Partial Differential Equations

Methods and Applications

Second Edition

Robert C. McOwen
Northeastern University

PEARSON EDUCATION, INC., Upper Saddle River, New Jersey 07458

Library of Congress Cataloging-in-Publication Data

McOwen, Robert C.
  Partial differential equations : methods and applications /
    Robert C. McOwen.–2nd ed.
      p.   cm.
  Includes bibliographical references and index.
  ISBN 0-13-009335-1
  1. Differential equations, Partial. I. Title.
QA377 .M323   2003
515.353–dc21   2002032251

Acquisition Editor: **George Lobell**
Editor-in-chief: **Sally Yagan**
Editorial Production/Supervision: **Bayani Mendoza de Leon**
Vice President/Director of Production and Manufacturing: **David W. Riccardi**
Senior Managing Editor: **Linda Mihatov Behrens**
Executive Managing Editor: **Kathleen Schiaparelli**
Manufacturing Buyer: **Michael Bell**
Manufacturing Manager: **Trudy Pisciotti**
Marketing Assistant: **Rachel Beckman**
Art Director: **Jayne Conte**
Cover Designer: **Bruce Kenselaar**
Editorial Assistant/Supplements Editor: **Jennifer Brady**

© 2003, 1996 by Pearson Education, Inc.
Pearson Education, Inc.
Upper Saddle River, NJ 07458

All rights reserved. No part of this book may be reproduced, in any form
or by any means, without permission in writing from the publisher.

Printed in the United States of America
10 9 8 7 6 5 4 3 2 1

ISBN 0-13-009335-1

Pearson Education LTD, *London*
Pearson Education Australia PTY, Limited, *Sydney*
Pearson Education Singapore, Pte, Ltd
Pearson Education North Asia Ltd, *Hong Kong*
Pearson Education Canada, Inc., *Toronto*
Pearson Education de Mexico, S.A. de C.V.
Pearson Education–Japan, Inc., *Tokyo*
Pearson Education Malaysia, Pte. Ltd

Contents

Preface   ix

Introduction   1
    Basic Definitions and Notation 7
    Organization and Numeration 10

Chapter 1. First-Order Equations   11
    1.1 The Cauchy Problem for Quasilinear Equations 11
       a. An Example: The Transport Equation   b. The Method of Characteristics   c. Semilinear Equations   d. Quasilinear Equations   e. General Solutions
    1.2 Weak Solutions for Quasilinear Equations 23
       a. Conservation Laws and Jump Conditions   b. Fans and Rarefaction Waves   c. Application to Traffic Flow
    1.3 General Nonlinear Equations 29
       a. The Method of Characteristics   b. Complete Integrals and General Solutions   c. Application to Geometrical Optics

Chapter 2. Principles for Higher-Order Equations   43
    2.1 The Cauchy Problem 43
       a. The Normal Form   b. Power Series and the Cauchy-Kovalevski Theorem   c. The Lewy Example
    2.2 Second-Order Equations in Two Variables 49
       a. Classification by Characteristics   b. Canonical Forms and General Solutions   c. First-Order Systems   d. Application to the Telegraph System
    2.3 Linear Equations and Generalized Solutions 59
       a. Adjoints and Weak Solutions   b. Transmission Conditions   c. Distributions   d. Convolutions and Fundamental Solutions

Chapter 3. The Wave Equation — 74

3.1 The One-Dimensional Wave Equation 74
    a. The Initial Value Problem   b. Weak Solutions   c. Initial/
    Boundary Value Problems   d. The Nonhomogeneous Equation

3.2 Higher Dimensions 83
    a. Spherical Means   b. Application to the Cauchy Problem
    c. The Three-Dimensional Wave Equation   d. The Two-
    Dimensional Wave Equation   e. Huygens's Principle

3.3 Energy Methods 91
    a. Conservation of Energy   b. The Domain of Dependence

3.4 Lower-order Terms 95
    a. Dispersion   b. Dissipation   c. The Domain of Dependence

3.5 Applications to Light and Sound 99
    a. Electromagnetism   b. Acoustics

Chapter 4. The Laplace Equation — 103

4.1 Introduction to the Laplace Equation 103
    a. Separation of Variables   b. Boundary Values and Physics
    c. Green's Identities and Uniqueness   d. Mean Values and the
    Maximum Principle

4.2 Potential Theory and Green's Functions 111
    a. The Fundamental Solution   b. Potential Theory and Electro-
    statics   c. Green's Function and the Poisson Kernel.   d. The
    Dirichlet Problem on a Half-Space   e. The Dirichlet Problem on
    a Ball   f. Properties of Harmonic Functions

4.3 Existence Theory 126
    a. Subharmonic Functions   b. Perron's Method

4.4 Eigenvalues of the Laplacian 132
    a. Eigenvalues and Eigenfunction Expansions   b. Application
    to the Wave Equation

4.5 Applications to Vector Fields 138
    a. The div-curl System   b. Helmholtz Decompositions

Chapter 5. The Heat Equation — 142

5.1 The Heat Equation in a Bounded Domain 142
    a. Existence by Eigenfunction Expansion   b. The Maximum
    Principle and Uniqueness

5.2 The Pure Initial Value Problem 146
    a. Fourier Transform   b. Solution of the Pure Initial Value
    Problem   c. The Fundamental Solution   d. The Nonhomoge-
    neous Equation

5.3 Regularity and Similarity 154
    a. Smoothness of Solutions   b. Scale Invariance and the Simi-
    larity Method

5.4 Application to Fluid Dynamics 159
    a. Slow Viscous Incompressible Flow   b. Brownian Motion

Chapter 6. Linear Functional Analysis 162

6.1 Function Spaces and Linear Operators 162
    a. Banach and Hilbert Spaces   b. Sobolev Spaces   c. Linear Operators and Functionals   d. The Hahn-Banach and Riesz Representation Theorems.

6.2 Application to the Dirichlet Problem 173
    a. Weak Solutions of the Poisson Equation   b. Weak Solutions of the Stokes Equations   c. More General Operators in Divergence Form   d. The Lax-Milgram Theorem

6.3 Duality and Compactness 181
    a. Dual Spaces   b. Weak Convergence   c. Compactness

6.4 Sobolev Imbedding Theorems 186
    a. The Sobolev Inequality for $p < n$ b. The Sobolev Imbedding Theorem for $p < n$ c. The Sobolev Inequality and Imbedding Theorem for $p > n$ d. Proof of the Sobolev Inequality for $p < n$

6.5 Generalizations and Refinements 193
    a. Hölder Continuity   b. Mollifiers and Smooth Approximations   c. Compact Imbeddings of $H_0^{1,p}(\Omega)$ d. Imbeddings of $H^{1,p}(\Omega)$ e. Higher-Order Sobolev Spaces and Strong Solutions

6.6 Unbounded Operators & Spectral Theory 208
    a. Densely-Defined and Closed Operators   b. Adjoints   c. Resolvents and Spectra   d. Symmetric and Self-Adjoint Operators

Chapter 7. Differential Calculus Methods 216

7.1 Calculus of Functionals and Variations 216
    a. The Dervative and Critical Points of a Functional   b. Coercive Functionals and Absolute Extrema   c. Weak Existence for Dirichlet & Neumann Problems   d. Convexity and Uniqueness   e. Mountain Passes and Saddle Points

7.2 Optimization with Constraints 227
    a. Lagrange Multipliers   b. Application to Eigenvalues of the Laplacian   c. The Maximin Characterization of Eigenvalues

7.3 Calculus of Maps between Banach Spaces 236
    a. The Method of Successive Approximations   b. The Inverse Function Theorem   c. The Implicit Function Theorem   d. C^1-maps on Sobolev spaces   e. Application to Small Mean Curvature

Chapter 8. Linear Elliptic Theory 244

8.1 Elliptic Operators on a Torus 244
    a. Fourier Analysis   b. A Priori Estimates and Regularity   c. L^p and Hölder Estimates

8.2 Estimates and Regularity on Domains 252
    a. Interior Estimates   b. Difference Quotients   c. Interior Regularity of Weak Solutions   d. Global Estimates and Regularity

8.3 Maximum Principles 260
  a. *The Weak Elliptic Maximum Principle*   b. *Application to a Priori Estimates*   c. *The Strong Elliptic Maximum Principle*   d. *Application to the Principal Eigenvalue.*
8.4 Solvability 267
  a. *Uniqueness and Solvability*   b. *Fredholm Solvability*   c. *Eigenvalues*

Chapter 9. Two Additional Methods   277

9.1 Schauder Fixed Point Theory 277
  a. *The Brouwer Fixed Point Theorem*   b. *The Schauder Fixed Point Theorem*   c. *The Leray-Schauder Fixed Point Theorem*   d. *Application to Stationary Navier-Stokes*
9.2 Semigroups and Dynamics 283
  a. *Finite-Dimensional Dynamics*   b. *Linear Evolution on Banach Spaces*   c. *The Nonhomogeneous Equation*   d. *Weak Solutions and Energy Methods*   e. *Nonlinear Dynamics*

Chapter 10. Systems of Conservation Laws   302

10.1 Local Existence for Hyperbolic Systems 302
  a. *Linear Systems*   b. *Nonlinear Systems*
10.2 Quasilinear Systems of Conservation Laws 306
  a. *Examples and Applications*   b. *Simple Waves and Rarefaction*   c. *Shocks and the Entropy Condition*   d. *Riemann Problems*
10.3 Systems of Two Conservation Laws 320
  a. *Riemann Invariants*   b. *The Hodograph Transformation*   c. *Application to Gas Dynamics*

Chapter 11. Linear and Nonlinear Diffusion   327

11.1 Parabolic Maximum Principles 327
  a. *The Weak Parabolic Maximum Principle*   b. *The Strong Parabolic Maximum Principle*   c. *Comparison Principles*
11.2 Local Existence and Regularity 333
  a. *Pure Initial Value Problems*   b. *Initial/Boundary Value Problems*   c. *Additional Smoothness*
11.3 Global Behavior 340
  a. *The Comparison Method*   b. *Energy Methods for Large Time Existence*   c. *Energy Methods for Asymptotic Behavior*
11.4 Applications to Navier-Stokes 346
  a. *Local Existence by the Semigroup Method*   b. *Weak Solutions of Navier-Stokes*   c. *A Priori Estimates*   d. *Existence of a Weak Solution: Galerkin's Method*   e. *Further Remarks*

Chapter 12. Linear and Nonlinear Waves   356

12.1 Symmetric Hyperbolic Systems 356
  a. *Energy Estimates for Linear Systems*   b. *Existence for Linear Systems*   c. *Local Existence for Quasilinear Systems*   d. *Application to Gas Dynamics*

12.2 Linear Wave Dynamics 369
    a. The Wave Equation in R^n b. The Klein-Gordon Equation
    in R^n c. Equations on Bounded Domains   d. The Schrödinger
    Equation
12.3 Semilinear Wave Dynamics 375
    a. Local Existence   b. Global Behavior for Conservative Systems

Chapter 13. Nonlinear Elliptic Equations — 381

13.1 Perturbations and Bifurcations 381
    a. Nonlinear Eigenvalue Problems   b. The Method of Lyapunov-
    Schmidt   c. Bifurcation from a Simple Eigenvalue
13.2 The Method of Sub- and Supersolutions 385
    a. Barriers for a Semilinear Equation   b. Monotone Iteration
    c. Application with Uniformly Bounded $f(x,u)$ d. Application
    with $f(x,u)$, Nondecreasing in u
13.3 The Variational Method 391
    a. A Semilinear Equation and Weak Solutions   b. Application
    of Lagrange Multipliers   c. Application of the Mountain Pass
    Theorem   d. Regularity and Positivity
13.4 Fixed Point Methods 398
    a. Semilinear Equations   b. Quasilinear Equations   c. Local
    Barriers and Boundary Gradient Estimates   d. Hölder Estimates
    of De Giorgi and Nash   e. Application to the Minimal Surface
    Equation

Appendix on Physics — 410

A.1 Physical Principles and PDEs 410
    a. Vibrating Strings and Membranes   b. Diffusion
A.2 Fields, Tensors, and Systems of PDEs 414
    a. Curl and Its Properties   b. Maxwell's Equations   c. Ideal Fluids and Euler's Equations   d. Cartesian Vector Fields and Tensors   e. Viscous Stress and Newtonian Fluids   f. The Navier-Stokes Equations

Hints & Solutions for Selected Exercises — 422

References — 441

Index — 447

Index of Symbols — 452

Preface to the Second Edition

I am grateful that so many individuals and institutions have chosen to use *Partial Differential Equations: Methods & Applications* since it first appeared in 1996. I have been even more grateful to the many individuals who have contacted me with suggestions and corrections for the first edition. I hope that this new edition will be much improved because of their interest and contributions.

The book originally evolved from a two-term graduate course in partial differential equations that I taught many times at Northeastern University. At that time, I felt there was an absence of textbooks that covered both the classical results of partial differential equations and more modern methods, such as functional analysis, which are used heavily in the current literature. Since I began to write the book, however, several other textbooks have appeared that also aspire to bridge the same gap: *An Introduction to Partial Differential Equations* by Renardy and Rogers (Springer-Verlag, 1993) and *Partial Differential Equations* by Lawrence C. Evans (AMS, 1998) are two good examples.

As with any book on such a broad and diverse subject as partial differential equations, I have had to make some difficult decisions concerning content and exposition. I make no apologies for these decisions, but I do acknowledge that other choices might have been made. For example, this text begins with the method of characteristics and first-order equations; although other texts often omit or slight this material in preference to the treatment of second-order equations, I have chosen to include it, and even emphasize its constructive aspects, because I feel it offers motivation and insights that are valuable in the study of higher-order equations. Indeed, the method of characteristics leads naturally to the Cauchy problem for higher-order equations, as well as the classification of second-order equations, which I treat in Chapter 2 (along with a discussion of generalized solutions). Following this momentum, I decided to treat the wave equation before Laplace's equation, even though this causes the use of eigenfunctions in a bounded domain to be delayed until the next chapter. Similarly, I have chosen to treat the heat equation after the Laplace equation for reasons of the maximum principle; of course, a bonus is that eigenfunction expansions are available for the heat equation in a bounded domain. Other texts treat these three second-order equations in different orders, and they all have their own reasons for doing so.

Exposure to the use of functional analysis begins in Chapter 6 with a rapid survey of the basic definitions and tools needed to study linear operators on Banach and Hilbert spaces. The Sobolev spaces are introduced as early as possible, as are their application to obtain weak solutions of the Dirichlet problems for the Poisson equation and the Stokes system, before encountering the more subtle issues of weak convergence, continuous imbeddings, compactness, unbounded operators, and spectral theory.

The theme of weak solutions is picked up again in Chapter 7, in the context of differential calculus on Banach spaces. The variational method of finding a weak solution by optimizing a functional, possibly with constraints, is applied to several problems, including the eigenvalues of the Laplacian. The forum of differential calculus also enables us to introduce, at this point, the contraction mapping principle, the inverse and implicit function theorems, a discussion of when they apply to Sobolev spaces, and an application to the prescribed mean curvature equation.

The issue of the regularity of weak solutions is taken up in Chapter 8, where the basic elliptic L^2-estimates are obtained by Fourier analysis on a torus, and transplantation to open domains. It is also natural, at this point, to discuss maximum principles for elliptic operators, and then the issues of uniqueness and solvability for linear elliptic equations.

Chapter 9 consists of two additional methods. The first of these, the Schauder fixed point theory, is presented and then illustrated with its application to the stationary Navier-Stokes equations; this application returns us to our theme of weak solutions in Sobolev spaces, and also builds on the discussion of the Stokes system in Chapter 6. The second additional method is the use of semigroups of operators on a Banach space to describe the dynamics of evolutionary partial differential equations. We first discuss systems of ordinary differential equations as a finite-dimensional example; this helps to motivate the ensuing discussion for partial differential equations, which is well seasoned with examples. This treatment of semigroups is very brief but serves the purpose of setting the stage for the hyperbolic and parabolic equations and systems that are studied in Chapters 10, 11, and 12.

Although Chapters 6 through 9 emphasize the development of tools and methods, I have tried to provide sufficient applications to motivate and illustrate the theory as it unfolds. However, beginning in Chapter 10, the focus switches from methods to applications, and developing the theories of hyperbolic systems conservation laws in one space dimension (Chapter 10), linear and nonlinear diffusion (Chapter 11), linear and nonlinear waves (Chapter 12), and nonlinear elliptic equations (Chapter 13) as far as possible in this limited space. I have, of course, needed to severely "limit the budget" in each of these last four chapters, but I hope I have given the flavor and some background on each topic, enough to enable the interested student to consult more detailed and comprehensive treatments.

Although I have made certain choices for the order, I have tried to make the exposition flexible enough to allow for the individual instructor to make changes without too much difficulty. For example, to enable the introduction of the spherical mean in connection with the Laplace equation instead of the wave equation, I have made Section 3.2a self-contained. This means that it is possible to reorder the material following Chapter 2: the one-dimensional wave equation, then Laplace's equation (with Section 3.2a added to Section 4.1d), and then the n-dimensional wave equation. Similarly, although I felt the need to collect all of the linear functional analysis and Sobolev space theory in Chapter 6, it is possible to discuss only the results for $H_0^{1,2}(\Omega)$ in order to study more quickly the Dirichlet problems in Chapters 7, 8, and 9. Another example would be to jump into Chapter 10 after only a minimal amount of Banach space theory and the contraction mapping principle.

I have tried to include a large number of exercises. Some of these exercises are fairly routine applications of the material covered in the text. Other exercises are designed to supply some steps that are omitted from the exposition in the text; this not only helps to streamline the exposition, but it also engages the student more actively in the learning experience. Still other exercises are intended to give the student a brief exposure to related topics that have been reluctantly omitted from the textual exposition, casualties of more hard choices of mine. When I teach this course, I usually assign many exercises, including some of each type. On the other hand, the instructor may choose to use lecture time to solve all omitted steps of proofs and/or pursue some of the omitted topics. In any case, hints and solutions of selected exercises are provided after Chapter 13; I hope the instructor and student find these useful.

Now let me list the major changes and additional topics that I have included in this second edition. To begin with, I have attempted to provide more details to some of the sketchier arguments in the first edition. Second, I have added sections with additional applications to Chapters 3, 4, and 5: respectively, applications to light and sound, applications to vector fields, and applications to fluid dynamics. In Chapter 6, I have added a section on unbounded operators and spectral theory that provides essential background for results in later chapters. I also have added an appendix on physics, in which the most important partial differential equations are derived from basic principles. Finally, I have made substantial changes to the Hints and Solutions for Selected Exercises.

ACKNOWLEDGMENTS

First let me thank the reviewers of the first edition: Christopher Grant (Brigham Young University), Peter Gingo (University of Akron), David Gurarie (Case Western Reserve University), Frank Jones (Rice University), Philip Mansfield (University of Toronto), Beny Neta (Naval Postgraduate), Peter Colwell (Iowa State University), Abdul Majid Wazwaz (Saint Xavier University), Chelluri Sastri (Dalhousie University), Steven Jay Smith (Northeast Missouri State University), and Enrique Thomann (Oregon State University). I'd also like to thank the reviewers of the second edition: Bo Guan (University of Tennessee), Frank Jones (Rice University),

Mark Kon (Boston University), Hailiang Liu (UCLA), and Shinshu Walter Wei (University of Oklahoma). Many of these reviewers offered comments that were very useful to me.

I would also like to thank many colleagues for their interest in this work: Bjorn Birnir (University of California, Santa Barbara), Walter Craig (Brown University), John D'Angelo (University of Illinois, Champaign-Urbana), Jerry Kazdan (University of Pennsylvania), and Nish Krikorian, Mikhail Shubin, and Marty Schwarz (all of Northeastern University). I am particularly indebted to mathematicians around the world who have used my book and offered corrections and suggestions: Mark Ashbaugh (University of Missouri), Jack Dockery (University of Montana), Faruk Gungor (Istanbul Technical University), Harald Hanche-Olsen (Trondheim, Norway), Helge Holden (Trondheim, Norway), Gregory C. Jones (University of Missouri-Columbia), Erik Talvila (University of Alberta), and Vaughan Weston (Purdue University).

Finally I would like to thank David Finn (Rose-Hulman Institute of Technology) for his assistance with the Hints and Solutions for Selected Exercises in the first edition, and my wife, Barbara, for supplying the graphics in the illustrations for both editions.

Robert McOwen
mcowen@neu.edu

Introduction

A *partial differential equation* (abbreviated *PDE*) is an equation involving a function u of several variables and its partial derivatives. For example, using subscripts to denote partial derivatives,

$$u_t = k u_{xx}, \tag{1}$$

$$u_{tt} = c^2 u_{xx}, \tag{2}$$

$$u_{xx} + u_{yy} = 0, \tag{3}$$

are all partial differential equations for functions of two variables that are familiar from undergraduate courses on differential equations: (1) is the one-dimensional *heat equation* in which u represents the temperature of a heat-conducting rod having k as its heat conductivity; (2) is the one-dimensional *wave equation* in which u represents the displacement of a vibrating string from its equilibrium position and c represents the speed of wave propagation; (3) is the two-dimensional *Laplace equation* that arises as a steady-state condition in heat conduction problems and occurs in many other problems of analysis and mathematical physics.

More generally, a PDE for a function $u(x_1, \ldots, x_n)$ is of the form

$$F(x_1, \ldots, x_n, u, u_{x_1}, \ldots, u_{x_n}, u_{x_1 x_1}, u_{x_1 x_2}, \ldots) = 0. \tag{4}$$

The *order* of (4) is the order of the highest derivative occuring in the equation. Moreover, the equation is *linear* if it depends linearly on u and its derivatives; if all derivatives of u occur linearly with coefficients depending only on x, then the equation is *semilinear*; and if all highest-order derivatives of u occur linearly with coefficients depending only on x, u, and lower-order derivatives of u, then the equation is *quasilinear*. Equations (1)–(3) are all second-order linear equations. A simple example of a first-order PDE is

$$u_t + a(u) u_x = 0. \tag{5}$$

When $a(u) \equiv a$ is a constant, (5) is a linear equation called the *transport equation*. In general, (5) is a quasilinear equation; for example, when $a(u) \equiv u$, the equation is called the *inviscid Burgers equation*, which arises in the study of a one-dimensional stream of particles or fluid having zero viscosity. An example of a first-order nonlinear PDE is

$$u_x^2 + u_y^2 = c^2, \tag{6}$$

which is the *eikonal equation of geometric optics*.

We can generalize equations (1)–(3) to higher dimensions if we introduce the *Laplacian* or *Laplace operator* $\Delta = \partial^2/\partial x_1^2 + \ldots + \partial^2/\partial x_n^2$. Then we can write

$$u_t = k\Delta u, \tag{7}$$

$$u_{tt} = c^2 \Delta u, \tag{8}$$

$$\Delta u = 0. \tag{9}$$

Equation (7) represents the diffusion of heat through an n-dimensional body; equation (8) represents surface waves if $n = 2$, and sound or light waves if $n = 3$; and equation (9) is the n-dimensional Laplace equation. Equations (7) and (8) are both *evolution equations* because they describe phenomena that may change with time; equation (9), on the other hand, is satisfied by *steady-state* (time independent) solutions of (7) and (8).

In order for a PDE to have a *unique* solution, we must impose additional conditions, sometimes called *side conditions,* on the solution; these are usually in the form of *initial conditions* or *boundary conditions* or some combination of the two. This is certainly familiar from ordinary differential equations (abbreviated *ODEs),* where a first-order equation requires an initial condition and a second-order equation requires either two initial conditions or a boundary condition at each end of a finite interval. The need for side conditions is also evident from the physical models. For example, we cannot know the temperature of a cooling body if we do not know its initial temperature; but even knowing the initial temperature will not be enough unless we also monitor what happens to the temperature on the boundary of the body. In this case, we must impose on (7) an initial condition and a boundary condition: the result is called an *initial/boundary value problem*. For other PDEs we may be able to consider a pure *initial value problem* or pure *boundary value problem*.

The values assigned to the side conditions are called the *data*. A PDE with side conditions is *well posed* if it admits a unique solution for any values assigned to the data. (Actually, well-posedness should also require the solution to depend continuously on the data, as we shall discuss later.)

Of course, equations (7)–(9) are very special, and we may wonder how to handle more complicated equations that arise in applications. To begin, we can add a function f to these equations to obtain *nonhomogeneous* equations. For example, the nonhomogeneous Laplace equation (sometimes called the *Poisson equation*)

(10) $$\Delta u = f$$

arises in various field theories such as electrostatics. Similarly, considerations of heat sources and external forcing terms in (7) and (8) respectively lead to the nonhomogeneous heat and wave equations

(11) $$u_t - k\Delta u = f,$$

(12) $$u_{tt} - c^2 \Delta u = f.$$

These equations may be modified further if additional considerations are in effect; for example, consideration of a restoring force in (8) leads to the *Klein-Gordon equation*

(13) $$u_{tt} - c^2 \Delta u + m^2 u = 0,$$

which arises in quantum field theory with m denoting mass, and consideration of a damping or *dissipation term* in (13) leads to

(14) $$u_{tt} - c^2 \Delta u + \alpha u_t + m^2 u = 0,$$

which in one space dimension is called the *telegrapher's equation* because it governs electrical transmission in a telegraph cable when current may leak to the ground.

Another important second-order linear equation that arises in quantum mechanics is *Schrödinger's equation*

(15) $$u_t = i(\Delta u + V(x)u)$$

in which $i = \sqrt{-1}$ indicates that the solution $u(x,t)$ must be complex-valued; the function $V(x)$ is called the *potential*.

So far, the second-order equations we have mentioned are all linear. This is not surprising, since the theory is simpler and for certain modeling purposes, a linear equation may suffice. But in other situations the nonlinear character of the problem is important and even essential. For example, if we allow f in (10)–(12) to depend on u, then we obtain the *semilinear Poisson equation*

(16) $$\Delta u = f(x,u),$$

the *semilinear heat equation*

(17) $$u_t - k\Delta u = f(x,t,u),$$

and the *semilinear wave equation*

(18) $$u_{tt} - c^2 \Delta u = f(x,t,u).$$

A specific instance of (16) is the *conformal scalar curvature equation*

(19a) $$\Delta u + K(x)e^{2u} = 0 \qquad (n=2),$$
(19b) $$\Delta u + K(x)u^{\frac{n+2}{n-2}} = 0 \qquad (n \geq 3),$$

which occurs in differential geometry when studying the scalar curvature of Riemannian metrics that are conformally Euclidean; for $n = 2$ the metric $e^{2u}(dx^2 + dy^2)$ will have Gauss curvature $K(x,y)$ if u satisfies (19a). Specific instances of (18) are the *semilinear Klein-Gordon equation*

(20) $$u_{tt} - c^2 \Delta u + m^2 u + \gamma u^p = 0 \qquad (p \text{ an integer } \geq 2),$$

which arises in quantum field theory with γ denoting a coupling constant, and the *sine-Gordon equation*

(21) $$u_{tt} - c^2 \Delta u + \sin u = 0,$$

which also arises in quantum field theory but was first studied in differential geometry in connection with surfaces of constant curvature. If we allow dissipation in (20) or (21), we get the *dissipative Klein-Gordon* or *dissipative sine-Gordon equations*

(22) $$u_{tt} - c^2 \Delta u + \alpha u_t + m^2 u + \alpha u^p = 0,$$

(23) $$u_{tt} - c^2 \Delta u + \alpha u_t + \sin u = 0.$$

A semilinear version of (15) is the *cubic Schrödinger equation*

(24) $$u_t = i(\Delta u + \sigma |u|^2 u) \qquad \sigma = \pm 1,$$

which arises in nonlinear optics, and also the study of deep water waves.

Equations also arise in applications that are not semilinear. For example, in differential geometry the *minimal surface equation*

(25) $$\text{div}\left(\frac{\nabla u}{(1+|\nabla u|^2)^{1/2}}\right) = 0$$

is a second-order quasilinear equation for a graph $z = u(x,y)$ that has the smallest surface area for a given boundary curve; for example, soap films are minimal surfaces. In (25), "div" denotes the *divergence* of the vector field $(1+|\nabla u|^2)^{-1/2}\nabla u$, and (25) is said to be in *divergence form*. A time-dependent quasilinear PDE that arises in applications is the *porous medium equation*

$$(26) \qquad u_t = k\,\Delta(u^\gamma),$$

where $k > 0$ and $\gamma > 1$ are constants; this equation governs the seepage of a fluid through a porous medium (e.g., water through soil). An example of a nonlinear PDE which is not quasilinear is the *Monge-Ampère equation*

$$(27) \qquad \det(u_{ij}) = f(x,u),$$

which arises in differential geometry; here the second-order derivatives $u_{ij} = \partial^2 u/\partial x_i \partial x_j$ occur in a nonlinear way.

Equations arising in applications need not be restricted to second-order. The *biharmonic equation*

$$(28) \qquad \Delta^2 u \equiv \Delta(\Delta u) = 0$$

is a fourth-order linear equation that occurs in elasticity theory, whereas the *Korteweg de Vries equation* (or *KdV equation*)

$$(29) \qquad u_t + cuu_x + u_{xxx} = 0$$

is a third-order quasilinear equation that was first encountered in the study of shallow water waves.

If u is replaced by a vector-valued function $\vec{u}(x_1,\ldots,x_n)$ and F is also vector valued, then (4) becomes a *system* of differential equations. For example, if $A(x,t)$ and $B(x,t)$ are $N \times N$ matrix-valued functions, and $\vec{c}(x,t)$ is a vector-valued function, then

$$(30) \qquad A(x,t)\vec{u}_x + B(x,t)\vec{u}_t = \vec{c}(x,t)$$

is a linear first-order system. An example of (30) is *Maxwell's equations* from electromagnetism theory in $\mathbf{R}^3$:

$$(31) \qquad \vec{E}_t - \operatorname{curl}\vec{H} = 0 \qquad \vec{H}_t + \operatorname{curl}\vec{E} = 0,$$

where $\vec{E}$ denotes the electric field and $\vec{H}$ the magnetic field. Notice that (31) is a system of six equations in the six unknowns $(\vec{E},\vec{H})$.

A natural example of a nonlinear first-order system occurs in fluid dynamics, when the balancing of forces according to Newton's law produces *Euler's equations*

$$(32) \qquad \vec{u}_t + (\vec{u}\cdot\nabla)\vec{u} + \frac{1}{\rho}\nabla p = 0$$

for an inviscid (no viscosity) fluid with velocity field $\vec{u}$, pressure p, and density ρ. In dimensions $n = 1, 2$, or 3, $\vec{u}$ has n components and is a function of the $n+1$ variables (x, t); moreover, $(\vec{u} \cdot \nabla)\vec{u}$ denotes the n vector whose jth component is $\sum_i u_i \partial u_j / \partial x_i$. Notice that (32) is a system of n equations in the $n+2$ unknowns $(\vec{u}, p, \rho)$ and so is underdetermined. One additional equation that must be coupled with (32) expresses the compressibility properties of the fluid. If the fluid is *incompressible*, this equation is div $\vec{u} = 0$, whereas for *compressible* fluids we must use $\rho_t + \text{div}\,(\rho\vec{u}) = 0$ which expresses *conservation of mass*. Of course, these equations coincide if the density ρ is constant, and coupling with (32) yields a well-posed system. More general than constant density is *isentropy*, which means that pressure is a known function of the density: $p = p(\rho)$, which is called an *equation of state*. Coupling (32) with an equation of state and either incompressibility or conservation of mass yields a well-posed system. For $n \geq 2$, the Euler equations are used to model vortices and turbulence. Natural questions to ask are whether smooth initial data produces a smooth solution at least for a short period of time, and whether an initially smooth solution can develop singularities.

If we consider a *viscous* but incompressible fluid, then we must replace (32) by the *Navier-Stokes equations*

$$\text{(33)} \qquad \vec{u}_t + (\vec{u} \cdot \nabla)\vec{u} + \frac{1}{\rho}\nabla p = \nu \Delta \vec{u},$$

where the density ρ is constant, ν is the kinematic viscosity, and Δ operates on $\vec{u}$ componentwise. Although (33) may seem more complicated than (32), the viscosity term actually has a tempering influence on the solutions that is not present in (32). A natural question to ask is whether the viscosity enables a smooth solution to exist for all time. This is known to be true for $n = 2$ but as yet has not been proved for $n = 3$, in spite of ample physical intuition! Another important question in fluid dynamics is how well (33) approximates (32) as $\nu \to 0$. This may be useful in modeling turbulence.

This comparison of (32) and (33) touches on some general themes that we shall encounter for evolution equations. Does a solution exist at least for a short time? If so, does it exist for all time, and can we describe its behavior asymptotically as $t \to \infty$? If the solution fails to exist globally in time, is this due to a blow-up (the values of the solution approaching infinity) or a gradient catastrophe (the values of the spatial derivatives becoming infinite) or some more complicated singularity at a finite time?

On the other hand, for time-independent or stationary PDEs such as the Laplace equation, the conformal scalar curvature equation, and the minimal surface equation, we may ask, What boundary conditions are appropriate? Does a solution exist? Is it unique? How smooth is the solution?

In this book, we shall develop some of the methods used to study such linear and nonlinear PDEs and apply these methods to obtain conclusions about their solvability and the behavior of the solutions.

Basic Definitions and Notation

Let $\mathbf{Z}$ denote the integers and $\mathbf{N}$ denote the natural numbers (i.e., the nonnegative integers). We shall denote n-dimensional Euclidean space by $\mathbf{R}^n$ and points by $x = (x_1, \ldots, x_n)$. However, we often use the more conventional notation (x, y) and (x, y, z) for points in $\mathbf{R}^2$ and $\mathbf{R}^3$, respectively. The interpretation of x as being a point or a coordinate should be clear from the context.

The open ball of radius r centered at $x \in \mathbf{R}^n$ is denoted by $B_r(x) = \{y \in \mathbf{R}^n : |x - y| < r\}$. A *domain* $\Omega \subset \mathbf{R}^n$ in $\mathbf{R}^n$ is a connected open subset: points $x, y \in \Omega$ can be connected by a curve in Ω, and for every $x \in \Omega$ we have $B_r(x) \subset \Omega$ provided r is sufficiently small. The domain Ω is *simply connected* if every closed path in Ω may be continuously contracted to a point (so Ω has no "holes"). Let $\overline{\Omega}$ denote the closure (all limit points) of Ω. The *boundary* $\partial \Omega$ of a domain Ω is the set of limit points of Ω that are not in Ω (i.e., $\partial \Omega = \overline{\Omega} \backslash \Omega$, where "\" denotes set subtraction). If $\Omega = \mathbf{R}^n$, then $\partial \Omega = \emptyset$, and if Ω is a bounded set, then so is $\partial \Omega$. In general, $\partial \Omega$ could be quite a complicated set, but for "nice" domains Ω, $\partial \Omega$ should be an $(n-1)$-dimensional hypersurface.

We shall denote the continuous functions on Ω by $C(\Omega)$, and those whose first-order derivatives are also all continuous by $C^1(\Omega)$. (We generally assume that functions are real valued; at times we shall consider complex-valued functions, but we shall not introduce additional notation for this.) Similarly, for $k \in \mathbf{N}$, $C^k(\Omega)$ denotes the functions having all derivatives up to the order k continuous on Ω. Let $C^1(\overline{\Omega})$ denote the functions $f \in C^1(\Omega)$ for which f and the first-order derivatives of f extend continuously to $\overline{\Omega}$; $C(\overline{\Omega})$ and $C^k(\overline{\Omega})$ are defined similarly. Occasionally we are interested in continuous functions that are *bounded* in Ω but may not extend continuously to $\overline{\Omega}$; we denote these by $C_B(\Omega)$. Notice that $C_B(\Omega) \subset C(\Omega)$. Similarly, we may define $C_B^k(\Omega)$. Various notations are used for differentiation, namely $\partial_x u$, $\partial u / \partial x$, and u_x all denote the partial derivative of u with respect to x.

We shall consider N-vector valued functions as maps from $\mathbf{R}^n$ to $\mathbf{R}^N$. For example, $C(\Omega, \mathbf{R}^N)$ denotes all continuous N-vector valued functions on the domain Ω.

The *support* of a continuous function $f(x)$ defined on $\mathbf{R}^n$ is the closure of the set of points where $f(x)$ is nonzero: $\operatorname{supp} f = \overline{\{x \in \mathbf{R}^n : f(x) \neq 0\}}$. A set in $\mathbf{R}^n$ is *bounded* if it is contained in a ball $B_R(0)$ with R sufficiently large. The closed, bounded sets in $\mathbf{R}^n$ are precisely the compact sets; so if $\operatorname{supp} f$ is bounded, we say f has *compact support* and denote such functions by $C_0(\mathbf{R}^n)$. Similarly, $C_0(\Omega)$ denotes the continuous functions on Ω whose support is a compact subset of Ω, and $C_0^k(\Omega)$ is defined analogously.

A function f defined on a domain Ω is *integrable* if $\int_\Omega |f(x)|\, dx$ is defined and finite; we denote all such functions by $L^1(\Omega)$. (Technically, this requires the notion of Lebesgue integration, but this will not be needed until Chapter 6.) Sometimes it is useful to consider the larger space $L^1_{\text{loc}}(\Omega)$ of functions that are locally integrable (i.e., integrable on any compact subset of Ω, but not necessarily integrable at the boundary of Ω or at infinity). Notice that the following are true: $C_0(\Omega) \subset L^1(\Omega)$ and $C(\Omega) \subset L^1_{\text{loc}}(\Omega)$. More generally, functions in $L^1(\Omega)$ or $L^1_{\text{loc}}(\Omega)$ may have discontinuities, including singularities, provided they are not too severe for the integral to converge; similarly, functions in $L^1(\Omega)$ may have noncompact support. [These conditions are discussed in more detail later, in connection with the $O(\cdot)$-notation. The concept of Lebesgue integration and the notation L^1 are discussed briefly in Chapter 6; at that point, additional function spaces will also be introduced.]

Suppose that Ω is a bounded domain with a C^1-*boundary:* this means that for every $\xi \in \partial\Omega$, there is an open ball $B_r(\xi)$ and a C^1-diffeomorphism $\psi : B_r(\xi) \to B_1(0)$ (i.e., ψ is invertible with ψ and ψ^{-1} being C^1 maps) such that $\psi(\partial\Omega \cap B_r(\xi)) \subset \{x \in \mathbf{R}^n : x_n = 0\}$ and $\psi(\Omega \cap B_r(\xi)) \subset \{x \in \mathbf{R}^n : x_n > 0\}$; see the figure.

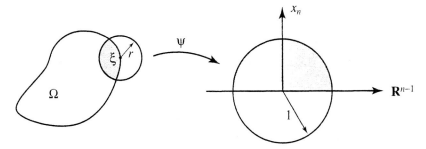

Figure. *The C^1-diffeomorphism $\psi: B_r(\xi) \to B_1(0)$.*

In particular, $\partial\Omega$ is an $(n-1)$-dimensional hypersurface. Suppose also that $\vec{V} = \langle V_1, \ldots, V_n \rangle$ is a vector field on Ω with $V_i \in C^1(\overline{\Omega})$ (i.e., $\vec{V} \in C^1(\Omega, \mathbf{R}^n)$). Then the *divergence theorem* holds:

$$(34) \qquad \int_{\partial\Omega} \vec{V} \cdot \nu \, dS = \int_\Omega \operatorname{div} \vec{V} \, dx,$$

where ν is the exterior unit normal on $\partial\Omega$, dS is the surface measure on $\partial\Omega$, and $\operatorname{div} \vec{V} = \sum_{i=1}^n (V_i)_{x_i}$. More generally, if $\partial\Omega$ is only *piecewise* C^1 (i.e., $\partial\Omega$ is a finite union of C^1 boundary parts) and the vector field components V_i are only in $C^1(\Omega) \cap C(\overline{\Omega})$, then (34) may still hold by approximating Ω by C^1-domains from the inside. This requires, in particular, the integral over Ω in (34) to be convergent.

Notice that we have used an arrow over the vector field to denote its being a vector, but not over the exterior unit normal. In general, vector functions that are considered as dependent variables will be given arrows whereas the independent variables will not. This is simply for convenience, and the meaning is usually clear from the context.

Another important classical result we shall use is the *Arzela-Ascoli theorem.* For our purposes here, the following formulation is sufficient: If Ω

is a bounded domain and $\{u_m\}_{m=1}^{\infty}$ is a sequence of functions $u_m \in C^1(\overline{\Omega})$ satisfying

$$\max_{x \in \overline{\Omega}} \left\{ u_m(x), \frac{\partial u_m(x)}{\partial x_1}, \ldots, \frac{\partial u_m(x)}{\partial x_n} \right\} \leq M,$$

where M is independent of m, then there is a subsequence $\{u_{m_k}\}$ that converges uniformly on $\overline{\Omega}$:

$$\max_{x \in \overline{\Omega}} |u_{m_k}(x) - u_{m_\ell}(x)| \to 0 \qquad \text{as } k, \ell \to \infty.$$

(A more general statement of the Arzela-Ascoli theorem will be found in Section 6.3.)

Notice that the preceeding use of "max" is consistent with the following: "max" is used when the maximum is known to be achieved, whereas "sup" is used when there is a least upper bound that is not necessarily attained. Similarly, we distinguish between "inf" and "min." We shall also use "log" to represent natural logarithm.

At times, we shall need to use the *inverse* and *implicit function theorems* of multivariable analysis. Recall that a map $f : U \to \mathbf{R}^n$ that is C^1 on the neighborhood U of $0 \in \mathbf{R}^n$ has a unique *inverse function* (i.e., $g : W \to \mathbf{R}^n$ is C^1 on a neighborhood W of $f(0)$ with $g(f(x)) = x$ for $x \in U$, *provided* the $n \times n$ matrix $A = (a_{ij}) = (\partial f_i(0)/\partial x_j)$ is invertible, i.e., $\det A \neq 0$.) More generally, a map $f : U \times V \to \mathbf{R}^n$, which is C^1 on the neighborhood $U \times V$ of $(0,0) \in \mathbf{R}^m_x \times \mathbf{R}^n_y$ and satisfies $f(0,0) = 0$, defines a unique *implicit function* $g : W \to \mathbf{R}^n$ which is C^1 on a neighborhood W of $0 \in \mathbf{R}^m$ and satisfies $f(x, g(x)) = 0$ for all $x \in W$, *provided* the $n \times n$ matrix $A = (a_{ij}) = (\partial f_i(0,0)/\partial y_j)$ is invertible (i.e., $\det A \neq 0$).

In some of the later chapters, the notations $O(\cdot)$ and $o(\cdot)$ are used. These are defined as follows: for $t > 0$ and a real number p

$$f(t) = O(t^p) \text{ as } t \to 0 \Leftrightarrow t^{-p}|f(t)| \text{ is bounded as } t \to 0$$

$$f(t) = o(t^p) \text{ as } t \to 0 \Leftrightarrow t^{-p}|f(t)| \to 0 \text{ as } t \to 0.$$

In addition we define

$$f(t) = g(t) + O(t^p) \Leftrightarrow f(t) - g(t) = O(t^p).$$

Similarly we may replace $t \to 0$ by $t \to \infty$. For example, $f(t) \equiv \sqrt{t^2+1} = O(t)$ as $t \to \infty$. In fact, using $(t^2+1)^{1/2} = t(1+t^{-2})^{1/2} = t(1+\frac{1}{2}t^{-2}+\cdots)$, we see that we can make the stronger statements $f(t) - t = o(1)$ or even $f(t) - t = O(t^{-1})$ as $t \to \infty$.

Let us use the $O(\cdot)$ notation to illustrate the integrability of functions. If $f(x)$ is a bounded measurable function on $\mathbf{R}^n$, then $f \in L^1_{\text{loc}}(\mathbf{R}^n)$. If, moreover, $f(x) = O(|x|^{-n-\epsilon})$ as $|x| \to \infty$ for $\epsilon > 0$, then $\int_{\mathbf{R}^n} |f(x)|\, dx$ is finite, so $f \in L^1(\mathbf{R}^n)$. If $g(x)$ is continuous on $0 < |x| \leq 1$, then $g \in$

$L^1_{\text{loc}}(B_1(0)\setminus\{0\})$. If, moreover, $g(x) = O(|x|^{-n+\epsilon})$ as $|x| \to 0$ for $\epsilon > 0$, then $\int_{|x|<1} |g(x)|\, dx$ is finite, so $g \in L^1(B_1(0))$. Notice that the condition $o(|x|^{-n})$ is *not* sufficient to imply integrability of either f on $\mathbf{R}^n$ or g on $B_1(0)$.

One last concept that will be useful in the later chapters is a localization tool called a *partition of unity*. Given an arbitrary set A in $\mathbf{R}^n$, and a countable collection $\{\Omega_k\}_{k=1}^{\infty}$ of open sets that cover A (i.e., $A \subset \cup_{k=1}^{\infty} \Omega_k$), a *locally finite partition of unity subordinate to* $\{\Omega_k\}_{k=1}^{\infty}$ is a collection of functions $\{\phi_k\}$ satisfying (i) $0 \le \phi_k(x) \le 1$, (ii) $\phi_k \in C_0^{\infty}(\Omega_k)$, (iii) every $x \in A$ has a neigborhood in which all but a finite number of the ϕ_k are zero, and (iv) $\sum_{k=1}^{\infty} \phi_k(x) = 1$ for every $x \in A$ [where (iii) implies that the infinite sum actually has only a finite number of nonzero terms]. If A is a compact set, then the open cover $\{\Omega_k\}_{k=1}^{\infty}$ admits a finite subcover, so we need only consider a finite collection of functions; this case, for example, is useful in localizing the analysis of $\overline{\Omega}$, as in the figure above. However, the more general case of an infinite open cover is important when A itself is an open set. The existence of a partition of unity is proved in many texts; for example, see [Adams] or [Yoshida].

Organization and Numeration

The material in this text is organized into chapters, with formulas numbered in sequence for each chapter. Each chapter is divided into sections, and the theorems, examples, and figures are numbered in sequence for each section; at the end of each section is a collection of exercises pertaining to the material of that section. Within each section, the exposition is divided into subsections that are labeled lexicographically: *a,b* and so on; this is for clarity and convenience but has no effect on the numeration of that section.

We shall use the symbol ♠ to denote the end of a proof, and ♣ to denote the end of an example.

Further References for Introduction

Most of the equations mentioned in this Introduction will be discussed later in this text, although the depth of treatment will vary quite a bit. However, for additional information on linear or nonlinear wave equations, consult [John, 2], [Strauss], and [Whitham]; on conformal scalar curvature and Monge-Ampére equations, consult [Aubin]; on the minimal surface equation, consult [Gilbarg-Trudinger] and [Giusti]; and on Navier-Stokes, consult [Aris], [Temam 1 & 2], and [Galdi].

For more information about the definitions and basic results of analysis that were briefly reviewed here, consult [Rudin, 1].

1 First-Order Equations

In this chapter, we study first-order equations using geometrical considerations and the method of characteristics. We begin with semilinear equations, and then show how to treat quasilinear and even general nonlinear equations. Along the way, we encounter weak solutions and several applications to physical examples.

1.1 The Cauchy Problem for Quasilinear Equations

Let us recall a simple fact from the theory of ordinary differential equations: the equation $du/dt = f(t, u)$ can be uniquely solved (at least for small values of t) for each initial condition $u(0) = u_0$, provided that f is continuous in t and Lipschitz continuous in the variable u. Recall that the solution may exist globally in time, or may blow up at some finite time.

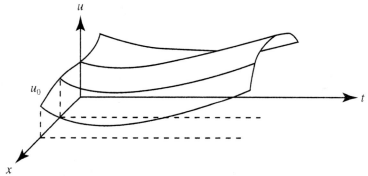

Figure 1. ODE depending on parameter x.

If we allow the equation and the initial condition to depend on a parameter x, then the solution u depends on x and may be written $u(x, t)$. In fact, u becomes a solution of $u_t = f(x, t, u)$, $u(x, 0) = u_0(x)$ that may be thought of as an initial value problem for a PDE in which u_x does not appear. Assuming f and u_0 are continuous functions of x, the solution $u(x, t)$ will be continuous in x (as well as t). Geometrically, the graph $z = u(x, t)$ is a surface in $\mathbf{R}^3$ that contains the curve $(x, 0, u_0(x))$ (see Figure 1). This surface may be defined for all $t > 0$, or may blow up at some finite t_0 (which may depend on x). However, if the surface remains bounded, then it will continue as a graph for all $t > 0$. In particular, the surface can*not* fold over on itself and thereby fail to be the graph of a function.

These elementary ideas from ODE theory lie behind the method of characteristics which applies to general quasilinear first-order PDEs, as we shall discover in this section.

a. An Example: The Transport Equation

Consider the initial value problem for the transport equation

$$\begin{cases} u_t + au_x = 0 \\ u(x,0) = h(x). \end{cases}$$

where a is a constant. Let us try to reduce this problem to an ODE along some curve $x(t)$; that is, find $x(t)$ so that

$$\frac{d}{dt}u(x(t),t) = au_x + u_t.$$

By the chain rule, we simply require $dx/dt = a$, i.e., $x = at + x_0$ where x_0 is the x-intercept of the curve. Along this curve we have $u_t = 0$ (i.e., $u \equiv \text{const}$); since u has the value $h(x_0)$ at the x-intercept, we must have $u(x,t) = h(x_0) = h(x - at)$. Indeed, if h is C^1, then we can check that $u(x,t) = h(x - at)$ satisfies the PDE and the initial condition. Notice that the solution corresponds to "transporting" (without change) the initial data $h(x)$ along the x-axis at a speed $dx/dt = a$ (see Figure 2). The lines $x = at + x_0$ are called the *characteristic curves* for $u_t + au_x = 0$.

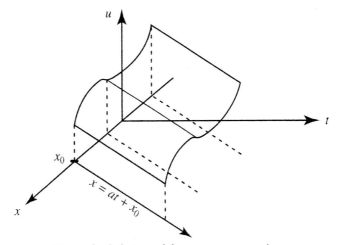

Figure 2. Solution of the transport equation.

The reduction of a PDE to an ODE along its characteristics is called the *method of characteristics* and applies to much more complicated equations. For example, the semilinear transport equation $u_t + au_x = f(u)$ reduces to the ODE $u_t = f(u)$ along the characteristic curve $x(t) = at + x_0$ and can therefore easily be integrated. Let us now see how and why this method applies to quasilinear PDEs.

b. The Method of Characteristics

Let us consider the quasilinear equation for a function u of two variables x and y

(1) $$a(x, y, u)u_x + b(x, y, u)u_y = c(x, y, u),$$

where the coefficient functions $a, b,$ and c are continuous in $x, y,$ and u. If $u(x, y)$ is a solution of (1), let us consider the graph $z = u(x, y)$. This surface has normal vector $N_0 = \langle -u_x(x_0, y_0), -u_y(x_0, y_0), 1 \rangle$ at the point $(x_0, y_0, u(x_0, y_0))$. But if we let $z_0 = u(x_0, y_0)$, then equation (1) implies that the vector $V_0 = \langle a(x_0, y_0, z_0), b(x_0, y_0, z_0), c(x_0, y_0, z_0) \rangle$ is perpendicular to this normal vector and hence must lie in the tangent plane to the graph of $z = u(x, y)$ at the point z_0 (see Figure 3).

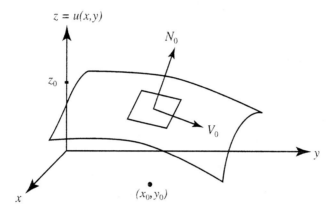

Figure 3. *The vector field V is tangent to the graph.*

In other words, $V(x, y, z) = \langle a(x, y, z), b(x, y, z), c(x, y, z) \rangle$ defines a vector field in $\mathbf{R}^3$, to which graphs of solutions must be tangent at each point. Surfaces that are tangent at each point to a vector field in $\mathbf{R}^3$ are called *integral surfaces* of the vector field, just as curves that are tangent to vector fields are called *integral curves*. Thus to find a solution of (1), we should try to find integral surfaces. Of course, there may be many integral surfaces of V, so we might try to be more specific and find the integral surface containing a given curve $\Gamma \subset \mathbf{R}^3$. Thus we are led to formulating the following:

The Cauchy Problem. *Given a curve Γ in $\mathbf{R}^3$, can we find a solution u of the first-order partial differential equation whose graph contains Γ?*

In the special case that Γ is the graph $(x, h(x))$ in the xz-plane of a function h, the Cauchy problem is just an initial value problem with the obvious interpretation of the variable y as "time."

How can we construct integral surfaces? We might try using the *characteristic curves* that are the integral curves of the vector field V. That is, $\chi = (x(t), y(t), z(t))$ is a characteristic if it satisfies the following system of ordinary differential equations called the *characteristic equations:*

(2) $\quad \dfrac{dx}{dt} = a(x,y,z), \qquad \dfrac{dy}{dt} = b(x,y,z), \qquad \dfrac{dz}{dt} = c(x,y,z).$

We can solve (2) uniquely for small $|t - t_0|$ if we are given initial conditions

$$x(t_0) = x_0, \qquad y(t_0) = y_0, \qquad z(t_0) = z_0,$$

and we assume that the functions a, b, c are all continuously differentiable in x, y, z. If the graph $z = u(x, y)$ is a smooth surface S that is a union of such characteristic curves, then at each point (x_0, y_0, z_0) the tangent plane contains the vector $V(x_0, y_0, z_0)$; hence S must be an integral surface. In other words, *a smooth union of characteristic curves is an integral surface.* So how can we obtain a smooth union of characteristic curves? If the given curve Γ is *noncharacteristic* (i.e., Γ is nowhere tangent to the vector field V), then a simple procedure for solving the Cauchy problem is to flow out from each point of Γ along the characteristic curve through that point, thereby sweeping out an integral surface (see Figure 4). This is the *method of characteristics.*

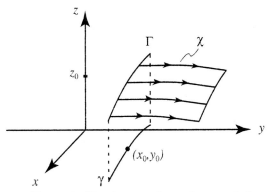

Figure 4. Flowing out along characteristics.

Analytically, this construction of an integral surface containing Γ can be achieved by first writing Γ as the graph of a curve $(f(s), g(s), h(s))$ parameterized by s and then solving the equations (2) for each s using $x_0 = f(s), y_0 = g(s), z_0 = h(s)$ as initial conditions. In this way, we obtain our integral surface S parameterized by s and t. In fact, it is also true that an integral surface S of the vector field $V = \langle a, b, c \rangle$ is *always* a union of characteristic curves: this can be proved using the uniqueness theorem for solutions of ordinary differential equations (see Exercise 1). Since the solutions of the characteristic equations are unique, we find that the integral surface S is unique. This means that we have proved the following.

Section 1.1: The Cauchy Problem for Quasilinear Equations   15

Theorem. *If Γ is noncharacteristic, then the vector field $V = \langle a, b, c \rangle$ admits a unique integral surface S containing Γ.*

To find the solution $u(x, y)$ of (1), it remains only to replace the variables s and t by expressions involving x and y. Theoretically, this can be achieved exactly when the integral surface S is the graph of a function. We shall discuss this in more detail in the next two subsections.

The method of characteristics is of more than theoretical value since it actually produces a formula for the solution, provided of course that we can explicitly solve the system of ordinary differential equations (2) and solve for s, t in terms of x, y. In the next three subsections we shall discuss several examples of semilinear and quasilinear equations in two variables, but first let us discuss how to generalize the foregoing procedure to an equation in n variables.

We now replace (1) with the equation

$$\text{(3)} \qquad \sum_{i=1}^{n} a_i(x_1, \ldots, x_n, u) u_{x_i} = c(x_1, \ldots, x_n, u).$$

The *characteristic curves* are now the integral curves of the system of $n+1$ equations in $n+1$ unknowns

$$\text{(4)} \qquad \frac{dx_i}{dt} = a_i(x_1, \ldots, x_n, z), \qquad \frac{dz}{dt} = c(x_1, \ldots, x_n, z),$$

which can be solved if we are given initial conditions on an $n-1$-dimensional manifold Γ:

$$x_i = f_i(s_1, \ldots, s_{n-1}), \qquad z = h(s_1, \ldots, s_{n-1}).$$

This generates an n-dimensional integral manifold M parameterized by $(s_1, \ldots, s_{n-1}, t)$. The solution $u(x_1, \ldots, x_n)$ is obtained by solving for $(s_1, \ldots, s_{n-1}, t)$ in terms of the variables $(x_1, \ldots, x_n)$. Thus we can apply the method of characteristics to solve initial value problems, or more generally Cauchy problems, for equations in n variables.

c. Semilinear Equations

Let us consider the Cauchy problem for the semilinear equation in two variables

$$\text{(5)} \qquad a(x, y) u_x + b(x, y) u_y = c(x, y, u),$$

with Γ parameterized by $(f(s), g(s), h(s))$. The characteristic equations become

$$\text{(6)} \qquad \frac{dx}{dt} = a(x, y), \qquad \frac{dy}{dt} = b(x, y), \qquad \frac{dz}{dt} = c(x, y, z),$$

with initial conditions

(7) $\quad x(s,0) = f(s), \qquad y(s,0) = g(s), \qquad z(s,0) = h(s).$

Notice that the first two equations form a system (decoupled from z), which may be solved to obtain a curve $(x(t), y(t))$ in the xy-plane; such a curve is sometimes called a *projected characteristic curve* since it is simply the projection into the xy-plane of the characteristic curve χ. If we first find the projected characteristics, we can then integrate the remaining characteristic equation to find z.

Moreover, regarding the problem of solving for s and t in terms of x and y, the inverse function theorem tells us that this can be achieved provided the Jacobian matrix is nonsingular:

(8) $$J \equiv \det \begin{pmatrix} x_s & y_s \\ x_t & y_t \end{pmatrix} \equiv x_s y_t - y_s x_t \neq 0.$$

Notice that this condition is independent of the behavior of z. In particular, at $t = 0$ we obtain the condition

(9) $\quad f'(s) \, b(f(s), g(s)) - g'(s) \, a(f(s), g(s)) \neq 0,$

which geometrically means that the projection of Γ into the xy-plane is a curve $\gamma = (f(s), g(s))$ that is nowhere parallel to the vector field $\langle a, b \rangle$. But (9) implies by continuity that (8) holds at least for small values of t, so we have the following result: *Provided the initial curve $\gamma = (f(s), g(s))$ satisfies (9), there exists a unique solution $u(x,y)$ of (5) in a neighborhood of γ.* However, away from γ (i.e. for larger values of t) the solution may develop a singularity where $J = 0$; in fact, even if (8) holds for *all* values of s and t, the solution may develop a blow up type of singularity if the equation for dz/dt is nonlinear.

Example 1. Let us solve the initial value problem $u_x + 2u_y = u^2$ with $u(x,0) = h(x)$. As an initial value problem, Γ is the graph $(x, h(x))$ in the xz-plane, so we may parameterize Γ by $(s, 0, h(s))$, and we see that (9) holds: $2 \neq 0$. The equations (6) become

$$\frac{dx}{dt} = 1, \qquad \frac{dy}{dt} = 2, \qquad \frac{dz}{dt} = z^2.$$

We may integrate the first two equations (treating s as a constant) to find $x(s,t) = t + c_1(s)$ and $y(s,t) = 2t + c_2(s)$, where the functions $c_1(s)$ and $c_2(s)$ may be determined from the initial conditions: $x(s,0) = c_1(s) = s$ and $y(s,0) = c_2(s) = 0$ so that $x = t + s$ and $y = 2t$. Notice that (8) holds for all s and t, and we can explicitly solve for s and t to find $s = x - \frac{y}{2}$ and

Section 1.1: The Cauchy Problem for Quasilinear Equations

$t = \frac{y}{2}$. We may integrate the equation for z to find $z(s,t) = -(t + c_3(s))^{-1}$ and use the initial condition $z(s,0) = h(s)$ to evaluate $c_3 = -1/h(s)$, so

$$z = \frac{h(s)}{1 - th(s)}.$$

Finally, we may eliminate s and t to express our solution as

$$u(x,y) = \frac{h(x - \frac{y}{2})}{1 - \frac{y}{2}h(x - \frac{y}{2})}.$$

Notice that $u(x,0) = h(x)$ and the solution $u(x,y)$ is certainly well defined for small enough values of y (assuming h is bounded); however, u may become infinite if y becomes large enough to cause the denominator to vanish. Even though the equations for dx/dt and dy/dt are linear and the solutions exist for all s and t, the equation for dz/dt is nonlinear and may produce a singularity. ♣

In the semilinear case, it is possible to take a different perspective on the Cauchy problem that will be important in Chapter 2. Consider the curve γ parameterized by $(f(s), g(s))$ as given and then impose *Cauchy data* $h(s)$ along γ. Provided γ is *noncharacteristic* [i.e. satisfies (9)], then there is a unique solution of (5) satisfying

(10) $$u(f(s), g(s)) = h(s).$$

This means that the Cauchy data $u|_\gamma = h$ also determines the derivatives $u_x|_\gamma$ and $u_y|_\gamma$; this is true *theoretically* because the solution u is uniquely defined in a neighborhood of γ, but in fact may be found *explicitly* by solving the system of algebraic equations along $\gamma(s) = (f(s), g(s))$:

(11) $$\begin{cases} a(\gamma(s))u_x(\gamma(s)) + b(\gamma(s))u_y(\gamma(s)) = c(f(s), g(s), h(s)) \\ f'(s)u_x(\gamma(s)) + g'(s)u_y(\gamma(s)) = h'(s), \end{cases}$$

which admit unique solutions by (9). Moreover, the Cauchy data also enables us to find *all* partial derivatives of u on γ, theoretically by differentiating the unique solution near γ, but explicitly by using equation (11) (see Example 2). Thus, in a well-posed Cauchy problem, the Cauchy data will determine *all* derivatives of the solution on the initial curve.

On the other hand, if $J \equiv x_s y_t - y_s x_t = 0$ along γ, then the method of characteristics breaks down. Is there any hope of solving the Cauchy problem? Yes, but *only* if $h(s)$ is chosen to make Γ a characteristic curve (which does not turn vertical). In fact, if $(f')^2 + (g')^2 \neq 0$ along γ, then the condition $J \equiv 0$ shows $dx/a = dy/b$ so that γ must be a projected characteristic. But this means the Cauchy data h cannot be chosen arbitrarily. Moreover, with this particular choice of h, there is an infinite number of solutions to the Cauchy problem since any smooth union of characteristics gives an integral surface, i.e., the solution is no longer unique. Thus, for characteristic γ the Cauchy problem is certainly not well posed.

Example 2. Consider the equation $u_x + xu_y = u^2$. We find the projected characteristics by solving $dx/1 = dy/x$ to find the parabolas

$$y = \frac{1}{2}x^2 + C.$$

Provided we take γ nowhere tangent to the parabolas, then the Cauchy problem is well posed, i.e., admits a unique solution with Cauchy data $u|_\gamma = h$ for arbitrary (smooth) h. For example, the Cauchy problem is well posed if the Cauchy data h is prescribed on the y-axis. Let us verify that $u|_{x=0} = h$ determines *all* derivatives of the solution on the y-axis. First of all, by solving (11), we can find $u_x|_{x=0}$ and $u_y|_{x=0}$. Differentiating these functions with respect to y will determine $u_{xy}|_{x=0}$ and $u_{yy}|_{x=0}$. To determine $u_{xx}|_{x=0}$ we differentiate the equation with respect to x:

$$u_{xx} = 2uu_x - u_y - xu_{xy}.$$

Now all terms on the right-hand side are known on the y-axis, so we have found u_{xx}. Similarly, we may proceed to find the other higher-order derivatives of u. (Exercise 9 illustrates this procedure.)

On the other hand, let us take γ to be the parabola $y = x^2/2$. Then in order for Γ to be characteristic, we must have

$$\frac{dz}{z^2} = dx.$$

We may integrate this equation to find $-1/z = x + c$ where c is an arbitrary constant. In fact the constant c is determined by picking a point over γ for Γ to pass through. For example, if Γ passes through $(0, 0, z_0)$, then $c = -1/z_0$ and Γ is given by

$$z = \frac{z_0}{1 - z_0 x}.$$

In order to find the infinite family of solutions of this Cauchy problem, let us use the general solution discussed in Section 1.1.e. If we let $\phi(x, y, z) = y - x^2/2$ and $\psi(x, y, z) = x + z^{-1}$, then the general solution may be written as $\phi = f(\psi)$ where f is an arbitrary function. Therefore

$$y - \frac{x^2}{2} = f(x + \frac{1}{z}).$$

Along Γ, $y - x^2/2 = 0$ and $x + z^{-1} = 1/z_0$, so the solution will pass through Γ provided $f(1/z_0) = 0$; with this sole restriction on f we see that there is an infinite number of solutions. ♣

Section 1.1: The Cauchy Problem for Quasilinear Equations

With obvious modifications, the preceding observations also pertain to semilinear equations in n variables. In particular, integrating (4) yields functions $x_i(s_1, ..., s_{n-1}, t)$ for $i = 1, ..., n$ and $z(s_1, ..., s_{n-1}, t)$ that may be used to determine the solution $u(x_1, ..., x_n)$ by the inverse function theorem provided

(12) $$\det\left(\frac{\partial x_i}{\partial(s_j, t)}\right) \neq 0 \quad (i = 1, \ldots, n;\ j = 1, \ldots, n-1).$$

Let us consider a linear example with $n = 3$.

Example 3. Let us solve the Cauchy problem $u_x + xu_y - u_z = u$ with $u(x, y, 1) = x + y$. To convert to the notation used in (3), let us replace x, y, z by x_1, x_2, x_3 so our equation becomes $u_{x_1} + x_1 u_{x_2} - u_{x_3} = u$ and the initial condition becomes $u(x_1, x_2, 1) = x_1 + x_2$ and we are now free to use the variable z for u and write the characteristic equations as

$$\frac{dx_1}{dt} = 1, \qquad \frac{dx_2}{dt} = x_1, \qquad \frac{dx_3}{dt} = -1, \qquad \frac{dz}{dt} = z.$$

The initial surface Γ is just the hyperplane $x_3 = 1$, $z = x_1 + x_2$, so it is certainly noncharacteristic (since any curve in Γ must have $dx_3/dt = 0$). We parameterize Γ by

$$x_1 = s_1, \qquad x_2 = s_2, \qquad x_3 = 1, \qquad z = s_1 + s_2$$

and use these as initial conditions to solve our characteristic equations (being careful to integrate x_1 before we attempt x_2) to find

$$x_1 = t + s_1, \qquad x_2 = \frac{1}{2}t^2 + s_1 t + s_2, \qquad x_3 = -t + 1, \qquad z = (s_1 + s_2)e^t.$$

We can then solve for s_1, s_2, and t and plug into z to find

$$u(x_1, x_2, x_3) = \left(x_1 + x_2 + (x_3 - 1)[1 + x_1 + \frac{1}{2}(x_3 - 1)]\right) e^{1-x_3}.$$

Notice that the solution exists for all values of x_1, x_2, and x_3. ♣

d. Quasilinear Equations

For simplicity we restrict our attention in this and the following section to $n = 2$. To solve the Cauchy problem for the quasilinear equation (1) with Γ parameterized by $(f(s), g(s), h(s))$, we solve the characteristic equations (2) with initial conditions (7) to find the integral surface S. The only difference from the semilinear case is that the characteristic equations for dx/dt and dy/dt need not decouple from the dz/dt equation; this means that we must take the z values into account even to find the projected charcteristic curves in the xy-plane. In particular, this allows for the possibility that the projected characteristics may cross each other.

The condition for solving for s and t in terms of x and y is again expressed as (8). In particular, at $t = 0$ this now takes the form

(13) $$f'(s)\, b(f(s), g(s), h(s)) - g'(s)\, a(f(s), g(s), h(s)) \neq 0,$$

which is geometrically the condition that the tangent to Γ and the vector field $\langle a, b, c\rangle$ along Γ project to vectors in the xy-plane that are nowhere parallel. By continuity, we have the following: *Provided Γ satisfies (13), there is a unique solution $u(x, y)$ of the Cauchy problem for (1), at least in a neighborhood of Γ.* As in the semilinear case, away from Γ (i.e. for larger values of t) the solution may develop singularities. Geometrically, this may be due to the integral surface folding over on itself at some point (x_1, y_1) (see Figure 5). In such a case, the solution experiences a "gradient catastrophe" (i.e., u_x becomes infinite) as $(x, y) \to (x_1, y_1)$, and thereafter the solution u cannot be both single valued and continuous. (We shall discuss discontinuous "weak" solutions in the next section.)

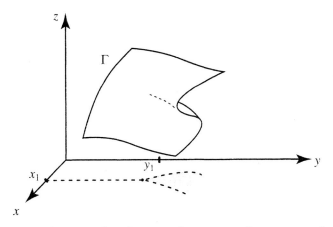

Figure 5. The integral surface experiences a gradient catastrophe.

Example 4. Let us solve the Cauchy problem $uu_x + yu_y = x$ with $u(x, 1) = 2x$. The characteristic equations are

$$\frac{dx}{dt} = z, \qquad \frac{dy}{dt} = y, \qquad \frac{dz}{dt} = x,$$

and Γ may be parameterized by $(s, 1, 2s)$. We easily check that (13) is satisfied: $1 \neq 0$. Notice that the characteristic equation for y happens to decouple and may be integrated to obtain $y = c(s)e^t$, and the initial condition then yields $y = e^t$. The equations for x and z form a 2×2

system, which may be solved by finding eigenvalues and eigenvectors, *or* we can simply observe that

$$\frac{d(x+z)}{dt} = x+z \quad \text{and} \quad \frac{d(x-z)}{dt} = -(x-z),$$

which yield

$$x+z = c_1(s)e^t \quad \text{and} \quad x-z = c_2(s)e^{-t}.$$

Using the initial conditions, we evaluate c_1 and c_2, then solve for x and z:

$$x = \frac{3}{2}se^t - \frac{1}{2}se^{-t}, \qquad y = e^t, \qquad z = \frac{3}{2}se^t + \frac{1}{2}se^{-t}.$$

Notice that z is defined for all s and t, but if we eliminate s and t in favor of x and y we obtain our solution

$$u(x,y) = x\frac{3y^2+1}{3y^2-1},$$

which exists for $|y| < 1/\sqrt{3}$: a blow-up singularity has developed at $y = 1/\sqrt{3}$, which is where $x_s y_t - y_s x_t$ vanishes. ♣

We shall encounter in the next subsection an example where the solution develops a gradient catastrophe type of singularity.

e. General Solutions

In ordinary differential equations, an initial value problem is often solved by finding a general solution that depends on an arbitrary constant and then using the initial condition to evaluate the constant. For quasilinear first-order PDEs, a similar process may be achieved by the *method of Lagrange*, which produces a general solution that depends on an arbitrary function; the Cauchy problem is then solved by evaluating the function.

In order to describe our general solution, let us consider a function $\phi(x,y,z)$ for which $\phi(x,y,z) = $ const is an integral surface of $V = \langle a,b,c \rangle$. This means, of course, that ϕ is constant along the characteristics (2), and so differentiating $\phi(x(t), y(t), z(t)) = $ const with respect to t shows that ϕ satisfies

(14) $$a\phi_x + b\phi_y + c\phi_z = 0.$$

Suppose we have another function $\psi(x,y,z)$ which is also constant along the characteristics (2), but is independent of ϕ (i.e., $\nabla_{x,y,z}\phi$ and $\nabla_{x,y,z}\psi$ are nowhere colinear). Consider the equation

(15) $$F(\phi, \psi) = 0,$$

where F is an arbitrary C^1 function with $F_\phi^2 + F_\psi^2 \neq 0$. Now (15) defines a curve $\mathcal{C}$ in (ϕ, ψ)-space. For each point $(\phi_0, \psi_0) \in \mathcal{C}$, consider the surfaces $S_\phi = \{(x, y, z) : \phi(x, y, z) = \phi_0\}$ and $S_\psi = \{(x, y, z) : \psi(x, y, z) = \psi_0\}$. Suppose S_ϕ and S_ψ intersect; by the independence of ϕ and ψ the intersection will be a curve $\chi \subset \mathbf{R}^3$. Moreover, the characteristic through any point of χ will lie in both S_ϕ and S_ψ and so must coincide with χ; i.e., χ is a characteristic. As we let (ϕ_0, ψ_0) vary over $\mathcal{C}$ we see that (15) defines a surface in $\mathbf{R}^3$ that is a union of characteristic curves χ of (2) and hence is an integral surface for V. Therefore, (15) implicitly defines a solution $z = u(x, y)$ of (1).

This means that the problem of finding a general solution of (1) is reduced to finding two independent functions ϕ and ψ that are constant along the characteristics, whose equations we shall write in nonparameterized form

(16) $$\frac{dx}{a} = \frac{dy}{b} = \frac{dz}{c}.$$

This may require some tricky manipulations, as illustrated in the following.

Example 4 (Revisited). Let us find a general solution for the equation $uu_x + yu_y = x$. We must find two independent functions constant along the characteristic curves
$$\frac{dx}{z} = \frac{dy}{y} = \frac{dz}{x}.$$

Notice that the first and third expressions do not contain a y term and so may be solved by integration to find $z^2 = x^2 + C$. The function $\phi = z^2 - x^2$ is clearly constant along the characteristic curve; in fact it is easy to check that $z\phi_x + y\phi_y + x\phi_z = 0$. Thus we have found one of our functions.

To find a second independent solution, let us use $\phi = z^2 - x^2 = C_1$ to write $z = \sqrt{x^2 + C_1}$. Substituting into the first equation, we have eliminated z and so we obtain the ordinary differential equation

$$\frac{dx}{\sqrt{x^2 + C_1}} = \frac{dy}{y}.$$

This may be integrated to find $\ln|x + \sqrt{x^2 + C_1}| + C_2 = \ln|y|$ or

$$C_3 = \left|\frac{y}{x + \sqrt{x^2 + C_1}}\right|.$$

But recalling $z = \sqrt{x^2 + C_1}$ we have found that $\psi(x, y, z) = y(x + z)^{-1}$ is a constant. It is easy to check that $z\psi_x + y\psi_y + x\psi_z = 0$, so we have found our second independent solution. The general solution can then be written as

$$F(u^2 - x^2, \frac{y}{x + u}) = 0,$$

Section 1.1: The Cauchy Problem for Quasilinear Equations

where $F(\phi, \psi)$ is an arbitrary C^1 function satisfying $F_\phi^2 + F_\psi^2 \neq 0$. Assuming $F_\phi \neq 0$, we may use the implicit function theorem to write (15) as $\phi = f(\psi)$, where f is an arbitrary C^1 function. In this example, then, we could also write the general solution as

$$u^2 = f\left(\frac{y}{x+u}\right) + x^2. \quad \clubsuit$$

We shall next consider an example of a quasilinear equation that arises in applications; it also gives the simplest example of a gradient catastrophe.

Example 5. *(Inviscid Burgers Equation.)* Suppose a one-dimensional stream of particles is in motion, each particle having constant velocity. We may interpret this as a velocity field $u(x, y)$: y denotes time, and $u(x, y)$ gives the velocity of the particle at position x at time y. If we follow an individual particle, we get a function $x(y)$ for which $u(x(y), y)$ remains constant. Differentiating this with respect to y, we obtain a quasilinear equation

(17) $$u\, u_x + u_y = 0.$$

Suppose the velocity field is known initially: $u(x, 0) = h(x)$. We want to solve the Cauchy problem for (17) with Γ parameterized as $(s, 0, h(s))$.

The characteristic equations are

$$\frac{dx}{dt} = z, \qquad \frac{dy}{dt} = 1, \qquad \frac{dz}{dt} = 0.$$

We may integrate these equations to find $z(s, t) = c_1(s)$, $y(s, t) = c_2(s) + t$, and $x(s, t) = c_1(s)t + c_3(s)$. Invoking the initial conditions, we find that $x = h(s)t + s$, $y = t$, and $z = h(s)$. Thus $s = x - zy$ so that

(18) $$z = h(x - zy)$$

defines our solution $z = u(x, y)$ implicitly.

Of course we can also approach the problem by finding the general solution of (17) and then invoking the initial condition to find the particular solution. In this approach, we must find ϕ and ψ constant on the curves

$$\frac{dx}{z} = \frac{dy}{1} = \frac{dz}{0}.$$

The last expression implies that z is constant along the characteristics, so we can take $\phi(x, y, z) = z$ (notice that $u\phi_x + \phi_y = 0$). Treating z as a constant in the first equation, $dx/z = dy$ implies $x = zy + C$. If we take $\psi(x, y, z) = x - zy$, we find that ψ is constant along characteristics. If we solve for ϕ in $F(\phi, \psi) = 0$, we can write $\phi = f(\psi)$ where f is an arbitrary function. In other words, our general solution is $z = f(x - zy)$, which defines u implicitly. The initial condition shows that $f(\psi) = h(\psi)$ so we again obtain (18).

Let us now investigate the possibility that the solution experiences a gradient catastrophe type of singularity. The projected characteristic curve beginning at the point s_1 on the x-axis is parameterized by $(h(s_1)t + s_1, t)$; this is a straight line in the xy-plane having "slope" $dx/dy = h(s_1)$; moreover, z has the constant value $h(s_1)$ along this line. If s_2 is another point on the x-axis with $s_1 < s_2$ but $h(s_1) > h(s_2)$, then the projected characteristic curves beginning from s_1 and s_2 will intersect at

$$y = t = \frac{s_2 - s_1}{h(s_1) - h(s_2)} > 0.$$

Since z is constant along both curves but has different initial values, the integral surface has folded over on itself. Thus if $h'(s_0) < 0$ for any s_0, the solution $u(x, y)$ fails to exist globally, and at the point where (8) fails, the solution suffers a gradient catastrophe type of singularity.

On the other hand, if $h'(s) \geq 0$ for all s, then the characteristics emanating from distinct points s_1 and s_2 on the x-axis will not intersect for positive values of $y = t$, and so the solution $u(x, y)$ will exist globally for $y > 0$.

Given the original physical model behind the equation (17), we can interpret these results as follows: If the initial velocities of the particles form a nondecreasing function of position, then the particles will spread out in a smooth fashion; but if the initial velocities are somewhere decreasing, then the stream of particles will undergo a "shock" that corresponds to collisions of converging particles. ♣

Remark. For nonlinear equations, the term *general solution* need not mean that *all* solutions are of this form. This phenomenon should be familiar from ordinary differential equations; an example of such a partial differential equation is given in Exercise 8.

Exercises for Section 1.1

1. Show that if $z = u(x, y)$ is an integral surface of $V = \langle a, b, c \rangle$ containing a point P, then the surface contains the characteristic curve χ passing through P. (Assume the vector field V is C^1.)

2. If S_1 and S_2 are two graphs [i.e., S_i is given by $z = u_i(x, y)$, $i = 1, 2$] that are integral surfaces of $V = \langle a, b, c \rangle$ and intersect in a curve χ, show that χ is a characteristic curve.

3. If Γ is a characteristic curve of V, show that there is an infinite number of solutions u of (1) containing Γ in the graph of u.

4. Solve the given initial value problem and determine the values of x and y for which it exists:
   (a) $xu_x + u_y = y$, $u(x,0) = x^2$
   (b) $u_x - 2u_y = u$, $u(0,y) = y$
   (c) $y^{-1}u_x + u_y = u^2$, $u(x,1) = x^2$
5. Solve the given initial value problem and determine the values of x, y, and z for which it exists:
   (a) $xu_x + yu_y + u_z = u$, $u(x,y,0) = h(x,y)$
   (b) $u_x + u_y + zu_z = u^3$, $u(x,y,1) = h(x,y)$
6. Solve the initial value problem and determine the values of x and y for which it exists:
   (a) $u_x + u^2 u_y = 1$, $u(x,0) = 1$
   (b) $u_x + \sqrt{u}\, u_y = 0$, $u(x,0) = x^2 + 1$
7. Find a general solution:
   (a) $(x+u)u_x + (y+u)u_y = 0$
   (b) $(x^2 + 3y^2 + 3u^2)u_x - 2xyu_y + 2xu = 0$
8. Consider the equation $u_x + u_y = \sqrt{u}$. Derive the general solution $u(x,y) = (x + f(x-y))^2/4$. Observe that the trivial solution $u(x,y) \equiv 0$ is not covered by the general solution.
9. Consider the equation $y^2 u_x + xu_y = \sin(u^2)$.
   (a) Describe all projected characteristic curves in the xy-plane.
   (b) For the solution u of the initial value problem with $u(x,0) = x$, determine the values of $u_x, u_y, u_{xx}, u_{xy}, u_{yy}$ on the x-axis.

1.2 Weak Solutions for Quasilinear Equations

In applications it is often necessary to consider solutions of quasilinear equations that are not continuous, but have certain mild singularities such as a jump discontinuity along a curve. For example, an initial condition may have such a jump discontinuity that may be propagated for positive time. On the other hand, a quasilinear equation with smooth initial condition may admit a solution that develops a jump discontinuity when the integral surface undergoes a gradient catastrophe, as discussed in the previous section. Such jump discontinuities are called *shocks*; in this section we investigate how to define shocks and locate their discontinuities.

a. Conservation Laws and Jump Conditions

Let us consider shocks for an equation

(19) $$(G(u))_x + u_y = 0,$$

where G is some smooth function of u. If we integrate (19) with respect to x for $a \leq x \leq b$, we obtain

(20) $$G(u(b,y)) - G(u(a,y)) + \frac{d}{dy}\int_a^b u(x,y)\,dx = 0.$$

This is an example of a *conservation law;* such formulas often arise in physics associated with such laws as conservation of momentum or conservation of energy. Notice that (20) implies (19) if u is C^1, but (20) makes sense for more general u; for example, if $\int_a^b u(x,y)\,dx$ is C^1 in y. A solution of (20) is called a *weak solution* of (19). (We shall further discuss weak solutions in Section 2.3.)

Let us consider a solution of (20) that, for fixed y, has a jump discontinuity at $x = \xi(y)$ (see Figure 1).

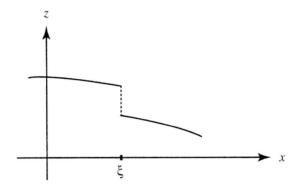

Figure 1. *Solution with a shock at $x = \xi(y)$.*

We shall assume that u, u_x, and u_y are all continuous up to ξ. Moreover, we shall assume that $\xi(y)$ is C^1 in y.

Taking $a < \xi(y) < b$ in (20), we obtain

$$G(u(b,y)) - G(u(a,y)) + \frac{d}{dy}\left(\int_a^\xi u\,dx + \int_\xi^b u\,dx\right)$$
$$= G(u(b,y)) - G(u(a,y)) + \xi'(y)u_\ell(\xi(y),y) - \xi'(y)u_r(\xi(y),y)$$
$$+ \int_a^\xi u_y(x,y)\,dx + \int_\xi^b u_y(x,y)\,dx = 0,$$

where u_ℓ and u_r denote the limiting values of u from the left and right sides of the shock. Letting $a \uparrow \xi(y)$ and $b \downarrow \xi(y)$, we get the *jump condition*

(21) $$\xi'(y) = \frac{G(u_r) - G(u_\ell)}{u_r - u_\ell}.$$

This means that the size of the jump discontinuity controls the propagation of the path of the shock: if we know the values of the solution on either side of the shock, then we can determine the path of the shock.

Example. Notice that (17) is of the form (19) if we take $G(u) = \frac{1}{2}u^2$. Suppose the one-dimensional velocity field has initial condition

(22) $$h(x) = \begin{cases} 0 & \text{for } x > 0 \\ u_0 & \text{for } x < 0, \end{cases}$$

where $u_0 > 0$ is a constant. Recall that u is constant along the characteristics that are of the form $(h(s)t + s, t)$, so $x = \xi(y)$ is a curve in the upper half-plane emanating from $(0,0)$ (where the initial jump discontinuity occurs). Moreover, to the left of this curve $u = u_0$ and to the right $u = 0$, as in Figure 2. Using $G(u) = \frac{1}{2}u^2$, $u_r = 0$, and $u_\ell = u_0$, the shock condition (21) shows that the velocity of the shock wave is one-half the field velocity:

$$\xi'(y) = \frac{1}{2}u_0.$$

We can integrate this to find $\xi(y) = \frac{1}{2}u_0 y + x_0$ and then use the initial condition $\xi(0) = 0$ to obtain $\xi(y) = \frac{1}{2}u_0 y$, as in Figure 2. ♣

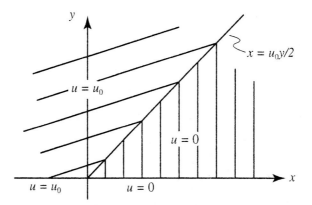

Figure 2. Solution of (15) with initial value (20).

Remark. In Exercise 2, we consider an example related to (22) in which the initial values are continuous but a shock develops after a finite time.

b. Fans and Rarefaction Waves

In (22) it was assumed that $u_0 > 0$, i.e., the initial condition has a larger constant value for $x < 0$ than for $x > 0$, in order for the characteristics to cross, and a shock to form. It is natural to wonder: What happens if $u_0 < 0$? This should have nothing to do with the formation of shocks, because the characteristics are spreading apart; in fact, this spreading has caused a wedge shaped region W to appear in which u is not defined by the characteristics (see Figure 3a). We ask if there is some way to define u in W so as to obtain a weak solution of (17). In fact, since the characteristics do not cross, we might hope our weak solution has no jump discontinuity (i.e., is continuous).

The way to do this is to pick $(x, y) \in W$, draw the line between (x, y) and the origin where the discontinuity occurs, and assign u the value of the "slope" of this line, namely x/y. This has the effect of filling W with a *fan* of lines along which u is constant (see Figure 3b) and allows us to define u by

$$u(x,y) = \begin{cases} 0 & \text{for } x \geq 0 \\ x/y & \text{for } u_0 < x/y < 0 \\ u_0 & \text{for } x/y \leq u_0. \end{cases}$$

Notice that $v(x,y) = x/y$ is a smooth solution of (17) provided $y > 0$; thus u is a smooth solution of (17) *except* along the patching lines $x = 0$ and $x = u_0 y$. Moreover, u is continuous for all values of (x,y), and it is easy to check that (21) is satisfied along $x = 0$ and $x = u_0 y$; for example, along $x = 0$ we have $\xi'(y) \equiv 0$, and if we fix $y > 0$ and let $x = a \uparrow 0$, we get $(G(u_r) - G(u_\ell))/(u_r - u_\ell) = a/2y \uparrow 0$. Hence u is a weak solution of (17).

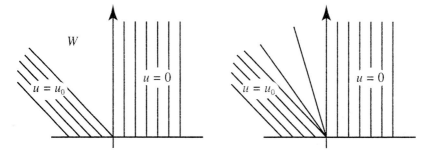

Figures 3a, 3b. Spreading characteristics and filling with a fan.

The critical ingredient in the construction of the weak solution was to recognize that $u(x,y) = x/y$ is a special solution of (17) called a *simple wave*, or *rarefaction wave* because its values appear to fan out. More generally, a solution of (19) of the form $u(x,y) = v(x/y)$ is called a *simple wave*.

Remark. Filling in the region W of Figure 3a by the rarefaction wave x/y is the most reasonable way to define the weak solution. There are other ways of defining a weak solution; for instance, by interposing a shock in the wedge W (see Exercise 6). This means that the weak solution of this initial value problem is not unique unless we somehow eliminate such unnecessary shocks. This is handled by imposing an *entropy condition*, which states that shocks are only allowed to form when the characteristics carrying the initial data cross each other; this will be discussed further in Chapter 10.

c. Application to Traffic Flow

Suppose that we model traffic flow on a highway as a continuous fluid. If the x-axis denotes the highway with traffic flowing in the positive x-direction, let $\rho(x,t)$ denote the density (number of cars per unit length) at the point x and time t, and let $q(x,t)$ denote the *flux* or *flow rate* (number of cars per unit time) at which cars pass the point x at the time t. Let

Section 1.2: Weak Solutions for Quasilinear Equations

us assume that no cars enter or leave the highway and that ρ and q are C^1-functions of x and t. Notice that at time t

$$\int_{x_1}^{x_2} \rho(x,t)\,dx$$

gives the number of cars between the two points x_1 and x_2 on the highway. If we differentiate this quantity with respect to t, we get the rate of change of the number of cars between x_1 and x_2, which also equals $q(x_1,t)-q(x_2,t)$ (since no cars can enter or leave between x_1 and x_2). Thus conservation of cars gives

$$(23) \qquad \frac{d}{dt}\int_{x_1}^{x_2} \rho(x,t)\,dx = q(x_1,t) - q(x_2,t)$$

as a particular case of (20). This can be rewritten as

$$\int_{x_1}^{x_2}\left(\frac{\partial \rho}{\partial t} + \frac{\partial q}{\partial x}\right)dx = 0.$$

Since this must hold for all values of x_1 and x_2, we obtain

$$(24) \qquad \frac{\partial \rho}{\partial t} + \frac{\partial q}{\partial x} = 0.$$

In order to obtain a partial differential equation involving only one unknown function, let us assume that the flow rate may be expressed as a function of the density, i.e., $q = G(\rho)$. In fact, if ρ is small we should expect q to be small and increase as ρ increases (cf. Exercise 7). On the other hand, if ρ is near the maximum possible density of cars (of the given length), then experience with traffic congestion suggests that q should decrease as ρ increases (cf. Exercise 8). In any case, equation (24) becomes $(G(\rho))_x + \rho_t = 0$, which is of the form (19). Notice that this equation can also be written

$$(25) \qquad \frac{dq}{d\rho}\rho_x + \rho_t = 0.$$

In the Exercises we shall apply (25) to some problems involving traffic flow.

Remark. In this application to traffic flow, we have derived the partial differential equation from a conservation law. In fact, this is quite common in physics, which makes it natural to call the partial differential equation itself a conservation law. In Chapter 10, we shall encounter *systems* of conservation laws.

Exercises for Section 1.2

1. Solve the initial value problem $a(u)u_x + u_y = 0$ with $u(x,0) = h(x)$, and show the solution becomes singular for some $y > 0$ unless $a(h(s))$ is a nondecreasing function of s.

2. Consider (17) with the initial condition
$$u(x,0) = h(x) = \begin{cases} u_0 & \text{for } x \leq 0 \\ u_0(1-x) & \text{for } 0 < x < 1 \\ 0 & \text{for } x \geq 1, \end{cases}$$
where $u_0 > 0$. Show that a shock develops at a finite time and describe the weak solution.

3. Consider (17) with the initial condition
$$u(x,0) = h(x) = \begin{cases} 0 & \text{for } x < 0 \\ u_0(x-1) & \text{for } x > 0, \end{cases}$$
where $u_0 > 0$. There is a weak solution $u(x,y)$ that has a jump discontinuity along a curve $x = \xi(y)$. Find this curve and describe the weak solution.

4. Consider (17) with initial condition
$$u(x,0) = h(x) = \begin{cases} u_\ell & \text{for } x < x_0 \\ u_r & \text{for } x > x_0, \end{cases}$$
where $u_\ell < u_r$. Find a continuous weak solution.

5. Find a simple wave solution $u(x,y) = v(x/y)$ for (19) when $G(u) = u^4/4$. Use this to define a continuous weak solution of (19) for $y > 0$ that satisfies (22) with $u_0 < 0$.

6. Consider (17) with (22) where $u_0 < 0$. In addition to the rarefaction solution described in the text, show that there is a weak solution with a shock along $x = u_0 y/2$.

7. A reasonable model for low-density traffic is (25) with $dq/d\rho = c$, where c is a constant.
   (a) Show that ρ is constant along the (characteristic) curves $x = ct + x_0$.
   (b) If a car is alone on the highway, what does $\rho(x,0)$ look like? What does $\rho(x,t)$ look like?
   (c) Explain why c represents the *free speed* of the highway.

8. If $\rho_{\max}$ denotes the maximum density of cars on a highway (i.e., under bumper-to-bumper conditions), then a reasonable relation between q

and ρ is given by $G(\rho) = c\rho(1 - \rho/\rho_{\max})$ where the constant c is the free speed of the highway (cf. Exercise 7). Suppose the initial density is

$$\rho(x,0) = \begin{cases} \frac{1}{2}\rho_{\max} & \text{for } x < 0 \\ \rho_{\max} & \text{for } x > 0. \end{cases}$$

Find the shock curve and describe the weak solution. Interpret your result for the traffic flow.

9. Using $G(\rho)$ as in Exercise 8, describe the traffic flow after a long red light turns green at $t = 0$; that is, the initial density is

$$\rho(x,0) = \begin{cases} \rho_{\max} & \text{for } x < 0 \\ 0 & \text{for } x > 0. \end{cases}$$

In particular, find the density at the green light, $\rho(0,t)$, while the light remains green.

1.3 General Nonlinear Equations

For simplicity, we again assume $n = 2$ (see Exercise 4 for $n \geq 3$). Let us write a general nonlinear equation $F(x, y, u, u_x, u_y) = 0$ as

(26) $$F(x, y, z, p, q) = 0,$$

where p and q represent u_x and u_y, respectively; we make the nondegeneracy assumption $F_p^2 + F_q^2 \neq 0$ to ensure that (26) is in fact a first-order equation. At any fixed point (x_0, y_0, z_0), (26) establishes a functional relationship between p and q. In fact, if we assume that $F_q(x_0, y_0, z_0, p, q) \neq 0$, then the implicit function theorem determines q as a function of p:

$$F(x_0, y_0, z_0, p, q(p)) = 0$$

for all p. [In the quasilinear case, we know $(p, q, -1)$ is perpendicular to a certain vector, so given p we can certainly determine q.]

The possible tangent planes to the graph $z = u(x, y)$ are given by

(27) $$(z - z_0) = p(x - x_0) + q(p)(y - y_0),$$

which, as p varies, describe a one-parameter family of planes through the point (x_0, y_0, z_0). These planes have as their "envelope" (which we shall discuss later) a surface C called the *Monge cone* (see Figure 1). Notice that, in general, the Monge cone C will not be a right circular cone, but rather a ruled surface, everywhere containing a line of tangency with one of the planes defined by (27).

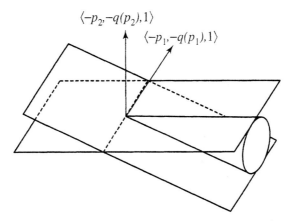

Figure 1. *The Monge cone and its family of tangent planes.*

An *integral surface* is now defined to be a surface S that, at each point (x_0, y_0, z_0), has a tangent plane P that is also tangent to the Monge cone C [i.e., P is of the form (27) for some choice of p]. The unique line of tangency between C and P determines a direction field on S. Integral curves of this field are again defined to be the *characteristic curves*, although they depend on the choice of tangent plane P [i.e., on a choice of p_0 since this determines $q_0 = q(p_0)$]. Notice that, in the quasilinear case $ap + bq = c$, the cone C degenerates to a line, so we need not specify p_0, as we saw in Section 1.1.

a. The Method of Characteristics

The preceding geometrical considerations also provide the means to construct solutions. As in the quasilinear case, we want to derive some ordinary differential equations that we can integrate to find the characteristics. But first let us describe the analytic process behind generating the Monge cone from the family of tangent planes (27).

Suppose S_a is a one-parameter family of surfaces in $\mathbf{R}^3$ given by $z = w(x, y; a)$ where w depends smoothly on x, y, and the real parameter a. Consider also the equation $\partial_a w(x, y; a) = 0$. For a fixed value of a, these two equations determine a curve γ_a in $\mathbf{R}^3$. The *envelope* $\mathcal{E}$ of the family of surfaces S_a is just the union of these curves γ_a. The equation for $\mathcal{E}$ is found simply by solving $\partial_a w(x, y; a) = 0$ for a as a function of x and y, $a = f(x, y)$ and then substituting into $z = w(x, y; a)$ to get $z = w(x, y; f(x, y))$. Moreover, along γ_a, a is constant and we have

$$dz = w_x dx + w_y dy \qquad 0 = w_{ax} dx + w_{ay} dy.$$

For example, if S_a is the one-parameter family of two-dimensional spheres in $\mathbf{R}^3$ of radius 1 and center $(a, 0, 0)$, then the envelope $\mathcal{E}$ of this family is the cylinder of radius 1 centered on the x-axis (see Exercise 5).

If we apply this to the family (27), where p is the parameter, we obtain

(28a) $$dz = p\,dx + q\,dy$$
(28b) $$0 = dx + \frac{dq}{dp}\,dy.$$

If we differentiate (26) with respect to p, we obtain $F_p + F_q \frac{dq}{dp} = 0$; combining this with (28b) we obtain

$$\frac{dx}{F_p} = \frac{dy}{F_q}.$$

With (28a), these equations can be written in parametric form as

(29) $$\frac{dx}{dt} = F_p(x,y,z,p,q) \qquad \frac{dy}{dt} = F_q(x,y,z,p,q)$$
$$\frac{dz}{dt} = p\frac{dx}{dt} + q\frac{dy}{dt} = pF_p(x,y,z,p,q) + qF_q(x,y,z,p,q).$$

However, this system of ordinary differential equations is underdetermined; we also need equations for dp/dt and dq/dt.

If we differentiate (26) with respect to x, we obtain $F_x + F_z p + F_p p_x + F_q q_x = 0$. Using $q_x = p_y$ and (29), we obtain $F_x + F_z p + \frac{dx}{dt} p_x + \frac{dy}{dt} p_y = 0$, so $dp/dt = -F_x - F_z p$. Similarly, differentiating (26) with respect to y yields $\frac{dq}{dt} = -F_y - F_z q$. Thus we may complete the system of ordinary differential equations (29) by adding

(30) $$\frac{dp}{dt} = -F_x - F_z p \qquad \frac{dq}{dt} = -F_y - F_z q.$$

Equations (29)–(30) are again called *characteristic equations*, and the solutions are now called *characteristic strips* because the specification of p and q gives infinitesimal pieces of the tangent planes along the curve $(x(t), y(t), z(t))$ (see Figure 2).

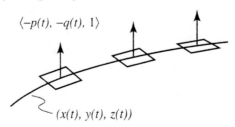

Figure 2. *A characteristic strip.*

Notice that in order to construct our integral surface S, we are really only interested in the *support* of the strip, namely the curve $(x(t), y(t), z(t))$, but to find it we also need to find the functions $p(t)$ and $q(t)$.

To solve the Cauchy problem for (26), we must assume the initial curve Γ is *noncharacteristic* in that, at each point of Γ, the Monge cone is not tangent to Γ. Even so, the Cauchy problem as it stands is unreasonable. *Geometrically* this is because (26) only determines a cone along Γ, and we do not know in which direction to flow along a characteristic; *analytically* we have a system of five ordinary differential equations to solve, but Γ only gives initial values for x, y, and z. The way to resolve this is to specify along Γ two functions ϕ and ψ to give initial conditions for p and q; geometrically this amounts to specifying a tangent plane at each point of Γ, and so replaces the curve Γ by a *strip* (see Figure 3).

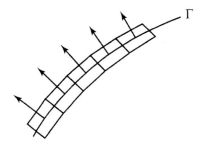

Figure 3. *An initial strip.*

Let us now suppose that Γ is parameterized by $(f(s), g(s), h(s))$. The choice of the tangent planes along Γ [i.e., the choice of the functions $\phi(s)$ and $\psi(s)$] is not arbitrary; they must meet two conditions. First, each tangent plane must be tangent to the Monge cone at that point; in other words the functions ϕ and ψ must satisfy the equation

(31) $$F(f(s), g(s), h(s), \phi(s), \psi(s)) = 0.$$

Second, the planes must fit together smoothly along Γ like the scales of a fish; from (26) we write $dz/ds = p\,(dx/ds) + q\,(dy/ds)$, which implies the *strip condition*

(32) $$h'(s) = \phi(s)f'(s) + \psi(s)g'(s).$$

Functions ϕ and ψ satisfying both (31) and (32) need not be unique and need not even exist! This means the Cauchy problem may not have any solution, or may have more than one solution. However, once ϕ and ψ satisfying (31) and (32) have been fixed, the integral surface S always exists and is unique; as in the quasilinear case, this is a consequence of the existence and uniqueness of solutions of ordinary differential equations. We may formulate these conclusions as the following.

Theorem. *If Γ is noncharacteristic for (26) and functions ϕ, ψ exist satisfying (31) and (32), then there is an integral surface S containing Γ (which is unique for the choice of ϕ and ψ).*

In order to obtain $u(x,y)$ from S, we need to have the Jacobian condition (8), just as in the quasilinear case. This means that, in order to solve the

Cauchy problem in a neighborhood of Γ, the following condition must be satisfied:

(33) $\qquad f'(s)\, F_p[f, g, h, \phi, \psi](s) - g'(s)\, F_q[f, g, h, \phi, \psi](s) \neq 0,$

where $F_p[f, g, h, \phi, \psi](s) \equiv F_p(f(s), g(s), h(s), \phi(s), \psi(s))$ (and analogously for F_q). Notice that (13) is a special case of (33).

Let us illustrate the theorem with a concrete example; other examples will be found in the Exercises and in the application to geometrical optics (see Section 1.3c).

Example. Solve the Cauchy problem $u_x u_y = u$ with $u(0, y) = y^2$. In this case we have $F = pq - z = 0$ and Γ parameterized as $(0, s, s^2)$. First we must complete Γ to a strip. From (31) and (32) we find that ϕ and ψ must satisfy

$$\phi(s)\psi(s) - s^2 = 0, \qquad 2s = \phi(s)0 + \psi(s)1.$$

We solve these algebraic equations to find $\phi(s) = s/2$ and $\psi(s) = 2s$. Now the characteristic equations are

$$\begin{cases} \dfrac{dx}{dt} = q, \qquad \dfrac{dy}{dt} = p, \qquad \dfrac{dz}{dt} = 2pq = 2z, \\ \qquad \dfrac{dp}{dt} = p, \qquad \dfrac{dq}{dt} = q. \end{cases}$$

We can easily integrate the third equation to obtain $z = C(s)e^{2t}$ and use the initial conditions to find $z = s^2 e^{2t}$. Similarly, the last two equations can be solved to find $p = se^t/2$ and $q = 2se^t$. The first two characteristic equations may now be written as $x_t = 2se^t$ and $y_t = se^t/2$. We integrate with initial conditions to obtain $x = 2se^t - 2s$ and $y = s(e^t + 1)/2$. Finally, to obtain $u(x, y)$, let us observe that $x/2 + 2y = 2se^t$, which implies $(x/4 + y)^2 = s^2 e^{2t} = z$; so the solution may be written as

$$u(x, y) = \left(\frac{x}{4} + y\right)^2. \quad \clubsuit$$

b. Complete Integrals and General Solutions

Recall that we considered general solutions for quasilinear equations in Section 1.1 by the method of Lagrange which produced a solution depending on an arbitrary function. Do such general solutions exist for fully nonlinear equations? The answer is yes, but the process is more complicated than in the quasilinear case.

Let us begin with some simple classes of solutions. A *complete integral* for (26) is a solution $u(x, y; a, b)$ that depends on independent parameters a and b. The independence of the parameters means that the mapping $(a, b) \to (u, u_x, u_y)$ should have rank 2 at each fixed x, y. This is not true, for example, if $u(x, y; a, b) = u(x, y; h(a.b))$ for some function $h(a, b)$. On the other hand, a and b are independent parameters if $(a, b) \to (u_x, u_y)$ is invertible; that is,

$$(34) \qquad D = \det \begin{pmatrix} u_{xa} & u_{ya} \\ u_{xb} & u_{yb} \end{pmatrix} \neq 0.$$

Example (Revisited). The equation $u_x u_y = u$ admits solutions of the form $u(x, y; a, b) = xy + ax + by + ab$. [It is possible to obtain these solutions by assuming separation of variables $u(x.y) = X(x)Y(y)$.] It is easy to calculate $D \equiv 1$, so $u(x, y; a, b)$ is indeed a complete integral.

We reiterate that condition (34) is a sufficient but not necessary condition that a and b occur independently; it may be necessary to consider the map $(a, b) \to (u, u_x)$ or $(a, b) \to (u, u_y)$. A simple example is the two-parameter family of solutions $u(x, y; a, b) = ax - ya^2 + b$ for the equation $u_x^2 + u_y = 0$ in which a and b occur independently even though $D \equiv 0$; see Exercise 7.

If S_a is a one-parameter family of integral surfaces for (26), then so is the envelope $\mathcal{E}$. This is obvious since every tangent plane of $\mathcal{E}$ coincides with a tangent plane of one of the S_a. We cannot take the envelope of the complete integral, since it depends on two parameters. But if we first choose an arbitrary function w relating the parameters a and b, $b = w(a)$, then we may consider the envelope $\mathcal{E}_w$ of the one-parameter family of surfaces given by the graphs of $u(x, y; a, w(a))$. This is our candidate for the *general solution* (depending on w).

Example (Revisited). The general solution of $u_x u_y = u$ is the envelope of the solutions $u(x, y; a, w(a)) = xy + ax + w(a)y + aw(a)$. To see that this expression is more general than the complete integral, let us take, for example, $w(a) = a$. Then $u(x, y; a, a) = xy + ax + ay + a^2$, and solving $\partial_a u(x, y; a, a) = x + y + 2a = 0$, we get $a = -(x + y)/2$. Substituting this into u, we find the solution $u(x, y) = -(x - y)^2/4$, which is certainly *not* included in the complete integral $xy + ax + by + ab$.

c. Application to Geometrical Optics

In geometrical optics, wave propagation is studied only in terms of the position of the wave front. In fact, if we have a light wave in the xy-plane, then its wave front at any time is a curve. Let us suppose that there is a function $u(x, y)$ such that the position of the wave front at time t is given by the level curve $u(x, y) = t$; see Figure 4.

Section 1.3: General Nonlinear Equations

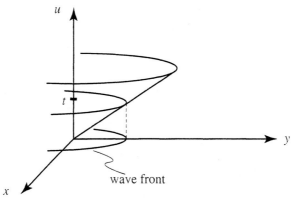

Figure 4. Geometric optics.

If $c(x, y)$ denotes the *propagation speed* of a wave at (x, y), then the curve $(x(t), y(t))$ is a *light ray* if at each point its speed coincides with $c(x, y)$ and its direction is orthogonal to the wave front:

(35)
$$\begin{cases} \sqrt{(\dot{x})^2 + (\dot{y})^2} = c(x, y), \\ \langle \dot{x}, \dot{y} \rangle \text{ and } \nabla u \text{ are colinear,} \end{cases}$$

where of course $\dot{}$ denotes differentiation with respect to t and we have used $\langle , \rangle$ to indicate that the quantity is being considered as a vector. Differentiating $u(x(t), y(t)) = t$, we obtain that the dot product is $(\nabla u) \cdot \langle \dot{x}, \dot{y} \rangle = 1$. But since $\langle \dot{x}, \dot{y} \rangle$ and ∇u are colinear, this implies

(36)
$$c^2(u_x{}^2 + u_y{}^2) = 1.$$

Equation (36) is called the *eikonal equation* of geometrical optics in two dimensions; let us study this equation as an example of a nonlinear partial differential equation. (The three-dimensional eikonal equation is obtained analogously; cf. Exercise 10.)

Let us assume that the propagation speed c is constant and take

$$F(x, y, z, p, q) = \frac{1}{2}(c^2 p^2 + c^2 q^2 - 1).$$

The relationhip $c^2(p^2 + q^2) = 1$ shows that the Monge cone is a right circular cone, making an angle $\theta = tan^{-1} c$ with the z-axis (see Figure 5). The characteristic equations are

(37)
$$\begin{cases} \dfrac{dx}{dt} = c^2 p, & \dfrac{dy}{dt} = c^2 q, & \dfrac{dz}{dt} = c^2 p^2 + c^2 q^2 = 1, \\ & \dfrac{dp}{dt} = 0, & \dfrac{dq}{dt} = 0, \end{cases}$$

where t is the parameter along the characteristic. Notice that p and q are constant along the characteristics. For initial values $(x_0, y_0, z_0, p_0, q_0)$ with $c^2(p_0^2 + q_0^2) = 1$, the characteristic is the straight line

$$x = c^2 p_0 t + x_0, \qquad y = c^2 q_0 t + y_0, \qquad z = t + z_0,$$

which lies on the Monge cone. It is therefore natural to identify t as time (elapsed from some intial time z_0). Moreover, $(x(t), y(t))$ satisfies the condition (35) for a light ray, so we identify the characteristics of (36) with light rays. Since the Monge cone is just the union of all light rays passing through the point (x_0, y_0, z_0), we call it the *light cone*.

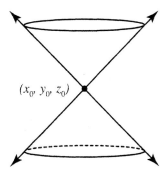

Figure 5. The light cone.

For a Cauchy problem, let us suppose Γ is given parametrically by

(38) $$x = f(s), \qquad y = g(s), \qquad z = h(s).$$

We must complete Γ to a strip by choosing ϕ and ψ to satisfy

(39) $$\begin{cases} c^2(\phi^2(s) + \psi^2(s)) = 1 \\ h'(s) = \phi(s) f'(s) + \psi(s) g'(s). \end{cases}$$

It is easily seen that (39) *cannot* be solved if

(40a) $$(f'(s))^2 + (g'(s))^2 < c^2(h'(s))^2,$$

whereas we will get two different solutions if

(40b) $$(f'(s))^2 + (g'(s))^2 > c^2(h'(s))^2.$$

When (40a) holds, we say Γ is *timelike* and the Cauchy problem *cannot* be solved (see Figure 6a); if (40b) holds we say Γ is *spacelike* and we can solve the Cauchy problem with two different solutions (see Figure 6b).

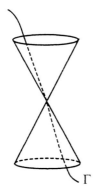
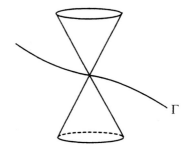

Figure 6a. Γ is timelike.   Figure 6b. Γ is spacelike.

Now let us suppose Γ lies in the xy-plane, so $h(s) \equiv 0$ and Γ is spacelike. Once we determine ϕ and ψ, we can use these as initial conditions in our solution of the characteristic equations (37) to find

(41)
$$x = c^2\phi(s)t + f(s), \quad y = c^2\psi(s)t + g(s), \quad z = t,$$
$$p = \phi(s), \quad q = \psi(s).$$

If we can solve for s, t in terms of x, y, then (41) will provide the desired solution.

For example, let us solve (36) with initial condition

(42) $\qquad \Gamma: \qquad x = \cos s, \qquad y = \sin s, \qquad z = 0.$

To complete Γ to a strip, we must find ϕ and ψ satisfying

$$c^2(\phi^2(s) + \psi^2(s)) = 1, \qquad \phi(s)(-\sin s) + \psi(s)(\cos s) = 0.$$

We find the two solutions $(\phi(s), \psi(s)) = \pm\frac{1}{c}(\cos s, \sin s)$ that lead to the values $x = \cos s\,(1 \pm ct)$ and $y = \sin s\,(1 \pm ct)$. In other words, we can write

(43) $\qquad\qquad\qquad x^2 + y^2 = (1 \pm ct)^2.$

These two solutions of the Cauchy problem correspond to two distinct physical situations. Taking the positive sign in (43) produces a wave front moving away from the origin, and taking the negative sign produces a wave front which will converge on the origin at time c^{-1} (see Figure 7).

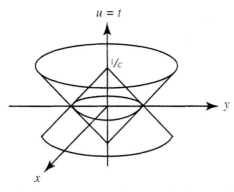

Figure 7. Two solutions to the initial value problem.

We may also study solutions of (36) by means of the complete integral and envelopes. Consider the family of solutions

(44) $$u(x, y; a, b) = c^{-1}(x \cos a + y \sin a) + b,$$

which are all planes making an angle $\theta = \tan^{-1} c$ with the z-axis; at any fixed time $z = t$, the wave front is just a straight line. The family (44) is a complete integral, even though condition (34) fails: $D \equiv 0$, but $(a, b) \to (u, u_x)$ is invertible away from $a = \pm n\pi$, and $(a, b) \to (u, u_y)$ is invertible near $a = \pm n\pi$.

We can take envelopes of (44) to generate additional solutions of (36). For example, with $b \equiv 0$ the family $u(x, y; a) = c^{-1}(x \cos a + y \sin a)$ are planes through the origin $(x, y) = (0, 0)$ corresponding to the possible tangent planes to solutions. If we take the envelope of these planes, we get $a = \tan^{-1}(y/x)$; this produces the solution $u(x, y) = c^{-1}(x^2 + y^2)$ which coincides with the light cone, as should be expected from its identification with the Monge cone.

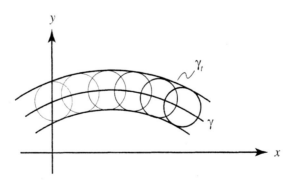

Figure 8. The wave front as the envelope of circular waves.

We can, of course, also use the light cone solutions to generate additional solutions as envelopes. For example, suppose we want to find the solution of (36) passing through a curve γ in the xy-plane. Let us parameterize γ by $(f(s), g(s))$ and let C_s denote the corresponding family of light cones along γ. If we form an envelope of the light cones, we shall get a solution of (36) passing through γ: for small t the wave front will be a curve γ_t that is everywhere at a distance ct from γ. (When γ is the unit circle, this can be seen in Figure 7; a general γ is depicted in Figure 8.) This process realizes the wave front as the envelope of circular wave fronts; this principle, attributed to Huygens, is in fact foundational to the theory of geometrical optics.

Exercises for Section 1.3

1. Solve $u_x^2 + yu_y - u = 0$ with initial condition $u(x,1) = \frac{x^2}{4} + 1$.

2. Solve $u = xu_x + yu_y + (u_x^2 + u_y^2)/2$ with initial condition $u(x,0) = (1-x^2)/2$. [This is an example of Clairaut's equation $u = xp + yq + f(p,q)$.]

3. Consider $u = u_x^2 + u_y^2$ with the initial condition $u(x,0) = ax^2$. For what positive constants a is there a solution? Is it unique? Find all solutions.

4. To generalize (26) to n variables, let $u(x) = u(x_1, \ldots, x_n)$ and $p_j = \partial u/\partial x_j$ for $j = 1, \ldots, n$. Then consider

$$F(x, u, p) = 0.$$

   (a) Find the characteristic equations [i.e., generalizations of (29),(30)].
   (b) Find the conditions on the initial values of p [i.e., generalizations of (31),(32)].

5. Show that the family of spheres S_a given by $(x-a)^2 + y^2 + z^2 = 1$ has as its envelope $\mathcal{E}$ the unit cylinder $y^2 + z^2 = 1$.

6. Consider *Clairaut's equation* $u = xu_x + yu_y + f(u_x, u_y)$.
   (a) Verify that $u(x,y;a,b) = ax + by + f(a,b)$ is a complete integral.
   (b) Use envelopes to generate a solution of $u = xu_x + yu_y + (u_x^2 + u_y^2)/2$ that is not linear in x and y.

7. The *Hamilton-Jacobi equation* $u_t + H(\nabla_x u) = 0$, where H depends only on $\nabla_x u = (u_{x_1}, u_{x_2}, \ldots, u_{x_n})$, arises frequently in physics, such as geometric optics (cf. the eikonal equation). Here we assume $n = 1$.
   (a) Verify that condition (34) fails for $u(x,t;a,b) = ax - tH(a) + b$, yet this is a complete integral.
   (b) Use envelopes to generate a solution of $u_t + u_x^2 = 0$ that is not linear in x and t.

8. To describe the wave front produced by an initial disturbance at a point, consider (36) with Γ being given by $f \equiv g \equiv h \equiv 0$. Describe the solution(s).

9. Instead of assuming the propagation speed is constant, consider $c = c(x,y)$ in (36) and derive the characteristic equations. In the special case $c = |x|$ with initial condition $u(x,0) = 0$, find the solution to be

$$u(x,y) = -\log \frac{\sqrt{x^2 + y^2} + y}{x} \quad \text{for } x > 0.$$

10. Consider the eikonal equation in three dimensions

$$u_x^2 + u_y^2 + u_z^2 = 1.$$

(a) Solve the initial value problem with $u = k = constant$ on the plane $\alpha x + \beta y + z = 0$.
(b) Find a complete integral.

Further References for Chapter 1

The issue of singularity development in solutions of first-order equations is connected with catastrophe theory; see [Arnold].

The discussion of scalar conservation laws in Section 1.2 only mentions the issue of uniqueness of weak solutions and does not consider their large time behavior. For information on these and other aspects of scalar conservation laws, consult [DiPerna], [Lax], or [Smoller].

In Section 1.3 we only mention (in Exercise 7) the rich subject of Hamilton-Jacobi theory, which includes issues of weak solutions, uniqueness, large time asymptotics, and many applications to physics; for more information, consult [Evans] or [Rund]. The application to geometric optics provides a nice blend of physics, geometry, and mathematics; for a detailed exposition, see [Luneburg].

2 Principles for Higher-Order Equations

In this chapter, we discuss some general principles for our study of higher-order equations and systems. We begin with some notation and a normal form for the Cauchy problem. We use this form to consider power series methods and the Cauchy-Kovalevski theorem, which holds for general orders and space dimensions when the equation and Cauchy data are real analytic. We next restrict our attention to the special case of two space dimensions and consider second-order equations and then first-order systems of equations. Finally, we generalize our notion of "solution" by introducing weak solutions and distributions.

2.1 The Cauchy Problem

Let us introduce multi-index notation in order to abbreviate partial derivatives. A *multi-index* is an n-tuple $\alpha = (\alpha_1, \alpha_2, \ldots, \alpha_n)$ where each $\alpha_i \in \mathbf{N}$ (i.e., $\alpha \in \mathbf{N}^n$). Let us also define the following expressions:

$$|\alpha| = \alpha_1 + \cdots + \alpha_n, \qquad \alpha! = (\alpha_1!)\cdots(\alpha_n!).$$

Using multi-index notation, we can abbreviate products and mixed partial derivatives by defining

$$x^\alpha = x_1^{\alpha_1} \cdots x_n^{\alpha_n}, \qquad D^\alpha = \left(\frac{\partial}{\partial x_1}\right)^{\alpha_1} \cdots \left(\frac{\partial}{\partial x_n}\right)^{\alpha_n},$$

where $x = (x_1, \ldots, x_n)$. We can now write a general mth-order equation in $\mathbf{R}^n$ as

(1) $$F(x, D^\alpha u) = 0, \qquad \text{where } |\alpha| \le m.$$

What is the Cauchy problem for this equation?

In Chapter 1, we discussed the Cauchy problem for first-order equations using the method of characteristics. In particular, in Section 1.1c we considered an initial curve $\gamma \subset \mathbf{R}^2$ and Cauchy data $u|_\gamma$; we found that $u|_\gamma$ determines the values of all derivatives of u on γ exactly when γ is noncharacteristic; in this case the Cauchy problem can be solved uniquely.

When $n \ge 3$, the Cauchy problem for (1) involves replacing the initial curve γ by an initial hypersurface $S \subset \mathbf{R}^n$. But if the order m is greater than one, what should we use for Cauchy data, what do we mean by S being noncharacteristic, and can the Cauchy problem be solved?

a. The Normal Form

If we consider the Cauchy problem as an initial value problem, then experience with ordinary differential equations suggests that the *Cauchy data* for the mth-order equation (1) should consist of the values along the initial hypersurface S of $u, \partial u/\partial \nu, \ldots, \partial^{m-1}u/\partial \nu^{m-1}$, where ν is the unit normal vector to S. If the Cauchy problem is well posed, then these values should determine all derivatives of u on S. It is natural, therefore, to take this as a definition: The hypersurface S is *noncharacteristic for* (1) if the values $u, \partial u/\partial \nu, \ldots, \partial^{m-1}u/\partial \nu^{m-1}$ on S determine all derivatives of the solution u on S. This will be true if we can express (1) in the form $\partial^m u/\partial \nu^m = G(x, D^\alpha u)$, where the right-hand side does not contain $\partial^m u/\partial \nu^m$. This is most easily verified if we make a change of variables $x \to \tilde{x}$ in which the hypersurface S is "straightened out" to coincide with the hyperplane where $\tilde{x}_n = 0$: $\tilde{S} = \mathbf{R}^{n-1} = \{\tilde{x} \in \mathbf{R}^n : \tilde{x}_n = 0\}$. The normal form becomes $\partial^m u/\partial \tilde{\nu}^m = \partial^m u/\partial \tilde{x}^m = \tilde{G}(\tilde{x}, \tilde{D}^\alpha u)$ and the condition that $\partial^m u/\partial \tilde{\nu}^m$ does not occur on the right-hand side is just the requirement that $\tilde{G}$ does not contain the term $\tilde{D}^\alpha u$ with $\alpha = (0, \ldots, 0, m)$.

For notational simplicity, we shall assume that our Cauchy problem has already been transformed to this normal form:

$$(2) \qquad \frac{\partial^m u}{\partial x_n^m} = G(x, D^\alpha u), \qquad \text{where } |\alpha| \leq m, \; \alpha \neq (0, \ldots, 0, m),$$

with Cauchy data on the initial surface $S = \{x \in \mathbf{R}^n : x_n = 0\} = \mathbf{R}^{n-1}$:

$$(3) \qquad u|_{x_n=0} = g_1, \; \frac{\partial u}{\partial x_n}\Big|_{x_n=0} = g_2, \ldots, \frac{\partial^{m-1} u}{\partial x_n^{m-1}}\Big|_{x_n=0} = g_m.$$

Now we can easily see that the Cauchy data indeed determines all derivatives of the solution u on S. In fact, to find $D^\alpha u$ on S,

(i) if $\alpha_n = 0$, we simply differentiate g_1 in S;
(ii) if $\alpha_n = 1$, we differentiate g_2 in S;
(iii) similarly, we can treat $\alpha_n \leq m-1$;
(iv) if $\alpha_n = m$, we can use (3) to evaluate (2) on S, and then differentiate in S;
(v) if $\alpha_n = k > m$, we first differentiate (2) with respect to x_n to get an equation of order k, and then follow (iv).

Example 1. Consider the Cauchy problem

$$\begin{cases} u_{xx} + uu_{yy} - u_y = u^2 \\ u(x,0) = 1 \\ u_y(x,0) = x. \end{cases}$$

Written in normal form, the equation becomes

$$u_{yy} = u + \frac{u_y - u_{xx}}{u}.$$

Let us assume u is a smooth solution. To see that we can determine all derivatives of u on $S = x$-axis, let us first observe that we can use $u(x,0) = 1$ to determine $u_x(x,0) = 0 = u_{xx}(x,0) = \cdots$. Next we can use $u_y(x,0) = x$ to determine $u_{yx}(x,0) = u_{xy}(x,0) = 1$, and $u_{yxx}(x,0) = u_{xyx}(x,0) = u_{xxy}(x,0) = 0 = u_{yxxx}(x,0) = \cdots$. To determine u_{yy} on the x-axis, we use the normal form of the equation together with the derivatives already determined:

$$u_{yy}(x,0) = u(x,0) + \frac{u_y(x,0) - u_{xx}(x,0)}{u(x,0)} = 1 + x.$$

Thus we also conclude $u_{yyx}(x,0) = u_{yxy}(x,0) = u_{xyy}(x,0) = 1$, and hence $u_{yyxx}(x,0) = 0 = u_{yxyx}(x,0) = \cdots$. If we differentiate the normal form of the equation we get

$$u_{yyy} = u_y + \frac{u_{yy} - u_{xxy}}{u} - \frac{(u_y - u_{xx})u_y}{u^2},$$

which we can use to determine $u_{yyy}(x,0) = 1 + 2x - x^2$. We may proceed similarly to obtain all derivatives of u on S. ♣

b. Power Series and the Cauchy-Kovalevski Theorem

As we saw in the previous section, (2) and (3) together determine all derivatives of the solution u on S, so we expect the Cauchy problem to be well posed. In fact, if we pick a point x_0 on S, we can use our knowledge of the derivatives of u at x_0 to obtain a formal power series representation for u (i.e., its Taylor series). The problem then becomes to show that this power series actually converges to u in a neighborhood of x_0. As the following example shows, this can only work in general when the functions in the problem are *real analytic* (i.e., have absolutely convergent Taylor series in a neighborhood of each point).

Example 2. Consider the first-order Cauchy problem for the Cauchy-Riemann equation:

$$\begin{cases} u_y - iu_x = 0 \\ u(x,0) = g(x). \end{cases}$$

Notice that u is a complex-valued function of the real variables x, y (although g can be real or complex valued). Notice that we could avoid complex values by separating u into real and imaginary parts to obtain a *system* involving real-valued functions. On the other hand, by the usual identification $z = x + iy$, we may consider u as a function of the complex variable z. Now if u is C^1 and satisfies the Cauchy-Riemann equation $u_y = iu_x$, then complex analysis tells us that u is an analytic (holomorphic) function of z. Therefore, $u \in C^\infty$ and has an absolutely convergent Taylor series about any point. If we restrict u to the real axis, we find that $u(x+i0) = g(x)$ has absolutely convergent Taylor series; hence g is real analytic. Conversely, if $g(x) = \sum a_n(x - x_0)^n$ is absolutely convergent for $|x - x_0| < R$, then we may define $u(z)$ by $u(z) = \sum a_n(z - x_0)^n$, which is absolutely convergent for $|z - x_0| < R$. Complex analysis tells us that u is analytic in $|z - x_0| < R$ and so must satisfy the Cauchy-Riemann equation $u_y = iu_x$. In this way we see that this Cauchy problem admits a C^1 solution if and only if $g(x)$ is real analytic.

Let us try to derive the convergent power series representation directly from the equation and the real analytic Cauchy data. Indeed, let us assume that

$$g(x) = \sum_{j=0}^{\infty} \frac{g^{(j)}(0)}{j!} x^j$$

converges absolutely for $|x| \leq R$. Since the series converges, we certainly have

$$\frac{|g^{(j)}(0)|R^j}{j!} \leq C \qquad \text{for all } j \geq 0.$$

Using $u_y = iu_x$, we obtain $u_{yy} = iu_{xy} = iu_{yx} = i(iu_{xx})$, and similarly

$$\partial_y^j \partial_x^k u = (i\partial_x)^j \partial_x^k u \qquad \text{for all } j, k \geq 0.$$

In terms of the Cauchy data, we may write

$$\partial_y^j \partial_x^k u(0,0) = (i)^j \partial_x^{j+k} u(0,0) = (i)^j g^{(j+k)}(0).$$

Thus our formal power series for u is

$$(4) \qquad u(x,y) \sim \sum_{j,k=0}^{\infty} \frac{\partial_y^j \partial_x^k u(0,0)}{j!\, k!} y^j x^k = \sum_{j,k=0}^{\infty} \frac{(i)^j g^{(j+k)}(0)}{j!\, k!} y^j x^k.$$

which is dominated by

$$(5) \qquad C \sum_{j,k=0}^{\infty} \frac{(j+k)!}{j!\, k!} \left(\frac{|y|}{R}\right)^j \left(\frac{|x|}{R}\right)^k.$$

Therefore, we must show that (5) converges for $|x|$ and $|y|$ sufficiently small. Let $|x| = aR$ and $|y| = bR$ and use

$$(a+b)^m = \sum_{j+k=m} \frac{m!}{j!\, k!} b^j a^k$$

to replace the series in (5) by

$$\sum_{m=0}^{\infty} \sum_{j+k=m} \frac{m!}{j!\, k!} b^j a^k = \sum_{m=0}^{\infty} (a+b)^m,$$

Section 2.1: The Cauchy Problem

which converges provided $a+b < 1$. So (5) converges provided $|x|+|y| < R$, and we see that (4) converges for (x,y) near $(0,0)$. ♣

This example suggests that (2)–(3) may be solved by power series methods provided the equation and the Cauchy data are all real analytic. In fact, since the power series is determined uniquely by the equation and the Cauchy data, we see that the solution must be unique. These are the conclusions of the following theorem, which is striking in its apparent generality.

Cauchy-Kovalevski Theorem. *If g_j for $j = 1, \ldots, m$ are real analytic in a neighborhood of $0 \in \mathbf{R}^{n-1}$, and G is real analytic in a neighborhood of $(0, D^\alpha u(0))$, then there exists a unique real analytic solution u of (2)–(3) defined in a neighborhood of $0 \in \mathbf{R}^n$.*

In particular, this theorem confirms our suspicion that, at least in the real analytic category, the Cauchy data for an mth-order equation in normal form should consist of m arbitrary functions.

The proof of this theorem follows the same outline as the argument in the preceding example: use the real analyticity of the equation and Cauchy data to majorize the formal power series for u, and show that it indeed converges to a solution (for details, see [Garabedian] or [John, 1]). Unfortunately, this procedure is not only lengthy in general but is not very useful for describing the solution.

Another shortcoming of the Cauchy-Kovalevski theorem is its requirement that the Cauchy data be real analytic; this means it fails to recognize *well posed* Cauchy problems. For example, if we take a sequence $g_k(x)$ of real analytic functions that converge uniformly to a continuous function $f(x)$, we cannot assert that the solutions u_k of the Cauchy-Riemann equation with data g_k will converge to a solution with data f. But for a physically meaningful problem, we expect that a small change in the data should induce only a small change in the solution.

Definition. A problem is *well posed (in the sense of Hadamard)* if a solution exists, is unique, and depends continuously on its data.

This is not a rigorous definition since we have not defined what we mean by "depends continuously on its data"; this needs to be specified for the given problem. We shall see in the next chapter that the Cauchy problem for the wave equation is well posed; this is attributable to the wave equation being *hyperbolic* (which we shall define in the next section). The Cauchy-Riemann and the Laplace equations, on the other hand, are *elliptic* equations for which the Cauchy problem is *not* well posed. In fact, in Exercise 7 we discuss *Hadamard's example* for Laplace's equation, which involves a sequence of Cauchy data converging uniformly to zero for which the solutions do not converge uniformly to zero.

The Cauchy-Kovalevski theorem also only asserts uniqueness of the real analytic solution; it does not preclude the existence of nonanalytic solutions. In fact, if we are to consider nonanalytic solutions, we should allow initial data to be nonanalytic. When the equation is *linear*, this is achieved by the *Holmgren uniqueness theorem*, which shows that a solution with vanishing Cauchy data must itself be zero. However, we shall not discuss the details of this result; see [John, 1], [Hörmander, 2, v. I], or [Treves, 1].

For these reasons, the Cauchy-Kovalevski theorem has limited value, in spite of its apparent generality. To obtain more useful techniques and results, we must specialize our study to the equation at hand, as we shall begin to discover in the next section.

c. The Lewy Example

Not only is the real analyticity of the Cauchy data a necessary condition of the Cauchy-Kovalevski theorem, but so is the real analyticity of the function G. This is illustrated by the famous and startling example of H. Lewy: There is a *linear* equation of the form

(6) $$-u_x - iu_y + 2i(x+iy)u_z = F(x,y,z)$$

with $F \in C^\infty(\mathbf{R}^3)$, which admits *no solutions in any open set* (cf. [John, 1], Chapter 8). This means that, regardless of our choice of Cauchy data, we shall be unable to obtain *any* solution (let alone a real analytic solution) in any neighborhood of $0 \in \mathbf{R}^3$.

Exercises for Section 2.1

1. Consider the initial value problem $u_{zz} = u^2 u_x + (u_{xy})^2$, $u(x,y,0) = x-y$, $u_z(x,y,0) = \sin x$. Find the values of u_{xz}, u_{yz}, u_{zz} when $z = 0$.
2. Is the heat equation $u_t = ku_{xx}$ in normal form for Cauchy data on the x-axis? On the t-axis? What form would the Cauchy data (3) take?
3. Find the solution of the initial value problem $u_{yy} = u_{xx} + u$, $u(x,0) = e^x$, $u_y(x,0) = 0$ in the form of power series expansion with respect to y [i.e., $\sum_0^\infty a_n(x)y^n$]. (*Note:* This is *not* a Taylor series.)
4. Find the Taylor series solution about $x,y = 0$ of the initial value problem $u_y = \sin u_x$, $u(x,0) = \pi x/4$.
5. Consider the initial value problem $u_t = u_{xx}$, $u(x,0) = g(x)$, where $g(x) = a_n x^n + \cdots + a_0$ is a polynomial. Find a Taylor series solution about $(0,0)$. Where does it converge?
6. Consider the same initial problem as in the preceding exercise, but with $g(x) = (1 - ix)^{-1}$, which is real analytic for $-\infty < x < \infty$. Derive the formal Taylor series solution $u(x,t)$, but show that it fails to converge for any x,t with $t \neq 0$. Why does this not violate the Cauchy-Kovalevski theorem?
7. Consider the Cauchy problem for Laplace's equation $u_{xx} + u_{yy} = 0$, $u(x,0) = 0$, $u_y(x,0) = k^{-1}\sin kx$, where $k > 0$. Use separation of

variables to find the solution explicitly. If we let $k \to \infty$, notice that the Cauchy data tends uniformly to zero, but the solution does not converge to zero for any $y \neq 0$. Therefore, a small change from zero Cauchy data [which has the solution $u(x,y) \equiv 0$] induces more than a small change in the solution; this means that the Cauchy problem for the Laplace equation is not well posed.

2.2 Second-Order Equations in Two Variables

In this section we shall consider second-order equations and first-order systems involving two independent variables, sometimes denoted x, y and sometimes x, t. For purposes of the Cauchy problem, we would like to know when a curve γ in the xy-plane is characteristic; i.e., when does the Cauchy data along γ *not* determine all derivatives of the solution along γ? For second-order equations, we shall find that the number of characteristic curves through a given point introduces a classification of the equation as one of three types: hyperbolic, parabolic, or elliptic. These three types are of fundamental importance in the theory of partial differential equations.

a. *Classification by Characteristics*

Let us consider second-order equations in which the derivatives of second-order all occur linearly, with coefficients only depending on the independent variables:

(7) $\qquad a(x,y)u_{xx} + b(x,y)u_{xy} + c(x,y)u_{yy} = d(x,y,u,u_x,u_y).$

We shall call an equation of this form *principally linear* and call the left-hand side the *principal part* of (7).

Given a curve γ in the xy-plane, the Cauchy data along γ is

(8) $\qquad\qquad\qquad u|_\gamma = h, \qquad \dfrac{\partial u}{\partial \nu}\big|_\gamma = h_1,$

where ν denotes a choice of unit normal vector along γ. But if τ denotes a unit tangent vector along γ, then we may find $\frac{\partial u}{\partial \tau}$ along γ by differentiating h; hence we may compute *any* directional derivative along γ by taking a linear combination of $\frac{\partial u}{\partial \nu}$ and $\frac{\partial u}{\partial \tau}$. This means that the Cauchy data can also be expressed as

(9) $\qquad\qquad u|_\gamma = h, \qquad \dfrac{\partial u}{\partial x}\big|_\gamma = \phi, \qquad \dfrac{\partial u}{\partial y}\big|_\gamma = \psi,$

provided the *compatibility condition* $h'(s) = \phi(s)f'(s) + \psi(s)g'(s)$ holds, where as usual we have parameterized γ by $(f(s), g(s))$. Notice that the compatibility condition implies that (9) actually involves only two degrees of freedom, as does (8).

Now, substituting $v = u_x$ and then $v = u_y$ into $dv = v_x dx + v_y dy$, we obtain
$$\begin{cases} \phi' = u_{xx} f' + u_{xy} g' \\ \psi' = u_{xy} f' + u_{yy} g'. \end{cases}$$

Thus we get a linear system of three equations in the three unknown functions u_{xx}, u_{xy}, and u_{yy}:
$$\begin{cases} f' u_{xx} + g' u_{xy} & = \phi' \\ \phantom{f' u_{xx} +} f' u_{xy} + g' u_{yy} & = \psi' \\ a u_{xx} + b u_{xy} + c u_{yy} & = d, \end{cases}$$

which can be solved uniquely provided that

$$D \equiv \begin{vmatrix} f' & g' & 0 \\ 0 & f' & g' \\ a & b & c \end{vmatrix} = a(g')^2 - bf'g' + c(f')^2 \neq 0.$$

In other words, the Cauchy data along γ determines all second-order derivatives along γ provided that $D \neq 0$. We therefore define γ to be *characteristic* if $D = 0$.

It is useful to express the characteristic condition $D = 0$ in more geometrical and algebraic terms. Let us associate to the principal part of (7) the *principal symbol*,

$$\sigma(\xi) = \sigma(x, y; \xi) = a(x, y)(\xi_1)^2 + b(x, y)\xi_1 \xi_2 + c(x, y)(\xi_2)^2,$$

where $\xi = (\xi_1, \xi_2)$ is a vector based at the point (x, y). Notice also that the curve γ has tangent vector (f', g'), so the vector $\xi = (g', -f')$ is normal to γ. Therefore, the curve γ is *characteristic* for (7) at (x, y) if and only if the principal symbol vanishes on its normal vector ξ: $\sigma(x, y; \xi) = 0$.

For further investigation, let us remove the parameter s by writing the characteristic condition as $a\, dy^2 - b\, dx\, dy + c\, dx^2 = 0$. If we solve for dy/dx, we obtain

(10) $$\frac{dy}{dx} = \frac{b \pm \sqrt{b^2 - 4ac}}{2a}$$

as an equation for the characteristic curve γ. We have three cases:
 (i) $b^2 > 4ac$, there are *two* characteristics, and (7) is called *hyperbolic*;
 (ii) $b^2 = 4ac$, there is *only one* characteristic, and (7) is called *parabolic*;
 (iii) $b^2 < 4ac$, there are *no* characteristics, and (7) is called *elliptic*.
These definitions are all taken at a point $x_0 \in \mathbf{R}^2$; unless a, b, and c are all constant, the *type* (hyperbolic, parabolic, or elliptic) may change with the point x_0.

Example 1. The equation $u_{xx} - u_{yy} = 0$ has $a = 1$, $b = 0$, and $c = -1$, so $b^2 > 4ac$ and the equation is *hyperbolic*; moreover, the characteristic equation $dy/dx = \pm 1$ has solutions $y = \pm x + c$ as the characteristic curves. (This equation is the one-dimensional wave equation with y representing time, and propagation speed $c = 1$.) ♣

Example 2. The equation $u_{xx} - u_y = 0$ has $a = 1$ and $b = 0 = c$, so $b^2 = 4ac$. The equation is *parabolic*, and the characteristic curves $y = c$ are found by solving $\frac{dy}{dx} = 0$. (This is the one-dimensional heat equation.) ♣

Example 3. The two-dimensional Laplace equation $u_{xx} + u_{yy} = 0$ has $a = 1 = c$ and $b = 0$, so $b^2 < 4ac$ and the equation is *elliptic;* there are no characteristics. ♣

Example 4. The Tricomi equation $u_{yy} - yu_{xx} = 0$ has $a = -y$, $b = 0$, and $c = 1$ so it is *elliptic* if $y < 0$, *parabolic* if $y = 0$, and *hyperbolic* if $y > 0$. Moreover, for $y > 0$ we solve $dy/dx = \pm 1/\sqrt{y}$ to find that the characteristic curves are $3x \pm 2y^{3/2} = c$ (see Figure 1). ♣

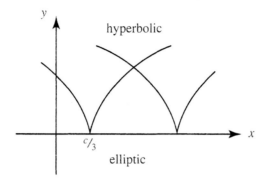

Figure 1. Characteristic curves for the Tricomi equation.

For quasilinear or fully nonlinear second-order equations, a similar classification can be made by *linearization* of the highest-order derivatives; but the resultant type (hyperbolic, parabolic, or elliptic) may now depend on the particular solution u being considered. To see this, consider the nonlinear equation
$$F(x, y, u, u_x, u_y, u_{xx}, u_{xy}, u_{yy}) = 0,$$
and then let
$$a = \frac{\partial F}{\partial u_{xx}}, \qquad b = \frac{\partial F}{\partial u_{xy}}, \qquad c = \frac{\partial F}{\partial u_{yy}},$$
all of which may depend on $x, y, u, u_x, u_y, u_{xx}, u_{xy}, u_{yy}$. In particular, for a quasilinear equation $au_{xx} + bu_{xy} + cu_{yy} = d$, where a, b, c, and d may depend on u, u_x, and u_y (in addition to x and y), the characteristics are again defined by (10) but now may depend on the solution u.

Example 5. The equation $u_{xx} - u\,u_{yy} = 0$ is hyperbolic where $u > 0$ and elliptic where $u < 0$. In particular, the equation is elliptic with respect to the solution $u \equiv c < 0$ and hyperbolic with characteristics $y = \pm\sqrt{c}\,x + k$ with respect to the solution $u \equiv c > 0$. ♣

b. Canonical Forms and General Solutions

By the introduction of new coordinates μ and η in place of x and y, the equation (7) may be transformed so that its principal part takes the form of either Example 1, 2, or 3; in this way these examples provide *canonical forms* for the three types of equations. For example, suppose (7) is *hyperbolic*. Then the characteristic curves are given by $\mu(x,y) = $ const and $\eta(x,y) = $ const, where μ, η define nondegenerate coordinates (i.e., the Jacobian determinant $\mu_x \eta_y - \mu_y \eta_x \neq 0$). Geometrically, of course, this means that the normal vectors $\nabla \mu = (\mu_x, \mu_y)$ and $\nabla \eta = (\eta_x, \eta_y)$ are not colinear; but since the curves are characteristic, the principal symbol must vanish on these vectors:

(11) $$\begin{cases} a\,\mu_x{}^2 + b\,\mu_x \mu_y + c\,\mu_y{}^2 = 0 \\ a\,\eta_x{}^2 + b\,\eta_x \eta_y + c\,\eta_y{}^2 = 0. \end{cases}$$

If we use the chain rule to change variables, we obtain

$$a u_{xx} + b u_{xy} + c u_{yy} = B u_{\mu\eta} + \{\text{lower order}\},$$

where

$$B = 2a\mu_x \eta_x + b(\mu_x \eta_y + \mu_y \eta_x) + 2c\mu_y \eta_y.$$

Using (11), the following may be verified by direct calculation:

$$B^2 = (b^2 - 4ac)(\mu_x \eta_y - \mu_y \eta_x)^2.$$

Hyperbolicity and nondegeneracy imply that $B \neq 0$, so we may divide by B to reduce (7) to

(12a) $$u_{\mu\eta} = \tilde{d}(\mu, \eta, u, u_\mu, u_\eta).$$

Finally, the substitution $\bar{x} = \mu + \eta$, $\bar{y} = \mu - \eta$ reduces (12a) to

(12b) $$u_{\bar{x}\bar{x}} - u_{\bar{y}\bar{y}} = \bar{d}(\bar{x}, \bar{y}, u, u_{\bar{x}}, u_{\bar{y}}).$$

Similarly, if (7) is *parabolic* or *elliptic*, then there exists a change of variables $\mu(x,y)$ and $\eta(x,y)$ under which (7) becomes

(13) $$u_{\mu\mu} = \tilde{d}(\mu, \eta, u, u_\mu, u_\eta),$$

or

(14) $$u_{\mu\mu} + u_{\eta\eta} = \tilde{d}(\mu, \eta, u, u_\mu, u_\eta),$$

respectively. We shall encounter a specific instance of the parabolic case in Example 7, but the elliptic case is quite complicated, requiring a solution of the *Beltrami equation* to define the new coordinates (cf. Exercise 10).

These canonical forms are of more than just theoretical interest, as they may sometimes be used to find the *general solution*.

Example 6. Consider the equation $x\,u_{xx} + 2x^2 u_{xy} = u_x - 1$. Here we have $a = x$, $b = 2x^2$, and $c = 0$, so $b^2 - 4ac = 4x^4 > 0$ and the equation is hyperbolic (provided $x \neq 0$). The characteristic curves are found by solving

$$\frac{dy}{dx} = \frac{2x^2 \pm \sqrt{4x^4}}{2x} = \begin{cases} 2x \\ 0 \end{cases}$$

to find $y = x^2 + c$ and $y = c$ (see Figure 2).

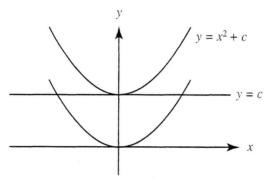

Figure 2. Characteristic curves for $xu_{xx} + 2x^2 u_{xy} = u_x - 1$.

Therefore, let $\mu(x,y) = x^2 - y$ and $\eta(x,y) = y$. We compute $\mu_x = 2x$, $\mu_y = -1$, $\eta_x = 0$, $\eta_y = 1$, $u_x = 2xu_\mu$, $u_{xx} = 2u_\mu + 4x^2 u_{\mu\mu}$, and $u_{xy} = -2xu_{\mu\mu} + 2xu_{\mu\eta}$. Inserting these expressions into our equation and simplifying, we obtain $4x^3 u_{\mu\eta} = -1$ or $u_{\mu\eta} = -\frac{1}{4}(\mu+\eta)^{-3/2}$, which is the desired canonical form. But we can integrate this last equation: $u_\mu = \frac{1}{2}(\mu+\eta)^{-1/2} + f(\mu)$ and so $u = (\mu+\eta)^{1/2} + F(\mu) + G(\eta)$. Converting to the variables x and y, we obtain our general solution

$$u(x,y) = x + F(x^2 - y) + G(y),$$

where F and G are arbitrary functions. ♣

Example 7. Consider the equation

$$x^2 u_{xx} + 2xu_{xy} + u_{yy} = u_y.$$

Now we have $a = x^2$, $b = 2x$, and $c = 1$, so $b^2 - 4ac = 0$ and the equation is parabolic. We find the characteristics by solving

$$\frac{dy}{dx} = \frac{1}{x}$$

to find $y = \ln|x| + c$ or $x = Ce^y$. Therefore, let $\mu(x,y) = xe^{-y}$; we must choose η to obtain new coordinates. A simple choice is $\eta(x,y) = y$, since the curves $y = const.$ are always transverse to the curves $xe^{-y} = const.$ (see Figure 3).

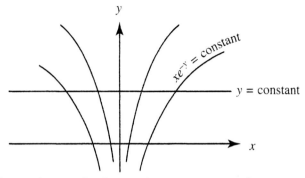

Figure 3. Level curves for the coordinates $\mu = xe^{-y}$ and $\eta = y$.

Expressing our equation in terms of these variables, we obtain our canonical form $u_{\eta\eta} = u_\eta$. Since we are fortunate enough to have only η-derivatives, we can integrate this equation to obtain $u_\eta - u = F(\mu)$. This first-order equation admits an integrating factor of $e^{-\eta}$, and we obtain $u = -F(\mu) + G(\mu)e^\eta$. Converting to x and y, we obtain our general solution

$$u(x,y) = -F(xe^{-y}) + G(xe^{-y})e^y,$$

where F and G are arbitrary functions. ♣

c. First-Order Systems

Recall that a second-order ODE of the form $y'' = f(x, y, y')$ can be reduced to a first-order system of ODEs; letting $u_1 = y$ and $u_2 = y'$ we obtain

$$\begin{pmatrix} u_1 \\ u_2 \end{pmatrix}' = \begin{pmatrix} u_2 \\ f(x, u_1, u_2) \end{pmatrix}.$$

A similar procedure reduces a second-order PDE to a first-order system.

Example 8. Consider $u_{xx} - u_{yy} = 0$ as in Example 1. Let $\vec{u}$ denote the column vector with components $u_1 = u_x$ and $u_2 = u_y$. (We could also let a third vector component be $u_0 = u$ in analogy with the ODE; but if we can find u_x and u_y, then we should be able to integrate to find u itself.) Since $(u_1)_y = u_{xy} = (u_2)_x$ and $(u_2)_y = u_{yy} = u_{xx} = (u_1)_x$, we can write the equation as the system

$$\vec{u}_y + \begin{pmatrix} 0 & -1 \\ -1 & 0 \end{pmatrix} \vec{u}_x = \begin{pmatrix} 0 \\ 0 \end{pmatrix}.$$

Section 2.2: Second-Order Equations in Two Variables

Since the second-order equation has two characteristics, we expect this system to have two characteristics. But what are "characteristics" for a first-order system? Are they of use in solving the system? ♣

We are therefore led to considering general systems of first-order PDEs on $\mathbf{R}^2$ of the form

(15) $$A(x,y)\vec{u}_x + B(x,y)\vec{u}_y = \vec{c}(x,y,\vec{u})$$

where $\vec{u}$ and $\vec{c}$ are N-vectors (functions of x, y and $x, y, \vec{u}$, respectively) and A and B are $N \times N$-matrices (functions of x, y); we shall assume that all functions depend continuously on their variables. What is a characteristic for this system? From our experience so far, we define a *characteristic* to be a curve γ in $\mathbf{R}^2$ for which the Cauchy data $\vec{u}|_\gamma$ does *not* determine the derivatives $\vec{u}_x$ and $\vec{u}_y$ along γ. To be more precise, let us suppose that γ is the graph of a function $\phi(x)$, so $(x, \phi(x))$ parameterizes γ and $\vec{u}(x, \phi(x)) = \vec{f}(x)$ is the Cauchy data. There are two equations to hold along γ:

(16) $$\begin{cases} A\vec{u}_x + B\vec{u}_y = \vec{c} \\ \vec{u}_x + \phi'\vec{u}_y = \vec{f}' \end{cases}$$

the latter of course arising by differentiating the Cauchy data with respect to x. Let us multiply the second equation by A and subtract it from the first:

$$(B - \phi' A)\vec{u}_y = \vec{c} - A\vec{f}'.$$

Thus we shall *not* be able to solve for $\vec{u}_y$ if

(17) $$\det(B - \frac{dy}{dx}A) = 0, \quad \text{where } \phi' = \frac{dy}{dx}.$$

The condition (17) is the *characteristic equation* for (15).

As for second-order equations, the characteristic condition may be expressed in more geometrical and algebraic terms. If we define the *principal part* of (15) to be the left-hand side and define the *principal symbol* to be the $N \times N$-matrix

$$\sigma(\xi) = \sigma(x, y; \xi) = A(x,y)\xi_1 + B(x,y)\xi_2,$$

which depends linearly on the vector $\xi = (\xi_1, \xi_2)$, then γ is *characteristic* for (15) at (x, y) if and only if the principal symbol matrix is singular on the vector ξ that is normal to γ.

The situation becomes somewhat simpler for an initial value problem (i.e., replacing y by t and letting γ be the curve $t = 0$). In this case the normal vector is $\xi = (0, \xi_2)$, so γ is noncharacteristic provided that

(18) $$\det B(x, 0) \neq 0 \quad \text{for all } x.$$

But continuity and condition (18) then guarantee that $\det B(x,t) \neq 0$ for t sufficiently small; hence the matrix $B(x,t)$ is invertible, at least for small values of t. If we multiply (15) by $B^{-1}(x,t)$ (and replace $B^{-1}A$ by A and $B^{-1}\vec{c}$ by $\vec{c}$), we may assume (15) is in the simpler form

$$\vec{u}_t + A(x,t)\vec{u}_x = \vec{c}(x,t,\vec{u}) \qquad \text{near } t = 0. \tag{19}$$

For (19) the condition (17) becomes simply

$$\det\left(\frac{dx}{dt} - A\right) = 0. \tag{20}$$

Thus to be a characteristic, the curve γ parameterized by $(x(t), t)$ should satisfy

$$\frac{dx}{dt} = \lambda(x,t), \tag{21}$$

where $\lambda(x,t)$ is an eigenvalue for $A(x,t)$.

What does "hyperbolicity" mean for (19)? Recall from the previous section that a second-order hyperbolic equation in $\mathbf{R}^2$ has two characteristics. If, as in Example 1, we reduce the equation to a 2×2 first-order system in the form (19), we should retain hyperbolicity by having two characteristics defined by (21). Thus hyperbolicity requires the existence of two distinct real eigenvalues λ_1, λ_2 for the matrix A.

We shall therefore define (19) to be *strictly hyperbolic* if the $N \times N$ matrix A has N distinct real eigenvalues $\lambda_1, \ldots, \lambda_N$. In this case, linear algebra guarantees that the associated eigenvectors $\vec{\mu}_1, \ldots, \vec{\mu}_N$ are linearly independent. As we shall see, this linear independence is the essential feature for our purposes, so it is natural to allow the following generalization: (19) is *hyperbolic* if A has all real eigenvalues and a basis of eigenvectors. For example, if the matrix A is symmetric (i.e., equal to its transpose), then linear algebra confirms that A has real eigenvalues and a basis of eigenvectors; in this case (19) is *symmetric hyperbolic*.

We can use the eigenvalues and eigenvectors to diagonalize the matrix A, thereby simplifying the left-hand side of (19). Let Γ denote the matrix having the eigenvectors $\vec{\mu}_1, \ldots, \vec{\mu}_N$ as column vectors. Then Γ^{-1} exists and $\Gamma^{-1}A\Gamma = \Lambda$, where Λ is the diagonal $N \times N$ matrix with $\lambda_1, \ldots, \lambda_N$ as diagonal entries. If we introduce the new dependent variable $\vec{v}$ by $\vec{u} = \Gamma\vec{v}$, we find that (19) becomes

$$\vec{v}_t + \Lambda\vec{v}_x = \vec{d}(x,t,\vec{v}), \tag{22}$$

where $\vec{d}(x,t,\vec{v}) = \Gamma^{-1}(\vec{c}(x,t,\vec{u}) - \Gamma_t \vec{v} - A\Gamma_x \vec{v})$. If we were also given the initial data $\vec{u}(x,0) = \vec{f}(x)$ for (19), then for (22) we have

$$\vec{v}(x,0) = \vec{g}(x), \tag{23}$$

where $\vec{g}(x) = \Gamma^{-1}\vec{f}(x)$.

Notice that the left-hand side of (22) "decouples": the kth equation in (22) depends on the derivatives of v_k only. If we should be so fortunate as to have the kth component of $d(x,t,\vec{v})$ depending only on v_k, then (22) decouples into N equations, each of which may be solved by the method of characteristics. On the other hand, even when (22) does not decouple, it may be used to show the initial value problem has a unique solution, at least for small values of t. This will be discussed in Section 10.1.

d. Application to the Telegraph System

In the study of electrical transmission along a linear cable in which the current $I = I(x,t)$ may leak to ground, the resultant decrease in current is governed by

$$(24) \qquad I_x = -GV - CV_t,$$

where $V = V(x,t)$ is the voltage, $G = G(x)$ is the conductance to ground, and $C = C(x)$ is the capacitance to ground. The change in voltage, on the other hand, is governed by

$$(25) \qquad V_x = -RI - LI_t,$$

where $R = R(x)$ is the resistance and $L = L(x)$ is the inductance of the cable. (See [Sokolnikoff-Redheffer] p. 465 for further explanation and the derivation of these equations.) If we introduce

$$\vec{u} = \begin{pmatrix} I \\ V \end{pmatrix},$$

then we can write (24) and (25) as in (19):

$$(26) \qquad \vec{u}_t + A(x)\vec{u}_x = \vec{c}(x,\vec{u}),$$

where

$$A(x) = \begin{pmatrix} 0 & 1/L(x) \\ 1/C(x) & 0 \end{pmatrix}, \quad \vec{c}(x,\vec{u}) = \begin{pmatrix} -R(x)/L(x) & 0 \\ 0 & -G(x)/C(x) \end{pmatrix}\vec{u}.$$

Notice that (26) is in fact a *linear* system, which we wish to solve with initial conditions

$$(27) \qquad I(x,0) = I_0(x), \qquad V(x,0) = V_0(x).$$

For simplicity, let us now assume that G, C, R, and L are all *positive constants*. Then the matrix A has two distinct real eigenvalues

$$\lambda_\pm = \pm \frac{1}{\sqrt{LC}},$$

with associated eigenvectors

$$\vec{\mu}_+ = \begin{pmatrix} \sqrt{C} \\ \sqrt{L} \end{pmatrix}, \qquad \vec{\mu}_- = \begin{pmatrix} \sqrt{C} \\ -\sqrt{L} \end{pmatrix}.$$

If we let

$$\Gamma = \begin{pmatrix} \sqrt{C} & \sqrt{C} \\ \sqrt{L} & -\sqrt{L} \end{pmatrix} \qquad \text{and} \qquad \vec{v} = \Gamma^{-1}\vec{u},$$

then (26) takes the form (22):

(28) $$\vec{v}_t + \Lambda \vec{v}_x = D\vec{v},$$

where

$$\Lambda = \begin{pmatrix} 1/\sqrt{LC} & 0 \\ 0 & -1/\sqrt{LC} \end{pmatrix},$$

and

$$D\vec{v} = \Gamma^{-1}\vec{c}(x, \Gamma \vec{u}) = -\frac{1}{2LC}\begin{pmatrix} RC+GL & RC-GL \\ RC-GL & RC+GL \end{pmatrix}\vec{v}.$$

In the special case where

(29) $$RC = GL,$$

then (28) decouples to two separate equations. It suffices to consider

(30) $$v_t \pm \frac{1}{\sqrt{LC}} v_x = -\frac{RC+GL}{2LC} v,$$

where $v = v_1$ if we take "+" and $v = v_2$ if we take "−." Of course, we also need initial conditions, which, from (27), may be written as

(31) $$v(x, 0) = \frac{1}{2\sqrt{LC}} \left[\sqrt{L} I_0(x) \pm \sqrt{C} V_0(x)\right],$$

where the ± corresponds to (30). The problem (30)–(31) may be solved by the method of characteristics (see Exercise 8). When (29) does *not* hold, then (28) is not so simple to solve. Nevertheless, a solution can be shown to exist (see Section 10.1).

Exercises for Section 2.2

1. Reduce to canonical form:
   (a) $u_{xx} + 5u_{xy} + 6u_{yy} = 0$
   (b) $x^2 u_{xx} - y^2 u_{yy} = 0$
2. Find the general solution:
   (a) $u_{xx} - 2u_{xy}\sin x - u_{yy}\cos^2 x - u_y \cos x = 0$
   (b) $y^2 u_{xx} - 2y\, u_{xy} + u_{yy} = u_x + 6y$

Section 2.3: Linear Equations and Generalized Solutions

3. Show that the function

$$u(x,y) = \begin{cases} 0 & \text{if } x \leq y \\ (x-y)^2 & \text{if } x > y \end{cases}$$

satisfies $u_{xx} - u_{yy} = 0$ for all x, y. Is $u \in C^1(\mathbf{R}^2)$? Where does u fail to be C^2?

4. Show that the *minimal surface equation* $(1 + u_y^2)u_{xx} - 2u_x u_y u_{xy} + (1 + u_x^2)u_{yy} = 0$ is everywhere elliptic.

5. Show that the *Monge-Ampère equation* $u_{xx}u_{yy} - u_{xy}^2 = f(x)$ is elliptic for a solution u exactly when $f(x) > 0$. [In this case the graph of $u(x,y)$ is *convex*.]

6. Reduce to the form (22) and solve the initial value problem

$$\vec{u}_t + \begin{pmatrix} -4 & -6 \\ 3 & 5 \end{pmatrix} \vec{u}_x = \begin{pmatrix} 1 \\ -1 \end{pmatrix}, \qquad \vec{u}(x,0) = \begin{pmatrix} x \\ 0 \end{pmatrix}.$$

7. Reduce the following systems to the form (22)

(a) $u_t + v_x = u$
$v_t + u_x = v$

(b) $u_t + v_x = u$
$v_t + u_x = 0$

Do they both decouple?

8. Solve (26),(27) under the condition (29) [by solving (30),(31)].

9. Solve the wave equation $u_{xx} - u_{yy} = 0$ with initial conditions $u(x,0) = g(x)$, $u_t(x,0) = h(x)$ by using a 3×3 system of first-order equations.

10. If $au_{xx} + 2bu_{xy} + cu_{yy} = d$ is elliptic (i.e., $ac - b^2 > 0$), let $W = \sqrt{ac - b^2}$. Show that solutions μ, η of the *Beltrami equations*

$$\mu_x = \frac{b\eta_x + c\eta_y}{W} \qquad \mu_y = -\frac{a\eta_x + b\eta_y}{W}$$

provide new coordinates transforming (7) to the form (14). [Note that $\mu(x,y) = \int_\gamma \mu_x dx + \mu_y dy$, where γ is a path joining (x,y) and a fixed point p_0; path independence is provided by the Beltrami equations.]

2.3 Linear Equations and Generalized Solutions

In this section, we consider some general remarks concerning the special case of (1) when F is linear in u and all its derivatives. In other words, we want to consider an mth-order linear differential operator

(32a) $$Lu = \sum_{|\alpha| \leq m} a_\alpha(x) D^\alpha u,$$

where each $a_\alpha(x)$ is a real-valued function; usually we assume $a_\alpha \in C^{|\alpha|}(\Omega)$, for reasons we shall soon discover. (The case of complex-valued or matrix-valued coefficients can also be treated with modifications; see Exercise 1.) As in the previous section, we define the *principal part* of (32a) to be the top-order terms

$$\text{(32b)} \qquad \sum_{|\alpha|=m} a_\alpha(x) D^\alpha,$$

and the *principal symbol* to be the function

$$\text{(32c)} \qquad \sigma_L(x,\xi) = \sum_{|\alpha|=m} a_\alpha(x) \xi^\alpha,$$

defined for $\xi = (\xi_1, \ldots, \xi_n) \in \mathbf{R}^n$. A vector $\xi \in \mathbf{R}^n$ is *characteristic* for L at x if $\sigma_L(x,\xi) = 0$, and a hypersurface S is *characteristic* for L at $x \in S$ if its normal vector ξ at x is characteristic.

We know from Section 2.1 that the Cauchy problem can be solved in the real-analytic category for noncharacteristic initial surfaces. In this section, however, we want to investigate other topics and properties that are peculiar to linear operators of the form (32a). These topics include weak solutions, distributions, convolutions, and fundamental solutions.

a. Adjoints and Weak Solutions

In Section 1.2 we encountered weak solutions to certain first-order equations. Although the equations were allowed to be nonlinear, they were of a special form that allowed evaluation of integration with respect to x; a function satisfying the resultant integro-differential equation was a weak solution. Integration is also an effective means of defining weak solutions for higher-order equations, again provided they are in an appropriate form. In particular, weak solutions may be defined for linear equations using integration by parts and a "test function"; this leads to the notion of the "adjoint" of a linear operator.

Let $\Omega \subset \mathbf{R}^n$ be a bounded domain with piecewise C^1 boundary so the divergence theorem holds [cf. (34) in the Introduction]. As a consequence, for $u, v \in C^1(\overline{\Omega})$ we obtain the *integration by parts* formula

$$\text{(33)} \qquad \int_\Omega \frac{\partial u}{\partial x_k} v \, dx = -\int_\Omega u \frac{\partial v}{\partial x_k} \, dx + \int_{\partial \Omega} u v \nu_k \, dS,$$

where ν_k is the kth component of the exterior unit normal on $\partial\Omega$. In particular, suppose v (or u) vanishes near $\partial\Omega$; then the boundary integral vanishes and we have

$$\text{(34)} \qquad \int_\Omega \frac{\partial u}{\partial x_k} v \, dx = -\int_\Omega u \frac{\partial v}{\partial x_k} \, dx.$$

Section 2.3: Linear Equations and Generalized Solutions

When v vanishes near $\partial\Omega$, we no longer require u and its first-order derivatives to be continuous on $\overline{\Omega}$, i.e., (34) holds for $u \in C^1(\Omega)$ and $v \in C_0^1(\Omega)$.

Of course we can repeat the integration by parts with any combination of derivatives, $D^\alpha = (\partial/\partial x_1)^{\alpha_1} \cdots (\partial/\partial x_n)^{\alpha_n}$, to obtain

$$(35) \qquad \int_\Omega (D^\alpha u)\, v\, dx = (-1)^m \int_\Omega u\, D^\alpha v\, dx \qquad (m = |\alpha|)$$

for all $u \in C^m(\Omega)$ and $v \in C_0^m(\Omega)$. Let us now consider the operator L of (32a); (35) shows that

$$(36) \qquad \int_\Omega (Lu)\, v\, dx = \int_\Omega u\, (L'v)\, dx$$

for all $u \in C^m(\Omega)$ and $v \in C_0^m(\Omega)$, where

$$(37) \qquad L'v = \sum_{|\alpha| \le m} (-1)^{|\alpha|} D^\alpha (a_\alpha(x) v).$$

The operator L' is called the *adjoint of* L and is an mth-order linear differential operator with continuous coefficients [involving the derivatives of $a_\alpha(x)$, which explains why we required $a_\alpha \in C^{|\alpha|}(\Omega)$].

Now, if u satisfies $Lu = f$ in Ω, then

$$(38) \qquad \int_\Omega u\, L'v\, dx = \int_\Omega f\, v\, dx$$

holds for *every* $v \in C_0^m(\Omega)$. But notice that (38) no longer requires u to have continuous derivatives; in general, u (and f) need only be *integrable on compact subsets of* Ω [i.e., belong to $L_{\text{loc}}^1(\Omega)$]. This leads us to define a function $u \in L_{\text{loc}}^1(\Omega)$ to be a *weak solution* of $Lu = f$ if (38) holds for every $v \in C_0^m(\Omega)$. In particular, if $L = \partial/\partial x_k$ and u is a weak solution of $\partial u/\partial x_k = f$, then we say f is the *weak derivative* of u. (Notice that we are *not* saying that every integrable function u has a weak derivative that is itself a function; in general, we may encounter *distributions*, which we shall discuss later.)

We should verify that this notion of a weak solution is consistent with our conventional notion of a solution: if u satisfies (38) in a domain Ω where u is C^m, then does $Lu = f$ in the classical sense? Consider a point $x \in \Omega$ and pick $\epsilon > 0$ so that $B = B_\epsilon(x)$ satisfies $\overline{B} \subset \Omega$. Consider a function $v \in C_0^m(B)$. Since $u \in C^m(B)$, we may use (36) to conclude $\int (Lu)\, v\, dx = \int f v\, dx$. But because v was arbitrary, this means $Lu \equiv f$ in B; and since x was arbitrary in Ω, this means $Lu \equiv f$ in Ω. The argument also illustrates the use of the functions $v \in C_0^m(\Omega)$ as *test functions* in an integral to determine the values of another function.

Example 1. Let L be the two-dimensional wave operator in characteristic coordinates

$$Lu = u_{\mu\eta},$$

which is *self-adjoint*: $L' = L$. The homogeneous equation $Lu = 0$ has many smooth solutions, including the trivial solution $u \equiv 0$ and $u(\mu, \eta) = \mu^2$. But let us define the function

$$\tilde{u}(\mu, \eta) = \begin{cases} \mu^2 & \text{if } \mu > 0 \\ 0 & \text{if } \mu \leq 0. \end{cases}$$

It is readily verified that $\tilde{u} \in C^1(\mathbf{R}^2)$, but $\tilde{u} \notin C^2$ since the second-order derivatives have a jump discontinuity along the line $\mu = 0$, which is a characteristic curve. However, let us verify that $\tilde{u}$ is a weak solution of $Lu = 0$. We must show $\int \tilde{u} Lv \, d\mu \, d\eta = 0$ for every $v \in C_0^2(\mathbf{R}^2)$. This is achieved by integrating by parts twice:

$$\int_{\mathbf{R}^2} \tilde{u} Lv \, d\mu \, d\eta = \int_{\mu > 0} \mu^2 v_{\mu\eta} \, d\mu \, d\eta = -\int_{\mu > 0} 2\mu \, v_\eta \, d\mu \, d\eta = 0$$

since the boundary terms all vanish. ♣

It is no coincidence that, in this example, the singularities of the weak solution occurred along a characteristic curve of the equation. Indeed, for *any* second-order linear equation in the xy-plane, suppose u is a weak solution that is C^2 *except* along a curve γ, where the second-order derivatives experience a jump discontinuity (but extend continuously to γ from either side, as in Example 1). Then γ must be a characteristic curve for the equation. The reason for this is that the Cauchy data on a noncharacteristic curve γ uniquely determines the second-order derivatives of u on γ, so the limiting values from either side of γ must agree.

This fact also implies that any singularity in the initial data of a Cauchy problem will propagate to the solution along characteristics. This is called the *propagation of singularities* and applies to stronger singularities than simple jumps in the highest-order derivatives. We cannot, however, indiscriminately patch together smooth solutions on either side of a characteristic curve $\gamma \subset \mathbf{R}^2$ (or, more generally, a hypersurface $S \subset \mathbf{R}^n$) and obtain a weak solution; there are conditions along γ that must be satisfied for the solution to be *transmitted* across the singularity. We shall investigate these in the next subsection.

b. Transmission Conditions

Let us use the notion of weak solution to investigate the propagation of singularities along characteristics for a linear second-order equation. Let us suppose the equation involves two independent variables and is hyperbolic, and so can be transformed to the canonical form

(39) $$u_{\mu\eta} = d(\mu, \eta, u, u_\mu, u_\eta),$$

where $d(\mu, \eta, u, u_\mu, u_\eta)$ is linear in u, u_μ, u_η. We wish to consider a weak solution that is singular along one of the characteristics, say along $\mu = 0$. Let Ω be an open domain intersecting the η-axis, and let $\Omega^\pm = \Omega \cap \mathbf{R}_\pm^2 = \{(\mu, \eta) \in \Omega : \pm\mu > 0\}$ (see Figure 1).

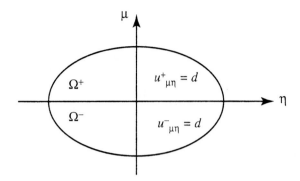

Figure 1. A jump discontinuity along $\mu = 0$ *in* Ω.

We suppose that $u^\pm = u|_{\Omega^\pm}$ and its derivatives extend continuously to $\gamma = \{(0, \eta) \in \Omega\}$. If u is a weak solution of (39) in Ω, then $u^\pm$ is a classical solution of (39) in $\Omega^\pm$, as observed previously, but u could presumably have a *jump discontinuity* along γ. Let us investigate the *transmission conditions* along γ in order for u to be a weak solution in all of Ω. If we use (39) to define $Lu = u_{\mu\eta} - d$, then $L'u = u_{\mu\eta} - d'$ for some expression d' that is also linear in u, u_μ, u_η. For a test function $v \in C_0^2(\Omega)$, we must have

$$(40) \quad \int_\Omega u\, L'v\, d\mu\, d\eta = \int_{\Omega^+} u^+ L'v\, d\mu\, d\eta + \int_{\Omega^-} u^- L'v\, d\mu\, d\eta = 0$$

in order for u to be a weak solution. If we integrate by parts, we obtain

$$(41) \quad \int_{\Omega^\pm} u^\pm L'v\, d\mu\, d\eta = \int_{\Omega^\pm} L(u^\pm)v\, d\mu\, d\eta + \int_{\partial\Omega^\pm} B^\pm(u^\pm, v)\, ds,$$

where ds is arclength and $B^\pm(u, v)$ is an expression involving u, v, and their normal derivatives on $\partial\Omega^\pm$. [Actually, $B^\pm(u, v)$ is not unique as it depends on the order in which the integrations are taken.] But $L(u^\pm) = 0$ in $\Omega^\pm$, and $v = 0$ in a neighborhood of $\partial\Omega$, so (40) holds if and only if

$$(42) \quad \int_\gamma B^+(u^+, v)\, d\eta + \int_\gamma B^-(u^-, v)\, d\eta = 0.$$

Recalling that v has arbitrary values along γ, (42) becomes the transmission conditions on the values $u^\pm$ and $u_\mu^\pm$ along γ; u is a weak solution if and only if these conditions are satisfied.

Let us compute the transmission condition (42) for the wave equation as in Example 1 (i.e., $Lu = u_{\mu\eta}$ with a jump singularity along $\mu = 0$). If we write

$$\text{(43)} \quad \int_\Omega u\, v_{\mu\eta}\, d\mu\, d\eta = \int_0^\infty \int_{-\infty}^\infty u^+ v_{\mu\eta}\, d\eta\, d\mu + \int_{-\infty}^0 \int_{-\infty}^\infty u^- v_{\mu\eta}\, d\eta\, d\mu$$

and integrate by parts first $d\eta$ and then $d\mu$, we get the transmission condition (cf. Exercise 2)

$$\text{(44)} \quad \int_{-\infty}^\infty [u_\eta^+(0,\eta) - u_\eta^-(0,\eta)] v(0,\eta)\, d\eta = 0.$$

Since this must be true for arbitrary $v(0,\eta) \in C_0^2(\gamma)$, we conclude that $u_\eta^+(0,\eta) - u_\eta^-(0,\eta) \equiv 0$, i.e., the jump discontinuity $u^+ - u^-$ along γ must be a constant. In other words, the solution u^+ in $\mu > 0$ may be continued to u^- in $\mu < 0$ as a weak solution "across $\mu = 0$" provided $u^+ - u^-$ is a constant along $\mu = 0$. We shall encounter exactly this type of singular solution for the one-dimensional wave equation in Section 3.1.

Transmission conditions along characteristics may be obtained analogously for higher-order equations as well as for more independent variables.

c. Distributions

The notion of a weak derivative introduced previously is closely related to *distribution theory*. Without getting too involved in the topological aspects of the theory, let us at least define distributions, especially the most important distribution, the *Dirac delta function* δ.

In fact, we first consider the *Heaviside function* of a single real variable:

$$\text{(45)} \quad H(x) = \begin{cases} 1 & \text{if } x \geq 0 \\ 0 & \text{if } x < 0. \end{cases}$$

This function is certainly not C^1, but if it had a weak derivative $H'(x)$, then (38) suggests that we use formal integration by parts: $H'(x)$ should satisfy

$$\int_{-\infty}^\infty H'(x) v(x)\, dx = -\int_{-\infty}^\infty H(x) v'(x)\, dx$$
$$= -\int_0^\infty v'(x)\, dx = v(0) - v(\infty) = v(0)$$

for every $v \in C_0^1(\mathbf{R})$. Since $H'(x) \equiv 0$ for $x \neq 0$, this suggests that "$H'(0) = \infty$" in such a way that $\int H'(x) v(x)\, dx = v(0)$. Of course, no true function behaves like this, so we call it a *generalized function* or *distribution*. Notice that this process enables us to take more derivatives of $H(x)$:

$$\int_{-\infty}^{\infty} H''(x)v(x)\,dx = \int_{-\infty}^{\infty} H(x)v''(x)\,dx = \int_{0}^{\infty} v''(x)\,dx = -v'(0),$$

$$\int_{-\infty}^{\infty} H'''(x)v(x)\,dx = -\int_{-\infty}^{\infty} H(x)v'''(x)\,dx = -\int_{0}^{\infty} v'''(x)\,dx = v''(0),$$

provided v is sufficiently smooth and has compact support, say $v \in C_0^{\infty}(\mathbf{R})$. In each case, the action of the derivative is computed by formally integrating by parts against a function in $C_0^{\infty}(\mathbf{R})$. This is called *distributional differentiation*, and the functions in $C_0^{\infty}(\mathbf{R})$ are called *test functions*.

Returning to $\mathbf{R}^n$, we define the *delta function* or, more accurately, the *delta distribution* to be the object $\delta(x)$ so that formally

$$(46) \qquad \int_{\mathbf{R}^n} \delta(x)v(x)\,dx = v(0)$$

for every test function $v \in C_0^{\infty}(\mathbf{R}^n)$; in the one-dimensional case we find $H'(x) = \delta(x)$. Notice that we can easily take any number of distributional derivatives of $\delta(x)$; in fact, $D^{\alpha}\delta(x)$ is the object satisfying $\int D^{\alpha}\delta(x)\,v(x)\,dx = (-1)^{|\alpha|} \int \delta(x) D^{\alpha}v(x)\,dx = (-1)^{|\alpha|} D^{\alpha}v(0)$. Also, we can translate the singularity in $\delta(x)$ to any point $\mu \in \mathbf{R}^n$ by letting $\delta_{\mu}(x) = \delta(x - \mu)$ so that a change of variables $y = x - \mu$ yields

$$\int_{\mathbf{R}^n} \delta_{\mu}(x)v(x)\,dx = \int_{\mathbf{R}^n} \delta(x - \mu)\,v(x)\,dx = \int_{\mathbf{R}^n} \delta(y)\,v(y + \mu)\,dy = v(\mu).$$

We have defined the delta function and its distributional derivatives through their actions on test functions; a distribution assigns to every function $v \in C_0^{\infty}(\mathbf{R}^n)$ a number. Also, notice that this action is linear: $\int \delta(\alpha v + \beta w)\,dx = \alpha v(0) + \beta w(0) = \alpha \int \delta v\,dx + \beta \int \delta w\,dx$, and similarly for the derivatives of δ and δ_μ. Therefore, a distribution should simply be thought of as a linear mapping $C_0^{\infty}(\mathbf{R}^n) \to \mathbf{R}$. In fact, generalized to an open domain $\Omega \subset \mathbf{R}^n$, we define distributions as follows.

Definitions. A *distribution F in Ω* is a linear mapping $F : C_0^{\infty}(\Omega) \to \mathbf{R}$ such that $F(v_j) \to 0$ for every sequence $\{v_j\} \subset C_0^{\infty}(\Omega)$ with support in a fixed compact set $K \subset \Omega$ and whose derivatives $D^{\alpha}v_j \to 0$ uniformly in K, as $j \to \infty$. If F and F_j are distributions in Ω, then $F_j \to F$ *as distributions* provided that $F_j(v) \to F(v)$ for every $v \in C_0^{\infty}(\Omega)$. The *support* of a distribution F in Ω is the smallest (relatively) closed set $K \subset \Omega$ such that $F(v) = 0$ whenever $v \equiv 0$ in a neighborhood of K.

Notation. The vector space $C_0^{\infty}(\Omega)$ of test functions is often denoted $\mathcal{D}(\Omega)$, and the space of distributions is denoted as its dual space, $\mathcal{D}'(\Omega)$. If

$v \in \mathcal{D}(\Omega)$ and $F \in \mathcal{D}'(\Omega)$, we sometimes denote the action of F on v by $\langle F, v \rangle$, or even $\int_\Omega F(x)v(x)\,dx$, which is called a *distributional integral*.

The definition of a distribution may seem complicated, but it has the important property that *distributions may always be differentiated:* if $F \in \mathcal{D}'(\Omega)$, then $D^\alpha F \in \mathcal{D}'(\Omega)$ is defined by $\langle D^\alpha F, v\rangle = (-1)^{|\alpha|}\langle F, D^\alpha v\rangle$ for all $v \in C_0^\infty(\Omega)$. The distributional integral notation has the advantage that this same formula looks like integration by parts: $\int_\Omega D^\alpha F(x)v(x)\,dx = (-1)^{|\alpha|}\int_\Omega F(x)D^\alpha v(x)\,dx$.

Examples of distributions are $\delta_\mu(x)$ and $f \in L^1_{\mathrm{loc}}(\Omega)$ [i.e., $\langle F_f, v\rangle = \int_\Omega f\,v\,dx$ (see Exercise 4)]; the latter explains the use of the distributional integral notation. Other examples of distributions and the convergence of distributions are discussed in the Exercises.

Let us now explain the role of distributions in PDE theory. Let L be given by (32a) and its adjoint L' by (37).

Definition. Distributions u and f satisfy $Lu = f$ in Ω in the *sense of distributions* (or u is a *distribution solution* of $Lu = f$) if $\langle u, L'v\rangle = \langle f, v\rangle$ for all $v \in C_0^\infty(\Omega)$.

In particular, if $u \in L^1_{\mathrm{loc}}(\Omega)$ is a weak solution of $Lu = f$, then it is a distribution solution, and the condition $\langle u, L'v\rangle = \langle f, v\rangle$ may be expressed in distributional integral notation to obtain (38).

d. Convolutions and Fundamental Solutions

The *convolution* of two functions f and g that are defined on $\mathbf{R}^n$ is a new function $h = f \star g$ defined on $\mathbf{R}^n$ by

$$(47) \qquad f \star g\,(x) = \int_{\mathbf{R}^n} f(x-y)\,g(y)\,dy,$$

assuming the integral converges. For example, if $f, g \in L^1_{\mathrm{loc}}(\mathbf{R}^n)$, and at least one has compact support, then (47) converges and $f \star g$ is well defined. A change of variables $z = x - y$ shows that

$$(48) \qquad f \star g = g \star f,$$

so convolution is commutative.

Let us view $f \star g$ as a distribution; that is, for $v \in C_0^\infty(\mathbf{R}^n)$, let us compute the value of the linear functional $f \star g$ on v:

$$(49) \qquad \begin{aligned} \langle f \star g, v \rangle &\equiv \int_{\mathbf{R}^n} f \star g\,(x)\,v(x)\,dx \\ &= \int_{\mathbf{R}^n}\int_{\mathbf{R}^n} f(x-y)\,g(y)\,v(x)\,dy\,dx \\ &= \int_{\mathbf{R}^n}\int_{\mathbf{R}^n} f(z)\,g(y)\,v(z+y)\,dy\,dz, \end{aligned}$$

Section 2.3: Linear Equations and Generalized Solutions   67

where the change in order of integration is justified by the compactness of the support of v and at least one of f or g. This shows us how to define the *convolution of distributions* F and G:

(50) $$\langle F \star G, v \rangle = \int_{\mathbf{R}^n} \int_{\mathbf{R}^n} F(z)\, G(y)\, v(z+y)\, dy\, dz,$$

which is well defined provided that F or G has compact support and we interpret the integrals in the distributional sense. The convolution of distributions is also commutative.

The following result will be used

Lemma. *If $f \in C(\mathbf{R}^n)$ and $g \in L^1_{\mathrm{loc}}(\mathbf{R}^n)$, one of which has compact support, then $f \star g \in C(\mathbf{R}^n)$.*

Notice that the result remains true for $g = \delta \in \mathcal{D}'(\mathbf{R}^n)$, since convolution by the delta distribution acts like the identity on continuous functions:

(51) $$f \star \delta(x) = \delta \star f(x) = \int_{\mathbf{R}^n} \delta(x-y)\, f(y)\, dy = f(x).$$

However, the result fails for general $g \in \mathcal{D}'(\mathbf{R}^n)$ (cf. Exervise 6b).

Proof of Lemma. Consider the case when $K = \operatorname{supp} g$ is compact. Let us show that $f \ast g$ is continuous at some fixed $x \in \mathbf{R}^n$; we restrict our attention to $y \in B_1(x)$. Let $\tilde{K} = \{y - z : |x - y| \le 1 \text{ and } z \in K\}$, which is also compact, so $f(w)$ is uniformly continuous for $w \in \tilde{K}$. Consequently, for any $\epsilon > 0$ there exists $1 > \epsilon' > 0$ such that $|f(x - z) - f(y - z)| < \epsilon$ whenever $z \in K$ with $|x - y| < \epsilon'$. This means

$$|f \ast g(x) - f \ast g(y)| = \left| \int_K (f(x-z) - f(y-z)) g(z)\, dz \right|$$
$$\le \int_K |f(x-z) - f(y-z)|\, |g(z)|\, dz \le \epsilon \int_K |g(z)|\, dz,$$

from which continuity follows. An analogous argument applies when f has compact support (cf. Exercise 6a). ♠

We can use (50) to take distributional derivatives of a convolution:

$$\langle D^\alpha(F \star G), v \rangle = (-1)^{|\alpha|} \langle F \star G, D^\alpha v \rangle$$
$$= (-1)^{|\alpha|} \int_{\mathbf{R}^n} \int_{\mathbf{R}^n} F(z)\, G(y)\, D^\alpha_z v(z+y)\, dy\, dz$$
$$= (-1)^{|\alpha|} \int_{\mathbf{R}^n} \int_{\mathbf{R}^n} F(z)\, G(y)\, D^\alpha_y v(z+y)\, dy\, dz.$$

But this implies

(52) $$D^\alpha(F \star G) = (D^\alpha F) \star G = F \star D^\alpha G,$$

so derivatives of a convolution can be taken on either factor. In particular, if $f \in C^k(\mathbf{R}^n)$ and $g \in L^1_{\text{loc}}(\mathbf{R}^n)$, one of which has compact support, then the preceding lemma implies $f \star g \in C^k(\mathbf{R}^n)$.

The convolution of distributions is useful in solving the nonhomogeneous equation

$$(53) \qquad Lu = f$$

when L has constant coefficients [i.e., the a_α in (32a) are all constants]. To achieve this, let us suppose that we can solve the particular case where we replace f by the delta distribution:

$$(54) \qquad Lu = \delta.$$

A solution $u = F$ of (54) is called a *fundamental solution* of L and must itself be understood as a distribution solution; that is,

$$(55) \qquad \langle F, L'v \rangle = \int_{\mathbf{R}^n} F(x)\, L'v(x)\, dx = v(0) \qquad \text{for all } v \in C_0^\infty(\mathbf{R}^n).$$

Notice that a fundamental solution is not unique since we can always add to it any solution of the homogeneous equation $Lu = 0$, and not disturb the property (54). Now let us assume f is a distribution with compact support and form the convolution with F:

$$(56) \qquad u(x) = F \star f(x) = \int_{\mathbf{R}^n} F(x - y)\, f(y)\, dy.$$

Then u is a distribution solution of (53) because $Lu = \sum_\alpha a_\alpha D^\alpha (F \star f) = \sum_\alpha a_\alpha D^\alpha F \star f = \delta \star f = f$.

Depending on the properties of F, it may be possible to extend (56) to functions f having noncompact support. In addition, if f is nice enough, we might be able to conclude *regularity* properties of u [e.g., we might like to conclude $u \in C^m(\mathbf{R}^n)$, so that u is a *classical solution* of (53)].

As examples of fundamental solutions, we have seen that the Heaviside function $H(x)$ is a fundamental solution for $L = d/dx$ when $n = 1$; in this case (56) is just the fundamental theorem of calculus. For $n = 2$, we can use complex analysis $z = x + iy$ to introduce the *Cauchy-Riemann operator* $\overline{\partial}_z = \frac{1}{2}(\partial_x + i\partial_y)$, which has $F(z) = (\pi z)^{-1}$ as a fundamental solution (for details, see [Treves, 1]). In Exercise 13, it is shown that

$$(57) \qquad F(x,t) = \frac{1}{2c} H(ct + x)\, H(ct - x)$$

is a fundamental solution of the one-dimensional wave operator $Lu = u_{tt} - c^2 u_{xx}$; among the many possible fundamental solutions, (57) is distinguished by the property of having its support equal to the *forward light cone* $\{(x,t) : ct \geq |x|\}$.

Section 2.3: Linear Equations and Generalized Solutions   69

Let us consider a simple example from ordinary differential equations to illustrate the role of a fundamental solution.

Example 2. Consider the ODE. $u'' = f$. To find a fundamental solution $F(x)$, we want to solve $F''(x) = \delta(x)$. Recall that the Heaviside function $H(x)$ satisfies $(H(x) + c)' = \delta(x)$ for any constant c. For convenience, let us take $c = -1/2$ and try to solve

$$F'(x) = \begin{cases} 1/2 & \text{if } x > 0 \\ -1/2 & \text{if } x < 0. \end{cases} \tag{58}$$

We may integrate (58) to obtain a particular solution

$$F(x) = \frac{1}{2}|x| \tag{59}$$

as our fundamental solution. [The choice $c = -1/2$ was made to simplify the resultant form (59) of the fundamental solution.] Assuming that $f \in L^1(\mathbf{R})$ has compact support, then the integral

$$u(x) = \frac{1}{2}\int_{-\infty}^{\infty} |x - y| f(y)\, dy \tag{60}$$

converges and defines a distribution solution of $u'' = f$.

Now let us address the regularity of u defined by (60); all the following claims are established in Exercise 11. If $f(x) \in L^1(\mathbf{R})$ has compact support, then $u \in C(\mathbf{R})$. In particular, u is a weak solution of $u'' = f$. If, in addition, $f(x)$ is continuous, then $u \in C^2$ (i.e., u is a classical solution). (This last fact, namely $f \in C^0 \Rightarrow u \in C^2$, does *not* hold for the Poisson equation $\Delta u = f$ in dimensions $n > 1$, as we shall see in later chapters.) ♣

In a sense, convolution by a fundamental solution "inverts" the operator L [i.e., $L(F \star f) = f$, at least for certain functions f]. It is important to observe, however, that this inverse is independent of any side conditions, such as initial or boundary conditions. When side conditions are present, we can take one of two approaches:

(i) First use (56) to define a particular solution u_p of (53), and then add an appropriate solution u_h of the homogeneous equation $Lu = 0$, so that $u_p + u_h$ satisfies the side conditions.
(ii) Try to adjust the inverse operator to take the side conditions into account.

The first approach should be familiar from elementary ordinary differential equations. We shall encounter the second approach when constructing Green's functions for boundary value problems in Chapter 4, but for now let us see how to define an inverse for an initial value problem.

Suppose we want to solve the Cauchy problem

(61)
$$\begin{cases} u_t = Lu & \text{for } x \in \mathbf{R}^n, t > 0 \\ u(x,0) = g(x) & \text{for } x \in \mathbf{R}^n, \end{cases}$$

where L is a differential operator in $\mathbf{R}^n$ with constant coefficients. Suppose $K(x,t)$ is a distribution in $\mathbf{R}^n$ for each value of $t \geq 0$, K is continuously differentiable in t (see Exercise 14) and satisfies

(62)
$$\begin{cases} K_t - LK = 0 & \text{as a distribution in } \mathbf{R}^n \text{ for each } t > 0 \\ K(x,0) = \delta(x) & \text{as a distribution in } \mathbf{R}^n. \end{cases}$$

We shall call K a *fundamental solution for the initial value problem*. The solution of (61) is then given by convolution in the space variables:

(63)
$$u(x,t) = \int_{\mathbf{R}^n} K(x-y,t) \, g(y) \, dy,$$

provided of course that this integral exists. We shall encounter this situation in connection with the heat equation in Chapter 5.

It is worth noting that the distribution $K(x,t)$ may be used to solve the initial value problem for the nonhomogeneous equation:

(64)
$$\begin{cases} u_t - Lu = f(x,t) & \text{for } x \in \mathbf{R}^n, t > 0 \\ u(x,0) = 0 & \text{for } x \in \mathbf{R}^n. \end{cases}$$

The solution is given by convolving the t variable as well:

(65)
$$u(x,t) = \int_0^t \int_{\mathbf{R}^n} K(x-y, t-s) \, f(y,s) \, dy \, ds.$$

Using the solution of the homogeneous equation with nonhomogeneous initial condition to solve the nonhomogeneous equation is called *Duhamel's principle*. We shall encounter this principle frequently for initial value problems.

Remark. For operators of the form $\partial_t - L$, the fundamental solution of the initial value problem, $K(x,t)$ as defined in (62), coincides with the "free space" fundamental solution, which satisfies $(\partial_t - L)K(x,t) = \delta(x,t)$, provided we extend $K(x,t)$ by zero to $t < 0$. This is true in general, but we shall see a proof in Section 5.2 for the heat equation.

Exercises for Section 2.3

1. (a) For complex-valued functions u and v on Ω, let $\langle u, v \rangle = \int_\Omega u\bar{v}\, dx$, where $\bar{v}$ denotes the complex conjugate of v. If the coefficients $a_\alpha(x)$ in (32a) are complex-valued functions, define the adjoint L^* so that $\langle Lu, v \rangle = \langle u, L^*v \rangle$ for all $u \in C^m(\Omega)$, $v \in C_0^m(\Omega)$.
   (b) For (complex) vector-valued functions $\vec{u}(x) = (u_1(x), \ldots, u_N(x))$ on Ω, let $\langle \vec{u}, \vec{v} \rangle = \int_\Omega (u_1\bar{v}_1 + \cdots u_N\bar{v}_N)\, dx$. If the coefficients $a_\alpha(x)$ in (32a) are $N \times N$ matrix-valued functions, then (32a) defines a *system* of mth-order operators. Define the adjoint L^* so that $\langle L\vec{u}, \vec{v} \rangle = \langle \vec{u}, L^*\vec{v} \rangle$ for all $\vec{u} \in C^m(\Omega, \mathbf{R}^N)$, $\vec{v} \in C_0^m(\Omega, \mathbf{R}^N)$.

2. Let $Lu = u_{\mu\eta}$ as in Example 1. Derive the transmission condition (44) along the characteristic $\mu = 0$.

3. Consider the first-order equation $u_t + cu_x = 0$.
   (a) If $f \in C(\mathbf{R})$, show that $u(x,t) = f(x - ct)$ is a weak solution.
   (b) Can you find any *discontinuous* weak solutions?
   (c) Is there a transmission condition for a weak solution with jump discontinuity along the characteristic $x = ct$?

4. If $f \in L^1_{\text{loc}}(\Omega)$, define $\langle F_f, v \rangle \equiv \int_\Omega f(x) v(x)\, dx$. Show that F_f is a distribution in Ω.

5. (a) If $\xi_1, \ldots, \xi_k$ are points in Ω and $a_1, \ldots, a_k$ are real numbers, show that $F \equiv a_1 \delta_{\xi_1} + \cdots + a_k \delta_{\xi_k}$ is a distribution in Ω.
   (b) For $\Omega = \mathbf{R}^n$, find an infinite sequence $\{\xi_k\}_{k=1}^\infty$ for which $f \equiv a_1 \delta_{\xi_1} + \cdots$ is a distribution. (This gives an example of a distribution that *cannot* be realized as a distributional derivative of an integrable function.)

6. (a) Prove the Lemma on p. 67 when $f \in C(\mathbf{R}^n)$ has compact support.
   (b) Find an example to show that the result fails to hold if g is replaced by $G \in \mathcal{D}'(\mathbf{R}^n)$.

7. Let Γ be a hypersurface (such as a sphere) in $\mathbf{R}^n$, let $a(z)$ be a continuous function of $z \in \Gamma$, and let dz denote the surface measure on Γ.
   (a) Show that $\langle F, v \rangle = \int_\Gamma a(z) v(z)\, dz$ is a distribution in $\mathbf{R}^n$. (It is natural to denote this distribution by $F = a\delta_\Gamma$.)
   (b) Suppose we can choose a unit normal ν along Γ. Formulate a definition for the "conormal distribution" $a\partial_\nu \delta_\Gamma$ in $\mathbf{R}^n$.

8. Let
$$f_n(x) = \begin{cases} \dfrac{n}{2} & \text{for } -\dfrac{1}{n} < x < \dfrac{1}{n} \\ 0 & \text{for } |x| \geq \dfrac{1}{n}. \end{cases}$$

   Show that $f_n(x) \to \delta(x)$ as distributions on $\mathbf{R}$.

9. If $f_n(x)$ and $f(x)$ are integrable functions such that for any compact set $K \subset \Omega$ we have $\int_K |f_n(x) - f(x)|\, dx \to 0$ as $n \to \infty$, then $f_n \to f$ as distributions.

10. Let $a \in \mathbf{R}$, $a \neq 0$.
    (a) Find a fundamental solution for $L = d/dx - a$ on $\mathbf{R}$ (i.e., solve $dF/dx - aF = \delta$).
    (b) Show that a fundamental solution for $L = d^2/dx^2 - a^2 = (d/dx + a)(d/dx - a)$ on $\mathbf{R}$ is given by
    $$F(x) = \begin{cases} a^{-1} \sinh ax & \text{if } x > 0 \\ 0 & \text{if } x < 0. \end{cases}$$

11. *Regularity of u defined by (60):*
    (a) If $f(x) \in L^1(\mathbf{R})$ has compact support, show that (60) defines a *continuous* weak solution of $u'' = f(x)$.
    (b) If, in addition to (a), $f(x)$ is a bounded function on $\mathbf{R}$, show $u \in C^1(\mathbf{R})$, and $u'(x) = \frac{1}{2}(\int_{-\infty}^{x} f(y)\,dy - \int_{x}^{\infty} f(y)\,dy)$.
    (c) If, in addition to (a), $f(x)$ is continuous, show that u is in fact a classical solution: $u \in C^2(\mathbf{R})$ with $u'' = f$.

12. (a) Use the fundamental solution (59) to solve the initial value problem $u'' = f(x)$ for $x > 0$ with $u(0) = u_0$ and $u'(0) = u'_0$, where $f \in C^\infty([0, \infty))$ and $f = O(|x|^{-2-\epsilon})$ as $|x| \to \infty$.
    (b) Do the same for the boundary value problem $u'' = f(x)$ for $0 < x < \ell$ with $u(0) = 0 = u(\ell)$.

13. (a) Using $\delta(\mu, \eta) = \delta(\mu)\delta(\eta)$, show that each of the following functions is a fundamental solution of $L = \partial^2/\partial\mu\partial\eta$:
    $$F_1(\mu, \eta) = H(\mu)\,H(\eta), \qquad F_2(\mu, \eta) = -H(\mu)\,H(-\eta),$$
    $$F_3(\mu, \eta) = -H(-\mu)\,H(\eta), \qquad F_4(\mu, \eta) = H(-\mu)\,H(-\eta).$$
    (b) Use part (a) with the change of variables $\mu = x + ct$, $\eta = x - ct$ to obtain four distinct fundamental solutions for the one-dimensional wave operator, each having support in one of the wedges determined by the lines $x = \pm ct$.

14. If $K(x, t)$ is a distribution in $\mathbf{R}^n$ depending on the parameter t, then we say K *depends continuously on* t if, for every $v \in C_0^\infty(\mathbf{R}^n)$, $f(t) \equiv \langle K(\cdot), v \rangle = \int K(x, t)v(x)\,dx$ defines a continuous function of t. Use this idea to define K *is continuously differentiable in* t.

15. According to Section 1.2, a *weak solution* of $u_y + (G(u))_x = 0$ in a rectangle $\Omega = (a, b) \times (c, d)$ satisfies
    $$\frac{d}{dy}\int_{x_1}^{x_2} u(x, y)\,dx + G(u(x, y))\Big|_{x=x_1}^{x=x_2} = 0$$
    for every $[x_1, x_2] \subset (a, b)$ and $c < y < d$. Show that this implies
    $$\int_\Omega uv_y + G(u)v_x\,dx\,dy = 0 \qquad \text{for every } v \in C_0^1(\Omega)$$

(i.e., the notion of weak solution of Section 1.2 is consistent with that of Section 2.3).

16. The mth-order operator (32a) is *elliptic* at x if its principal symbol (32c) has no nonzero characteristics [i.e., $\sigma_L(x,\xi) \neq 0$ for all $\xi \in \mathbf{R}^n\setminus\{0\}$]. In this case, show that m must be an *even* integer.

Further References for Chapter 2

For more on the Cauchy problem for linear equations, see [John, 1], [Hörmander, 1] or [Hörmander, 2, v. I], or [Treves, 1].

The theory of generalized functions or distributions is rich and beautiful and has extensive applications; we have only given a taste of the theory that is sufficient for our purposes. For more information, consult [Gelfand-Shilov], [Schwartz], or [Treves, 2].

For the construction of fundamental solutions for some familiar constant coefficient PDEs, see [Treves, 1]. In fact, it is true that *any* constant coefficient PDE admits a fundamental solution; this result is called the *Malgrange-Ehrenpreis theorem;* see [Hörmander, 1] or [Hörmander, 2, v. II].

3 The Wave Equation

We study the wave equation in all dimensions, but with particular detail for the physical cases of dimensions one, two, and three. First we consider the one-dimensional wave equation, using characteristics and separation of variables to study initial and boundary value problems for the homogeneous equation, and then invoke Duhamel's principle to study the inhomogeneous equation. Next, the initial value problem for the three-dimensional wave equation is solved by spherical mean reduction to the one-dimensional case, and then the two-dimensional wave equation is solved by "descent" from three dimensions; a similar procedure is followed for all $n > 1$. The concept of energy is introduced and then used to study the domain of dependence in n-dimensions. We also study the effect of lower-order derivative terms in the one-dimensional wave equation, including dispersion, dissipation, and the domain of dependence. Finally, we show that the wave equation provides a reasonable model for sound and light waves, not just vibratine strings and membranes.

3.1 The One-Dimensional Wave Equation

The *one-dimensional wave equation*

$$(1) \qquad u_{tt} - c^2 u_{xx} = 0$$

governs wave motion in a string, at least for small displacements u from its equilibrium position (cf. Appendix, Section A.1a). We shall assume that the *propagation speed* c is a constant. The characteristics for (1) are $x \pm ct = \text{const}$, and the change of variable $\mu = x + ct$, $\eta = x - ct$ transforms (1) to

$$(2) \qquad u_{\mu\eta} = 0.$$

The general solution of (2) is $u(\mu, \eta) = F(\mu) + G(\eta)$, where F and G are C^1 functions. Returning to the variables x, t we find that

$$(3) \qquad u(x, t) = F(x + ct) + G(x - ct)$$

solves (1). Moreover, u is C^2 provided that F and G are C^2.

Notice that the solution u given by (3) can be realized as the superposition of two waves propagating without change of shape with velocity c in opposite directions along the x-axis (see Figure 1).

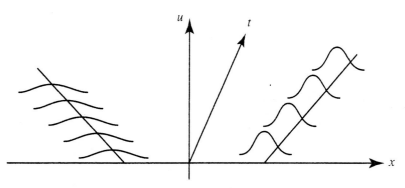

Figure 1. The solution as a superposition of traveling waves

In fact, if $F \equiv 0$, then u has constant values along the lines $x - ct = $ const, so may be described as a wave moving in the positive x-direction with speed $dx/dt = c$; similarly, if $G \equiv 0$, then u is a wave moving in the negative x-direction with speed c.

We shall now investigate several consequences of (3).

a. The Initial Value Problem

Let us consider the *Cauchy problem* for (1) as an *initial value problem*. That is, the Cauchy data is given by the initial conditions

(4) $$u(x,0) = g(x), \qquad u_t(x,0) = h(x),$$

where g and h are arbitrary functions. Using (3) in (4), we find that F and G must satisfy

(5) $$F(x) + G(x) = g(x), \qquad cF'(x) - cG'(x) = h(x).$$

If we integrate the second equation in (5), we get $cF(x) - cG(x) = H(x)$, where $H(x) = \int_0^x h(\xi)\,d\xi + C$. Combining this with the first equation in (5), we can solve for F and G to find

$$\begin{cases} F(x) = \tfrac{1}{2}g(x) + \tfrac{1}{2c}\int_0^x h(\xi)\,d\xi + C_1 \\ G(x) = \tfrac{1}{2}g(x) - \tfrac{1}{2c}\int_0^x h(\xi)\,d\xi - C_1, \end{cases}$$

where $C_1 = (2c)^{-1}C$. Using these expressions in (3), we obtain *d'Alembert's formula* for the solution of the initial value problem (1),(4):

(6) $$u(x,t) = \frac{1}{2}(g(x+ct) + g(x-ct)) + \frac{1}{2c}\int_{x-ct}^{x+ct} h(\xi)\,d\xi.$$

We can formulate this result as the following.

Theorem 1. *If $g \in C^2$ and $h \in C^1$, then (6) defines a C^2 solution of (1),(4).*

Let us draw some further conclusions from the formula (6). First, the explicit construction shows that the solution of (1),(4) is *unique*. (Uniqueness is also a consequence of energy methods as discussed in Section 3.3.) Second, small changes in the Cauchy data g, h will affect the solution u only by a small amount, so the solution *depends continuously on the data*. We conclude that the Cauchy problem for the wave equation is *well posed*.

There are additional important properties of the solution that can be seen from (6). Notice that, at any point (x, t), u is determined by the values of g and h in the interval $[x - ct, x + ct]$ of the x-axis. This interval is called the *domain of dependence* for (x, t). Conversely, any point ξ on the x-axis lies only in the domain of dependence of points (x, t) in a wedge-shaped region called the *range of influence* of ξ; see Figure 2. Physically, these concepts express the *finite propagation speed* of disturbances or signals: an initial disturbance near $x = \xi$ will not be felt at a point x_1 until time $t_1 = |x_1 - \xi|/c$.

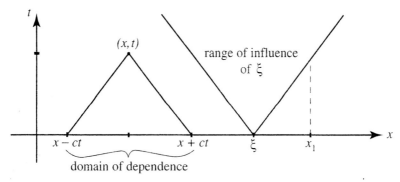

Figure 2. The domain of dependence and range of influence.

b. Weak Solutions

It is natural to expect that (3) defines a *weak* solution of (1) when we no longer require F and G to be C^2-functions. In fact, if $F(\mu)$ and $G(\eta)$ are both piecewise C^2-functions, then it is a simple matter to use the transmission conditions obtained in Section 2.3b to confirm that (3) is a weak solution as defined in Section 2.3a. However, we shall use an algebraic property of (3) to extend our definition of a solution of (1).

Consider a rectangle $ABCD$ in the $\mu\eta$-plane whose sides are parallel to the coordinate axes, as in Figure 3a. Since F is constant along vertical lines and G is constant along horizontal lines, we have $F(A) = F(D)$, $F(B) = F(C)$, $G(A) = G(B)$, and $G(C) = G(D)$. Using $u(\mu, \eta) = F(\mu) + G(\eta)$, we find

(7) $$u(A) + u(C) = u(B) + u(D)$$

(i.e., the sums of the values of u at opposite vertices are equal). Translated to the xy-plane, we view (7) as a *parallelogram rule* that holds for every parallelogram whose sides are all segments of characteristics, as in Figure 3b.

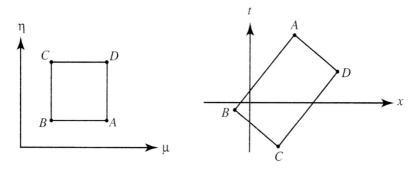

Figures 3a, 3b. Rectangles of characteristics.

This enables us to define a *weak solution* of (1) as any function $u(x,t)$ satisfying (7) for every such parallelogram in its domain. As observed, this generalizes C^2-solutions and the weak solutions as defined in Section 2.3. But the real reason for this characterization of weak solutions is its usefulness for solving the wave equation with both initial and boundary conditions.

c. Initial/Boundary Value Problems

In order to consider oscillations of a string of finite length, we would like to be able to solve (1) on a finite x-interval, say $[0, L]$, with Cauchy data for u at $t = 0$ and boundary conditions for u at $x = 0$ and $x = L$. For example, let us consider

(8) $$\begin{cases} u_{tt} - c^2 u_{xx} = 0 & \text{for } 0 < x < L \text{ and } t > 0 \\ u(x,0) = g(x), \ u_t(x,0) = h(x) & \text{for } 0 < x < L \\ u(0,t) = \alpha(t), \ u(L,t) = \beta(t) & \text{for } t \geq 0. \end{cases}$$

This means that we want to determine the motion of the string from the position $(\alpha(t), \beta(t))$ of its ends and from its initial position $g(x)$ and velocity $h(x)$. (Alternatively, we could impose boundary conditions involving u_x; cf. Exercise 3.)

One approach to this problem, familiar from undergraduate courses in differential equations, is the use of *separation of variables* to obtain an *expansion in eigenfunctions*. For example, if $\alpha \equiv 0 \equiv \beta$, then we can try to find $u(x,t)$ in the form

(9) $$u(x,t) = \sum_{n=1}^{\infty} a_n(t) \sin \frac{n\pi x}{L}.$$

Formally, u satisfies the boundary conditions $u(0,t) = 0 = u(L,t)$. Moreover, if we substitute (9) in the equation $u_{tt} = c^2 u_{xx}$ and equate coefficients of $\sin(n\pi x/L)$, we find that the functions $a_n(t)$ must satisfy the ordinary differential equation $a_n'' + (n\pi c/L)^2 a_n = 0$, whose general solution is

$$a_n(t) = c_n \sin \frac{n\pi ct}{L} + d_n \cos \frac{n\pi ct}{L}. \tag{10}$$

The constants c_n and d_n are determined by the initial conditions; namely, if $g(x)$ and $h(x)$ admit Fourier sine series expansions, then use $u(x,0) = g(x)$ and $u_t(x,0) = h(x)$ to find

$$g(x) = \sum_{n=1}^{\infty} d_n \sin \frac{n\pi x}{L}, \qquad h(x) = \sum_{n=1}^{\infty} c_n \frac{n\pi c}{L} \sin \frac{n\pi x}{L}.$$

By orthogonality, we may multiply by $\sin(m\pi x/L)$ and integrate to find

$$d_n = \frac{2}{L} \int_0^L g(x) \sin \frac{n\pi x}{L}\, dx, \qquad c_n = \frac{2}{n\pi c} \int_0^L h(x) \sin \frac{n\pi x}{L}\, dx. \tag{11}$$

The formulas (9), (10), and (11) define the solution; the only disadvantage is that the values of $u(x,t)$ must be found by summing the infinite series.

Another approach to the problem (8) is to use d'Alembert's formula and the parallelogram rule to *piece together* the solution. To do this, let us divide the region $R = \{(x,t) : 0 < x < L,\ t > 0\}$ into subregions defined by the characteristics from the corners $(0,0)$ and $(L,0)$, reflected at the boundaries (see Figure 4): $R = R_1 \cup R_2 \cup \ldots$, where each R_j contains the characteristic(s) along its upper edge(s), but not along its lower edge(s).

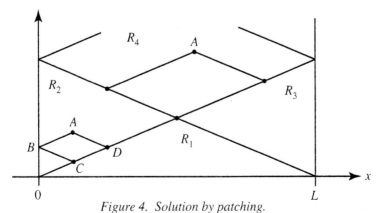

Figure 4. Solution by patching.

In region R_1 the solution u is defined by d'Alembert's formula (6). For $A = (x,t)$ in region R_2 let us form the parallelogram with B on the t-axis and C and D on the characteristic emanating from the corner $(0,0)$;

by (7) we have $u(A) = -u(C) + u(B) + u(D)$. But $u(B)$ is known by the boundary condition $u(0,t) = \alpha(t)$, and $u(C)$ and $u(D)$ are both known from the solution in R_1, so we have defined u in R_2. How smoothly does u fit together along the boundary between R_1 and R_2? Letting $A \to D$ we must have $B, C \to 0$, so that $u(A) \to -g(0) + \alpha(0) + u(D)$. Thus, to have u continuous in $R_1 \cup R_2$ [i.e., $u(A) \to u(D)$], we need a *compatibility condition* between the initial and boundary values to hold: $\alpha(0) = g(0)$. Similarly we require $\alpha'(0) = h(0)$ in order to have $u \in C^1$, and $\alpha''(0) = c^2 g''(0)$ in order to have $u \in C^2$.

The construction may be repeated in R_3 yielding compatibility conditions at $(L, 0)$ to ensure that u is continuous, C^1, or C^2 along the boundary with R_1. Using the solution in R_2 and R_3, (7) can be used to extend the solution to R_4, and so on.

Notice that, if the initial and boundary conditions are all smooth, then the solution will be smooth *except* along the characteristics forming the boundaries between the regions. Nevertheless, the transmission conditions of Section 2.3.b are satisfied, so the solution is a weak solution as defined in Section 2.3.a with singularities propagating along characteristics.

We shall compare the Fourier series solution with that obtained by d'Alembert's formula and the parallelogram rule in the following example.

Example 1. Consider the initial/boundary value problem

$$\begin{cases} u_{tt} - u_{xx} = 0 & \text{for } 0 < x < \pi \text{ and } t > 0 \\ u(x,0) = 1, \quad u_t(x,0) = 0 & \text{for } 0 < x < \pi \\ u(0,t) = 0, \quad u(\pi,t) = 0 & \text{for } t \geq 0. \end{cases}$$

If we look for a Fourier series solution, we compute

$$c_n = 0, \qquad d_n = \frac{2}{\pi} \int_0^\pi \sin nx \, dx = \begin{cases} 0 & \text{if } n \text{ is even} \\ \frac{4}{n\pi} & \text{if } n \text{ is odd}. \end{cases}$$

This means that the initial condition $u(x,0) = 1$ may be written in series form as

(12) $$u(x,0) = 1 = \frac{4}{\pi} \sum_{n=0}^{\infty} \frac{\sin(2n+1)x}{(2n+1)} \quad \text{for } 0 < x < \pi,$$

and the solution may be written as

(13) $$u(x,t) = \frac{4}{\pi} \sum_{n=0}^{\infty} \frac{\sin(2n+1)x \cos(2n+1)t}{(2n+1)}.$$

Next let us use (7) to find the solution in the different regions R_1, R_2, and so on (as in Figure 4 with $c = 1$ and $L = \pi$). In R_1 we use (6) to compute $u \equiv 1$. For A in R_2, select B on the t-axis and C and D on the boundary with R_1; then $u(A) = -u(C) + u(B) + u(D) = -1 + 0 + 1 = 0$, so $u \equiv 0$ in R_2. Similarly, $u \equiv 0$ in R_3. For A in R_4, select B and D on the boundaries with R_2 and R_3 and C at the intersection of these characteristics; then $u(A) = -u(C) + u(B) + u(D) = -1 + 0 + 0 = -1$. We can repeat this procedure to see that $u \equiv 0$ in the triangular shaped boundary regions R_2, R_3, R_5, R_6, and so on; $u \equiv 1$ in the regions R_1, R_7, and so on; and $u \equiv -1$ in the regions R_4, R_{10}, and so on.

In order to verify that this step function u agrees with the series solution (13), let us use the trigonometric identity $2 \sin A \cos B = \sin(A - B) + \sin(A + B)$ to rewrite (13) as

$$(14) \quad u(x,t) = \frac{2}{\pi} \left(\sum_{n=0}^{\infty} \frac{\sin[(2n+1)(x-t)]}{2n+1} + \sum_{n=0}^{\infty} \frac{\sin[(2n+1)(x+t)]}{2n+1} \right).$$

Now in R_1 we have $(x \pm t) \in (0, \pi)$, so we can use (12) to sum the series in (14) and obtain $u \equiv 1/2 + 1/2 = 1$. In R_2 we have $-\pi < x - t < 0$ and $0 < x + t < \pi$, so we may sum the series in (14) to obtain $u \equiv -1/2 + 1/2 = 0$. Similarly, in R_3 we have $\pi > x - t > 0$ and $2\pi > x + t > \pi$ so $u \equiv 1/2 - 1/2 = 0$. Continuing in this manner, we see that we can sum the series (14) and recover the function $u(x,t)$ obtained by using (7). ♣

We have used Fourier series rather formally here, avoiding issues of convergence. We shall return to separation of variables, Fourier series, and more general expansions in eigenfunctions in Chapter 4.

d. The Nonhomogeneous Equation

Next, let us consider the *nonhomogeneous wave equation*

$$(15) \quad u_{tt} - c^2 u_{xx} = f(x,t),$$

which occurs when an external force is driving the motion. To begin with, suppose we want to solve (15) with *homogeneous initial conditions*

$$(16) \quad u(x,0) = u_t(x,0) = 0.$$

One simple approach would be to first find any solution u_p of (15) and then add a solution u_h of (1) to achieve (16). Instead, we shall solve (15),(16) by Duhamel's principle, which reduces the problem to the special homogeneous equation with nonhomogeneous initial conditions

$$(17) \quad \begin{cases} U_{tt} - c^2 U_{xx} = 0 & \text{for } x \in \mathbf{R} \text{ and } t > 0 \\ U(x,0,s) = 0 & \text{for } x \in \mathbf{R} \\ U_t(x,0,s) = f(x,s) & \text{for } x \in \mathbf{R}, \end{cases}$$

which is to be solved for each value of the parameter $s > 0$.

Section 3.1: The One-Dimensional Wave Equation

Duhamel's Principle. If $U(x,t,s)$ is C^2 in x and t, C^0 in s, and solves the problem (17), then

$$(18) \qquad u(x,t) = \int_0^t U(x, t-s, s)\, ds$$

solves the problem (15),(16).

Proof. Differentiate (18) with respect to t to obtain

$$u_t(x,t) = U(x, 0, t) + \int_0^t U_t(x, t-s, s)\, ds = \int_0^t U_t(x, t-s, s)\, ds.$$

Differentiate again with respect to t to obtain

$$u_{tt}(x,t) = U_t(x, 0, t) + \int_0^t U_{tt}(x, t-s, s)\, ds = f(x,t) + \int_0^t U_{tt}(x, t-s, s)\, ds.$$

Differentiate (18) twice with respect to x to obtain

$$u_{xx}(x,t) = \int_0^t U_{xx}(x, t-s, s)\, ds.$$

Thus

$$u_{tt} - c^2 u_{xx} = f(x,t) + \int_0^t (U_{tt} - c^2 U_{xx})(x, t-s, s)\, ds = f(x,t),$$

so u solves (15). Since u clearly satisfies (16), the proof is complete. ♠

Now we may use d'Alembert's formula (6) to solve (17):

$$U(x,t,s) = \frac{1}{2c} \int_{x-ct}^{x+ct} f(\xi, s)\, d\xi.$$

Provided that $f(x,t)$ is C^1 in x and C^0 in t, then U will be C^2 in x, t and C^0 in s. Duhamel's principle then provides the solution of (15),(16):

$$(19) \qquad u(x,t) = \frac{1}{2c} \int_0^t \left(\int_{x-c(t-s)}^{x+c(t-s)} f(\xi, s)\, d\xi \right) ds.$$

Theorem 2. If $f(x,t)$ is C^1 in x and C^0 in t, then (19) is a C^2 solution of (15),(16).

We can also solve (15) with nonhomogeneous initial conditions,

$$(20) \qquad \begin{cases} u_{tt} - c^2 u_{xx} = f(x,t) \\ u(x,0) = g(x) \\ u_t(x,0) = h(x), \end{cases}$$

simply by adding together the formulas (6) and (19).

Remark. If we consider a driving force on a string of finite length L, we encounter a boundary value problem for a nonhomogeneous wave equation. If we first find a particular solution of the nonhomogeneous equation, this reduces the problem to a boundary value problem for the homogeneous equation as discussed in subsection c. (See Exercise 6.)

Exercises for Section 3.1

1. Solve the initial value problems:
    (a) $u_{tt} - c^2 u_{xx} = 0$, with $u(x,0) = x^3$ and $u_t(x,0) = \sin x$.
    (b) $u_{tt} - c^2 u_{xx} = 2t$, with $u(x,0) = x^2$ and $u_t(x,0) = 1$.

2. Solve the initial/boundary value problem

$$\begin{cases} u_{tt} - u_{xx} = 0 & \text{for } 0 < x < \pi \text{ and } t > 0 \\ u(x,0) = 0, \; u_t(x,0) = 1 & \text{for } 0 < x < \pi \\ u(0,t) = 0, \; u(\pi,t) = 0 & \text{for } t \geq 0 \end{cases}$$

using a Fourier series. Using the parallelogram rule, find the values of the solution in the various regions. Is the resulting solution continuous? Is it C^1?

3. Consider the initial/boundary value problem

$$\begin{cases} u_{tt} - u_{xx} = 0 & \text{for } 0 < x < \pi \text{ and } t > 0 \\ u(x,0) = x, \; u_t(x,0) = 0 & \text{for } 0 < x < \pi \\ u_x(0,t) = 0, \; u_x(\pi,t) = 0 & \text{for } t \geq 0. \end{cases}$$

   (a) Find a Fourier series solution, and sum the series in regions bounded by characteristics. Do you think the solution is unique?
   (b) Use the parallelogram rule to solve this problem; is the resulting solution unique? continuous? C^1?

4. Consider the initial boundary value problem

$$\begin{cases} u_{tt} - c^2 u_{xx} = 0 & \text{for } x, t > 0 \\ u(x,0) = g(x), \; u_t(x,0) = h(x) & \text{for } x > 0 \\ u(0,t) = 0 & \text{for } t \geq 0, \end{cases}$$

where $g(0) = 0 = h(0)$. If we extend g and h as *odd* functions on $-\infty < x < \infty$, show that d'Alembert's formula (6) gives the solution.

5. Find in closed form (similar to d'Alembert's formula) the solution $u(x,t)$ of

$$\begin{cases} u_{tt} - c^2 u_{xx} = 0 & \text{for } x, t > 0 \\ u(x,0) = g(x), \; u_t(x,0) = h(x) & \text{for } x > 0 \\ u(0,t) = \alpha(t) & \text{for } t \geq 0, \end{cases}$$

where $g, h, \alpha \in C^2$ satisfy $\alpha(0) = g(0)$, $\alpha'(0) = h(0)$, and $\alpha''(0) = c^2 g''(0)$. Verify that $u \in C^2$, even on the characteristic $x = ct$.

6. Solve the initial/boundary value problem

$$\begin{cases} u_{tt} - u_{xx} = 1 & \text{for } 0 < x < \pi \text{ and } t > 0 \\ u(x,0) = 0 \, , \, u_t(x,0) = 0 & \text{for } 0 < x < \pi \\ u(0,t) = 0 \, , \, u(\pi,t) = -\pi^2/2 & \text{for } t \geq 0. \end{cases}$$

Describe the singularities (i.e., is u C^2? If not, where does it fail? Is u C^1? etc.).

7. (a) Use Fourier series to solve the initial/boundary value problem for the Klein-Gordon equation

$$\begin{cases} u_{tt} - c^2 u_{xx} + m^2 u = 0 & \text{for } x, t > 0 \\ u(x,0) = g(x) \, , \, u_t(x,0) = h(x) & \text{for } x > 0 \\ u(0,t) = 0 = u(\pi, t) & \text{for } t > 0. \end{cases}$$

Notice that the solution $u(x,t)$ is bounded as $t \to \infty$.
(b) Do the same for the equation $u_{tt} - c^2 u_{xx} - m^2 u = 0$. If $c^2 < m^2$, show that the solution could be unbounded as $t \to \infty$.

8. Derive a formula similar to (6) for the pure initial value problem for the Klein-Gordon equation $u_{tt} - c^2 u_{xx} + m^2 u = 0$.

3.2 Higher Dimensions

For more than one space dimension, the solutions of the wave equation no longer have a simple form like (3). In this section we introduce the method of spherical means and use it to solve the Cauchy problem for the three-dimensional wave equation. The method of descent enables us to obtain the solution in the two-dimensional case. Finally we shall discuss the generalization to all even and odd dimensions greater than one. We shall consider only the pure initial value problem here. To consider initial/boundary value problems in a bounded domain $\Omega \subset \mathbf{R}^n$, we need to use an expansion in eigenfunctions of the Laplacian; this shall be discussed in Chapter 4.

a. Spherical Means

For a continuous function $h(x)$ on $\mathbf{R}^n$, let us introduce its *spherical mean* or *average on a sphere of radius r and center x* (see Figure 1):

(21) $$M_h(x,r) = \frac{1}{\omega_n} \int_{|\xi|=1} h(x + r\xi) \, dS_\xi,$$

where ω_n denotes the area of the unit sphere $S^{n-1} = \{\xi \in \mathbf{R}^n : |\xi| = 1\}$ and dS_ξ denotes surface measure. Since h is continuous in x, $M_h(x, r)$ is continuous in x and $r \geq 0$. In fact, letting $r \to 0$, we find

(22) $$M_h(x, 0) = h(x).$$

Furthermore, if $h \in C^k(\mathbf{R}^n)$, then we may differentiate under the integral in (21) to conclude that $M_h \in C^k(\mathbf{R}^n \times [0, \infty))$.

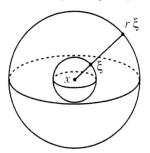

Figure 1. A sphere of radius r, centered at x.

We shall be interested in formulas for the derivatives of M_h with respect to r. Using the chain rule, we find

(23) $$\frac{\partial}{\partial r} M_h(x, r) = \frac{1}{\omega_n} \int_{|\xi|=1} \sum_{i=1}^n h_{x_i}(x + r\xi) \xi_i \, dS_\xi.$$

To compute the right-hand side of (23), we shall apply the divergence theorem in the domain $\Omega = \{\xi \in \mathbf{R}^n : |\xi| < 1\}$, which has boundary $\partial\Omega = S^{n-1}$ and exterior unit normal vector $\nu(\xi) = \xi$. The integrand in (23) is just $V \cdot \nu$ where $V(\xi) = r^{-1} \nabla_\xi h(x + r\xi) = \nabla_x h(x + r\xi)$. Computing the divergence of V, we obtain

$$\operatorname{div} V(\xi) = r \sum_{i=1}^n h_{x_i x_i}(x + r\xi) = r \Delta_x h(x + r\xi),$$

so the divergence theorem implies

(24) $$\frac{\partial}{\partial r} M_h(x, r) = \frac{r}{\omega_n} \Delta_x \int_{|\xi|<1} h(x + r\xi) \, d\xi.$$

Now if we change variables by $\xi' = r\xi$ so that $d\xi' = r^n d\xi$, then the last integral in (24) becomes

$$\int_{|\xi|<1} h(x + r\xi) \, d\xi = \frac{1}{r^n} \int_{|\xi'|<r} h(x + \xi') \, d\xi'.$$

In turn, this term may be written in spherical coordinates as

$$\frac{1}{r^n}\int_{|\xi'|<r} h(x+\xi')\,d\xi' = \frac{1}{r^n}\int_0^r \rho^{n-1}\int_{|\xi|=1} h(x+\rho\xi)\,dS_\xi d\rho$$

$$= \frac{\omega_n}{r^n}\int_0^r \rho^{n-1} M_h(x,\rho)\,d\rho.$$

Substituting this back into (24), we obtain

$$\frac{\partial}{\partial r} M_h(x,r) = \frac{1}{r^{n-1}}\Delta_x \int_0^r \rho^{n-1} M_h(x,\rho)\,d\rho.$$

If we multiply by r^{n-1}, differentiate with respect to r, and then divide by r^{n-1}, this equation may be rewritten as

(25) $$\left(\frac{\partial^2}{\partial r^2} + \frac{n-1}{r}\frac{\partial}{\partial r}\right) M_h(x,r) = \Delta_x M_h(x,r)$$

which is known as the *Darboux equation*. Notice that for a *radial* function $h = h(r)$, we have $M_h = h$, so (25) provides the Laplacian of h in spherical coordinates.

b. Application to the Cauchy Problem

We want to solve the equation

(26) $$u_{tt} = c^2 \Delta u \qquad \text{for} \quad x \in \mathbf{R}^n, t > 0$$

with initial conditions

(27) $$u(x,0) = g(x), \qquad u_t(x,0) = h(x) \qquad \text{for} \quad x \in \mathbf{R}^n.$$

We shall use *Poisson's method of spherical means* to reduce this problem to a partial differential equation in the two variables r and t.

Suppose that $u(x,t)$ solves (26). We can view t as a parameter and take the spherical mean to obtain $M_u(x,r,t)$, which satisfies

$$\frac{\partial^2}{\partial t^2} M_u(x,r,t) = \frac{1}{\omega_n}\int_{|\xi|=1} u_{tt}(x+r\xi,t)\,dS_\xi$$

$$= \frac{1}{\omega_n}\int_{|\xi|=1} c^2 \Delta u(x+r\xi,t)\,dS_\xi$$

$$= c^2 \Delta M_u(x,r,t).$$

Then we can invoke (25) to conclude that

(28) $$\frac{\partial^2}{\partial t^2} M_u(x,r,t) = c^2 \left(\frac{\partial^2}{\partial r^2} + \frac{n-1}{r}\frac{\partial}{\partial r}\right) M_u(x,r,t).$$

which is called the *Euler-Poisson-Darboux equation*. We can obtain initial conditions by taking the spherical means in (27):

(29) $$M_u(x, r, 0) = M_g(x, r), \qquad \frac{\partial M_u}{\partial t}(x, r, 0) = M_h(x, r).$$

If we can solve (28),(29) to find $M_u(x, r, t)$, we can then recover $u(x, t)$ by using (22):

(30) $$u(x, t) = \lim_{r \to 0} M_u(x, r, t).$$

It is easiest to solve (28),(29) when $n = 3$, so we shall consider this physically important case in detail.

c. The Three-Dimensional Wave Equation

When $n = 3$, we can rewrite (28) as

(31) $$\frac{\partial^2}{\partial t^2} r M_u(x, r, t) = c^2 \frac{\partial^2}{\partial r^2} r M_u(x, r, t).$$

This means that at each point x, we may consider the function $V^x(r, t) = r M_u(x, r, t)$ as a solution of a one-dimensional wave equation in $r, t > 0$:

(32) $$\frac{\partial^2}{\partial t^2} V^x(r, t) = c^2 \frac{\partial^2}{\partial r^2} V^x(r, t).$$

Moreover, we may use (29) to find the Cauchy data at $t = 0$

(33) $$\begin{aligned} V^x(r, 0) &= r M_g(x, r) \equiv G^x(r) \\ V^x_t(r, 0) &= r M_h(x, r) \equiv H^x(r), \end{aligned}$$

and use (22) to find the boundary condition at $r = 0$

(34) $$V^x(0, t) = \lim_{r \to 0} r M_u(x, r, t) = 0 \cdot u(x, t) = 0.$$

Notice that $G^x(0) = 0 = H^x(0)$, so by Exercise 4 of Section 3.1 we may extend G^x and H^x as odd functions of r and use d'Alembert's formula for $V^x(r, t)$:

(35) $$V^x(r, t) = \frac{G^x(r + ct) + G^x(r - ct)}{2} + \frac{1}{2c} \int_{r-ct}^{r+ct} H^x(\rho) \, d\rho.$$

Since G^x and H^x are odd functions, we have for $r < ct$

$$G^x(r - ct) = -G^x(ct - r) \quad \text{and} \quad \int_{r-ct}^{r+ct} H^x(\rho) \, d\rho = \int_{ct-r}^{ct+r} H^x(\rho) \, d\rho,$$

and therefore

$$M_u(x,r,t) = \frac{1}{r}V^x(r,t) = \frac{G^x(ct+r) - G^x(ct-r)}{2r} + \frac{1}{2cr}\int_{ct-r}^{ct+r} H^x(\rho)\,d\rho$$

$$= \frac{(ct+r)M_g(x,ct+r) - (ct-r)M_g(x,ct-r)}{2r} + \frac{1}{2cr}\int_{ct-r}^{ct+r} \rho M_h(x,\rho)\,d\rho.$$

Letting $r \to 0$, the difference quotients become derivatives and we obtain

(36)
$$u(x,t) = \frac{\partial}{\partial \tau}(\tau M_g(x,\tau))|_{\tau=ct} + tM_h(x,ct)$$
$$= \frac{\partial}{\partial t}(tM_g(x,ct)) + tM_h(x,ct).$$

We therefore have found that the solution of (26),(27) is given by

(37) $$u(x,t) = \frac{1}{4\pi}\frac{\partial}{\partial t}\left(t\int_{|\xi|=1} g(x+ct\xi)\,dS_\xi\right) + \frac{t}{4\pi}\int_{|\xi|=1} h(x+ct\xi)\,dS_\xi.$$

This is called *Kirchhoff's formula*, and we have proved the following.

Theorem 1. *If $g \in C^3(\mathbf{R}^3)$ and $h \in C^2(\mathbf{R}^3)$, then (37) defines a C^2-solution of (26),(27).*

Let us analyze the solution given by (37). As in the one-dimensional case, we find that the Cauchy problem is well posed. Moreover, the *domain of dependence* for a point (x,t) is the surface of the sphere $\{x+ct\xi : |\xi| = 1\}$ in $\mathbf{R}^3$ (see Figure 2). Similarly, the *range of influence* of a point $x_0 \in \mathbf{R}^3$ is the set of points $\{(x,t) : |x - x_0| = ct\}$ called the *forward light cone*.

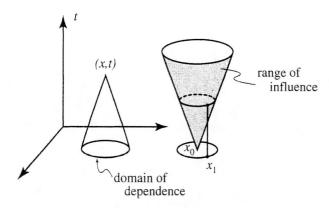

Figure 2. The domain of dependence and range of influence.

Physically, this is experienced as a *finite propagation speed*, but more specifically as the existence of *sharp signals* for three-dimensional waves, such as light or sound waves (cf. Section 3.5). For example, a large initial disturbance very near $x = 0$ will be perceived at another point $x_1 \in \mathbf{R}^3$ only at the time $t_1 = |x_1|/c$, and not after that time. This is in contrast to the one-dimensional case, as we saw in the preceding section; it also differs from the two-dimensional case, as we shall see in the next section. Another difference from the one-dimensional case is the *loss of regularity*: If g is C^2 and h is C^1, then (37) guarantees only that u is C^1. This is due to the *focusing* of initial irregularities onto a smaller set; the conservation of energy discussed in Section 3.3 will help elucidate this phenomenon.

d. The Two-Dimensional Wave Equation

Now let us consider the Cauchy problem (26),(27) when $n = 2$, which governs surface waves (e.g., water waves) in a two-dimensional medium with propagation speed c. We shall solve this problem by Hadamard's *method of descent*, namely, view (26),(27) as a special case of a three-dimensional problem with initial conditions independent of x_3.

We need to convert surface integrals in $\mathbf{R}^3$ to domain integrals in $\mathbf{R}^2$. Specifically, we need to express the surface measure on the upper half of the unit sphere S_+^2 in terms of the two variables ξ_1 and ξ_2. To do this, consider S_+^2 as the graph of the function $\psi(\xi_1, \xi_2) = \sqrt{1 - \xi_1^2 - \xi_2^2}$ over the unit disk $\xi_1^2 + \xi_2^2 < 1$, so the surface measure is

$$(38) \qquad dS_\xi = \sqrt{1 + (\psi_{\xi_1})^2 + (\psi_{\xi_2})^2}\, d\xi_1 d\xi_2 = \frac{d\xi_1 d\xi_2}{\sqrt{1 - \xi_1^2 - \xi_2^2}}.$$

If we use (37), we may perform the integrals over S_+^2 and multiply by 2 (since the initial conditions are independent of x_3); using (38), we obtain

$$(39) \qquad \begin{aligned} u(x_1, x_2, t) = {} & \frac{1}{4\pi} \frac{\partial}{\partial t} \left(2t \int_{\xi_1^2 + \xi_2^2 < 1} \frac{g(x_1 + ct\xi_1, x_2 + ct\xi_2)\, d\xi_1 d\xi_2}{\sqrt{1 - \xi_1^2 - \xi_2^2}} \right) \\ & + \frac{t}{4\pi} \left(2 \int_{\xi_1^2 + \xi_2^2 < 1} \frac{h(x_1 + ct\xi_1, x_2 + ct\xi_2)\, d\xi_1 d\xi_2}{\sqrt{1 - \xi_1^2 - \xi_2^2}} \right). \end{aligned}$$

Theorem 2. *If $g \in C^3(\mathbf{R}^2)$ and $h \in C^2(\mathbf{R}^2)$, then (39) defines a C^2-solution of (26),(27).*

The formula (39) shows that, like the one-dimensional case, but unlike the three-dimensional case, the *domain of dependence* for a point (x, t) is the *interior* of the circle $\{x + ct\xi : |\xi| = 1\}$ in $\mathbf{R}^2$ (see Figure 3). Similarly, the *range of influence* of a point $x_0 \in \mathbf{R}^2$ is the interior of the light cone [i.e., the set of points $\{(x, t) : |x - x_0| \leq ct\}$].

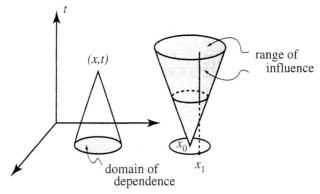

Figure 3. The domain of dependence and range of influence.

Physically, this is experienced as a *finite propagation speed* and the *absence of sharp signals* for two-dimensional waves; for example, dropping a pebble in a pond very near $x = 0$ will produce expanding circular waves that will continue to be felt at another point $x_1 \in \mathbf{R}^2$ after the time $t_1 = |x_1|/c$. Notice that, like the three-dimensional case, there is a possible *one-order loss of regularity* in the solution u.

e. Huygens's Principle

The pattern of using spherical means to solve (26),(27) for odd values of n, and the method of descent for even values of n, persists for all $n \geq 2$. In fact, for $n = 2k + 1$ the substitution

$$(40) \qquad V^x(r,t) = \left(\frac{1}{r}\frac{\partial}{\partial r}\right)^{k-1} \left(r^{2k-1} M_u(x,r,t)\right)$$

yields (32) (see Exercise 7). Similarly, we define the Cauchy data at $t = 0$

$$(41) \qquad \begin{aligned} V^x(r,0) &= \left(\frac{1}{r}\frac{\partial}{\partial r}\right)^{k-1} \left(r^{2k-1} M_g(x,r)\right) \equiv G^x(r) \\ V_t^x(r,0) &= \left(\frac{1}{r}\frac{\partial}{\partial r}\right)^{k-1} \left(r^{2k-1} M_h(x,r)\right) \equiv H^x(r), \end{aligned}$$

and check the boundary condition at $r = 0$: $V^x(0,t) = 0$. This means that the function $V^x(r,t)$ is again given by (35). Proceeding exactly as in the case $n = 3$, we find (see Exercise 7) with $c_n = (n-2)(n-4)\cdots 1$:

$$(42) \qquad \begin{aligned} u(x,t) = &\frac{1}{c_n \omega_n} \frac{\partial}{\partial t} \left(\left(\frac{1}{t}\frac{\partial}{\partial t}\right)^{\frac{n-3}{2}} t^{n-2} \int_{|\xi|=1} g(x + ct\xi)\, dS_\xi \right) \\ &+ \frac{1}{c_n \omega_n} \left(\frac{1}{t}\frac{\partial}{\partial t}\right)^{\frac{n-3}{2}} t^{n-2} \int_{|\xi|=1} h(x + ct\xi)\, dS_\xi. \end{aligned}$$

Notice that $u(x,t)$ depends only on the Cauchy data g, h on the surface of the hypersphere $\{x + ct\xi : |\xi| = 1\}$ in $\mathbf{R}^n$; in other words, we have *sharp signals*.

If we use the method of descent to obtain the solution for $n = 2k$, the hypersurface integrals in (42) become domain integrals, exactly as in the case $n = 2$. This means that there are *no sharp signals*.

The fact that sharp signals exist only for *odd dimensions* $n > 1$ is known as *Huygens's principle*. Notice that the loss of regularity in the solution also persists for all dimensions $n > 1$.

Exercises for Section 3.2

1. (a) If $F(s)$ is a C^2-function of the one-variable s, find a condition on the vector $\alpha = (\alpha_1, \alpha_2, \alpha_3)$ so that
$$u(x_1, x_2, x_3, t) = F(\alpha_1 x_1 + \alpha_2 x_2 + \alpha_3 x_3 - t)$$
is a solution of (26). (Such solutions are called *plane waves* and are constant on the planes $\alpha \cdot x - t = $ constant.)
(b) Find the relationship which must hold between the initial data $g(x)$ and $h(x)$ for a plane wave solution.
(c) Find all plane wave solutions of (26) with the initial condition $u(x_1, x_2, x_3, 0) = x_1 - x_2 + 1$.

2. Find the solution of the initial value problem
$$\begin{cases} u_{tt} = u_{xx} + u_{yy} + u_{zz} \\ u(x,y,z,0) = x^2 + y^2, \quad u_t(x,y,z,0) = 0 \end{cases}$$
(a) by using (37) and (b) by using (39).

3. Use Duhamel's principle to find the solution of the nonhomogeneous wave equation for three space dimensions $u_{tt} - c^2 \Delta u = f(x,t)$ with initial conditions $u(x,0) = 0 = u_t(x,0)$. What regularity in $f(x,t)$ is required for the solution u to be C^2?

4. Let $\Omega = \{(x,y) \in \mathbf{R}^2 : 0 < x < a \text{ and } 0 < y < b\}$, and use separation of variables to solve the initial/boundary value problem
$$\begin{cases} u_{tt} = u_{xx} + u_{yy} & \text{for } (x,y) \in \Omega \text{ and } t > 0 \\ u(x,y,t) = 0 & \text{for } (x,y) \in \partial\Omega \text{ and } t > 0 \\ u(x,y,0) = \sin\dfrac{\pi x}{a} \sin\dfrac{2\pi y}{b}, & \text{and} \quad u_t(x,y,0) = 0 \quad \text{for } (x,y) \in \Omega. \end{cases}$$

5. Find a formula for the solution $v(x,t) = v(x_1, x_2, t)$ of the Cauchy problem for the two-dimensional Klein-Gordon equation:
$$\begin{cases} v_{tt} = c^2 \Delta v - m^2 v & \text{for } x \in \mathbf{R}^2 \text{ and } t > 0 \\ v(x,0) = g(x), \quad v_t(x,0) = h(x). \end{cases}$$

6. (a) For $n = 3$, suppose that $g, h \in C_0^\infty(\mathbf{R}^3)$ and consider the solution of (26),(27) given by (37). Show that there is a constant C so that $|u(x,t)| \leq C/t$ for all $x \in \mathbf{R}^3$ and $t > 0$.
(b) Is a similar result true for $n = 2$?

7. Suppose $n = 2k + 1$ for a positive integer k and u satisfies (26).
(a) Show that $V^x(r,t)$ given by (40) satisfies (32).
(b) Derive (42) from (35) and (41); note $c_n = 1 \cdot 3 \cdots (n-2)$.

3.3 Energy Methods

Given a function $u(x,t)$, let us define its *energy* at time t by

$$(43) \qquad \mathcal{E}(t) = \frac{1}{2} \int_{\mathbf{R}^n} (u_t^2 + c^2 |\nabla u|^2)\, dx.$$

The energy gives an integral measure of the first-order regularity of a function: if the solution develops a singularity so that the first-order derivatives become unbounded, we might expect the energy to become unbounded. On the other hand, if the energy is constant, then such singularities must become concentrated on smaller and smaller sets; this explains the focusing of singularities mentioned in Section 3.2.

In this section we shall show that solutions of the n-dimensional wave equation exhibit such *conservation of energy*. This also provides proofs of uniqueness and the domain of dependence and will be an important property in the consideration of weak solutions of the wave equation.

a. Conservation of Energy

Suppose that $u \in C^2(\mathbf{R}^n \times (0,\infty))$ solves (26),(27), where g and h both have compact support [say g and h are both zero outside of the large ball $\{x \in \mathbf{R}^n : |x| \leq R\}$]. By finite propagation speed (see also Section 3.3b), for fixed t, $u(x,t)$ is zero outside the ball $\{x \in \mathbf{R}^n : |x| \leq R + ct\}$. In particular, the energy $\mathcal{E}(t)$ of the solution is finite for each time t. If we differentiate this function of t, we obtain

$$(44) \qquad \frac{d\mathcal{E}}{dt} = \int_{\mathbf{R}^n} \left(u_t u_{tt} + c^2 \sum_{i=1}^n u_{x_i} u_{x_i t}\right) dx.$$

Integration by parts then yields

$$(45) \qquad \frac{d\mathcal{E}}{dt} = \int_{\mathbf{R}^n} u_t (u_{tt} - c^2 \Delta u)\, dx = 0.$$

In other words, $\mathcal{E}(t)$ must be a constant.

Theorem 1. *If $u \in C^2(\mathbf{R}^n \times (0,\infty))$ solves (26),(27), where g and h have compact support, then $\mathcal{E}(t) \equiv \mathcal{E}(0)$.*

In particular, if u_1 and u_2 are two solutions of (26),(27), then $w = u_1 - u_2$ has zero Cauchy data and hence $\mathcal{E}_w(0) = 0$. By Theorem 1, $\mathcal{E}_w(t) \equiv 0$, which implies $w(x,t) \equiv$ const. But $w(x,0) = 0$ then implies $w(x,t) \equiv 0$, so the solution is unique.

Corollary. *The solution of (26),(27) is unique.*

The concept of energy may also be defined for boundary value problems, and used to show uniqueness of solutions; see Exercises 1 and 2. (Existence theory for the wave equation on bounded domains in dimension $n \geq 2$ will be studied in Section 4.4b using eigenfunction expansions.)

b. The Domain of Dependence

In the preceding discussion of conservation of energy and the uniqueness of solutions, we used the property of finite propagation speed to integrate by parts without encountering a boundary term in (45). With a little more care, however, we may use a similar calculation with localized energy to establish the domain of dependence for all n and obtain the finite propagation speed as a *consequence* of energy integrals.

Theorem 2. *Suppose u is a C^2 solution of (26),(27). For any $x_0 \in \mathbf{R}^n$ and $t_0 > 0$, if $g \equiv 0 \equiv h$ in $\overline{B}_0 = \{x \in \mathbf{R}^n : |x - x_0| \leq ct_0\}$ then $u(x_0, t_0) = 0$.*

Proof. For any time $\tau \in [0, t_0]$, let $\overline{B}_\tau = \{x \in \mathbf{R}^n : |x - x_0| \leq c(t_0 - \tau)\}$ (see Figure 1).

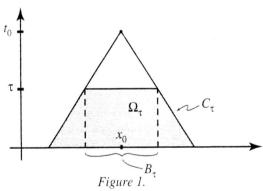

Figure 1.

Consider the local energy function

$$(46) \quad \mathcal{E}_{x_0, t_0}(\tau) = \frac{1}{2} \int_{B_\tau} (u_t^2 + c^2 |\nabla u|^2)\big|_{t=\tau} \, dx \quad \text{for} \quad 0 \leq \tau \leq t_0.$$

We claim that (46) is a nonincreasing function of τ; that is, the following *energy inequality* holds:

$$(47) \quad \mathcal{E}_{x_0, t_0}(\tau) \leq \mathcal{E}_{x_0, t_0}(0) \quad \text{for} \quad 0 \leq \tau \leq t_0.$$

To prove (47), let us introduce the following notation (see Figure 1):

$$\Omega_\tau = \{(x,t) : |x - x_0| < c(t_0 - t), 0 < t < \tau\}$$
$$C_\tau = \{(x,t) : |x - x_0| = c(t_0 - t), 0 < t < \tau\}.$$

Notice that $\partial\Omega_\tau = C_\tau \cup (\overline{B}_0 \times \{0\}) \cup (\overline{B}_\tau \times \{\tau\})$, where the unions are disjoint. Moreover, the exterior unit normal ν on $\partial\Omega_\tau$ is given on $\overline{B}_\tau \times \{\tau\}$ by $\nu = \langle 0, \ldots, 0, 1 \rangle$, and on $\overline{B}_0$ by $\nu = \langle 0, \ldots, 0, -1 \rangle$. On C_τ, the normal $\nu = \langle \nu_1, \ldots, \nu_n, \nu_{n+1} \rangle$ satisfies $c^2(\nu_1^2 + \cdots + \nu_n^2) = \nu_{n+1}^2$; together with the unit length condition $\nu_1^2 + \cdots + \nu_n^2 + \nu_{n+1}^2 = 1$, this implies

(48) $$\nu_1^2 + \cdots + \nu_n^2 = \frac{\nu_{n+1}^2}{c^2} = \frac{1}{1+c^2}.$$

Given a solution u of (26), let us define the vector field

$$\vec{V} = \langle 2c^2 u_t u_{x_1}, \ldots, 2c^2 u_t u_{x_n}, -(c^2|\nabla u|^2 + u_t^2) \rangle.$$

If we calculate the divergence in (x,t), we find

$$\operatorname{div}\vec{V} = 2c^2(u_{tx_1}u_{x_1} + u_t u_{x_1 x_1} + \cdots + u_{tx_n}u_{x_n} + u_t u_{x_n x_n})$$
$$- 2c^2(u_{tx_1}u_{x_1} + \cdots + u_{tx_n}u_{x_n}) - 2u_t u_{tt} = 0.$$

The divergence theorem therefore implies

(49) $$\int_{\partial\Omega_\tau} \vec{V} \cdot \nu \, dS = 0.$$

Now, on C_τ the following inequality holds (see Exercise 3):

(50) $$2u_t(u_{x_1}\nu_1 + \cdots + u_{x_n}\nu_n) \leq \frac{c}{\sqrt{1+c^2}}|\nabla u|^2 + \frac{1}{c\sqrt{1+c^2}}u_t^2.$$

Using (50), we may compute on C_τ

$$\vec{V} \cdot \nu = 2c^2 u_t(u_{x_1}\nu_1 + \cdots + u_{x_n}\nu_n) - (c^2|\nabla u|^2 + u_t^2)\nu_{n+1} \leq 0,$$

so in particular

$$\int_{C_\tau} \vec{V} \cdot \nu \, dS \leq 0.$$

From (49) we have

$$0 \leq \int_{\overline{B}_0} \vec{V} \cdot \nu \, dS + \int_{\overline{B}_\tau \times \{\tau\}} \vec{V} \cdot \nu \, dS$$
$$= \int_{\overline{B}_0} (c^2|\nabla u|^2 + u_t^2)|_{t=0}\, dx - \int_{\overline{B}_\tau} (c^2|\nabla u|^2 + u_t^2)|_{t=\tau}\, dx,$$

which proves (47).

Now to prove Theorem 2, use (47) to conclude $u(x,t) \equiv$ cont in Ω_{t_0}. But $u(x,t) \equiv 0$ in $\overline{B_0}$ then implies $u(x,t) \equiv 0$ in Ω_{t_0}; in particular, $u(x_0, t_0) = 0$. ♠

Exercises for Section 3.3

1. Let Ω be a smooth, bounded domain in $\mathbf{R}^n$. For a C^2 solution $u(x,t)$ of the wave equation $u_{tt} = c^2 \Delta u$ for $x \in \Omega, t > 0$, define the energy to be $\mathcal{E}_\Omega(t) = \frac{1}{2} \int_\Omega (u_t^2 + c^2 |\nabla u|^2) \, dx$. If u satisfies either the boundary condition $u(x,t) = 0$ or $\partial u / \partial \nu(x,t) = 0$ for $x \in \partial\Omega$, where ν is the exterior unit normal, then show that $\mathcal{E}_\Omega(t)$ is constant.

2. Use the previous exercise to show uniqueness of the solution for the (nonhomogeneous) wave equation $u_{tt} = \Delta u + f(x,t)$ in a smooth, bounded domain $\Omega \subset \mathbf{R}^n$ with either (a) Dirichlet condition $u = g$ on $\partial \Omega$, or (b) Neumann condition $\partial u / \partial \nu = h$ on $\partial \Omega$.

3. Use (48) to prove (50).

4. The partial differential equation $u_{tt} = c^2 \Delta u - q(x) u$ arises in the study of wave propagation in a nonhomogeneous elastic medium: $q(x)$ is nonnegative and proportional to the coefficient of elasticity at x.
   (a) Define an appropriate notion of energy for solutions.
   (b) Verify the corresponding energy inequality.
   (c) Use the energy method to prove that solutions are uniquely determined by their Cauchy data.

5. Consider the n-dimensional wave equation with dissipation
$$\begin{cases} u_{tt} - c^2 \Delta u + \alpha u_t = 0 \\ u(x,0) = g(x), \quad u_t(x,0) = h(x), \end{cases}$$
   where g and h have compact support and $\alpha \geq 0$ is a constant. Define the energy $\mathcal{E}(t)$ by (43).
   (a) Prove a domain of dependence result to conclude that solutions have a finite propagation speed.
   (b) Show that $\mathcal{E}(t)$ is *nonincreasing* in $t > 0$.
   (c) Use the energy method to prove that solutions are uniquely determined by their Cauchy data.

6. Consider a flexible beam with clamped ends at $x = 0$ and $x = 1$. Small wave motion in the beam satisfies
$$\begin{cases} u_{tt} + \gamma^2 u_{xxxx} = 0 \\ u(0,t) = 0 = u(1,t) \\ u_x(0,t) = 0 = u_x(1,t), \end{cases}$$
   where γ^2 is a constant depending on the shape and the material of the beam. Show that the energy $\mathcal{E} = \frac{1}{2} \int_0^1 (u_t^2 + \gamma^2 u_{xx}^2) \, dx$ is conserved.

3.4 Lower-Order Terms

In this section we shall consider the effect of lower-order terms on the wave equation. For simplicity, we shall just consider the one-dimensional case

(51) $\quad u_{tt} - u_{xx} + \alpha u_t + \beta u_x + \gamma u = f(x,t) \quad$ for $\quad x \in \mathbf{R}, \; t > 0,$

where α, β, γ are functions of x and t, possibly constants.

a. Dispersion

First, let us see the effect of the zero-order term in (51) by investigating a phenomenon called *dispersion* for the equation

(52) $\quad u_{tt} - u_{xx} + \lambda u = 0 \quad x \in \mathbf{R}, \; t > 0,$

where λ is a constant. Let us look for a special type of solution, called a *uniform wave*, of the form

(53) $\quad u(x,t) = U(kx - \omega t),$

where k and ω are real numbers; k is called the *wave number* and ω is the *frequency*. The function (53) is called a uniform wave because it represents a wave propagating with constant velocity

(54) $\quad c = \dfrac{\omega}{k},$

but without change of shape. If $\lambda = 0$, then $\omega = \pm k$ yields the type of solutions given in (3). If $\lambda \neq 0$, then substitution of (53) in (52) yields $(\omega^2 - k^2)U'' + \lambda U = 0$. Solutions that are *bounded* require $\lambda^{-1}(\omega^2 - k^2) > 0$, in which case U may be expressed as a linear combination of sines and cosines. However, it is simpler and more general to write

(55) $\quad U(kx - \omega t) = A e^{i(kx - \omega t)}$

and substitute into (52) to obtain the *dispersion relation*

(56) $\quad \omega^2 - k^2 = \lambda \quad \Leftrightarrow \quad \omega(k) = \pm\sqrt{k^2 + \lambda},$

which guarantees that (55) solves (52). Euler's formula $e^{i\theta} = \cos\theta + i\sin\theta$ may then be used to replace the complex wave (55) by real waves when $\lambda^{-1}(\omega^2 - k^2) > 0$.

As (56) suggests, the dispersion relation defines the frequency ω as a function of the wave number k. Similarly, the velocity (54) becomes a function of k; $c = c(k)$ is called the *phase velocity*. When the velocity is a nonconstant function of k (i.e., $dc/dk \neq 0$), then we say the solution (55) is a *dispersive wave*. Notice that the uniform wave (55) is dispersive when $\lambda \neq 0$ but nondispersive for the pure wave equation ($\lambda = 0$). Thus one consequence of the addition of lower-order terms to the wave equation is the possible introduction of dispersive waves.

The significance of dispersive waves can be seen if we use superposition to form linear combinations of (56) with different $k_1, \ldots, k_N$ to obtain a *discrete wave train*

$$(57) \qquad u(x,t) = \sum_{j=1}^{N} A_j e^{i(k_j x - \omega_j t)}.$$

This function satisfies (52) provided that each $w_j = w_j(k_j)$ satisfies the dispersion relation $w_j(k_j) = \pm\sqrt{k_j^2 + \lambda}$. In the dispersive case $\lambda \neq 0$, the terms in (58) corresponding to different k_j will propagate at different speeds, i.e., they will *disperse*.

It is possible to generalize (58) to infinite series, provided the coefficients A_j decay sufficiently as $j \to \infty$ to guarantee convergence of the series. In fact, in the continuous limit, we obtain Fourier integrals such as

$$(58) \qquad u(x,t) = \int_{-\infty}^{\infty} A(k) e^{i(kx - \omega(k)t)} \, dk.$$

These can be analyzed (cf. [Whitham]) to show that such solutions also disperse when $\lambda \neq 0$. This means that initial conditions may have their support in a small neighborhood of a single point, say $x = 0$, but as time evolves, the solution disperses into separate waves.

It is important to note that, even in the dispersive case, solutions of (52) satisfy *conservation of energy*, where the energy is defined by

$$(59) \qquad \mathcal{E}(t) = \frac{1}{2} \int_{-\infty}^{\infty} u_t^2 + u_x^2 + \lambda u^2 \, dx$$

(assuming the Cauchy data has compact support; cf. Exercise 4 in the previous section). Therefore, even when the solution disperses into separate waves, the total energy remains constant.

Remark. Dispersion also occurs for other linear and nonlinear equations. In general, a *dispersive wave* is a solution of the form (55) for which the dispersion relation establishes ω as (i) a *real-valued* function of k, (ii) having *nonconstant* velocity $c = \omega/k$. An equation admitting dispersive wave solutions is called *dispersive*. Examples will be given in the Exercises.

b. Dissipation

Now let us investigate the behavior of solutions of

$$(60) \qquad u_{tt} - u_{xx} + \alpha u_t + \beta u_x + \gamma u = 0,$$

when α, β, and γ are constants; in particular, we are interested in the case $\alpha > 0$, which corresponds to dissipation.

By a change of dependent variable, we can reduce this problem to (52) of the previous subsection. In fact, first introduce the characteristic coordinates $\mu = x + t$ and $\eta = x - t$ as in Section 3.1. The equation (60) becomes
$$u_{\mu\eta} - \frac{\alpha+\beta}{4}u_\mu + \frac{\alpha-\beta}{4}u_\eta - \frac{\gamma}{4}u = 0.$$

Now let

(61) $$w(\mu,\eta) = u(\mu,\eta)\exp\left(\frac{\alpha-\beta}{4}\mu - \frac{\alpha+\beta}{4}\eta\right)$$

to find that w satisfies

(62) $$w_{\mu\eta} + \frac{\lambda}{4}w = 0, \quad \text{where} \quad \lambda = \frac{\alpha^2 - \beta^2 - 4\gamma}{4}.$$

Converting back to x, t coordinates, we have

(63) $$u(x,t) = w(x,t)\exp\left(\frac{\beta}{2}x - \frac{\alpha}{2}t\right),$$

where w satisfies (62).

By the previous subsection, w satisfies a conservation of energy law, but if $\alpha > 0$, then $u(x,t)$ is exponentially decreasing in t. In particular, if $\beta = 0$, then this decrease is uniform. In fact, when $\beta = 0$ we see that the energy
$$\mathcal{E}(t) = \frac{1}{2}\int (u_t^2 + u_x^2 + \gamma u^2)\,dx$$
of a solution u is nonincreasing:
$$\frac{d}{dt}\mathcal{E}(t) = \int (u_t u_{tt} + u_x u_{xt} + \gamma u u_t)\,dx$$
$$= \int u_t(u_{tt} - u_{xx} + \gamma u)\,dx = -\alpha \int u_t^2\,dx \leq 0,$$

assuming, for example, that the initial data has compact support.

c. The Domain of Dependence

Recall from Section 3.1 that the solution of the Cauchy problem for the nonhomogenous wave equation

(64) $$\begin{aligned}u_{tt} - u_{xx} &= f(x,t)\\ u(x,0) &= g(x), \qquad u_t(x,0) = h(x)\end{aligned}$$

can be expressed as

$$(65) \quad u(P) = \frac{g(Q) + g(R)}{2} + \frac{1}{2} \int_Q^R h(x)\, dx + \frac{1}{2} \int\!\!\int_D f(x,t)\, dx\, dt,$$

where $P = (x,t)$, $Q < R$ are the points where the characteristic through P intersects the x-axis, and D is the domain bounded by these characteristics and the x-axis (see Figure 1). The closure $\overline{D}$ is the domain of dependence of (64).

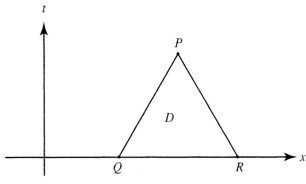

Figure 1.

Now suppose we want to study solutions of (51) with the same Cauchy data as in (64). We shall invoke a technique that occurs frequently in partial differential equations, namely replace the PDE by an *integro-differential* equation. To achieve this, let us rewrite (51) as

$$(66) \quad u_{tt} - u_{xx} = F(x,t,u,u_x,u_t),$$

where

$$(67) \quad F(x,t,u,u_x,u_t) = f(x,t) - \alpha u_x - \beta u_t - \gamma u.$$

Then let us introduce $\tilde{f}(x,t) = F(x,t,u,u_x,u_t)$ and invoke (65):

$$(68) \quad u(P) = \frac{g(Q) + g(R)}{2} + \frac{1}{2} \int_Q^R h(x)\, dx + \frac{1}{2} \int\!\!\int_D F(x,t,u,u_x,u_t)\, dx\, dt.$$

Of course, (68) does not *define* the solution of (51) since F depends on u itself; however, it may be used to draw conclusions about the solution. In particular, we see that u only depends on the behavior of the Cauchy data on $[Q,R]$ and the functions α, β, γ, and f in D. In this way, we conclude that the *domain of dependence of* (51) coincides with $\overline{D}$, the domain of dependence of (64).

Another important observation to make about (68) is that, although it cannot be used to *define* a solution u of (51), it can be used to show that a solution *exists*. This is achieved by means of the *contraction mapping principle* (also known as *successive approximations;* cf. Section 7.3). This method is especially important because it can yield numerical approximations to the solution. However, we shall not discuss it at this time.

Exercises for Section 3.4

1. Find dispersive wave solutions of the n-dimensional linear Klein-Gordon equation $u_{tt} - c^2 \Delta u + m^2 u = 0$.
2. Show that each of the following linear equations has dispersive wave solutions $u = A \exp[i(kx - \omega t)]$:
   (a) The flexible beam equation $u_{tt} + \gamma^2 u_{xxxx} = 0$
   (b) The linearized KdV equation $u_t + c u_x + u_{xxx} = 0$
   (c) The Boussinesq equation $u_{tt} - c^2 u_{xx} = \gamma^2 u_{ttxx}$
   (d) The Schrödinger equation $u_t = i \Delta u$
3. Show that the heat equation $u_t = u_{xx}$ admits uniform wave solutions of the form (55) in which $\omega(k)$ is complex-valued and the wave is exponentially decaying in t. (Such uniform waves are called *diffusive.* Thus the heat equation is diffusive and not dispersive.)
4. Find two uniform wave solutions of (52) with $\lambda > 0$, satisfying the initial condition $u(x,0) = 3 \cos 2x$.
5. Find a condition on g and h that is necessary for the existence of a uniform wave solution of (52) satisfying the initial condition $u(x,0) = g(x)$ and $u_t(x,0) = h(x)$.
6. Find the solution of the telegrapher's equation $u_{tt} - u_{xx} + u_t + m^2 u = 0$ satisfying the initial conditions $u(x,0) = g(x)$ and $u_t(x,0) = 0$ where g is an arbitrary C^2 function.

3.5 Applications to Light and Sound

In this section we use principles of electromagnetism and acoustics to show how the wave equation applies to the propagation of light and sound.

a. *Electromagnetism*

Let us consider Maxwell's equations for the propagation of the electric field $\vec{E}(x,t)$ and magnetic field $\vec{H}(x,t)$ with electric charge density $\rho(x,t)$ (cf. Section A.2b in the Appendix):

(69a) $\qquad \partial_t \vec{E} = c \operatorname{curl} \vec{H} - 4\pi \sigma \vec{E},$

(69b) $\qquad \partial_t \vec{H} = -c \operatorname{curl} \vec{E},$

(69c) $\qquad \operatorname{div} \vec{E} = 4\pi \rho,$

(69d) $\qquad \operatorname{div} \vec{H} = 0,$

where c is the velocity of light and σ is another (known) constant. We want to solve these equations for $\vec{E}$ and $\vec{H}$ in terms of given initial conditions

(70) $\qquad \vec{E}(x,0) = \vec{E}_0(x) \quad \text{and} \quad \vec{H}(x,0) = \vec{H}_0(x),$

which of course must satisfy (69c) and (69d), respectively. [As observed in Section A.2b, the function $\rho(x,t)$ is determined by its initial values $\rho_0(x) = (4\pi)^{-1} \text{div}\, \vec{E}_0(x)$.]

Take the curl of both sides of (69a), using (69b) to evaluate curl $\vec{E}$, (14b) of the Appendix to evaluate curl(curl $\vec{H}$), and finally (69d):

$$\text{curl}\,(\partial_t \vec{E}) = \partial_t \text{curl}\, \vec{E} = -c^{-1} \partial_t^2 (\vec{H}),$$

$$\text{curl}\,(c\,\text{curl}\, \vec{H} - 4\pi\sigma \vec{E}) = c[\nabla(\text{div}\,\vec{H}) - \Delta \vec{H}] - 4\pi\sigma[-\frac{1}{c}\partial_t \vec{H}]$$

$$= -c\Delta \vec{H} + \frac{4\pi\sigma}{c} \partial_t \vec{H}.$$

This means that $\vec{H}$ satisfies a damped wave equation

(71a) $\qquad \partial_t^2 \vec{H} + 4\pi\sigma \partial_t \vec{H} = c^2 \Delta \vec{H}.$

Moreover, we know appropriate initial conditions for (71b):

(71b) $\qquad \vec{H}(x,0) = \vec{H}_0(x) \quad \text{and} \quad \partial_t \vec{H}(x,0) = -c\,\text{curl}\, \vec{E}_0(x).$

It is important to emphasize that (71a) is really three separate scalar equations for the components H_i of $\vec{H}$, and therefore the solution methods of this chapter are applicable. However, we might be concerned whether the equation (69d), which was assumed to hold at $t = 0$, continues to hold for all $t > 0$. But div $\vec{H}(x,t)$ satisfies the damped wave equation $\partial_t^2(\text{div}\,\vec{H}) + 4\pi\sigma \partial_t(\text{div}\,\vec{H}) = c^2 \Delta(\text{div}\,\vec{H})$, with the trivial initial conditions div$\vec{H}(x,0) = 0$ and $\partial_t(\text{div}\,\vec{H}(x,0)) = -c\,\text{div}\,\text{curl}\, E_0(x) = 0$ by (14b) of the Appendix, so by the uniqueness (cf. Exercise 5 in Section 3.3) we know that div$\vec{H}(x,t) = 0$ for all $t > 0$ so that (69d) is indeed satisfied.

A similar calculation performed on (69b) using (69c) shows that $\vec{E}$ satisfies an inhomogeneous damped wave equation

(72a) $\qquad \partial_t^2 \vec{E} + 4\pi\sigma \partial_t \vec{E} = c^2 \Delta \vec{E} - c^2 \nabla \rho,$

with initial conditions

(72b) $\qquad \vec{E}(x,0) = \vec{E}_0(x) \quad \text{and} \quad \partial_t \vec{E}(x,0) = c\,\text{curl}\, \vec{H}_0(x) - 4\pi\sigma \vec{E}_0(x).$

Moreover, the formula $\rho(x,t) = \rho(x,0) \exp(-4\pi\sigma t)$ obtained in Section A.2b of the Appendix guarantees that (69c) holds for all $t > 0$. In the special case $\rho(x,0) = 0$ we have $\rho(x,t) \equiv 0$, and (72a) becomes a homogeneous damped wave equation that in fact agrees with (71a).

In conclusion, we have shown that Maxwell's equations in the absence of a charge density (i.e. when $\rho \equiv 0$) decouple into damped wave equations for the components of $\vec{E}$ and $\vec{H}$. Moreover, in a complete vacuum we also have $\sigma = 0$; in this case, the damping factor disappears, so the components of $\vec{E}$ and $\vec{H}$ satisfy the wave equation and hence may be found explicitly (in terms of Cauchy data) using the Kirchhoff formula (37). In particular, this justifies the use of the wave equation as a model for light propagation.

b. Acoustics

The propagation of sound through air or any gas is caused by small displacements of the molecules. Such displacemnts may be described by a velocity field $\vec{u}(x,t)$ and a density function $\rho(x,t)$ that satisfy Euler's equations (conservation of momentum) $\vec{u}_t + (\vec{u} \cdot \nabla)\vec{u} + \rho^{-1}\nabla p = 0$, where p denotes pressure, and conservation of mass $\rho_t + \text{div}(\rho\vec{u}) = 0$; see also Section A.2c in the Appendix. These are nonlinear equations and therefore difficult to solve. But since the displacements are small, the equations are well approximated by a linearization process that we shall now describe.

Let us assume that the gas is isentropic, so that pressure is a (smooth) function of density: $p = p(\rho)$ and $\nabla p = p'(\rho)\nabla \rho$. With no disturbance, the gas has a constant density $\rho_0 > 0$ and zero velocity. To linearize, let us consider a small perturbation: $\rho = \rho_0 + \epsilon\tilde{\rho}$ and $\vec{u} = \epsilon\vec{v}$, where $\epsilon > 0$ is small. Using Taylor series at ρ_0, we find

$$\rho^{-1} = \rho_0^{-1} - \rho_0^{-2}\epsilon\tilde{\rho} + O(\epsilon^2) \quad \text{as } \epsilon \to 0,$$
$$\nabla p = (p'(\rho_0) + O(\epsilon))\epsilon\nabla\tilde{\rho} \quad \text{as } \epsilon \to 0.$$

Substituting in Euler's equations, we obtain

$$\epsilon\vec{v}_t + (\epsilon\vec{v} \cdot \nabla)\epsilon\vec{v} + (\rho_0^{-1} + O(\epsilon))(p'(\rho_0)\epsilon\nabla\tilde{\rho} + O(\epsilon^2)) = 0.$$

Selecting just the leading-order (linear) terms in ϵ, we obtain

$$\vec{v}_t + \rho_0^{-1}p'(\rho_0)\nabla\tilde{\rho} = 0.$$

Similarly, we may substitute our perturbation into the conservation of mass equation to obtain

$$\epsilon(\tilde{\rho}_t + \rho_0 \text{div}\,\vec{v}) + \epsilon^2 \text{div}(\tilde{\rho}\vec{v}) = 0$$

and then select the leading-order terms in ϵ:

$$\tilde{\rho}_t + \rho_0 \text{div}\,\vec{v} = 0.$$

Returning to our original notation (replacing $\tilde{\rho}$ by ρ and $\vec{v}$ by $\vec{u}$), we have the following *linearized equations of acoustics*:

(73a) $$\vec{u}_t + \frac{c_0^2}{\rho_0}\nabla\rho = 0,$$

(73b) $$\rho_t + \rho_0 \text{div}\,\vec{u} = 0,$$

where $c_0^2 = p'(\rho_0)$; as we shall see, c_0 represents the *speed of sound*.

The system (73ab) is first-order but coupled; we can replace it with a decoupled second-order system as follows. Differentiate (73b) with respect to t and then use (73a):

$$0 = \rho_{tt} + \rho_0 \operatorname{div} \vec{u}_t = \rho_{tt} + \rho_0 \operatorname{div}(-\frac{c_0^2}{\rho_0}\nabla \rho) = \rho_{tt} - c_0^2 \Delta \rho.$$

We see that ρ satisfies the wave equation with propagation speed c_0. Similarly, differentiating (73a) with respect to t and then using (73b) yields

(74) $$\vec{u}_{tt} = c_0^2 \nabla(\operatorname{div} \vec{u}).$$

Now we would like to use (74) to conclude that the components of $\vec{u}$ also satisfy a wave equation, but we must exercise some caution: if u is scalar valued, then $\nabla(\operatorname{div} u) = \Delta u$, but this may not be the case for vector-valued $\vec{u}$. However, let us add an initial condition, namely that the gas is initially *irrotational* (i.e. its *vorticity* $\operatorname{curl} \vec{u}$ is zero at $t = 0$). Using (73a), we can conclude that the vorticity is zero for $t > 0$:

$$\partial_t(\operatorname{curl} \vec{u}) = \operatorname{curl} \vec{u}_t = -\frac{c_0^2}{\rho_0}\operatorname{curl}(\nabla \rho) = 0.$$

But using $\operatorname{curl} \vec{u} \equiv 0$ and (14c) in Section A.2a of the Appendix, we obtain $\nabla(\operatorname{div} \vec{u}) = \Delta \vec{u}$ and consequently we may conclude from (74) that $\vec{u}_t = c_0^2 \Delta \vec{u}$ (i.e. each compenent of $\vec{u}$ satisfies the wave equation with propagation speed c_0). This justifies the use of the wave equation as an approximate model for the propagation of sound.

Further References for Chapter 3

For additional details about Fourier series, including issues of orthogonality, completeness, and convergence, see [Churchill] or [Pinsky].

We know the one-dimensional wave equation is hyperbolic. Similarly, the n-dimensional wave equation is included in a class of equations called hyperbolic (see Exercise 1 in Section 11.1). For these more general hyperbolic equations, it is natural to investigate the well-posedness of the Cauchy problem, propagation of singularities, and so on. This naturally leads to a refinement of distribution theory called the wave front set, and to generalizations of partial differential equations called pseudo-differential operators and Fourier-integral operators; these are related to the representation (58) and the Fourier transform, which will be discussed in Section 5.2. For more information, consult [Taylor] or [Hörmander, 2, v. II].

For more information on dispersive and nondispersive waves, an excellent reference is [Whitham].

4 The Laplace Equation

In this chapter we study boundary value problems, especially Dirichlet and Neumann problems, for the n-dimensional Laplace and Poisson equations. In doing so, we encounter maximum principles, fundamental solutions, Green's functions, and subharmonic functions. We also discuss the eigenvalues of the Laplacian and their application to the wave equation in bounded domains. Finally, we discuss applications to vector fields.

4.1 Introduction to the Laplace Equation

The *Laplace equation*

$$(1) \qquad \Delta u = 0 \quad \text{in} \quad \Omega \subset \mathbf{R}^n$$

and its nonhomogeneous partner, the *Poisson equation*

$$(2) \qquad \Delta u = f \quad \text{in} \quad \Omega \subset \mathbf{R}^n,$$

arise quite frequently in the physical sciences. For example, solutions of (2) correspond to steady states for time evolutions such as heat flow or wave motion, with f corresponding to external driving forces (heat sources or wave generators). When no external forces are present, such steady states correspond to solutions of (1); solutions of (1) are called *harmonic functions* in Ω.

The Laplace and Poisson equations also play an important role in *field theories* in which a field (e.g., electric, magnetic, gravitational force, or fluid velocity field) is given as the gradient of a *potential function* u. For example, suppose $\vec{E}$ is an electric field in a domain Ω of $\mathbf{R}^3$. If the components of $\vec{E}$ are C^1, then $\operatorname{div} \vec{E}$ has a natural interpretation as a continuous distribution of electric charge in Ω; in standard units, Maxwell's equations imply that $\operatorname{div} \vec{E} = 4\pi\rho$ where ρ is the charge density. Now suppose $\vec{E}$ is a gradient field with a potential function u; this is usually written $\vec{E} = -\nabla u$ (the minus sign is a convention connected with the interpretation of u as "potential energy"). Then $\Delta u = -\operatorname{div} \vec{E} = -4\pi\rho$, so u satisfies the Poisson equation. On the other hand, if the electric field is induced by charged particles lying *outside* of Ω, then $\rho \equiv 0$ in Ω and u is *harmonic* in Ω. (The interpretation of (1) and (2) in electrostatics will prove particularly useful in Section 4.2; for a discussion of force fields and fluid velocity fields defined by potential functions, see [Kellogg].)

Cauchy problems for equations (1) and (2) are not well posed (cf. Exercise 7 in Section 2.1). What sort of side conditions should we impose? In this section, we use separation of variables for some special domains Ω to find the boundary conditions that are appropriate for (1) and (2), and we discuss the significance of such boundary value problems in physics; then we use the divergence theorem and the spherical mean to investigate the uniqueness of solutions and their continuous dependence on the data for general domains Ω.

a. Separation of Variables

If the domain Ω has natural symmetries, then we can try to separate the variables and use Fourier series to solve the equation.

Example 1. Consider the problem

$$\begin{cases} u_{xx} + u_{yy} = 0 & \text{for } 0 < x, y < \pi \\ u(0,y) = 0 = u(\pi,y) & \text{for } 0 \leq y \leq \pi \\ u(x,0) = 0, \quad u(x,\pi) = g(x) & \text{for } 0 \leq x \leq \pi, \end{cases}$$

where g is a continuous function satisfying $g(0) = 0 = g(\pi)$.

If we assume $u(x,y) = X(x)Y(y)$, then substitution in the PDE gives $X''Y + XY'' = 0$. We can algebraically "separate the variables" x and y in this equation, so both sides must be equal to some constant:

$$\frac{X''}{X} = -\frac{Y''}{Y} = -\lambda.$$

But the boundary conditions $u(0,y) = 0 = u(\pi,y)$ require $X(0) = 0 = X(\pi)$, for which the solutions of the eigenvalue problem $X'' + \lambda X = 0$ are $\lambda_n = n^2$ with $n = 1, 2, \ldots$ and $X_n(x) = a_n \sin nx$. With these values of λ_n we solve $Y'' - n^2 Y = 0$ to find $Y_n(y) = b_n \sinh ny + c_n \cosh ny$, but the boundary condition $u(x,0) = 0$ requires $c_n = 0$.

By superposition we write

$$u(x,y) = \sum_{n=1}^{\infty} a_n \sin nx \sinh ny,$$

which formally satisfies the equation and the three homogeneous boundary conditions. The boundary condition at $y = \pi$ now yields

$$u(x,\pi) = \sum_{n=1}^{\infty} a_n \sin nx \sinh n\pi = g(x).$$

But since g is continuous with $g(0) = 0 = g(\pi)$, the Fourier sine series converges absolutely and uniformly to g, and so the coefficients a_n may be obtained from

$$a_n \sinh n\pi = \frac{2}{\pi} \int_0^{\pi} g(x) \sin nx\, dx. \clubsuit$$

Section 4.1: Introduction to the Laplace Equation

The method of separation of variables can, of course, be used for more general boundary conditions; this leads to general boundary value problems for ODEs or *Sturm-Liouville theory* (cf. [Pinsky]). We encounter some additional examples in the Exercises. In fact, the next example involves a different type of boundary condition on a different domain.

Example 2. Let Ω be the unit disk in $\mathbf{R}^2$ and consider the problem involving the normal derivative of the solution on the boundary

$$\begin{cases} \Delta u = 0 & \text{in } \Omega \\ \frac{\partial u}{\partial \nu} = h & \text{on } \partial \Omega, \end{cases}$$

where h is a continuous function. It is natural to use polar coordinates (r, θ) in which the problem becomes

$$\begin{cases} \dfrac{\partial^2 u}{\partial r^2} + \dfrac{1}{r} \dfrac{\partial u}{\partial r} + \dfrac{1}{r^2} \dfrac{\partial^2 u}{\partial \theta^2} = 0 & \text{for } 0 \leq r < 1,\, 0 \leq \theta < 2\pi \\ \dfrac{\partial u}{\partial r}(1, \theta) = h(\theta) & \text{for } 0 \leq \theta < 2\pi. \end{cases}$$

If we write $r = e^{-t}$ and $u(r, \theta) = X(t)Y(\theta)$, then

$$r^2 \partial_r^2 u + r \partial_r u + \partial_\theta^2 u = \partial_t^2 u + \partial_\theta^2 u = X''(t)Y(\theta) + X(t)Y''(\theta) = 0.$$

Separating the variables, we obtain

$$\frac{X''(t)}{X(t)} = -\frac{Y''(\theta)}{Y(\theta)} = \lambda.$$

But $Y''(\theta) + \lambda Y(\theta) = 0$ has solutions $\lambda_n = n^2$ and $Y_n(\theta) = a_n \cos n\theta + b_n \sin n\theta$; notice $Y_0(\theta) = a_0 = \text{const}$. The equation $X'' + n^2 X = 0$ has solutions $X_0(t) = c_0 t + d_0$ and $X_n(t) = c_n e^{nt} + d_n e^{-nt}$ for $n = 1, \ldots$. This means that $u_0(r, \theta) = -c_0 \log r + d_0$ and $u_n(r, \theta) = (a_n \cos n\theta + b_n \sin n\theta)(c_n r^{-n} + d_n r^n)$ for $n = 1, \ldots$ But u must be finite at $r = 0$, so $c_n = 0$. By superposition we may write (after relabeling coefficients)

$$u(r, \theta) = a_0 + \sum_{n=1}^{\infty} r^n (a_n \cos n\theta + b_n \sin n\theta).$$

But then

$$u_r(1, \theta) = \sum_{n=1}^{\infty} n(a_n \cos n\theta + b_n \sin n\theta) = h(\theta),$$

which shows that the coefficients a_n, b_n for $n \geq 1$ are determined from the Fourier series for $h(\theta)$. Notice, however, that a_0 is not determined by $h(\theta)$ and therefore may take an arbitrary value. Moreover, the constant term in the Fourier series for $h(\theta)$ must be zero [i.e., $\int_0^{2\pi} h(\theta)\, d\theta = 0$]. Therefore, the problem is *not* solvable for an arbitrary function $h(\theta)$, and when it *is* solvable, the solution is *not* unique. ♣

b. Boundary Values and Physics

From the preceding examples, we see that the Cauchy problem for the Laplace equation in a bounded domain Ω is *overdetermined*: we cannot specify both $u = g$ and $\frac{\partial u}{\partial \nu}|_{\partial\Omega} = h$ on $\partial\Omega$. We must therefore choose to specify *either* the value of u on $\partial\Omega$ *or* the value of the normal derivative of u on $\partial\Omega$. This leads to two different boundary value problems. The *Dirichlet problem for the Laplace equation* involves finding a solution of (1) that satisfies

$$(3) \qquad u(x) = g(x) \qquad \text{for} \quad x \in \partial\Omega,$$

whereas the *Neumann problem for Laplace's equation* involves finding a solution of (1) that satisfies

$$(4) \qquad \frac{\partial u(x)}{\partial \nu} = h(x) \qquad \text{for} \quad x \in \partial\Omega.$$

Similarly, we may define the *Dirichlet and Neumann problems for the Poisson equation* (2). However, we generally consider boundary value problems in which either the equation is homogeneous (i.e., Laplace's equation) or the boundary condition is homogeneous (i.e., $g = 0$ or $h = 0$). By superposition, there is no loss of generality in doing so; for example, to solve $\Delta u = f$ in Ω with $u = g$ on $\partial\Omega$, we write $u = u_1 + u_2$, where $\Delta u_1 = f$ and $\Delta u = 0$ in Ω, and $u_1 = 0$ and $u_2 = g$ on $\partial\Omega$.

The Dirichlet and Neumann problems both have important physical significance. In electrostatics, for example, the Dirichlet condition (3) specifies the values of the potential function u on $\partial\Omega$, which induces the electric field $\vec{E} = -\nabla u$ in Ω. If we can show that (2),(3) is well posed, this means that the electric field is completely determined by the charge distribution inside Ω together with the values of the potential function u on $\partial\Omega$.

The physical significance of (3) and (4) is even more evident in the application to heat diffusion. Suppose the values of a temperature distribution in a domain Ω are specified at the boundary $\partial\Omega$ by $g(x)$. Assume that Ω contains no heat sources, and the temperature is allowed to achieve its steady state, $u(x)$. Then $u(x)$ is a solution of the Dirichlet problem (1),(3). Similarly, the Neumann problem (1),(4) arises for the steady-state temperature distribution when the amount of heat that flows into (or out of Ω) is controlled by $h(x)$ on $\partial\Omega$. We could also consider more complicated physical problems where the temperature is controlled on one part of $\partial\Omega$, while the heat flow is controlled on the rest of $\partial\Omega$ (for example, a rod with one insulated end). To find an equilibrium solution, we must find a harmonic function satisfying *mixed* Dirichlet and Neumann conditions.

Section 4.1: Introduction to the Laplace Equation   107

Another kind of boundary condition arises naturally from the physical consideration of heat flow. Suppose the space around Ω is maintained at a constant temperature v. According to Newton's law of cooling, the heat flow across $\partial\Omega$ should be proportional to the difference $u - v$. This leads to the condition

(5) $$\frac{\partial u}{\partial \nu} + \alpha u = \beta \quad \text{on } \partial\Omega.$$

Finding a harmonic function u in Ω satisfying (5) is called the *Robin* or *third boundary value problem*. However, we shall concern ourselves mostly with the Dirichlet and Neumann problems.

c. Green's Identities and Uniqueness

Suppose Ω is a smooth, bounded domain in $\mathbf{R}^n$ and $u, v \in C^2(\overline{\Omega})$. The following are called *Green's identities*:

(6) $$\int_{\partial\Omega} v \frac{\partial u}{\partial \nu} dS = \int_\Omega (v\Delta u + \nabla v \cdot \nabla u)\, dx$$

(7) $$\int_{\partial\Omega} (v\frac{\partial u}{\partial \nu} - u\frac{\partial v}{\partial \nu})\, dS = \int_\Omega (v\Delta u - u\Delta v)\, dx,$$

where dS denotes the surface measure on $\partial\Omega$. In fact, (6) and (7) hold more generally for $u, v \in C^2(\Omega) \cap C^1(\overline{\Omega})$, provided the integrals over Ω converge.

To prove (6), use the divergence theorem with vector field $\vec{V} = v\nabla u$ so that $\vec{V} \cdot \nu\, dS = v\frac{\partial u}{\partial \nu} dS$ and $\text{div } \vec{V} = v\Delta u + \nabla v \cdot \nabla u$. To prove (7), interchange the roles of u and v and subtract the result from (6). The generalization to $u, v \in C^2(\Omega) \cap C^1(\overline{\Omega})$ is proved by approximating Ω from the inside. (In fact, this approximation procedure shows that the domain Ω need only be piecewise smooth.)

Let us obtain several useful consequences from these identities. If we take $v = 1$ in (6), we obtain

(8) $$\int_{\partial\Omega} \frac{\partial u}{\partial \nu} dS = \int_\Omega \Delta u\, dx.$$

From this we obtain important *solvability conditions for the Neumann problem*. For example, in order for (1),(4) to admit a solution $u \in C^2(\Omega) \cap C^1(\overline{\Omega})$, we must have $\int_{\partial\Omega} h\, dS = 0$. Notice that we already encountered this phenomenon in Example 2.

If we let $u = v$ in (6), we obtain

(9) $$\int_{\partial\Omega} u \frac{\partial u}{\partial \nu} dS = \int_\Omega (u\Delta u + |\nabla u|^2)\, dx.$$

From this we see that if $u \in C^2(\Omega) \cap C^1(\overline{\Omega})$ satisfies $\Delta u = 0$ in Ω, and either $u = 0$ or $\frac{\partial u}{\partial \nu} = 0$ on $\partial\Omega$, then $u \equiv \text{const}$ in Ω; of course, $u = 0$ on $\partial\Omega$ additionally implies $u \equiv 0$ in Ω. Taking the difference of $u = u_1 - u_2$ of two possible solutions u_1, u_2 of the given problem yields the following uniqueness result.

108   Chapter 4: The Laplace Equation

Theorem 1. *Any two solutions in $C^2(\Omega) \cap C^1(\overline{\Omega})$ of the Dirichlet problem (2),(3) must agree; any two such solutions of the Neumann problem (2),(4) must differ by a constant.*

Since the Neumann problem requires $\frac{\partial u}{\partial \nu}$ to exist on $\partial \Omega$, it is natural to require $u \in C^1(\overline{\Omega})$ for uniqueness in the Neumann problem. However, we would prefer *not* to require $u \in C^1(\overline{\Omega})$ for uniqueness in the Dirichlet problem. We investigate next this stronger uniqueness property for the Dirichlet problem as a consequence of a maximum principle.

d. Mean Values and the Maximum Principle

In one dimension, the harmonic functions are simply the linear functions, and the value of a linear function at the midpoint of a finite interval is just the average of the values at the endpoints. If we wish to generalize this property to higher dimensions, we might try to show that the value of a harmonic function u at the center ξ of a ball of radius r is simply the average of u on the surface of the sphere. We therefore recall the notion of spherical mean from Section 3.2 in order to obtain the following. Notice that we only require u to be C^2 on the *interior* of Ω.

Theorem 2 (Gauss Mean Value Theorem). *If $u \in C^2(\Omega)$ is harmonic in Ω, let $\xi \in \Omega$ and pick $r > 0$ so that $\overline{B_r(\xi)} = \{x : |x - \xi| \leq r\} \subset \Omega$. Then*

$$(10) \qquad u(\xi) = M_u(\xi, r) \equiv \frac{1}{\omega_n} \int_{|x|=1} u(\xi + rx)\, dS_x,$$

where ω_n is the measure of the $(n-1)$-dimensional sphere in $\mathbf{R}^n$.

Proof. Using (8) we have
$$(11) \qquad 0 = \int_{\partial B_r(\xi)} \frac{\partial u}{\partial \nu}\, dS = r^{n-1} \int_{|x|=1} \frac{\partial u}{\partial r}(\xi + rx)\, dS_x = r^{n-1} \omega_n \frac{\partial}{\partial r} M_u(\xi, r),$$

so that $M_u(\xi, r)$ is independent of r. By continuity, $M_u(\xi, r) \to u(\xi)$ as $r \to 0$, so (10) holds. ♠

A continuous function u satisfying (10) is said to satisfy the *mean value property*, so the theorem simply states that a harmonic function satisfies the mean value property. The analogy with linear functions in one dimension suggests that perhaps the converse is true: if u satisfies the mean value property, then u is harmonic. This is true but requires solving the Dirichlet problem on balls (cf. Exercise 7 in Section 4.2).

If we replace r in (10) by ρr, multiply by ρ^{n-1}, and integrate $0 \leq \rho \leq 1$, we obtain the following.

Corollary. With u, Ω, ξ, and r as in the preceding theorem, we have

(12) $$u(\xi) = \frac{n}{\omega_n} \int_{|x|\leq 1} u(\xi + rx)\, dx.$$

If we replace the condition that $u \in C^2(\Omega)$ be harmonic by the condition that $\Delta u \geq 0$ in Ω, then a calculation as in (11) shows that $M_u(\xi, r)$ is increasing in $r \geq 0$, and hence

(13) $$u(\xi) \leq M_u(\xi, r) \equiv \frac{1}{\omega_n} \int_{|x|=1} u(\xi + rx)\, dS_x.$$

Moreover, in place of (12) we obtain

(14) $$u(\xi) \leq \frac{n}{\omega_n} \int_{|x|\leq 1} u(\xi + rx)\, dx.$$

From this latter inequality we obtain the following.

Theorem 3 (Maximum Principle). If $u \in C^2(\Omega)$ satisfies $\Delta u \geq 0$ in Ω, then either u is a constant, or

(15) $$u(\xi) < \sup_{x \in \Omega} u(x)$$

for all $\xi \in \Omega$.

Proof. We may assume $A = \sup_{x \in \Omega} u(x) < \infty$, so by continuity of u we know that $\{x \in \Omega : u(x) = A\}$ is relatively closed in Ω. But by (14), if $u(\xi) = A$ at an interior point ξ, then $u(x) = A$ for all x in a ball about ξ, so $\{x \in \Omega : u(x) = A\}$ is open. The connectedness of Ω implies $u(\xi) < A$ or $u(\xi) \equiv A$ for all $\xi \in \Omega$. ♠

The maximum principle shows that $u \in C^2(\Omega)$ with $\Delta u \geq 0$ can attain an interior maximum only if u is a constant. (The possibility that u attains its maximum at a boundary point of Ω is discussed in the Exercises; see also the elliptic maximum principle in Chapter 8.) In particular, if $\overline{\Omega}$ is compact, and $u \in C^2(\Omega) \cap C(\overline{\Omega})$ satisfies $\Delta u \geq 0$ in Ω, then

(16) $$\max_{x \in \overline{\Omega}} u(x) = \max_{x \in \partial \Omega} u(x).$$

This is sometimes called the *weak maximum principle*, since it is not as strong as Theorem 3 [although (15) required the connectivity of Ω and (16) does not]. As an immediate consequence of (16) we obtain the continuous dependence on the boundary data in the Dirichlet problem for either Laplace's equation or Poisson's equation.

Corollary 1. Suppose Ω is a bounded domain and $u, v \in C^2(\Omega) \cap C(\overline{\Omega})$ both satisfy (2) in Ω. If $|u(x) - v(x)| \leq \epsilon$ for $x \in \partial\Omega$, then $|u(x) - v(x)| \leq \epsilon$ for $x \in \overline{\Omega}$.

When $u = v$ on $\partial\Omega$, we also obtain uniqueness of solutions to the Dirichlet problem without requiring them to be C^1 on $\partial\Omega$:

Corollary 2. If Ω is a bounded domain and $u, v \in C^2(\Omega) \cap C(\overline{\Omega})$ both satisfy (2) in Ω with $u = v$ on $\partial\Omega$, then $u \equiv v$ on $\overline{\Omega}$.

These two corollaries show that the Dirichlet problem (2),(3) for $g \in C(\partial\Omega)$ will be well posed, *if* we can show a solution always exists. The next two sections are devoted to the study of this question of existence.

Exercises for Section 4.1

1. Let $\Omega = \{(x, y) \in \mathbf{R}^2 : x^2 + y^2 < 1\} = \{(r, \theta) : 0 \leq r < 1, 0 \leq \theta < 2\pi\}$, and use separation of the variables (r, θ) to solve the Dirichlet problem

$$\begin{cases} \Delta u = 0 & \text{in } \Omega \\ u(1, \theta) = g(\theta) & \text{for } 0 \leq \theta < 2\pi. \end{cases}$$

2. Let $\Omega = (0, \pi) \times (0, \pi)$, and use separation of the variables to solve the mixed boundary value problem

$$\begin{cases} \Delta u = 0 & \text{in } \Omega \\ u_x(0, y) = 0 = u_x(\pi, y) & \text{for } 0 < y < \pi \\ u(x, 0) = 0, \quad u(x, \pi) = g(x) & \text{for } 0 < x < \pi. \end{cases}$$

3. Prove that the solution of the Robin or third boundary value problem (5) for the Laplace equation is unique when $\alpha > 0$ is a constant.

4. Let Ω be the unit disk as in Exercise 1. (a) Solve the Robin problem (5) for the Laplace equation when $\alpha > 0$ is a constant. (b) When $\alpha = -1$, show that uniqueness fails.

5. Suppose $q(x) \geq 0$ for $x \in \Omega$ and consider solutions $u \in C^2(\Omega) \cap C^1(\overline{\Omega})$ of $\Delta u - q(x)u = 0$ in Ω. Establish uniqueness theorems for (a) the Dirichlet problem, and (b) the Neumann problem.

6. By direct calculation, show that $v(x) = |x - x_0|^{2-n}$ is harmonic in $\mathbf{R}^n \setminus \{x_0\}$ for $n \geq 3$. Do the same for $v(x) = \log|x - x_0|$ if $n = 2$.

7. (a) If Ω is a bounded domain and $u \in C^2(\Omega) \cap C(\overline{\Omega})$ satisfies (1), then $\max_{\overline{\Omega}} |u| = \max_{\partial\Omega} |u|$.
(b) If $\Omega = \{x \in \mathbf{R}^n : |x| > 1\}$ and $u \in C^2(\Omega) \cap C(\overline{\Omega})$ satisfies (1) and $\lim_{|x| \to \infty} u(x) = 0$, then $\max_{\overline{\Omega}} |u| = \max_{\partial\Omega} |u|$.

8. Suppose $u \in C^2(\Omega)$ and $\Delta u \geq 0$ in Ω where the boundary $\partial\Omega$ has the following property: For every $x_0 \in \partial\Omega$ there is a ball $B_\epsilon(x_1) \subset \Omega$ such

that $\partial\Omega \cap \overline{B_\epsilon(x_1)} = \{x_0\}$. Prove that if u is not a constant and for some $x_0 \in \partial\Omega$ we have $u(x_0) = \sup\{u(x) : x \in \Omega\}$, then the outward normal derivative of u is positive at x_0: $\frac{\partial u}{\partial \nu}(x_0) > 0$, where ν is the exterior unit normal to the ball.

9. Suppose $u \in C(\Omega)$ satisfies the mean value property in Ω.
   (a) Show that u satisfies the maximum principle (15).
   (b) Show that boundary values on $\overline{B_r(\xi)} \subset \Omega$ uniquely determine u: If $v \in C(\overline{B_r(\xi)})$ satisfies the mean value property in $B_r(\xi)$ and $u = v$ on $\partial B_r(\xi)$, then $u \equiv v$ in $B_r(\xi)$.

4.2 Potential Theory and Green's Functions

In this section we describe the fundamental solution of the Laplace operator and use it to introduce domain and boundary potentials as well as the Green's function, which is associated with the Dirichlet problem. We then calculate the Green's functions explicitly for a half-space and a ball.

a. The Fundamental Solution

Recall from Section 2.3 that a fundamental solution $K(x)$ for the Laplace operator is a distribution satisfying

$$(17) \qquad \Delta K(x) = \delta(x)$$

where δ is the delta distribution supported at $x = 0$. In order to solve (17), we should first observe that Δ is symmetric in the variables $x_1, \ldots, x_n$, and $\delta(x)$ is also radially symmetric (i.e., its value only depends on $r = |x|$). It is therefore natural to try to solve (17) with a radially symmetric function, $K(x) = \Psi(r)$. Since $\delta(x) = 0$ for $x \neq 0$, we see that (17) requires Ψ to be harmonic for $r > 0$. For the radially symmetric function Ψ, Laplace's equation becomes

$$(18) \qquad \Psi'' + \frac{n-1}{r}\Psi' = 0 \quad \text{for} \quad r > 0$$

[cf. (25) in Section 3.2]. The general solution of (18) is

$$\Psi(r) = \begin{cases} c_1 + c_2 \log r & \text{if } n = 2 \\ c_1 + c_2 r^{2-n} & \text{if } n \geq 3. \end{cases}$$

Now we must determine c_2 so that (17) holds.

For $v \in C_0^\infty(\mathbf{R}^n)$, we want to show

$$(19) \qquad \int_{\mathbf{R}^n} \Psi(|x|)\, \Delta v(x)\, dx = v(0).$$

112 Chapter 4: The Laplace Equation

Suppose $v(x) \equiv 0$ for $|x| \geq R$ and let $\Omega = B_R(0)$; for small $\epsilon > 0$ let $\Omega_\epsilon = \Omega \setminus \overline{B_\epsilon(0)}$. By the integrability of $\Psi(r)$ at $x = 0$, $\int_{\Omega_\epsilon} \Psi \Delta v \, dx \to \int_\Omega \Psi \Delta v \, dx$ as $\epsilon \to 0$. Moreover, Ψ is harmonic in Ω_ϵ, so by (7) of Section 4.1

$$\text{(20)} \qquad \int_{\Omega_\epsilon} \Psi \Delta v \, dx = \int_{|x|=\epsilon} \left(\Psi \frac{\partial v}{\partial \nu} - v \frac{\partial \Psi}{\partial \nu} \right) dS,$$

where ν is the unit normal pointing *toward* 0 on the sphere $|x| = \epsilon$ (i.e., $\nu = -x/|x|$). But for $|x| = \epsilon$ we have $\Psi(|x|) = \Psi(\epsilon)$ so that, as $\epsilon \to 0$,

$$\text{(21)} \qquad \int_{|x|=\epsilon} \Psi \frac{\partial v}{\partial \nu} dS = -\Psi(\epsilon) \int_{|x|<\epsilon} \Delta v \, dx = \Psi(\epsilon) O(\epsilon^n) \to 0,$$

where the estimate of the integral being $O(\epsilon^n)$ is a consequence of the continuity of Δv at $x = 0$. We also have on $|x| = \epsilon$

$$\frac{\partial \Psi}{\partial \nu}(|x|) = -\frac{\partial \Psi}{\partial r}(\epsilon) = \bar{c}_2 \epsilon^{1-n}, \quad \text{where } \bar{c}_2 = \begin{cases} -c_2 & \text{if } n = 2 \\ (n-2)c_2 & \text{if } n \geq 3. \end{cases}$$

Therefore, we may compute

$$\text{(22)} \qquad \int_{|x|=\epsilon} v \frac{\partial \Psi}{\partial \nu} dS = \bar{c}_2 \epsilon^{1-n} \int_{|x|=\epsilon} v \, dS = \bar{c}_2 \omega_n M_v(0, \epsilon) \to \bar{c}_2 \omega_n v(0),$$

since v is continuous at $x = 0$. Taking c_2 so that $\bar{c}_2 \omega_n = -1$, we find that (19) is satisfied; that is, a fundamental solution is given by

$$\text{(23)} \qquad K(x) = \begin{cases} \dfrac{1}{2\pi} \log r & \text{if } n = 2 \\ \dfrac{1}{(2-n)\omega_n} r^{2-n} & \text{if } n \geq 3. \end{cases}$$

We shall call (23) *the fundamental solution for the Laplace operator*.

b. Potential Theory and Electrostatics

By the results of Section 2.3, we know that $u(x) = \int K(x-y) f(y) \, dy$ is a distribution solution of $\Delta u = f$ when f is integrable and has compact support. In particular, we have

$$\text{(24)} \qquad u(x) = \int_{\mathbf{R}^n} K(x-y) \Delta u(y) \, dy \qquad \text{whenever } u \in C_0^\infty(\mathbf{R}^n).$$

The following result generalizes (24) by allowing boundary terms.

Theorem 1. If Ω is a smooth bounded domain in $\mathbf{R}^n$, $u \in C^2(\overline{\Omega})$, and $x \in \Omega$, then

$$
\begin{aligned}
(25) \quad u(x) = & \int_\Omega K(x-y) \Delta u(y) \, dy \\
& + \int_{\partial \Omega} \left(u(y) \frac{\partial K(x-y)}{\partial \nu_y} - K(x-y) \frac{\partial u(y)}{\partial \nu} \right) dS_y.
\end{aligned}
$$

More generally, (25) holds if $u \in C^2(\Omega) \cap C^1(\overline{\Omega})$ and the integral over Ω converges.

Proof. For $u \in C^2(\overline{\Omega})$ and $x \in \Omega$, choose $0 < \epsilon < \text{dist}(x, \partial \Omega)$ and let $\Omega_\epsilon = \Omega \setminus \overline{B_\epsilon(x)}$; then apply (7) to conclude

$$
\int_{\Omega_\epsilon} K(x-y) \Delta u(y) \, dy = \int_{\partial \Omega_\epsilon} \left(K(x-y) \frac{\partial u(y)}{\partial \nu} - u(y) \frac{\partial K(x-y)}{\partial \nu_y} \right) dS_y,
$$

where $\partial \Omega_\epsilon = \partial \Omega \cup \partial B_\epsilon(x)$. Letting $\epsilon \to 0$, the integral over Ω_ϵ becomes an integral over Ω [since $K(x-y)$ is integrable at $x = y$], and the integral over $|x - y| = \epsilon$ converges to $u(x)$ exactly as in (21),(22).

If we only have $u \in C^2(\Omega) \cap C^1(\overline{\Omega})$, then we can introduce a sequence of smooth domains $\Omega_j = \{x \in \Omega : \text{dist}(x, \partial \Omega) > 1/j\}$ and apply (25) in each Ω_j. Letting $j \to \infty$ yields (25) for Ω. ♠

Let us interpret the preceding results in the context of electrostatics in $\mathbf{R}^3$. A single unit charge placed at a point $\xi \in \mathbf{R}^3$ is known to induce an electric field $\vec{E}_\xi(x) = (x - \xi)|x - \xi|^{-3}$, and it is easy to check that $u_\xi(x) = |x - \xi|^{-1}$ is a potential function for $\vec{E}$ (i.e., $\vec{E} = -\nabla u$). Let us make two observations. First, $u_\xi(x) = |x - \xi|^{-1}$ represents the *potential energy* that a unit charge of the same sign would have if brought from infinity to the point x; the positivity of the potential energy explains the sign convention $\vec{E} = -\nabla u$. Second, from $u_\xi(x) = -4\pi K(x - \xi)$ we see that the fundamental solution $K(x - \xi)$ is just a multiple of the potential function for the electric field induced by placing a unit point charge at ξ. Another way to write this is

$$
(26) \quad u_\xi(x) = \int_{\mathbf{R}^3} \frac{\delta_\xi(y)}{|x-y|} \, dy = -4\pi \int_{\mathbf{R}^3} K(x-y) \delta_\xi(y) \, dy.
$$

Now let us replace $\delta_\xi(y)$ by $\rho(y)$, a distribution of electric charges in $\Omega \subset \mathbf{R}^3$, and let us define the *domain potential*

$$
(27) \quad u(x) = \int_\Omega \frac{\rho(y)}{|x-y|} \, dy = -4\pi \int_\Omega K(x-y) \rho(y) \, dy
$$

to be the *potential function induced by the charge density* ρ. If $\rho \in L^1(\Omega)$ and we extend ρ to all of $\mathbf{R}^3$ by zero outside of Ω, then we recognize (27) as the convolution on $\mathbf{R}^3$ of the fundamental solution with the compactly supported distribution ρ; by the results of Section 2.3, u is a distribution solution of $\Delta u = -4\pi\rho$. In particular, u is harmonic in the complement of $\overline{\Omega}$, $\mathbf{R}^3\setminus\overline{\Omega}$, and provided ρ is sufficiently regular, we might *expect* that $u \in C^2(\Omega)$ with $\Delta u = -4\pi\rho$ in Ω; in other words, u is indeed the potential for the electric field in Ω induced by ρ.

There are also electric potentials associated with charge distributions lying only on the surface $\partial\Omega$. The functions

$$(28) \qquad \int_{\partial\Omega} \frac{\rho(y)}{|x-y|} \, dS_y \quad \text{and} \quad \int_{\partial\Omega} \frac{\partial}{\partial\nu_y}(|x-y|^{-1})\mu(y) \, dS_y$$

are called, respectively, a *single layer potential (with charge density ρ)* and a *double layer potential (with dipole moment density μ)*; the latter represents the potential induced by a double layer of charges (of opposite sign) on $\partial\Omega$. Notice that these potential functions are harmonic in Ω and $\mathbf{R}^3\setminus\overline{\Omega}$; in fact, they define distribution solutions of the Poisson equations $\Delta u = -4\pi\rho\delta_{\partial\Omega}$ and $\Delta u = 4\pi\mu\frac{\partial}{\partial\nu}\delta_{\partial\Omega}$, respectively (where the distributions $\delta_{\partial\Omega}$ and $\frac{\partial}{\partial\nu}\delta_{\partial\Omega}$ were defined in Exercise 7 of Section 2.3).

Returning to n-dimensions, it is natural to say that the formula (25) gives a *representation theorem* for any function $u \in C^2(\overline{\Omega})$ as a sum of a domain potential (with charge density Δu), a double layer potential (with moment density u), and a single layer potential (with charge density $\partial u/\partial\nu$).

It is clearly important to investigate in what sense the domain potential

$$(29a) \qquad u(x) = \int_\Omega K(x-y) \, f(y) \, dy$$

defines a solution of Poisson's equation $\Delta u = f$. When f is an integrable function on the bounded domain Ω, we know from Section 2.3 that u is at least a distribution solution of $\Delta u = f$ in $\mathbf{R}^n$ (extend f by zero outside Ω). The following proposition expresses ways in which additional regularity of f improves the regularity of u.

Proposition. *For a bounded domain Ω and $f \in L^1(\Omega)$, define u by (29a).*
(i) *u is C^∞ and harmonic in $\mathbf{R}^n\setminus\overline{\Omega}$;*
(ii) *if f is bounded on Ω, then $u \in C^1(\mathbf{R}^n)$;*
(iii) *if $f \in C^1(\overline{\Omega})$, then $u \in C^2(\Omega)$.*

Proof. To prove (i), we simply pass derivatives under the convergent integral and use the fact that $K(x-y)$ is C^∞ and harmonic for $x \neq y$.

To prove (ii), for any $j = 1, \ldots, n$ let

$$(29b) \qquad u_{(j)}(x) \equiv \int_\Omega \frac{\partial K(x-y)}{\partial x_j} f(y) \, dy,$$

which is well defined because $\partial K/\partial x_j$ is $O(|x-y|^{1-n})$ as $|x-y| \to 0$ and hence locally integrable. To show $u_{(j)} = \partial u/\partial x_j$, we first approximate u

Section 4.2: Potential Theory and Green's Functions

by a smooth function as follows. Let $\eta \in C^\infty(\mathbf{R})$ satisfy $\eta(t) = 0$ for $t < 1$, $\eta(t) = 1$ for $t > 2$, and $0 \leq \eta(t) \leq 1$ and $0 \leq \eta'(t) \leq 2$ for all t; then define

(29c)
$$u_\epsilon(x) \equiv \int_\Omega K_\epsilon(x-y) f(y)\, dy, \quad \text{where}$$
$$K_\epsilon(z) \equiv K(z)\, \eta\left(\frac{|z|}{\epsilon}\right).$$

The function $K_\epsilon(z)$ is smooth, hence so is u_ϵ (cf. the lemma in Section 2.3). It is straightforward to check that $u_\epsilon \to u$ uniformly on compact subsets of $\mathbf{R}^n$, but let us show that $\partial_j u_\epsilon \to u_{(j)}$, where $\partial_j = \partial/\partial x_j$. We may write

$$u_{(j)}(x) - \partial_j u_\epsilon(x) = \int_{|x-y|<2\epsilon} \partial_j \left[\left(1 - \eta\left(\frac{|x-y|}{\epsilon}\right)\right) K(x-y)\right] f(y)\, dy$$
$$= \int_{|z|<2\epsilon} \left[\frac{\eta'(|z|/\epsilon)}{\epsilon} \frac{z_j}{|z|} K(z) - \left(1 - \eta\left(\frac{|z|}{\epsilon}\right)\right) \frac{\partial K(z)}{\partial z_j}\right] f(x-z)\, dz.$$

Now $\int_{|z|<2\epsilon} \epsilon^{-1} K(z)\, dz = (\epsilon(n-2))^{-1} \int_0^{2\epsilon} r\, dr = 2\epsilon/(n-2)$ when $n > 2$, and $\int_{|z|<2\epsilon} \epsilon^{-1} K(z)\, dz = \epsilon^{-1} \int_0^{2\epsilon} (\log r)\, r\, dr = 2\epsilon \log(2\epsilon) - \epsilon$ when $n = 2$. Moreover, for $n \geq 2$ we can easily estimate $\int_{|z|<2\epsilon} \partial |K(z)/\partial z_j|\, dz \leq 2\epsilon$. Recalling $0 \leq \eta' \leq 2$ and $0 \leq (1 - \eta) \leq 1$, we conclude

(29d)
$$|u_{(j)}(x) - \partial_j u_\epsilon(x)| \leq \sup_{y \in \Omega} |f(y)| \cdot C(\epsilon),$$

where $C(\epsilon) \to 0$ as $\epsilon \to 0$. But this shows that on any compact subset of $\mathbf{R}^n$, $u_{(j)}$ is the uniform limit of the continuous functions $\partial_j u_\epsilon$. We conclude that $u \in C^1(\mathbf{R}^n)$ and $\partial_j u = u_{(j)}$.

To prove (iii), again let $\partial_j = \partial/\partial x_j$ and $v = \partial_j u = u_{(j)}$; then introduce

$$v_{(k)}(x) \equiv \int_\Omega \partial_k \partial_j K(x-y) f(y)\, dy$$
$$= \int_\Omega \partial_k \partial_j K(x-y)(f(y) - f(x))\, dy + f(x) \int_\Omega \partial_k \partial_j K(x-y)\, dy.$$

Now observe that

(29e)
$$\partial K(x-y)/\partial x_j = -\partial K(x-y)/\partial y_j,$$

which we use with the divergence theorem to conclude

$$\int_\Omega \partial_k \partial_j K(x-y)\, dy = -\int_\Omega \frac{\partial}{\partial y_k} \partial_j K(x-y)\, dy = -\int_{\partial\Omega} \partial_j K(x-y)\nu_k\, dS_y.$$

To show $\partial_k v = v_{(k)}$, we use the smoothing trick,

$$v_\epsilon(x) \equiv \int_\Omega \partial_j K(x-\xi)\,\eta\left(\frac{|x-y|}{\epsilon}\right) f(y)\,dy,$$

and calculate [similarly as for $v_{(k)}$]

$$\partial_k v_\epsilon(x) = \int_\Omega \partial_k \left[\partial_j K(x-y)\,\eta\left(\frac{|x-y|}{\epsilon}\right)\right] (f(y) - f(x))\,dy$$
$$+ f(x) \int_\Omega \partial_k \left[\partial_j K(x-y)\eta\left(\frac{|x-y|}{\epsilon}\right)\right] dy.$$

Again using the divergence theorem, we obtain

$$\int_\Omega \partial_k \left[\partial_j K(x-y)\,\eta\left(\frac{|x-y|}{\epsilon}\right)\right] dy = -\int_{\partial\Omega} \partial_j K(x-y)\,\eta\left(\frac{|x-y|}{\epsilon}\right)\nu_k\,dS_y$$
$$= -\int_{\partial\Omega} \partial_j K(x-y)\nu_k\,dS_y,$$

provided we pick $\epsilon > 0$ sufficiently small that (for fixed $x \in \Omega$) we have $2\epsilon < \mathrm{dist}(x,\partial\Omega)$, and hence $\eta(|x-y|/\epsilon) \equiv 1$ for all $y \in \partial\Omega$. Thus we find

$$v_{(k)}(x) - \partial_k v_\epsilon(x) = \int_{|x-y|<2\epsilon} \partial_k \left[\left(1 - \eta\left(\frac{|x-y|}{\epsilon}\right)\right)\partial_j K(x-y)\right](f(y) - f(x))\,dy,$$

since the boundary integrals cancel each other. Similar to (29d), we can estimate this integral to show

$$|v_{(k)}(x) - \partial_k v_\epsilon(x)| \leq \sup_{y\in\Omega} \frac{|f(x) - f(y)|}{|x-y|} \cdot C(\epsilon),$$

where $C(\epsilon) \to 0$ as $\epsilon \to 0$. But $f \in C^1(\overline{\Omega})$ ensures that the supremum is finite, so we conclude that $\partial_k v_\epsilon \to v_{(k)}$ uniformly on compact neighborhoods of x as $\epsilon \to 0$. In particular, $v \in C^1(\Omega)$ and hence $u \in C^2(\Omega)$. ♠

Remarks. We might expect that $f \in C(\overline{\Omega})$ implies $u \in C^2(\Omega)$ or that $f \in C^1(\overline{\Omega})$ implies $u \in C^2(\overline{\Omega})$, but in fact neither is true in general. However, the notion of Hölder continuity enables us to weaken substantially the hypothesis $f \in C^1(\overline{\Omega})$ (cf. Section 6.5), and the conclusion $u \in C^2(\overline{\Omega})$ may be obtained for sufficiently smooth domains (cf. Section 8.2). On the other hand, $f \in C^k(\overline{\Omega})$ does imply $u \in C^{k+1}(\Omega)$ (cf. Exercise 2).

Notice that (29a) does not allow us to specify the boundary values of u. However, to solve a boundary value problem for the Poisson equation, a domain potential may be used to reduce the problem to a similar boundary value problem for the Laplace equation. For example, to solve $\Delta w = f$ in

Ω with $w = 0$ on $\partial\Omega$, let u be defined by (29a) and then let $w = u + v$, where v is harmonic in Ω with $v = -u$ on $\partial\Omega$. Consequently, the rest of this chapter will focus on the Laplace equation with the understanding that these results have obvious implications for the Poisson equation.

c. Green's Function and the Poisson Kernel

With a slight change in notation, Theorem 1 has the following special case.

Theorem 2. *If Ω is a smooth bounded domain in $\mathbf{R}^n$, $u \in C^2(\Omega) \cap C^1(\overline{\Omega})$ is harmonic, and $\xi \in \Omega$, then*

$$(30) \qquad u(\xi) = \int_{\partial\Omega} \left(u(x) \frac{\partial K(x-\xi)}{\partial \nu_x} - K(x-\xi) \frac{\partial u(x)}{\partial \nu} \right) dS_x.$$

Observe that (30) suggests the solution of the Dirichlet problem be sought as a double layer potential, and the solution of the Neuman problem be sought as a single layer potential. However, in order to achieve this, we must first modify the fundamental solution (23).

Let $\omega(x)$ be any harmonic function in Ω, and for $x, \xi \in \Omega$ consider $G(x,\xi) = K(x-\xi) + \omega(x)$. If we use (7) to observe

$$\int_\Omega \omega \Delta u \, dx + \int_{\partial\Omega} (u \frac{\partial \omega}{\partial \nu} - \omega \frac{\partial u}{\partial \nu}) dS = 0,$$

then (25) immediately generalizes to

$$(31) \quad u(\xi) = \int_\Omega G(x,\xi) \Delta u \, dx + \int_{\partial\Omega} \left(u(x) \frac{\partial G(x,\xi)}{\partial \nu_x} - G(x,\xi) \frac{\partial u(x)}{\partial \nu} \right) dS_x.$$

Any function $G(x,\xi)$ satisfying (31) for any $u \in C^2(\Omega) \cap C^1(\overline{\Omega})$ is called a *fundamental solution in Ω with pole at ξ*.

Now suppose that *for each* $\xi \in \Omega$ we can find a function $\omega_\xi(x)$ that is harmonic in Ω *and* satisfies $\omega_\xi(x) = -K(x-\xi)$ for all $x \in \partial\Omega$. Then $G(x,\xi) = K(x-\xi) + \omega_\xi(x)$ is a fundamental solution with the distinguished property that

$$(32) \qquad\qquad G(x,\xi) = 0 \qquad \text{for all} \quad x \in \partial\Omega.$$

Such a function G is called the *Green's function* and is useful in satisfying Dirichlet boundary conditions. [A fundamental solution $N(x,\xi)$ that satisfies the boundary condition $\partial N(x,\xi)/\partial \nu = 0$ for all $x \in \partial\Omega$ is called *Neumann's function* and is useful in satisfying Neumann boundary conditions; however, we shall not consider this here.]

The Green's function is difficult to construct for a general domain Ω since it requires solving the Dirichlet problem $\Delta \omega_\xi = 0$ in Ω, $\omega_\xi(x) =$

$-K(x-\xi)$ for $x \in \partial\Omega$, for *each* $\xi \in \Omega$. There is a "physical" argument from electrostatics for the existence of $G(x,\xi)$. Suppose $\partial\Omega$ is a perfectly conducting surface, and a positive unit charge is placed at the point $\xi \in \Omega$. Then negative charges are induced on $\partial\Omega$, and the potential function for the resultant electric field has the stated properties of the Green's function.

In the next section, we explicitly construct $G(x,\xi)$ when Ω is a half-space and when Ω is a ball. The latter is used in the subsequent section to show that, in theory, $G(x,\xi)$ exists for most domains Ω. But for now, let us suppose that $G(x,\xi)$ is known for Ω. Then from (31) we find

$$(33) \qquad u(\xi) = \int_\Omega G(x,\xi)\,\Delta u\,dx + \int_{\partial\Omega} u(x)\frac{\partial G(x,\xi)}{\partial \nu_x}\,dS_x$$

for every $u \in C^2(\overline{\Omega})$. In particular, if $\Delta u = 0$ in Ω and $u = g$ on $\partial\Omega$ then we obtain the *Poisson integral formula*

$$(34) \qquad u(\xi) = \int_{\partial\Omega} H(x,\xi)g(x)\,dS_x, \qquad \text{where} \quad H(x,\xi) = \frac{\partial G(x,\xi)}{\partial \nu_x}.$$

The function $H(x,\xi)$ is called the *Poisson kernel*.

Thus *if* we know that the Dirichlet problem has a solution $u \in C^2(\overline{\Omega})$, then we can calculate u from the Poisson integral formula [provided of course that we can compute $G(x,\xi)$].

d. The Dirichlet Problem on a Half-Space

We use the *method of reflection* to compute the Green's function and the corresponding Poisson integral formula for the special case when the domain is a half-space.

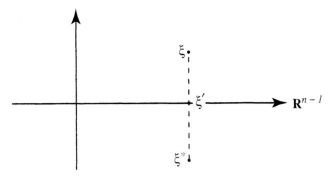

Figure 1. The method of reflection on a half-space.

Let us take $\Omega = \mathbf{R}^n_+ = \{(x_1,...,x_n) \in \mathbf{R}^n : x_n > 0\}$, where $n \geq 2$, and let $K(x)$ denote the fundamental solution as defined in (23). Introduce the notation $x = (x', x_n)$, where $x' \in \mathbf{R}^{n-1}$, and then identify $(x', 0)$ with x'. For $\xi = (\xi_1,...,\xi_n) \in \Omega$, define its *reflection* $\xi^* = (\xi_1,...,-\xi_n)$, which is *not* in Ω (see Figure 1). Notice that $K(x-\xi^*)$ is harmonic in $x \in \Omega$; moreover, for $x' \in \mathbf{R}^{n-1}$ we have $|x'-\xi| = |x'-\xi^*|$ and hence $K(x'-\xi) = K(x'-\xi^*)$. Thus

(35) $$G(x,\xi) = K(x-\xi) - K(x-\xi^*)$$

is the Green's function on Ω!

To compute the Poisson kernel, we must differentiate $G(x,\xi)$ in the *negative* x_n direction. For $n \geq 2$, a simple calculation shows

$$\frac{\partial}{\partial x_n} K(x-\xi) = \frac{x_n - \xi_n}{\omega_n} |x-\xi|^{-n},$$

so that the Poisson kernel is given by

(36) $$H(x',\xi) = -\frac{\partial}{\partial x_n} G(x,\xi)|_{x_n=0} = \frac{2\xi_n}{\omega_n} |x'-\xi|^{-n},$$

for $x' \in \mathbf{R}^{n-1}$. Thus we obtain the formula for the solution of the Dirichlet problem (3) for Laplace's equation when $\Omega = \mathbf{R}^n_+$:

(37) $$u(\xi) = \frac{2\xi_n}{\omega_n} \int_{\mathbf{R}^{n-1}} \frac{g(x')}{|x'-\xi|^n} dx'.$$

As a formula for the solution, (37) is very satisfactory. But we have not made clear our assumptions on the boundary function g. To guarantee that the integral in (37) converges, it is reasonable to assume that g is bounded, and since our uniqueness theorems require continuity of the solution on the boundary, it is reasonable to require g to be continuous. The following theorem proves that (37) indeed solves the Dirichlet problem in this case.

Theorem 3. *If $g(x')$ is bounded and continuous for $x' \in \mathbf{R}^{n-1}$, then the function $u(\xi)$ defined by (37) is C^∞ and harmonic in $\mathbf{R}^n_+$ and extends continuously to $\overline{\mathbf{R}^n_+}$ such that $u(\xi') = g(\xi')$.*

Proof. We first observe that for fixed $\xi \in \mathbf{R}^n_+$ we have

$$H(x',\xi) = O(|x'|^{-n}) \qquad \text{as } |x'| \to \infty.$$

Notice that the decay $O(|x'|^{-n})$ as $|x'| \to \infty$ is sufficient for an integral over $\mathbf{R}^{n-1}$ to converge. In fact, by taking $u \equiv 1$ and $\Omega = B_R(0) \cap \mathbf{R}^n_+$ in (33), then letting $R \to \infty$, we can show (cf. Exercise 5)

(38) $$\int_{\mathbf{R}^{n-1}} H(x',\xi) \, dx' = 1 \qquad \text{for every } \xi \in \mathbf{R}^n_+.$$

In particular, if g is bounded, then the integral in (37) converges absolutely, and the function u is bounded by $M = \sup_{x'}\{|g(x')|\}$. Moreover, for fixed $x' \in \mathbf{R}^{n-1}$, $H(x', \xi)$ is C^∞ in $\xi \in \mathbf{R}^n_+$, and all ξ-derivatives of $H(x', \xi)$ will continue to be at least $O(|x'|^{-n})$ if we fix $\xi \in \mathbf{R}^n_+$ and let $|x'| \to \infty$. Thus we may differentiate u by differentiating under the integral sign, thereby concluding $u \in C^\infty(\mathbf{R}^n_+)$. Since $\Delta_\xi H(x', \xi) = -\partial_{x_n} \Delta_\xi G(x, \xi)|_{x_n=0} = 0$ for $x' \in \mathbf{R}^{n-1}$ and $\xi \in \mathbf{R}^n_+$, we conclude that u is harmonic in $\mathbf{R}^n_+$.

It only remains to verify that u extends continuously to $\overline{\mathbf{R}^n_+}$ and $u = g$ on $\mathbf{R}^{n-1}$. Let $\xi' \in \mathbf{R}^{n-1}$; we must show $u(\xi) \to g(\xi')$ as $\xi \to \xi'$ (i.e. as $\xi_n \to 0$). But, using (38), we may write

$$g(\xi') = \int_{\mathbf{R}^{n-1}} H(x', \xi) g(\xi')\, dx',$$

so that

(39)
$$u(\xi) - g(\xi') = I_1 + I_2 \equiv \left(\int_{|x'-\xi'|<\delta} + \int_{|x'-\xi'|>\delta} \right) H(x', \xi)[g(x') - g(\xi')]\, dx'.$$

By the continuity of g, given $\epsilon > 0$ we may choose $\delta > 0$ so that $|g(x') - g(\xi')| < \epsilon$ provided $|x' - \xi'| < \delta$; invoking (38), we find $|I_1| < \epsilon$. Fix $\zeta \in \mathbf{R}^n_+$ and observe that if $\xi \to \xi'$, then $\xi_n \to 0$ and $\xi_n |x' - \zeta|^n |x' - \xi|^{-n} \to 0$ uniformly for $|x' - \xi'| > \delta$. We may therefore choose $\delta' > 0$ so that $H(x', \xi) < \epsilon(2M)^{-1} H(x', \zeta)$ whenever $|x' - \xi'| > \delta$ and $|\xi - \xi'| < \delta'$. Then, using $|g(x') - g(\xi')| \leq 2M$, we find

$$|I_2| = \left| \int_{|x'-\xi'|>\delta} H(x', \xi)[g(x') - g(\xi')]\, dx' \right| < \epsilon \int_{\mathbf{R}^{n-1}} H(x', \zeta)\, dx' = \epsilon.$$

From (39) we conclude $|u(\xi) - g(\xi')| < 2\epsilon$, provided $|\xi - \xi'| < \delta'$. ♠

e. The Dirichlet Problem on a Ball

We again use the method of reflection to construct the Green's function. For $\xi \in \Omega = B_a(0) = \{x \in \mathbf{R}^n : |x| < a\}$, define $\xi^* = a^2 \xi/|\xi|^2$ as its *reflection in* $\partial\Omega$; notice that ξ^* is *not* in Ω (see Figure 2).

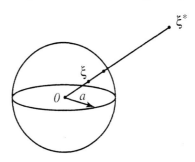

Figure 2. *The method of reflection on a ball.*

A calculation shows that for $|x| = a$

$$|x - \xi^*|^2 = a^2 - 2\frac{a^2}{|\xi|^2}\sum_{i=1}^n x_i\xi_i + \frac{a^4}{|\xi|^2}$$

$$= \frac{a^2}{|\xi|^2}\left(a^2 - 2\sum_{i=1}^n x_i\xi_i + |\xi|^2\right) = \frac{a^2}{|\xi|^2}|x - \xi|^2.$$

Thus we obtain

(40) $$\frac{|x - \xi^*|}{|x - \xi|} = \frac{a}{|\xi|} \quad \text{for} \quad |x| = a.$$

From (23) and (40) we conclude that for $x \in \partial\Omega$ (i.e., $|x| = a$),

(41) $$K(x - \xi) = \begin{cases} \frac{1}{2\pi}\log\left(\frac{|\xi|}{a}|x - \xi^*|\right) & \text{if } n = 2 \\ \left(\frac{a}{|\xi|}\right)^{n-2}K(x - \xi^*) & \text{if } n > 2. \end{cases}$$

Let us define for $x, \xi \in \Omega$:

(42) $$G(x, \xi) = \begin{cases} K(x - \xi) - \frac{1}{2\pi}\log\left(\frac{|\xi|}{a}|x - \xi^*|\right) & \text{if } n = 2 \\ K(x - \xi) - \left(\frac{a}{|\xi|}\right)^{n-2}K(x - \xi^*) & \text{if } n > 2. \end{cases}$$

Since ξ^* is *not* in Ω, the term $\left(\frac{a}{|\xi|}\right)^{n-2}K(x - \xi^*)$ (or $\log\frac{|\xi|}{a}|x - \xi^*|$ is $n = 2$) is harmonic in $x \in \Omega$; moreover, by (41) we have $G(x, \xi) = 0$ if $x \in \partial\Omega$; thus G defined by (42) is the Green's function for Ω!

Now let us calculate the Poisson kernel on $\partial\Omega$. We first compute the gradient of $G(x, \xi)$ for $n \geq 2$:

$$\frac{\partial}{\partial x_i}G(x, \xi) = \frac{1}{\omega_n}[|x - \xi|^{-n}(x_i - \xi_i) - \left(\frac{a}{|\xi|}\right)^{n-2}|x - \xi^*|^{-n}(x_i - \xi_i^*)].$$

If we let $|x| = a$ and use (40), we obtain

$$\frac{\partial G}{\partial x_i} = \frac{x_i}{\omega_n}|x - \xi|^{-n}\left(1 - \frac{|\xi|^2}{a^2}\right) \quad \text{for} \quad x \in \partial\Omega.$$

The exterior unit normal on $\partial\Omega$ is $\nu = x/a$, so the Poisson kernel is

(43) $$H(x, \xi) = \nu \cdot \nabla G = \frac{a^2 - |\xi|^2}{a\omega_n|x - \xi|^n} \quad \text{for} \quad x \in \partial\Omega,$$

and the solution of the Dirichlet problem (3) when $\Omega = B_a(0)$ is given by the Poisson integral formula

(44) $$u(\xi) = \frac{a^2 - |\xi|^2}{a\omega_n}\int_{|x|=a}\frac{g(x)}{|x - \xi|^n}\,dS_x.$$

Theorem 4. *If $g(x)$ is continuous on $\partial\Omega = \{x \in \mathbf{R}^n : |x| = a\}$, then the function $u(\xi)$ defined by (44) is C^∞ and harmonic in $\Omega = B_a(0)$ and extends continuously to $\overline{\Omega}$ such that $u(\xi) = g(\xi)$ for $|\xi| = a$.*

Proof. Taking $u \equiv 1$ in (33) shows that
$$\int_{|x|=a} H(x,\xi)\,dS_x = 1 \qquad \text{for every } \xi \in \Omega = B_a(0).$$
This is analogous to (38) in the half-space case. In fact, the rest of the proof of Theorem 4 parallels that of Theorem 3, so we omit the details. ♠

f. Properties of Harmonic Functions

In this subsection we gather some properties of harmonic functions which are consequences of the Poisson integral formula (44).

Theorem 5 (Smoothness). *If $u \in C^2(\Omega)$ is harmonic, then $u \in C^\infty(\Omega)$.*

Proof. For $\xi \in \Omega$, choose $a > 0$ small enough that $\overline{B_a(\xi)} \subset \Omega$ and then let $g = u|_{\partial B_a(\xi)} \in C(\partial B_a(\xi))$. Applying (44) to $B_a(\xi)$, we can solve $\Delta u_1 = 0$ in $B_a(\xi)$, $u_1 = g$ on $\partial B_a(\xi)$ with $u_1 \in C^\infty(B_a(\xi)) \cap C(\overline{B_a(\xi)})$. But uniqueness for the Dirichlet problem implies $u = u_1$, and so $u \in C^\infty(B_a(\xi))$. Since ξ was arbitrary in Ω, we conclude that $u \in C^\infty(\Omega)$. ♠

Using Exercise 7, we see that the hypotheses "$u \in C^2(\Omega)$ and harmonic" can be weakened to "$u \in C(\Omega)$ and satisfies the mean value property."

Another consequence of (44) is the following result.

Theorem 6 (Harnack Inequality). *Suppose $u \in C^2(\Omega)$ is harmonic and nonnegative, and Ω_1 is a bounded domain satisfying $\overline{\Omega}_1 \subset \Omega$. Then there is a constant C_1 depending only on Ω_1 such that*

$$(45) \qquad \sup_{x \in \Omega_1} u(x) \le C_1 \inf_{x \in \Omega_1} u(x).$$

Because C_1 does not depend on u, the inequality (45) implies that nonnegative harmonic functions cannot oscillate wildly on bounded sets; for the proof of (45), see Exercise 10.

The Poisson integral formula (44) may also be used to estimate the derivatives of harmonic functions. First, we may calculate

$$\frac{\partial}{\partial \xi_j} H(x,\xi) = \frac{1}{a\omega_n}\left(\frac{-2\xi_j}{|x-\xi|^n} + n(a^2 - |\xi|^2)\frac{(x_j - \xi_j)}{|x-\xi|^{n+2}} \right)$$

to conclude that $H_{\xi_j}(x,0) = nx_j/(\omega_n a^{n+1})$ for $|x| = a$. If u is harmonic in Ω and $\overline{B_a(0)} \subset \Omega$, then (44) implies

$$(46) \qquad \left|\frac{\partial u}{\partial \xi_j}(0)\right| = \frac{n}{\omega_n a^{n+1}}\left|\int_{|x|=a} x_j u(x)\,dS_x\right| \le \frac{n}{a}\max_{|x|=a}|u(x)|.$$

Of course, (46) may be generalized to any ball $\overline{B_a(\xi)} \subset \Omega$.

One consequence of (46) is the following (cf. Exercise 11).

Theorem 7 (Liouville's Theorem). *A bounded harmonic function defined on all of $\mathbf{R}^n$ must be a constant.*

We can extend (46) to compact subsets Γ of a domain Ω in which u is harmonic. Let $d > 0$ represent the distance between Γ and $\partial\Omega$, and pick an increasing sequence $d_k \to d$. We can apply (46) to $B_{d_k}(\xi)$ for every $\xi \in \Gamma$:

$$\left|\frac{\partial u}{\partial \xi_j}(\xi)\right| \leq \frac{n}{d_k} \max_{\partial B_{d_k}(\xi)} |u(x)| \leq \frac{n}{d_k} \sup_{x \in \Omega} |u(x)|.$$

Taking the maximum over $\xi \in \Gamma$ and letting $k \to \infty$, we obtain

$$(47) \qquad \max_{\xi \in \Gamma} \left|\frac{\partial u}{\partial \xi_j}(\xi)\right| \leq \frac{n}{d} \sup_{x \in \Omega} |u(x)|.$$

A simple consequence of this inequality is the following important convergence property of sequences of harmonic functions.

Theorem 8. *If $u_k \in C^2(\Omega)$ is a uniformly bounded sequence of harmonic functions in Ω and $\overline{B_a(\xi)} \subset \Omega$, then u_k contains a subsequence that converges uniformly on $B_a(\xi)$.*

Proof. The estimate (47) gives a uniform bound for u_k and $D_\xi u_k$ on $\overline{B_a(\xi)}$. The Arzela-Ascoli theorem then gives the uniformly convergent subsequence. ♠

We would like to conclude that the uniform limit of a sequence of harmonic functions is itself harmonic, but this requires estimates on the second-order derivatives. These are obtained from (46) by iteration. Namely, suppose u is harmonic in Ω and $\overline{B_a(\xi)} \subset \Omega$. Invoke (46) with u replaced by $\partial u / \partial \xi_i$ and a replaced by $a/2$:

$$\left|\frac{\partial^2 u}{\partial \xi_j \partial \xi_i}(0)\right| \leq \frac{2n}{a} \max_{|x| = \frac{a}{2}} \left|\frac{\partial u}{\partial \xi_i}(x)\right|.$$

But for $|\bar{x}| = a/2$ we can apply (46) to the ball $B_{a/2}(\bar{x})$. Combined with the preceding estimate, we get

$$\left|\frac{\partial^2 u}{\partial \xi_j \partial \xi_i}(0)\right| \leq \left(\frac{2n}{a}\right)^2 \sup_{x \in \Omega} |u(x)|.$$

Arguing as we did for (47), this extends to compact subsets $\Gamma \subset \Omega$:

$$(48) \qquad \max_{\xi \in \Gamma} \left|\frac{\partial^2 u}{\partial \xi_j \partial \xi_i}(\xi)\right| \leq \left(\frac{2n}{d}\right)^2 \sup_{x \in \Omega} |u(x)|.$$

In particular, if u_k is a sequence of harmonic functions converging uniformly on $\Gamma = \overline{B_a(\xi)} \subset \Omega$ to a function u, then applying (48) to $u_k - u_\ell$ shows that the second-order derivatives of the u_k also converge uniformly on $\overline{B_a(\xi)}$. In particular, $\Delta u_k \to \Delta u$. But $\Delta u_k = 0$ implies $\Delta u = 0$, so u is harmonic in $B_a(\xi)$. We have proved the following:

Theorem 9. *Suppose Ω is a bounded domain and $u_k \in C^2(\Omega) \cap C(\overline{\Omega})$ are harmonic in Ω and converge uniformly to $u \in C(\overline{\Omega})$. Then $u \in C^\infty(\Omega)$ and u is harmonic in Ω.*

Iteration of (47) may also be used to bound higher-order derivatives of a harmonic function u in a domain Ω:

$$(49a) \qquad |D^\alpha u(0)| \leq \left(\frac{n|\alpha|}{a}\right)^{|\alpha|} \max_{|x|=a} |u(x)| \quad \text{provided} \quad \overline{B_a(0)} \subset \Omega$$

and

$$(49b) \qquad \max_{\xi \in \Gamma} |D^\alpha u(\xi)| \leq \left(\frac{n|\alpha|}{d}\right)^{|\alpha|} \sup_{x \in \Omega} |u(x)|. \quad \text{where } d = \text{dist}(\Gamma, \partial\Omega).$$

The estimate (49a) may be used to show that harmonic functions are not only smooth, but in fact have convergent Taylor series.

Theorem 10 (Real Analyticity). *If $u \in C^2(\Omega)$ is harmonic, then u is real analytic in Ω.*

Proof. Pick any point $x_0 \in \Omega$ and $a > 0$ so that $\overline{B_a(x_0)} \subset \Omega$. For notational convenience, we assume $x_0 = 0$. We want to show that for $a_1 > 0$ sufficiently small, the Maclaurin series for u converges to u on the ball $\overline{B_{a_1}(0)}$. To verify this, fix x with $|x| \leq a_1 = \epsilon a$, and consider the remainder term

$$(50) \qquad R_N(x) \equiv u(x) - \sum_{0 \leq |\alpha| < N} \frac{D^\alpha u(0)}{\alpha!} x^\alpha = \sum_{|\alpha| = N} \frac{D^\alpha u(tx)}{\alpha!} x^\alpha,$$

for some $0 \leq t \leq 1$. If we let $M = \max\{|u(x)| : x \in B_a(0)\}$, then by (49b) we have

$$|D^\alpha u(tx)| \leq M \left(\frac{nN}{(1-\epsilon)a}\right)^N.$$

Thus

$$\left|\sum_{|\alpha|=N} \frac{D^\alpha u(tx)}{\alpha!} x^\alpha\right| \leq M \sum_{|\alpha|=N} \frac{(nN\epsilon a)^N}{(1-\epsilon)^N a^N \alpha!} \leq CM \sum_{|\alpha|=N} \frac{n^{2N} e^N \epsilon^N}{(1-\epsilon)^N},$$

where we have used the inequality $|\alpha|^{|\alpha|} \leq C\alpha!(en)^{|\alpha|}$ (cf. Exercise 12). If ϵ is small enough, we can make $\epsilon e n^2/(1-\epsilon) < 1/2$, so $|R_N(x)| \leq CM2^{-N} \sum_{|\alpha|=N} 1 \leq CM2^{-N} \sum_{|\alpha|\leq N} 1 \leq CM2^{-N}(N+1)^n \to 0$ as $N \to \infty$. ♠

In the next section, we use some of these properties of harmonic functions when we discuss the existence theory of Perron.

Exercises for Section 4.2

1. (a) If $n = 2$ and $a = 1$, show that (44) is equivalent to
$$u(r,\theta) = \frac{1-r^2}{2\pi} \int_0^{2\pi} \frac{g(\phi)\,d\phi}{1+r^2-2r\cos(\theta-\phi)}.$$
   (b) Use (a) and Exercise 1 in Section 4.1 to verify the formula
$$r^k \cos k\theta = \frac{1-r^2}{2\pi} \int_0^{2\pi} \frac{\cos(k\phi)\,d\phi}{1+r^2-2r\cos(\theta-\phi)},$$
   where k is an integer and $0 \le r < 1$.

2. Let Ω be a bounded domain and $f \in C^k(\overline{\Omega})$. Show that in fact the domain potential (29a) satisfies $u \in C^{k+1}(\Omega)$. Conclude that $f \in C^\infty(\overline{\Omega})$ implies $u \in C^\infty(\Omega)$. [In Section 8.2 we show that $f \in C^\infty(\overline{\Omega})$ implies $u \in C^\infty(\overline{\Omega})$, provided $\partial\Omega$ is C^∞.]

3. The *symmetry* of the Green's function [i.e., $G(x,\xi) = G(\xi,x)$ for all $x, \xi \in \Omega$] is an important fact connected with the self-adjointness of Δ.
   (a) Verify the symmetry of $G(x,\xi)$ by direct calculation when $\Omega = \mathbf{R}_+^n$ and $\Omega = B_a(0)$.
   (b) Prove the symmetry of $G(x,\xi)$ when Ω is any smooth, bounded domain.

4. (a) Use the weak maximum principle (16) to prove that $G(x,\xi) \le 0$ for $x, \xi \in \Omega$ with $x \ne \xi$.
   (b) Use the strong maximum principle (15) to prove that $G(x,\xi) < 0$ for $x, \xi \in \Omega$ with $x \ne \xi$.

5. Use (33) to prove (38).

6. For $n = 2$, use the method of reflections to find the Green's function for the first quadrant $\Omega = \{(x,y) : x, y > 0\}$.

7. If $u \in C(\Omega)$ satisfies the mean value property of Section 4.1.d, then u is harmonic in Ω.

8. Let $\Omega = B_a(0)$, $\Omega_+ = \Omega \cap \mathbf{R}_+^n$, and $\Omega_0 = \{x \in \Omega : x_n = 0\}$. If $u \in C^2(\Omega) \cap C(\Omega_+ \cup \Omega_0)$ is harmonic in Ω_+, and $u = 0$ on Ω_0, prove that u may be extended to a harmonic function on all of Ω. (This is called a *reflection principle*.)

9. Show that the *bounded* solution of the Dirichlet problem in a half-space is unique. Give unbounded counterexamples.

10. Suppose $u \in C^2(\Omega)$ is harmonic, $u \ge 0$, and $\overline{B_a(0)} \subset \Omega$.
    (a) Use (44) to show
$$\frac{a^{n-2}(a-|\xi|)}{(a+|\xi|)^{n-1}} u(0) \le u(\xi) \le \frac{a^{n-2}(a+|\xi|)}{(a-|\xi|)^{n-1}} u(0) \quad \text{for } |\xi| < a.$$
    (b) Prove (45).

11. Use (46) to prove Liouville's theorem.

12. Use *Stirling's formula* $k! \sim (2\pi k)^{1/2} k^k e^{-k}$ as $k \to \infty$ to show there is a constant C so that $|\alpha|^{|\alpha|} \le C \alpha! e^{|\alpha|} n^{|\alpha|}$ holds for all multi-indices α.

4.3 Existence Theory

In this section we consider the existence of solutions to the Dirichlet and Neumann problems for the Laplace and Poisson equations in domains other than a ball or half-space. There are several methods to prove existence, each having certain advantages and disadvantages. Let us mention a few of them. One method is to use formulas like (30) to reduce the problem to *integral equations on the boundary* (cf. [Folland], [Garabedian], [Guenther-Lee], and [Kevorkian]). This method applies to both the Neumann and Dirichlet problems and yields a representation for the solution; it also generalizes to higher-order equations and is closely related to modern techniques of singular-integral or pseudo-differential operators (cf. [Taylor] or [Hörmander, 2]). However, the method is limited by requiring the boundary of the domain to be fairly smooth and does not easily generalize to nonlinear equations.

A second method for proving existence is the *variational method*, which is discussed in Chapter 7. It has the advantages that it applies to more general domains and generalizes to nonlinear equations, as we see in Chapter 13; it has the disadvantage that it does not yield a concrete representation for the solution as did the integral equation method.

In this section, however, we use a third method to study the Dirichlet problem, originally due to O. Perron. The *method of subharmonic functions*, like the variational method, yields little concrete information about the solution but applies to very general domains (only requiring a "barrier condition" on the boundary) and can be generalized to nonlinear equations, as we see in Chapter 13. We also state an existence theorem for the Neumann problem that follows from the results of Chapters 7 and 8.

a. Subharmonic Functions

We found in Section 4.1d that harmonic functions in Ω satisfy the mean value property: $u(\xi) = M_u(\xi, r)$ for every $\xi \in \Omega$ provided r is sufficiently small. We now define $u \in C(\Omega)$ to be *subharmonic* in Ω if

$$(51) \qquad u(\xi) \leq M_u(\xi, r) \equiv \frac{1}{\omega_n} \int_{|x|=1} u(\xi + rx)\, dS_x$$

for every $\xi \in \Omega$ provided r is sufficiently small. By (13) we know that if $u \in C^2(\Omega)$ satisfies $\Delta u \geq 0$ in Ω, then u is subharmonic in Ω; conversely it may be shown (see Exercise 1) that, if $u \in C^2(\Omega)$ is subharmonic, then $\Delta u \geq 0$ in Ω. However, it is important to emphasize that we do *not* require subharmonic functions to be in $C^2(\Omega)$. In fact, a glance at the maximum principle in Section 4.1.d shows that condition (51) instead of $\Delta u \geq 0$ is necessary for the proof. Therefore, we immediately obtain the following.

Theorem 1 (Maximum Principle for Subharmonic Functions). *If Ω is connected and $u \in C(\Omega)$ is subharmonic in Ω, then either u is a constant or*

$$(52) \qquad u(\xi) < \sup_{x \in \Omega} u(x)$$

for all $\xi \in \Omega$.

In particular, if Ω is a bounded domain and $u \in C(\overline{\Omega})$ is subharmonic in Ω, then

$$(53) \qquad \max_{x \in \overline{\Omega}} u(x) = \max_{x \in \partial\Omega} u(x).$$

We see next that subharmonic functions are very useful in proving the existence of harmonic functions.

b. Perron's Method for the Dirichlet Problem

The idea behind Perron's proof of the existence of a solution to the Dirichlet problem for Laplace's equation is the following. If $u \in C^2(\Omega) \cap C(\overline{\Omega})$ is harmonic in Ω with $u = g$ on $\partial\Omega$ and v is a subharmonic function in Ω with $v \leq g$ on $\partial\Omega$, then we may use the Gauss mean value theorem to confirm that $v - u$ is subharmonic in Ω. But $v - u \leq 0$ on $\partial\Omega$, so by (53) we conclude that $v - u \leq 0$ in Ω (i.e., $u \geq v$ in Ω). On the other hand, u itself is a candidate for v, so we see that u is the "largest" subharmonic function in Ω that is majorized by g on $\partial\Omega$. This suggests taking as a candidate for our solution the following function w_g (see Figure 1).

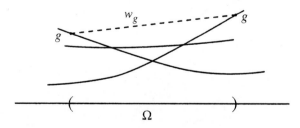

Figure 1. Taking the supremum of subfunctions.

Definition. For $g \in C(\partial\Omega)$, let $w_g(x) = \sup\{u(x) : u \in S_g\}$, where $S_g = \{u \in C(\overline{\Omega}) : u \text{ is subharmonic in } \Omega, \text{ and } u \leq g \text{ on } \partial\Omega\}$.

Notice that the function $w_g(x)$ is defined pointwise by taking the supremum of $u(x)$ as u varies over the class S_g (which are called *subfunctions* for the problem). Since S_g is nonempty (e.g., S_g contains any constant less than the minimum of g on $\partial\Omega$), and since by the maximum principle (53) we have $u(x) \leq M = \max\{|g(y)| : y \in \partial\Omega\}$ for every $x \in \Omega$ and $u \in S_g$, we conclude that w_g is well defined. However, it is not immediately clear that w_g is, for example, a continuous function. In fact, we show that w_g is harmonic in Ω and (under an additional assumption on Ω) satisfies $w_g = g$ on $\partial\Omega$.

Given $u \in C(\Omega)$ and $\overline{B_r(\xi)} \subset \Omega$, we may apply the results of the previous section to find the unique solution of the Dirichlet problem in $B_r(\xi)$ with boundary condition $g = u|_{\partial B_r(\xi)}$: let us extend this function to coincide with u on $\Omega\setminus\overline{B_r(\xi)}$. In this way we have modified u to make it harmonic on $B_r(\xi)$ but still continuous in Ω.

Definition. For $u \in C(\Omega)$ and $\overline{B_r(\xi)} \subset \Omega$, define

$$u_{\xi,r}(x) = \begin{cases} u(x) & \text{for } x \in \Omega\setminus\overline{B_r(\xi)} \\ \text{harmonic for} & x \in B_r(\xi). \end{cases}$$

Our first result shows that if u is subharmonic, then this modification process does not decrease the values of the function and does not disturb the subharmonic property (see Figure 2).

Lemma 1. If u is subharmonic in Ω and $\overline{B_r(\xi)} \subset \Omega$, then
(i) $u(x) \le u_{\xi,r}(x)$ for $x \in \Omega$, and
(ii) $u_{\xi,r}$ is subharmonic in Ω.

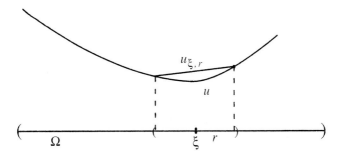

Figure 2. Making subharmonic functions harmonic on $\overline{B_r(\xi)}$.

Proof. Since $u - u_{\xi,r}$ is subharmonic in $B_r(\xi)$ and vanishes on $\partial B_r(\xi)$, we find $u(x) \le u_{\xi,r}(x)$ for $x \in B_r(\xi)$ by (53); since $u = u_{\xi,r}$ on $\Omega - \overline{B_r(\xi)}$ we have established (i). To prove (ii), we must show for any $\zeta \in \Omega$ that $u_{\xi,r}(\zeta) \le M_{u_{\xi,r}}(\zeta, \rho)$ provided ρ is sufficiently small. If $\zeta \in B_r(\xi)$ this follows from the harmonicity of $u_{\xi,r}$ in $B_r(\xi)$, and if $\zeta \in \Omega - \overline{B_r(\xi)}$ this follows from the subharmonicity of $u_{\xi,r} = u$ in Ω; thus we need only consider $\zeta \in \partial B_r(\xi)$. But then (i) implies $M_u(\zeta, \rho) \le M_{u_{\xi,r}}(\zeta, \rho)$ and the subharmonicity of u implies $u_{\xi,r}(\zeta) = u(\zeta) \le M_u(\zeta, \rho)$ for ρ sufficiently small; together these establish (ii). ♠

Section 4.3: Existence Theory   129

Lemma 2. *If the functions $u_1, \ldots, u_k \in C(\overline{\Omega})$ are subharmonic in Ω, define $v(x) = \max(u_1(x), \ldots, u_k(x))$. Then $v \in C(\overline{\Omega})$ and v is subharmonic in Ω.*

Proof. Clearly $v \in C(\overline{\Omega})$. For $\xi \in \Omega$ pick r sufficiently small that $u_j(\xi) \leq M_{u_j}(\xi, r)$ for $1 \leq j \leq k$. On the other hand, $M_{u_j}(\xi, r) \leq M_v(\xi, r)$ for $1 \leq j \leq k$, so

$$v(\xi) = \max(u_1(\xi), \ldots, u_k(\xi))$$
$$\leq \max(M_{u_1}(\xi, r), \ldots, M_{u_k}(\xi, r)) \leq M_v(\xi, r),$$

proving that v is subharmonic. ♠

Proposition 1. *If $g \in C(\partial\Omega)$, then w_g is harmonic in Ω.*

Proof. It suffices to show w_g is harmonic in $B_{r/2}(\xi) \subset \overline{B_r(\xi)} \subset \Omega$. We proceed in several steps.

1. Let $x_1, x_2, \ldots$ be a sequence of points in $B_{r/2}(\xi)$. By the definition of w_g, for each x_k we can find functions $u_k^j \in S_g$, and such that

(54) $$w_g(x_k) = \lim_{j \to \infty} u_k^j(x_k).$$

2. Let $u^j(x) = \max(u_1^j(x), \ldots, u_j^j(x))$ for $x \in \overline{\Omega}$. By Lemma 2, $u^j \in S_g$. Moreover, $u_k^j \leq u^j$ for $x \in \overline{\Omega}$ and $k \leq j$; hence

(55) $$w_g(x_k) = \lim_{j \to \infty} u^j(x_k) \quad \text{for all} \quad k.$$

3. Let $m = \min\{g(x) : x \in \partial\Omega\}$ and $M = \max\{g(x) : x \in \partial\Omega\}$. We may assume $u^j(x) \geq m$ for all $x \in \Omega$ [otherwise replace $u^j(x)$ by $\max(u^j(x), m)$ and (55) still holds]. Thus we have

(56) $$m \leq u^j(x) \leq M \quad \text{for} \quad x \in \Omega.$$

4. We may assume u^j is harmonic in $B_r(\xi)$ [otherwise replace u^j by $u_{\xi,r}^j$, which is subharmonic in Ω by Lemma 1 and clearly continues to satisfy (55) and (56)].

5. By Theorems 8 and 9 of Section 4.2, we may extract a subsequence that converges uniformly on $\overline{B_{r/2}(\xi)}$ to a function w that is harmonic in $B_{r/2}(\xi)$. In particular,

(57) $$w_g(x_k) = w(x_k) \quad \text{for all} \quad x_k \in B_{r/2}(\xi).$$

6. We may now conclude that w_g is continuous on $B_{r/2}(\xi)$: For $x_k \to x_0 \in B_{r/2}(\xi)$ we use the continuity of w to obtain $w_g(x_k) = w(x_k) \to w(x_0) = w_g(x_0)$.

7. If we pick a sequence $\{x_k\}$ that is *dense* in $B_{r/2}(\xi)$, then (57) shows that $w_g = w$ in $B_{r/2}(\xi)$; in particular, w_g is harmonic in $B_{r/2}(\xi)$ as was to be shown. ♠

We have therefore associated to $g \in C(\partial\Omega)$ a function w_g that is harmonic in Ω. In order to have $w_g = g$ on $\partial\Omega$, however, we need to make an additional assumption on Ω.

Definitions. A *barrier function* at $z \in \partial\Omega$ is a function $Q_z \in C(\overline{\Omega})$ that is subharmonic in Ω such that (i) $Q_z(z) = 0$, and (ii) $Q_z(x) < 0$ for $x \in \partial\Omega - \{z\}$. A point $z \in \partial\Omega$ is *regular* if there exists a barrier function at z.

Proposition 2. *If $g \in C(\partial\Omega)$ and $z \in \partial\Omega$ is regular, then*

$$\lim_{\Omega \ni x \to z} w_g(x) = g(z). \tag{58}$$

Proof. Consider the function $u_-(x) = g(z) - \epsilon + \kappa Q_z(x)$, where $\epsilon, \kappa > 0$. This function is in $C(\overline{\Omega})$ and is subharmonic in Ω. Since g is continuous on $\partial\Omega$, we can find $\delta > 0$ such that for $x \in \partial\Omega$

$$|g(x) - g(z)| < \epsilon \quad \text{if} \quad |x - z| < \delta, \tag{59}$$

and then choose $\kappa > 0$ such that for $x \in \partial\Omega$

$$\kappa Q_z(x) \leq -2M \quad \text{if} \quad |x - z| \geq \delta, \tag{60}$$

where $M = \max\{|g(x)| : x \in \partial\Omega\}$. Thus $u_- \leq g$ on $\partial\Omega$, so by the definition of w_g we have

$$u_-(x) = g(z) - \epsilon + \kappa Q_z(x) \leq w_g(x) \quad \text{for} \quad x \in \Omega.$$

Similarly, consider the function $u_+(x) = g(z) + \epsilon - \kappa Q_z(x)$. This function is in $C(\overline{\Omega})$ and $-u_+$ is subharmonic in Ω (such a function u_+ is called *superharmonic*); using (59) and (60), we conclude that $-u_+ \leq -g$ on $\partial\Omega$. Now for any $u \in S_g$ we find that $u - u_+ \in C(\overline{\Omega})$ is subharmonic on Ω, and $u - u_+ \leq 0$ on $\partial\Omega$. Hence by (53) we have $u(x) \leq u_+(x)$ for $x \in \Omega$, which by the definition of w_g implies

$$w_g(x) \leq u_+(x) = g(z) + \epsilon - \kappa Q_z(x) \quad \text{for} \quad x \in \Omega.$$

Thus we have established the inequality

$$g(z) - \epsilon + \kappa Q_z(x) \leq w_g(x) \leq g(z) + \epsilon - \kappa Q_z(x). \tag{61}$$

Since $Q_z(x) \to 0$ as $x \to z$ and since ϵ was arbitrary, we obtain (58). ♠

Propositions 1 and 2 immediately imply the following existence result for the Dirichlet problem.

Theorem 2. *If Ω is a bounded domain with regular boundary, then the Dirichlet problem (1),(3) admits a (unique) solution $u \in C^\infty(\Omega) \cap C(\overline{\Omega})$ for every $g \in C(\partial\Omega)$.*

This theorem, however, is only useful if we can recognize a "regular boundary." Perhaps the simplest such domains are those that are *strictly convex*, that is, for each point $z \in \partial\Omega$ there is a hyperplane in $\mathbf{R}^n$ that intersects $\partial\Omega$ only at the point z; the barrier function Q_z can be taken to be a linear function that vanishes along the hyperplane and is negative on the side containing Ω. Another, more general, class of regular domains are those that satisfy the *exterior sphere condition* on $\partial\Omega$, that is, for each $z \in \partial\Omega$ there is a ball $B_\epsilon(\xi)$ satisfying $\overline{B_\epsilon(\xi)} \cap \overline{\Omega} = \{z\}$; the fundamental solution is useful for constructing the barrier function at z (see Exercise 2). In particular, a domain with C^2 boundary satisfies the exterior sphere condition (see Exercise 3), so the Dirchlet problem may be solved for domains that are sufficiently smooth. But even the smoothness is not required: if Ω satisfies an *exterior cone condition* at $z \in \partial\Omega$, then z is a regular point (see Exercise 4 when $n = 2$). So the domain must be pretty wild to be nonregular. The other methods for proving existence of a solution to the Dirichlet problem do not apply to such general domains.

c. The Neumann Problem

The following result is established using the variational theory of Chapter 7 together with the regularity theory of Chapter 8. We record it here for convenience.

Theorem 3. *Let Ω be a bounded domain with smooth boundary $\partial\Omega$. Suppose $f \in C^\infty(\overline{\Omega})$ and $h \in C^\infty(\partial\Omega)$ satisfy the compatibility condition*

$$\int_{\partial\Omega} h \, dS = \int_\Omega f \, dx.$$

Then there is a solution $u \in C^\infty(\overline{\Omega})$ of $\Delta u = f$ in Ω and $\partial u / \partial \nu = h$ on $\partial\Omega$ that is unique up to an additive constant.

Exercises for Section 4.3

1. Show that if $u \in C^2(\Omega)$ is subharmonic then $\Delta u \geq 0$ in Ω.
2. Show that if Ω satisfies an exterior sphere condition at $z \in \partial\Omega$, then z is a regular point.
3. Show that a bounded domain Ω with C^2 boundary $\partial\Omega$ satisfies an exterior sphere condition and hence is regular.
4. If $n = 2$ and Ω satisfies an *exterior cone condition* at $z \in \partial\Omega$, that is, there is a cone $C = \{x \in \mathbf{R}^2 : |x - z| < \epsilon, |\arg(x - z) - \theta_0| < \delta\}$ such that $C \cap \bar{\Omega} = \{z\}$, then z is a regular point. (This is similarly true for $n \geq 3$.)

4.4 Eigenvalues of the Laplacian

In this section we consider the equation

(62)
$$\begin{cases} \Delta u + \lambda u = 0 & \text{in } \Omega \\ u = 0 & \text{on } \partial\Omega, \end{cases}$$

where Ω is a bounded domain and λ is a (complex) number. Of course, (62) always admits the trivial solution $u \equiv 0$; but the values of λ for which (62) admits a nontrivial solution u are called the *eigenvalues* of Δ in Ω, and the solution u is an *eigenfunction associated to the eigenvalue* λ. (The convention $\Delta u + \lambda u = 0$ is chosen instead of $\Delta u = \lambda u$ so that all eigenvalues λ will be positive.)

More properly, the solutions to (62) should be called the *Dirichlet* eigenvalues and eigenfunctions, due to the boundary condition involved. Similarly, there are *Neumann* eigenvalues and eigenfunctions (cf. Exercise 5), or other eigenvalues and eigenfunctions, depending on a choice of boundary condition.

We first discuss the collection of eigenvalues and how their associated eigenfunctions may be used to expand a given function in an infinite series. We then explore applications of these eigenfunctions to a representation of the Green's function for Δ on Ω, and to solving an initial boundary value problem for the n-dimensional wave equation.

a. Eigenvalues and Eigenfunction Expansions

We begin with an elementary example.

Example 1. Let Ω be the two-dimensional rectangle $(0, a) \times (0, b)$ in the xy-plane. The problem (62) becomes

(63)
$$\begin{cases} u_{xx} + u_{yy} + \lambda u = 0 & \text{in } \Omega \\ u(0, y) = 0 = u(a, y) & \text{for } 0 \leq y \leq b \\ u(x, 0) = 0 = u(x, b) & \text{for } 0 \leq x \leq a. \end{cases}$$

We can solve this problem easily by separation of variables. If we let $u(x, y) = X(x)Y(y)$, we find the separate equations for X and Y:

$$\begin{array}{ll} X'' + \mu^2 X = 0 & \quad Y'' + \nu^2 Y = 0 \\ X(0) = 0 = X(a) & \quad Y(0) = 0 = Y(b), \end{array}$$

where $\lambda = \mu^2 + \nu^2$. The solutions of these one-dimensional eigenvalue problems are

$$\begin{array}{ll} \mu_m = \frac{m\pi}{a} & \quad \nu_n = \frac{n\pi}{b} \\ X_m(x) = \sin\frac{m\pi x}{a} & \quad Y_n(y) = \sin\frac{n\pi y}{b}, \end{array}$$

Section 4.4: Eigenvalues of the Laplacian

where $m, n = 1, 2, \ldots$. Thus we obtain solutions of (63) of the form

(64) $\qquad \lambda_{mn} = \pi^2 \left(\dfrac{m^2}{a^2} + \dfrac{n^2}{b^2} \right), \qquad u_{mn}(x, y) = \sin \dfrac{m\pi x}{a} \sin \dfrac{n\pi y}{b},$

where $m, n = 1, 2, \ldots$.

Let us observe that the eigenvalues $\{\lambda_{mn}\}_{m,n=1}^{\infty}$ of (63) are positive a form a countable set of real numbers accumulating only at infinity. T smallest eigenvalue λ_{11} has only one (up to a scalar multiple) eigenfunct $u_{11}(x, y) = \sin(\pi x/a) \sin(\pi y/b)$; notice that u_{11} is positive in Ω. Otl eigenvalues λ may correspond to more than one choice of m and n; example, in the case $a = b$ we have $\lambda_{nm} = \lambda_{mn}$, so if $\lambda = \lambda_{nm}$ with $n \neq$ then there are two linearly independent eigenfunctions. However, fo particular value of λ there are (at most) finitely many linearly independ eigenfunctions. Moreover, a calculation easily shows that

(65) $\qquad \displaystyle\int_0^a \int_0^b u_{mn}(x, y) u_{m'n'}(x, y) \, dx \, dy = \begin{cases} 0 & \text{if } m \neq m' \text{ or } n \neq n' \\ \dfrac{ab}{4} & \text{if } m = m' \text{ and } n = n'. \end{cases}$

In particular, the $\{u_{mn}\}$ are pairwise orthogonal. In fact, we could n malize each u_{mn} by a scalar multiple so that $\frac{ab}{4}$ in (65) becomes 1.

Let us also recall that we may associate to a function $f(x, y)$ on $[0, a$ $[0, b]$ its double Fourier sine series

(66a) $\qquad f(x, y) \sim \displaystyle\sum_{m=1}^{\infty} \sum_{n=1}^{\infty} A_{mn} \sin \dfrac{m\pi x}{a} \sin \dfrac{n\pi y}{b},$

where the coefficients may be computed by

(66b) $\qquad A_{mn} = \dfrac{4}{ab} \displaystyle\int_0^a \int_0^b f(x, y) \sin \dfrac{m\pi x}{a} \sin \dfrac{n\pi y}{b} dx \, dy.$

As we shall see, the sense in which the sine series represents the funct $f(x, y)$ depends on the properties of f. However, the point we wish make here is that the double Fourier series (66ab) may be considered as *expansion in eigenfunctions* of (63). ♣

Now let us turn to the general eigenvalue problem (62), which may studied using a couple of methods. One is to use the Green's function rewrite (62) as an integral equation

(67) $\qquad u(x) + \lambda \displaystyle\int_\Omega G(x, y) \, u(y) \, dy = 0$

and then use properties of the Green's operator defined by the symmet integral kernel $G(x, y)$ (cf. [Garabedian]). Another is to look for w solutions of (62) using variational techniques (cf. Section 7.2b). Both

these methods require some functional analysis, which we develop later in this book (cf. Chapter 6), but for now let us summarize the properties of the eigenvalues and eigenfunctions for (62) that we found in Example 1:
1. The eigenvalues of (62) form a countable set $\{\lambda_n\}_{n=1}^{\infty}$ of positive numbers with $\lambda_n \to \infty$ as $n \to \infty$.
2. For each eigenvalue λ_n there is a finite number (called the *multiplicity* of λ_n) of linearly independent eigenfunctions u_n.
3. The first (or *principal*) eigenvalue, λ_1, has multiplicity 1 (i.e., λ_1 is *simple*) and u_1 does not change sign in Ω.
4. Eigenfunctions corresponding to distinct eigenvalues are orthogonal.
5. The eigenfunctions may be used to expand certain functions on Ω in an infinite series.

The collection of eigenvalues $\{\lambda_n\}$ is called the *spectrum* of the Laplacian (with Dirichlet boundary conditions); the study of the eigenvalues and eigenvectors of an operator is called *spectral theory* (see Section 6.6).

Let us change our notation somewhat so that each eigenvalue λ_n corresponds to a particular eigenfunction $\phi_n(x)$; that is, we may write

$$(68) \qquad 0 < \lambda_1 < \lambda_2 \leq \lambda_3 \ldots \to \infty,$$

where we repeat each eigenvalue according to its multiplicity. If we choose an orthonormal basis of eigenfunctions in each eigenspace, we may arrange that $\{\phi_n\}_{n=1}^{\infty}$ is pairwise orthonormal:

$$(69) \qquad \int_\Omega \phi_n(x)\phi_m(x)\,dx = \begin{cases} 0 & \text{if } m \neq n \\ 1 & \text{if } m = n. \end{cases}$$

In this notation, the eigenfunction expansion of $f(x)$ defined on Ω becomes

$$(70) \qquad f(x) \sim \sum_{n=1}^{\infty} a_n \phi_n(x), \qquad \text{where} \quad a_n = \int_\Omega f(x)\phi_n(x)\,dx.$$

An important consequence of Property 4 is the *uniqueness* of the eigenfunction expansion: if $\sum a_n \phi_n(x) \sim 0$, then $a_n = 0$ for all n.

These properties will be established in detail in Section 7.2b. In particular, the sense in which "$f(x) = \sum a_n \phi_n(x)$" will be found to be *convergence in square mean*: $\int_\Omega |f(x) - \sum_{n=1}^{N} a_n \phi_n(x)|^2 \, dx \to 0$ as $N \to \infty$.

Remark. Actual pointwise convergence of the eigenfunction expansion to $f(x)$ requires additional regularity. For example, if $f \in C^2(\overline{\Omega})$ and $f = 0$ on $\partial\Omega$, then the eigenfunction expansion converges absolutely and uniformly on $\overline{\Omega}$ to $f(x)$ (cf. the expansion theorem in [Courant-Hilbert], vol. I).

Let us consider a couple of immediate applications of eigenfunction expansions.

Section 4.4: Eigenvalues of the Laplacian

Example 2. *(Poisson Equation)* Suppose we want to find the eigenfunction expansion of the solution of (2) with $u = 0$ on $\partial\Omega$, when f has the expansion (70) in the orthonormal Dirichlet eigenfunctions ϕ_n. Then, writing $u = \sum c_n \phi_n$, insertion in (2) and comparision of coefficients yields $c_n = -\lambda_n^{-1} a_n$; that is,

$$(71) \qquad u(x) \sim -\sum_{n=1}^{\infty} \frac{a_n \phi_n(x)}{\lambda_n}.$$

Assuming sufficient regularity of f for (70) to converge absolutely and uniformly on $\overline{\Omega}$, then (71) certainly converges absolutely and uniformly on $\overline{\Omega}$ since the factors λ_n^{-1} only improve the convergence assumed of (70). ♣

Example 3. *(Eigenfunction Expansion of the Green's Function)* Suppose we fix x and attempt to expand the Green's function $G(x, y)$ in the orthonormal eigenfunctions $\phi_n(y)$:

$$G(x, y) \sim \sum_{n=1}^{\infty} a_n(x)\phi_n(y), \qquad \text{where} \quad a_n(x) = \int_{\Omega} G(x, z)\phi_n(z)\, dz.$$

But if we use (67) and (69), we discover that $a_n(x) = -\lambda_n^{-1} \phi_n(x)$, so this formula becomes

$$(72) \qquad G(x, y) \sim -\sum_{n=1}^{\infty} \frac{\phi_n(x)\phi_n(y)}{\lambda_n}.$$

The sense in which the infinite series converges to $G(x, y)$ is convergence in mean square:

$$(73) \qquad \int_{\Omega}\int_{\Omega} \left| G(x, y) - \sum_{n=1}^{N} \frac{\phi_n(x)\phi_n(y)}{\lambda_n} \right|^2 dx\, dy \to 0 \qquad \text{as} \quad N \to \infty.$$

Since we are interested in $G(x, y)$ as an integral kernel, the convergence in mean is satisfactory. ♣

b. Application to the Wave Equation

In Section 3.1 we considered an initial boundary value problem for the one-dimensional wave equation on an interval, and we found that the solution could be obtained using Fourier series. If we replace the Fourier series by an expansion in eigenfunctions, we can consider an initial/boundary value problem for the n-dimensional wave equation. For example, let us consider

$$(74) \qquad \begin{cases} u_{tt} = \Delta u & \text{for } x \in \Omega \text{ and } t > 0 \\ u(x, 0) = g(x), \quad u_t(x, 0) = h(x) & \text{for } x \in \Omega \\ u(x, t) = 0 & \text{for } x \in \partial\Omega \text{ and } t > 0, \end{cases}$$

where the initial conditions $g(x)$ and $h(x)$ are sufficiently "nice" functions. From our experience in Section 3.1, we suspect that the compatibility condition $g(x) = 0 = h(x)$ on $\partial\Omega$ will be required to ensure $u \in C^2(\Omega)$; thus it is not unreasonable to assume that $g, h \in C^2(\overline{\Omega})$ with $g = h = 0$ on $\partial\Omega$. From the Remark in the previous subsection, this means that the eigenfunction expansions

$$(75) \qquad g(x) = \sum_{n=1}^{\infty} a_n \phi_n(x) \quad \text{and} \quad h(x) = \sum_{n=1}^{\infty} b_n \phi_n(x)$$

hold pointwise, with each series converging absolutely and uniformly on $\overline{\Omega}$.

Let us assume the solution $u(x,t)$ may be expanded in the eigenfunctions with coefficients depending on t: $u(x,t) = \sum_{n=1}^{\infty} u_n(t)\phi_n(x)$. Assuming sufficient convergence to pass the derivatives inside the summation, this implies $u_n''(t) + \lambda_n u_n(t) = 0$ for each n. Since $\lambda_n > 0$, this ordinary differential equation has general solution $u_n(t) = A_n \cos\sqrt{\lambda_n}\,t + B_n \sin\sqrt{\lambda_n}\,t$. At $t = 0$ we obtain $u_n(0) = A_n$ and $u_n'(0) = B_n\sqrt{\lambda_n}$, so

$$u(x,0) = \sum_{n=1}^{\infty} A_n \phi_n(x) = g(x) \quad \text{and} \quad u_t(x,0) = \sum_{n=1}^{\infty} B_n\sqrt{\lambda_n}\,\phi_n(x) = h(x).$$

Comparing with (75) we obtain

$$(76) \quad A_n = a_n = \int_\Omega g(x)\phi_n(x)\,dx \quad \text{and} \quad B_n = \frac{b_n}{\sqrt{\lambda_n}} = \frac{1}{\sqrt{\lambda_n}} \int_\Omega h(x)\phi_n(x)\,dx.$$

Putting this together, we find that the solution of (74) is given by

$$(77) \qquad u(x,t) = \sum_{n=1}^{\infty} (A_n \cos\sqrt{\lambda_n}\,t + B_n \sin\sqrt{\lambda_n}\,t)\,\phi_n(x)$$

with the coefficients given by (76).

To analyze this solution, let us consider one term $u_n(t)\phi_n(x)$ in the series, and for simplicity (which can be achieved by a phase shift) assume $B_n = 0$ so that

$$(78) \qquad u_n(t)\phi_n(x) = A_n \cos(\sqrt{\lambda_n}\,t)\,\phi_n(x).$$

We see that this term exhibits simple harmonic motion with *frequency* $f_n = \sqrt{\lambda_n}/2\pi$. Moreover, points where $\phi_n(x) = 0$ are called *nodal sets*. Notice that the solution (78) remains constantly zero on the nodal sets.

When $\Omega = (0, L)$ is a one-dimensional interval, then (74) is just the initial boundary value problem of Section 3.1 that models a vibrating string. Notice that $\sqrt{\lambda_n} = n\pi/L$ so that all the frequencies f_n are integral multiples of each other. This accounts for the coherent perception of a single tone corresponding to a vibrating string, say on a musical instrument. Moreover, the eigenfunction $\phi_n(x) = \sin(n\pi x/L)$ will have exactly $n-1$ zeros, so the solution (78) will have $n-1$ stationary nodal points.

When Ω is a two-dimensional region, the frequencies will no longer occur as integral multiples of each other, and the nodal sets will be curves (cf. Exercise 1). In particular, we observe that a vibrating drumhead does not create a single musical tone unless the vibration corresponds to one of the special solutions (78) for which the vibration leaves certain points on the drumhead stationary.

Exercises for Section 4.4

1. Consider (74) with $\Omega = (0, a) \times (0, b)$, the rectangle of Example 1.
   (a) Use the eigenvalues and eigenfunctions computed in the text to express the solution (77).
   (b) Find the frequencies of the special solutions (78), and show that they are *not* integral multiples of each other.
   (c) Assume $a = b/2$. Find the two smallest frequencies and their corresponding nodal curves.

2. Let $\Omega = (0, a) \times (0, b) \times (0, c) \subset \mathbf{R}^3$. Find the Dirichlet eigenvalues and eigenfunctions for Δ in Ω.

3. Consider the initial/boundary value problem with forcing term

$$\begin{cases} u_{tt} = \Delta u + f(x,t) & \text{for } x \in \Omega \text{ and } t > 0 \\ u(x,0) = 0, \quad u_t(x,0) = 0 & \text{for } x \in \Omega \\ u(x,t) = 0 & \text{for } x \in \partial\Omega \text{ and } t > 0. \end{cases}$$

Use Duhamel's principle and an expansion of f in eigenfunctions to obtain a (formal) solution.

4. Suppose the forcing term in the previous exercise is $f(x,t) = A(x)\sin\omega t$. Find the (formal) solution when (a) $\omega^2 \neq \lambda_n$ for any of the eigenvalues λ_n, and (b) $\omega^2 = \lambda_k$ for some k (resonance).

5. Let $\Omega = (0,a) \times (0,b)$ and consider the initial/ boundary value problem

$$\begin{cases} u_{tt} = \Delta u & \text{for } x \in \Omega \text{ and } t > 0 \\ u(x,0) = g(x), \quad u_t(x,0) = h(x) & \text{for } x \in \Omega \\ \frac{\partial u}{\partial \nu}(x,t) = 0 & \text{for } x \in \partial\Omega \text{ and } t > 0. \end{cases}$$

(a) Find the eigenvalues and eigenfunctions for the associated Neumann problem on Ω.
(b) Find the solution as an expansion similar to (77).

6. The ordinary differential equation $r^2 y'' + r y' + (r^2 - n^2)y = 0$ for $y(r)$ in $r > 0$ is called *Bessel's equation of order n*, where $n = 0, 1, 2, \ldots$ This equation has a solution $y = J_n(r)$, which behaves like r^n as $r \to 0$, and has an infinite number of zeros $J_n(r) = 0$ at $r = \rho_{n1}, \rho_{n2}, \ldots \to \infty$.

(a) If $\lambda > 0$, show that $y = J_n(\sqrt{\lambda}r)$ solves the ordinary differential equation
$$r^2 y'' + ry' + (\lambda r^2 - n^2)y = 0.$$

(b) If $\Omega = \{(x_1, x_2) \in \mathbf{R}^2 : x_1^2 + x_2^2 < a\}$, then use separation of variables and part (a) to obtain eigenvalues and eigenfunctions for
$$\begin{cases} \Delta u + \lambda u = 0 & \text{in } \Omega \\ u = 0 & \text{on } \partial\Omega. \end{cases}$$

(c) Assuming the functions $g(r,\theta)$ and $h(r,\theta)$ may be expanded in the eigenfunctions of (b), find the solution of the initial/boundary value problem
$$\begin{cases} u_{tt} = \Delta u & \text{in } \Omega \\ u(r,\theta,0) = g(r,\theta) & \text{for } 0 \leq r < a \text{ and } 0 \leq \theta < 2\pi \\ u_t(r,\theta,0) = h(r,\theta) & \text{for } 0 \leq r < a \text{ and } 0 \leq \theta < 2\pi \\ u(a,\theta,t) = 0 & \text{for } 0 \leq \theta < 2\pi, t > 0. \end{cases}$$

7. If $\lambda \leq 0$ and $\partial\Omega$ is smooth, show that the only solution $u \in C^2(\overline{\Omega})$ of (62) is the trivial solution $u \equiv 0$.

4.5 Applications to Vector Fields

In this section we apply some of the results obtained in this chapter to problems involving vector fields on Euclidean domains.

a. The div-curl System

In applications in $\mathbf{R}^3$, it is often useful to recover a vector field from its divergence and curl. In other words, given a scalar function $\rho(x)$ and a vector field $\vec{v}(x)$ in $\mathbf{R}^3$, we want to solve the following system:

(79a) $\qquad \operatorname{div} \vec{u} = \rho$

(79b) $\qquad \operatorname{curl} \vec{u} = \vec{v}.$

Obviously, there is a compatibility condition that must be imposed on $\vec{v}$ in order to solve (79b) [cf. (14a) in the Appendix]:

(80) $\qquad \operatorname{div} \vec{v} = 0.$

Moreover, the solution of (79ab) is certainly not unique, since $\vec{u}(x) + \vec{c}$ is also a solution whenever $\vec{u}(x)$ is. (Here $\vec{c}$ denotes any constant vector.) To have a unique solution, we need to impose some sort of side condition such as a decay condition at infinity.

Section 4.5: Applications to Vector Fields

It is natural to split the problem (79ab) up by letting $\vec{u} = \vec{u}_1 + \vec{u}_2$, where $\vec{u}_1$ satisfies $\operatorname{div} \vec{u}_1 = \rho$ and $\operatorname{curl} \vec{u}_1 = 0$ and $\vec{u}_2$ satisfies $\operatorname{div} \vec{u}_2 = 0$ and $\operatorname{curl} \vec{u}_2 = \vec{v}$. Let us first find $\vec{u}_1$. By vector calculus, we know that $\operatorname{curl} \vec{u}_1 = 0$ implies the existence of a potential function ϕ for $\vec{u}_1$; this means $\vec{u}_1 = \nabla \phi$, and consequently ϕ satisfies $\Delta \phi = \rho$. Since we are on $\mathbf{R}^3$, we may use the fundamental solution $K(x) = -(4\pi|x|)^{-1}$ to find ϕ: if $\rho \in C_0^\infty(\mathbf{R}^3)$, then

$$(81) \qquad \phi(x) = -\int_{\mathbf{R}^3} \frac{\rho(y)}{4\pi|x - y|} \, dy$$

is well defined, smooth, $O(|x|^{-1})$ as $|x| \to \infty$, and satisfies $\Delta \phi = \rho$; we can then let $\vec{u}_1 = \nabla \phi$, which is clearly $O(|x|^{-2})$ as $|x| \to \infty$.

To find $\vec{u}_2$, let us take the curl of $\operatorname{curl} \vec{u}_2 = \vec{v}$ and use (14c) of the Appendix to obtain $\nabla \operatorname{div} \vec{u}_2 - \Delta \vec{u}_2 = \operatorname{curl} \vec{v}$. But using $\operatorname{div} \vec{u}_2 = 0$, this becomes $\Delta \vec{u}_2 = -\operatorname{curl} \vec{v}$, which may be solved componentwise using the fundamental solution: if $\operatorname{curl} \vec{v} \in C_0^\infty(\mathbf{R}^3)$, then we can let

$$(82a) \qquad \vec{u}_2(x) = \int_{\mathbf{R}^3} \frac{\operatorname{curl} \vec{v}(y)}{4\pi|x - y|} \, dy,$$

which is well defined, smooth, and $O(|x|^{-1})$ as $|x| \to \infty$. However, if we integrate by parts to throw y-derivatives of $\vec{v}$ onto $|x - y|^{-1}$ and then use (29e), we find

$$(82b) \qquad \vec{u}_2(x) = \operatorname{curl} \left(\int_{\mathbf{R}^3} \frac{\vec{v}(y)}{4\pi|x - y|} \, dy \right),$$

which shows that in fact $\vec{u}_2 = O(|x|^{-2})$ as $|x| \to \infty$, as well as $\operatorname{div} \vec{u}_2 = 0$ and $\operatorname{curl} \vec{u}_2 = \vec{W}$.

Combining these formulas for $\vec{u}_1$ and $\vec{u}_2$ suggests that we try to solve (79ab) in $\mathbf{R}^3$ by the formula

$$(83) \qquad \vec{u}(x) = -\nabla_x \left(\int_{\mathbf{R}^3} \frac{\rho(y)}{4\pi|x - y|} \, dy \right) + \operatorname{curl} \left(\int_{\mathbf{R}^3} \frac{\vec{v}(y)}{4\pi|x - y|} \, dy \right).$$

The function ϕ as in (81) is called the *scalar potential*, and the vector field $\int_{\mathbf{R}^3} (4\pi|x-y|)^{-1} \vec{v}(y) \, dy$ is called the *vector potential* for $\vec{u}$. It is now simply a matter of imposing more general conditions on ρ and $\vec{v}$ under which (83) indeed gives a well-defined solution.

Theorem 1. *Suppose ρ and $\vec{v}$ are C^1 on $\mathbf{R}^3$ and $O(|x|^{-2-\epsilon})$ as $|x| \to \infty$ for some $\epsilon > 0$; also suppose $\operatorname{div} \vec{v} = 0$. Then (83) defines a C^1-solution of (79ab) and is the unique solution satisfying $\vec{u}(x) = o(1)$ as $|x| \to \infty$.*

Proof. The order of decay $O(|x|^{-2-\epsilon})$ on ρ and $\vec{v}$ ensures that the scalar and vector potentials in (83) are well-defined C^1-functions. But of course we need to know that the potentials are C^2 in order to conclude that $\vec{u}$ is C^1, and then we need to show that $\vec{u}(x) = o(1)$ as $|x| \to \infty$. The proof that the potentials are C^2 is essentially that of the claim (iii) of the Proposition in Section 4.2b; Exercise 1 at the end of this section not only shows that the decay hypotheses allow the replacement of the bounded domain Ω by $\mathbf{R}^3$, but that the resultant $\vec{u}$ is $O(|x|^{-2})$. To verify uniqueness, let $\vec{u}$ be the difference between two solutions of (79ab) so that $\vec{u}$ satisfies (79ab) with $\rho \equiv 0$ and $\vec{v} \equiv 0$. Taking curl of curl $\vec{u} = 0$ and using (14c) of the Appendix, we find $\Delta \vec{u} = 0$ (i.e. each component of $\vec{u}$ is harmonic). Using $\vec{u} = o(1)$ as $|x| \to \infty$ and Theorem 7 of Section 4.2, $\vec{u} \equiv 0$. ♠

b. Helmholtz Decompositions

Notice that (83) states that $\vec{u}$ is realized as the sum of a *gradient field* $\vec{u}_1$ (i.e., $\vec{u}_1 = \nabla \phi$) and a *divergence-free* vector field $\vec{u}_2$ (i.e., div $\vec{u}_2 = 0$); this is known as the *Helmholtz decomposition* of $\vec{u}$. In fact, substituting (79ab) in (83), we obtain the explicit representation

$$(84) \qquad \vec{u}(x) = -\nabla_x \left(\int_{\mathbf{R}^3} \frac{\operatorname{div} \vec{u}(y)}{4\pi |x-y|} \, dy \right) + \operatorname{curl} \left(\int_{\mathbf{R}^3} \frac{\operatorname{curl} \vec{u}(y)}{4\pi |x-y|} \, dy \right).$$

Corollary. *Suppose that $\vec{u}$ is a C^1-vector field on $\mathbf{R}^3$ with $\vec{u} = o(1)$, div $\vec{u} = O(|x|^{-2-\epsilon})$, and curl $\vec{u} = O(|x|^{-2-\epsilon})$ as $|x| \to \infty$. Then (84) provides a Helmholtz decomposition $\vec{u} = \vec{u}_1 + \vec{u}_2$, where $\vec{u}_1$ and $\vec{u}_2$ are orthogonal: $\int_{\mathbf{R}^3} \vec{u}_1 \cdot \vec{u}_2 \, dx = 0$.*

Proof. Apply (83) with $\rho = \operatorname{div} \vec{u}$ and $\vec{v} = \operatorname{curl} \vec{u}$. To verify orthogonality, consider a large ball $|x| \leq R$ and apply the divergence theorem. In fact, if we use $\phi(x)$ as in (81) and use div $\vec{u}_2 = 0$, we obtain

$$\int_{|x| \leq R} \vec{u}_1 \cdot \vec{u}_2 \, dx = \int_{|x| \leq R} \operatorname{div}(\phi \vec{u}_2) \, dx = \int_{|x|=R} \phi \vec{u}_2 \cdot \vec{\nu} \, dS.$$

But using $\phi = O(|x|^{-1})$ and $\vec{u}_2 = O(|x|^{-2})$ as $|x| \to \infty$ (cf. Exercise 1), we find that the last term vanishes as $R \to \infty$, and so $\int_{\mathbf{R}^3} \vec{u}_1 \cdot \vec{u}_2 \, dx = 0$. ♠

Remark. Although $\vec{u}$ was assumed to be C^1, it was not claimed that $\vec{u}_1$ or $\vec{u}_2$ is C^1. In fact, this need not be the case, since we cannot conclude (for example) from div $\vec{u} \in C(\mathbf{R}^3)$ that $\phi(x) = -\int (4\pi|x-y|)^{-1} \operatorname{div} \vec{u}(y) \, dy \in C^2(\mathbf{R}^3)$. However, if $\vec{u} \in C^\infty(\mathbf{R}^3)$, then it is true that $\vec{u}_1, \vec{u}_2 \in C^\infty(\mathbf{R}^3)$.

There is also a Helmholtz decomposition of vector fields on bounded domains; its existence may be reduced to a Neumann problem.

Theorem 2. *If $\Omega \subset \mathbf{R}^3$ is a smooth bounded domain and $\vec{u}$ is a smooth vector field on $\overline{\Omega}$, then there is a unique orthogonal decomposition $\vec{u} = \vec{u}_1 + \vec{u}_2$, where $\vec{u}_1$ is a smooth gradient field and $\vec{u}_2$ is both divergence-free and parallel to $\partial \Omega$ (i.e., $\vec{u}_2 \cdot \vec{\nu} = 0$ on $\partial \Omega$, where $\vec{\nu}$ is the exterior unit normal).*

Proof. We seek a potential function $\phi(x)$ such that $\vec{u} = \nabla \phi + \vec{u}_2$, where $\vec{u}_2$ is divergence-free and parallel to $\partial \Omega$. If we take the divergence of this equation, we find that $\Delta \phi = f$ in Ω, where $f = \text{div}\, \vec{u}$; if we dot the equation with $\vec{\nu}$, we find that $\partial \phi / \partial \nu = h$ on $\partial \Omega$, where $h = \vec{u} \cdot \vec{\nu}$. Thus ϕ must satisfy a Neumann problem. But

$$\int_{\partial \Omega} h\, dS = \int_{\partial \Omega} \vec{u} \cdot \vec{\nu}\, dS = \int_{\partial \Omega} \vec{u}_1 \cdot \vec{\nu}\, dS = \int_{\partial \Omega} \nabla \phi \cdot \vec{\nu}\, dS$$
$$= \int_\Omega \text{div} \nabla \phi\, dx = \int_\Omega \Delta \phi\, dx = \int_\Omega f\, dx$$

shows that the compatibility condition in Theorem 3 of Section 4.3 is satified, so we can solve the Neumann problem to obtain $\phi \in C^\infty(\overline{\Omega})$ and then let $\vec{u}_1 = \nabla \phi$. By construction we find that $\vec{u}_2 = \vec{u} - \vec{u}_1$ is divergence-free.

We verify orthogonality much as in the free space case:

$$\int_\Omega \vec{u}_1 \cdot \vec{u}_2\, dx = \int_\Omega \text{div}(\phi \vec{u}_2)\, dx = \int_{\partial \Omega} \phi \vec{u}_2 \cdot \vec{\nu}\, dS = 0.$$

Finally, we verify uniqueness. If we also have $\vec{u} = \vec{v}_1 + \vec{v}_2$ with $\vec{v}_1 = \nabla \psi$ and $\vec{v}_2$ divergence-free and parallel to $\partial \Omega$, then $0 = \vec{u}_1 - \vec{v}_1 + \vec{u}_2 - \vec{v}_2 = \nabla(\phi - \psi) + \vec{u}_2 - \vec{v}_2$, and so we may dot with $\vec{u}_2 - \vec{v}_2$ and integrate over Ω to obtain

$$0 = \int_\Omega \left((\vec{u}_2 - \vec{v}_2) \cdot \nabla(\phi - \psi) + |\vec{u}_2 - \vec{v}_2|^2 \right) dx.$$

But we may use the divergence theorem and the properties of $\vec{u}_2 - \vec{v}_2$ to conclude that $\int (\vec{u}_2 - \vec{v}_2) \cdot \nabla(\phi - \psi)\, dx = 0$. This means $\int |\vec{u}_2 - \vec{v}_2|^2\, dx = 0$, and hence $\vec{u}_2 \equiv \vec{v}_2$; but this also implies $\vec{u}_1 \equiv \vec{v}_1$, so uniqueness is established. ♠

Exercises for Section 4.5

1. Let $K(x) = -(4\pi |x|)^{-1}$, $f \in C(\mathbf{R}^3)$ with $f(x) = O(|x|^{-2-\epsilon})$ as $x \to \infty$, and $u(x) = \int_{\mathbf{R}^3} K(x - y) f(y)\, dy$.
   (a) Show $u \in C^1(\mathbf{R}^n)$ and $\nabla u(x) = O(|x|^{-2})$ as $|x| \to \infty$.
   (b) If additionally $f \in C^1(\mathbf{R}^3)$, show $u \in C^2(\mathbf{R}^3)$.

Further References for Chapter 4

For additional details about separation of variables and the resultant Sturm-Liouville theory for ODEs, see [Churchill] or [Pinsky]. For more on eigenfunction expansions, including issues of orthogonality, completeness, and convergence, see [Courant-Hilbert, vol. I] or Chapter 7 of this book.

For $n = 2$, the theory of the Laplace and Poisson equations are closely allied with complex analysis and conformal mappings; a good reference is [Nehari]. For more about potential theory and electrostatics, see [Kellogg] or [Wermer]; for applications of potential theory to magnetic fields, Newtonian gravity, and incompressible fluid flow, see [Kellogg]. A very thorough treatment of modern potential theory is given in [Landkof].

5 The Heat Equation

In this chapter we study the equation that governs the conduction of heat through a body. We begin with the heat equation in a bounded domain, using the eigenfunctions of the Laplacian to study existence of a solution and a maximum principle to study uniqueness of the solution. Next we study the pure initial value problem, discussing the role of the Fourier transform in $\mathbf{R}^n$, deriving the heat kernel, and exploring its role as a fundamental solution. We also discuss the regularity of solutions of the heat equation and the similarity method for constructing special solutions of the heat equation and other equations having sufficient symmetry. Finally, we give two applications to fluid dynamics.

5.1 The Heat Equation in a Bounded Domain

If Ω is a bounded domain in $\mathbf{R}^n$ that we view as a physical body with constant heat conductivity k, then the *heat* or *diffusion equation*

$$(1) \qquad u_t = k\Delta u \qquad \text{for } x \in \Omega \text{ and } t > 0$$

governs the *propagation* or *diffusion of heat*, where $u(x,t)$ represents the temperature of the body at the point x and time t (cf. Section A.1b in the Appendix). We generally take $k = 1$ in (1) since otherwise this may be achieved by rescaling the time variable: $t \to kt$. The appropriate side conditions are the initial temperature

$$(2) \qquad u(x,0) = g(x)$$

and some sort of boundary condition: Dirichlet if the temperature is controlled on the boundary $\partial\Omega$, Neumann if the heat flow across $\partial\Omega$ is controlled, or Robin if the heat flow obeys Newton's law of cooling (cf. Section 4.1b).

In this section we discuss the existence and uniqueness of solutions to these initial/boundary value problems.

a. Existence by Eigenfunction Expansion

For an initial/boundary value problem, the solution may be found explicitly in the form of an eigenfunction expansion, involving the eigenfunctions of the Laplacian for the appropriate boundary conditions.

Section 5.1: The Heat Equation in a Bounded Domain   143

Example 1. Consider the initial value problem with homogeneous Dirichlet condition:

(3)
$$\begin{cases} u_t = \Delta u & \text{for } x \in \Omega \text{ and } t > 0 \\ u(x,0) = g(x) & \text{for } x \in \overline{\Omega} \\ u(x,t) = 0 & \text{for } x \in \partial\Omega \text{ and } t > 0. \end{cases}$$

If we assume the initial temperature distribution g is in $C^2(\overline{\Omega})$ with $g = 0$ on $\partial\Omega$, then we may write

(4) $$g(x) = \sum_{n=1}^{\infty} a_n \phi_n(x), \quad \text{where } a_n = \int_{\Omega} g(x)\,\phi_n(x)\,dx.$$

In (4), (λ_n, ϕ_n) denote the eigenvalues and normalized eigenfunctions of the Laplacian on Ω with Dirichlet boundary conditions that we discussed in Section 4.4; by the Remark in Section 4.4a, the series in (4) converges absolutely and uniformly on $\overline{\Omega}$.

Let us write $u(x,t) = \sum_{n=1}^{\infty} u_n(t)\phi_n(x)$ and insert this expression in the heat equation. Assuming we can pass derivatives through the summation, the coefficients of ϕ_n yield $u'_n(t) + \lambda_n u_n(t) = 0$, which has the general solution $u_n(t) = A_n e^{-\lambda_n t}$. At $t = 0$ we find $u_n(0) = A_n$, so $u(x,0) = \sum A_n \phi_n(x)$ and comparison with (4) shows $a_n = A_n$. Thus the solution of (3) is given by

(5) $$u(x,t) = \sum_{n=1}^{\infty} a_n e^{-\lambda_n t} \phi_n(x),$$

where the a_n are given in (4). Notice that the series (5) converges absolutely and uniformly on $\overline{\Omega} \times [0,\infty)$ since the factor $e^{-\lambda_n t}$ only improves the convergence of (4); this justifies the interchange of differentiation and summation that we performed in the derivation.

If we combine (4) and (5) and interchange the integration and summation, we obtain

(6) $$u(x,t) = \int_{\Omega} \sum_{n=1}^{\infty} e^{-\lambda_n t} \phi_n(x)\phi_n(y) g(y)\,dy.$$

This means that, if we formally define

(7) $$K(x,y,t) = \sum_{n=1}^{\infty} e^{-\lambda_n t} \phi_n(x)\phi_n(y),$$

we can write the solution as an integral transform of the initial condition:

(8) $$u(x,t) = \int_{\Omega} K(x,y,t)\,g(y)\,dy.$$

The integral kernel $K(x,y,t)$ is called the *heat kernel*. ♣

The method of eigenfunction expansion may also be applied to cover more general problems than (3), namely

(9) $$\begin{cases} u_t = \Delta u + f(x,t) & \text{for } x \in \Omega \text{ and } t > 0 \\ u(x,0) = g(x) & \text{for } x \in \overline{\Omega} \\ u(x,t) = h(x,t) & \text{for } x \in \partial\Omega \text{ and } t > 0, \end{cases}$$

in which $f(x,t)$ represents a forcing term, such as a heat source, and $h(x,t)$ represents a nonhomogeneous temperature control on $\partial\Omega$. This is discussed in Exercises 2, 3, and 4. It is also possible to replace the Dirichlet boundary condition $u(x,t) = h(x,t)$ by a Neumann or Robin condition, provided we replace (λ_n, ϕ_n) by the eigenvalues and eigenfunctions for the appropriate boundary value problem (cf. Exercise 5).

Let us next consider the uniqueness of solutions.

b. The Maximum Principle and Uniqueness

The heat equation resembles the Laplace equation in that its solutions satisfy a maximum principle. To formulate this principle, let us introduce the "cylinder" $U = U_T = \Omega \times (0,T)$. We know from Chapter 4 that harmonic (and subharmonic) functions achieve their maximum on the boundary of the domain. For the heat equation, the result is improved in that the maximum is achieved on a certain part of the boundary, which we call the *parabolic boundary*:

(10) $$\Gamma = \{(x,t) \in \overline{U} : x \in \partial\Omega \quad \text{or} \quad t = 0\}$$

(see Figure 1). Let us also denote by $C^{2;1}(U)$ functions satisfying $u_t, u_{x_i x_j} \in C(U)$.

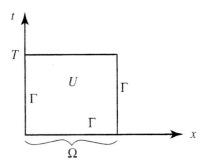

Figure 1. The cylinder U and its parabolic boundary Γ.

Theorem 1 (Weak Maximum Principle). Let $u \in C^{2;1}(U) \cap C(\overline{U})$ satisfy $\Delta u \geq u_t$ in U. Then u achieves its maximum on the parabolic boundary of U:

(11) $$\max_{(x,t) \in \overline{U}} u(x,t) = \max_{(x,t) \in \Gamma} u(x,t).$$

Proof. We proceed in two steps. First, assume $\Delta u > u_t$ in U. For $0 < \tau < T$ consider

$$U_\tau = \Omega \times (0, \tau), \qquad \Gamma_\tau = \{(x,t) \in \overline{U}_\tau : x \in \partial\Omega \text{ or } t = 0\}.$$

If the maximum of u on $\overline{U}_\tau$ occurs at $x \in \Omega$ and $t = \tau$, then $u_t(x, \tau) \geq 0$ and $\Delta u(x, \tau) \leq 0$, violating our assumption; similarly, u cannot attain an interior maximum on U_τ. Hence (11) holds for U_τ: $\max_{\overline{U}_\tau} u = \max_{\Gamma_\tau} u$. But $\max_{\Gamma_\tau} u \leq \max_\Gamma u$ and by continuity of u, $\max_{\overline{U}} u = \lim_{\tau \to T} \max_{\overline{U}_\tau} u$. This establishes (11).

Second, we consider the general case of $\Delta u \geq u_t$ in U. Introduce $v = u - kt$ for $k > 0$. Notice that $v \leq u$ on $\overline{U}$ and $\Delta v - v_t = \Delta u - u_t + k > 0$ in U. Thus we may apply (11) to v:

$$\max_{\overline{U}} u = \max_{\overline{U}}(v + kt) \leq \max_{\overline{U}} v + kT = \max_\Gamma v + kT \leq \max_\Gamma u + kT.$$

Letting $k \to 0$ establishes (11) for u. ♠

By applying the maximum principle to the difference of two solutions of (9), we obtain the following.

Corollary (Uniqueness). *If $u, v \in C^{2;1}(U) \cap C(\overline{U})$ both satisy (9), then $u \equiv v$.*

The weak maximum principle also shows that the solution of (9) *depends continuously on the data* (i.e., on the initial and boundary values). Thus the initial/boundary value problem (9) is *well posed*.

When a Neumann condition or Robin condition replaces the Dirichlet condition in (9), then an appropriate expansion may be available for existence (see Exercise 5); but uniqueness requires a maximum principle that considers the behavior of $\partial u/\partial \nu$ when a maximum is achieved at $x \in \partial\Omega$; this is the *strong parabolic maximum principle,* which is discussed in Section 11.1.

Exercises for Section 5.1

1. A square two-dimensional plate of side length a is heated to a uniform temperature $U_0 > 0$. Then at $t = 0$ all sides are reduced to zero temperature. (Assume that the top and bottom of the plate are insulated so that the heat flow can be restricted to two dimensions.) Describe the heat diffusion $u(x, y, t)$.

2. Describe how you would solve the initial/boundary value problem (3) with the homogeneous Dirichlet condition replaced by $u(x, t) = h(x)$ for $x \in \partial\Omega$ and $t > 0$. What happens to $u(x, t)$ as $t \to \infty$?

3. Suppose $f(x, t) = \sum_{n=1}^\infty a_n(t)\phi_n(x)$ converges absolutely and uniformly for $(x, t) \in \overline{U}_T$ for every $T > 0$. Let $g(x) \equiv 0$ and $h(x, t) \equiv 0$, and describe the solution of (9).

4. More generally than in Exercise 2, describe how you would solve the initial boundary value problem (3) with the homogeneous Dirichlet condition replaced by $u(x,t) = h(x,t)$ for $x \in \partial\Omega$ and $t > 0$.

5. Suppose the square plate of Exercise 1 is given an initial temperature distribution $u(x,y,0) = g(x,y)$ and then insulated so that heat cannot flow across $\partial\Omega$ (i.e., $\partial u/\partial \nu = 0$). Use an expansion in appropriate eigenfunctions to describe the heat diffusion $u(x,y,t)$.

6. Suppose $U = \Omega \times (0,T)$ and $u \in C^{2;1}(U) \cap C(\overline{U})$ satisfies $u_t \leq \Delta u + cu$ in U, where $c \leq 0$ is a constant. If $u \geq 0$, show that (11) holds for u. Give a counterexample without the condition $u \geq 0$.

7. If u satisfies (1), define its *heat energy* by $\mathcal{E}(t) = \int_\Omega u^2(x,t)\,dx$.
   (a) If $U = \Omega \times (0,\infty)$ and $u \in C^{2;1}(\overline{U})$ satisfies (1) and either (i) $u = 0$ on $\partial\Omega$, or (ii) $\partial u/\partial \nu = 0$ on $\partial\Omega$, then $\mathcal{E}(t)$ is nonincreasing in t.
   (b) Use (a) to conclude the uniqueness of a solution $u \in C^{2;1}(\overline{U})$ for either the nonhomogeneous Dirichlet problem (9) or the corresponding nonhomogeneous Neumann problem.

5.2 The Pure Initial Value Problem

Instead of restricting the x-variables to lie in a bounded domain Ω, suppose we consider the pure initial value problem

(12) $$\begin{cases} u_t = \Delta u & \text{for } t > 0, x \in \mathbf{R}^n \\ u(x,0) = g(x) & \text{for } x \in \mathbf{R}^n. \end{cases}$$

Although of limited significance for the diffusion of heat, (12) is a natural first-order time evolution that involves second-order spatial derivatives. The natural tool to use is the Fourier transform in the x variables: we find that this enables us to define the *heat kernel*, which enables us to solve (12) by an integral transform. The heat kernel will also be identified as the fundamental solution of the heat equation as defined in Section 2.3. (For the purpose of using the Fourier transform, throughout this section we assume that all functions are complex valued.)

a. Fourier Transform

If $g \in C_0^\infty(\mathbf{R}^n)$, we define its *Fourier transform* $\widehat{g}$ by

(13) $$\widehat{g}(\xi) = \frac{1}{(2\pi)^{n/2}} \int_{\mathbf{R}^n} e^{-ix\cdot\xi} g(x)\,dx \quad \text{for} \quad \xi \in \mathbf{R}^n.$$

The function $\widehat{g}(\xi)$ is well defined since the integral converges by the compactness of the support of g. Moreover, we can differentiate $\widehat{g}$:

$$\frac{\partial}{\partial \xi_j}\widehat{g}(\xi) = \frac{1}{(2\pi)^{n/2}} \int_{\mathbf{R}^n} e^{-ix\cdot\xi}(-ix_j)g(x)\,dx,$$

which is well defined since $(-ix_j)g(x)$ also has compact support; in fact, $\widehat{g}$ is C^1. Iterating this computation, we obtain $\widehat{g}$ is C^∞ and

(14a) $$\left(\frac{\partial}{\partial \xi_j}\right)^k \widehat{g}(\xi) = [\widehat{(-ix_j)^k g}](\xi).$$

Similarly, integrating by parts shows

(14b) $$\left(\widehat{\frac{\partial^k g}{\partial x_j^k}}\right)(\xi) = (i\xi_j)^k \widehat{g}(\xi).$$

The formulas (14a,b) express the fact that *Fourier transform interchanges differentiation and multiplication by the coordinate function.*

Notice that the Fourier transform $\widehat{g}$ is also well defined when $g \in L^1(\mathbf{R}^n)$. However, the properties (14a,b) make it very natural to consider the Fourier transform on a subspace of $L^1(\mathbf{R}^n)$ called the *Schwartz class functions*, $\mathcal{S}$, which are defined to be the smooth functions whose derivatives of all orders decay faster than any polynomial, i.e.,

$$\mathcal{S} = \{u \in C^\infty(\mathbf{R}^n) : \text{ for every } k \in \mathbf{N} \text{ and } \alpha \in \mathbf{N}^n,$$
$$|x|^k |D^\alpha u(x)| \text{ is bounded on } \mathbf{R}^n\}.$$

For $g \in \mathcal{S}$, the Fourier transform $\widehat{g}$ exists since g decays rapidly at ∞.

Lemma. *(i) If $g \in L^1(\mathbf{R}^n)$, then $\widehat{g}$ is bounded. (ii) If $g \in \mathcal{S}$, then $\widehat{g} \in \mathcal{S}$.*

Proof. For $g \in L^1(\mathbf{R}^n)$, we immediately see that $\widehat{g}$ is bounded, since

$$|\widehat{g}(\xi)| \le \frac{1}{(2\pi)^{n/2}} \int_{\mathbf{R}^n} |g(x)|\, dx = M < \infty.$$

For $g \in \mathcal{S}$, we can still differentiate (13) under the integral sign to conclude that $\widehat{g} \in C^\infty$ and (14a) holds. Similarly, we find that (14b) holds for $g \in \mathcal{S}$. Combining (14a) and (14b) shows $|(D_x^\alpha x^\beta g)\widehat{}(\xi)| = |\xi^\alpha D_\xi^\beta \widehat{g}(\xi)|$. Since $D_x^\alpha x^\beta g \in L^1(\mathbf{R}^n)$, this means by (i) that $\xi^\alpha D_\xi^\beta \widehat{g}$ is bounded for $\xi \in \mathbf{R}^n$; in other words, $\widehat{g} \in \mathcal{S}$. ♠

We may also define the *inverse Fourier transform* for $g \in L^1(\mathbf{R}^n)$:

(15) $$g^\vee(\xi) = \frac{1}{(2\pi)^{n/2}} \int_{\mathbf{R}^n} e^{ix\cdot\xi} g(x)\, dx \quad \text{for} \quad \xi \in \mathbf{R}^n.$$

By the lemma, if $g \in \mathcal{S}$, then $\widehat{g} \in \mathcal{S}$, so we can define $(\widehat{g})^\vee$, and a proof similar to that of (ii) in the lemma shows that $(\widehat{g})^\vee \in \mathcal{S}$. We might wonder, especially given the name "inverse Fourier transform," about the relationship of g to $(\widehat{g})^\vee$. The proof of the following result may be found in most books on functional analysis (e.g., [Stein-Weiss] or [Yoshida]).

Fourier Inversion Theorem. If $g \in \mathcal{S}$, then $(\widehat{g})^\vee = g$; that is,

$$g(x) = \frac{1}{(2\pi)^{n/2}} \int_{\mathbf{R}^n} e^{ix \cdot \xi} \widehat{g}(\xi)\, d\xi = \frac{1}{(2\pi)^n} \int \int_{\mathbf{R}^{2n}} e^{i(x-y)\cdot \xi} g(y)\, dy\, d\xi.$$

Remark. The Fourier inversion theorem holds for more functions than those in $\mathcal{S}$, but it does *not* hold for general functions in $L^1(\mathbf{R}^n)$. In many applications, including the one in the next subsection, it suffices to apply the theorem on $\mathcal{S}$ or even $C_0^\infty(\mathbf{R}^n)$.

b. Solution of the Pure Initial Value Problem

Let us consider the pure initial value problem (12) where we first assume $g \in C_0^\infty(\mathbf{R}^n)$. The idea is to take the Fourier transform of the heat equation in the x-variables. Of course, we do not know at this point that $u(x,t)$ is nice enough for the integrals to converge, so we must stress that we are only operating *formally* in order to obtain a formula for the solution; once the formula is obtained it must be verified that it indeed solves the problem. Formally, we obtain

$$\widehat{(u_t)}(\xi, t) = \frac{1}{(2\pi)^{n/2}} \int_{\mathbf{R}^n} e^{-ix\cdot\xi} u_t(x,t)\, dx = \frac{\partial}{\partial t}\widehat{u}(\xi, t)$$

and $\quad \widehat{\Delta u}(\xi, t) = \sum_{j=1}^n (i\xi_j)^2 \widehat{u}(\xi, t) = -|\xi|^2 \widehat{u}(\xi, t),$

where we have treated t as a parameter in the first formula and used (14b) in the second formula. The heat equation therefore becomes

(16) $$\frac{\partial}{\partial t}\widehat{u}(\xi, t) = -|\xi|^2\, \widehat{u}(\xi, t),$$

which may be thought of as an ordinary differential equation in t. Clearly the solution is $\widehat{u}(\xi, t) = Ce^{-|\xi|^2 t}$, where the constant C may be determined by transforming the initial condition: $\widehat{u}(\xi, 0) = \widehat{g}(\xi)$. Hence $\widehat{u}(\xi, t) = \widehat{g}(\xi) e^{-|\xi|^2 t}$, and we can recover u itself by applying the inverse Fourier transform:

$$u(x, t) = \frac{1}{(2\pi)^{n/2}} \int_{\mathbf{R}^n} e^{ix\cdot\xi} \widehat{g}(\xi) e^{-|\xi|^2 t}\, d\xi$$

$$= \frac{1}{(2\pi)^n} \int_{\mathbf{R}^{2n}} e^{i(x-y)\cdot\xi - |\xi|^2 t}\, g(y)\, dy\, d\xi$$

$$= \frac{1}{(2\pi)^n} \int_{\mathbf{R}^n} \left(\int_{\mathbf{R}^n} e^{i(x-y)\cdot\xi - |\xi|^2 t}\, d\xi \right) g(y)\, dy.$$

Now if we can evaluate the integral

(17) $$K(x, y, t) = \frac{1}{(2\pi)^n} \int_{\mathbf{R}^n} e^{i(x-y)\cdot\xi - |\xi|^2 t}\, d\xi$$

Section 5.2: The Pure Initial Value Problem

then we will have represented our solution as an integral transform of the initial condition.

Let us first consider the special case $n = 1$, where we must evaluate

$$\int_{-\infty}^{\infty} e^{i(x-y)\xi - \xi^2 t} \, d\xi.$$

By contour integration in the complex plane, this integral gives the same answer as integrating along $\xi = ai + \zeta$ where $-\infty < \zeta < \infty$ and $a \in \mathbf{R}$ is fixed. Since x, y, t are fixed, we may let $a = (x - y)/2t$ and introduce the change of variables $\zeta = \eta/\sqrt{t}$: $d\zeta = \frac{1}{\sqrt{t}} d\eta$ and

$$i(x-y)\xi - \xi^2 t = i(x-y)\left(\frac{x-y}{2t}i + \frac{\eta}{\sqrt{t}}\right) - \left(\frac{x-y}{2t}i + \frac{\eta}{\sqrt{t}}\right)^2 t$$

$$= -\frac{(x-y)^2}{4t} - \eta^2.$$

Therefore,

$$\frac{1}{2\pi}\int_{-\infty}^{\infty} e^{i(x-y)\xi - \xi^2 t}\, d\xi = \frac{e^{-(x-y)^2/4t}}{2\pi\sqrt{t}} \int_{-\infty}^{\infty} e^{-\eta^2}\, d\eta = \frac{e^{-(x-y)^2/4t}}{\sqrt{4\pi t}}.$$

Since the integral over $\mathbf{R}^n$ in (17) can be expressed as a product of one-dimensional integrals, we obtain

(18) $$K(x, y, t) = \frac{1}{(4\pi t)^{n/2}} e^{-\frac{|x-y|^2}{4t}}.$$

The function $K(x, y, t)$ is called the *Gaussian* or *heat kernel*. Having computed $K(x, y, t)$, we can use it to solve (12) for more general initial values than $g \in C_0^\infty(\mathbf{R}^n)$. The proof of the following result is similar to that for the Dirichlet problem on a half-space and is outlined in Exercise 1.

Theorem 1. *If $g(x)$ is bounded and continuous for $x \in \mathbf{R}^n$, then*

(19) $$u(x,t) = \int_{\mathbf{R}^n} K(x, y, t) g(y)\, dy = \frac{1}{(4\pi t)^{n/2}} \int_{\mathbf{R}^n} e^{-\frac{|x-y|^2}{4t}} g(y)\, dy$$

is a C^∞-function satisfying $u_t = \Delta u$ for $x \in \mathbf{R}^n$ and $t > 0$ and extends continuously to $t \geq 0$ such that $u(x, 0) = g(x)$.

This theorem establishes the *existence* of a solution to the pure initial value problem, but *uniqueness fails*: there are nontrivial solutions of (12) with $g = 0$ (see Exercise 4). Thus the pure initial value problem for the heat equation is *not* well posed, as it was for the wave equation. However, the nontrivial solutions are unbounded as functions of x when $t > 0$ is fixed; uniqueness can be regained by adding a boundedness condition on the solution (see Exercise 5b).

Another distinguishing feature of the solution (19) is the fact that $u(x,t)$ for *any* $t > 0$ depends on the initial values $g(x)$ for all $x \in \mathbf{R}^n$. This means that initial conditions display *infinite propagation speed* in their effect on the solution. This contrasts with the finite propagation speed, which applies to solutions of the wave equation.

c. The Fundamental Solution

In this subsection we use the distribution theory of Section 2.3 to see that the heat kernel defined in the previous subsection may be realized as the fundamental solution of the heat operator. As such, it is useful in solving the nonhomogeneous heat equation.

To begin with, observe that the heat kernel $K(x, y_0, t)$ of (18) may be thought of as the temperature at (x, t) caused by an initial burst of heat at $(y_0, 0)$. If we replace $g(y)$ by the Dirac distribution $\delta_{y_0}(y)$ in (19), then formally

$$u(x,t) = \int_{\mathbf{R}^n} K(x, y, t) \delta_{y_0}(y) \, dy$$

$$= \frac{1}{(4\pi t)^{n/2}} \int_{\mathbf{R}^n} e^{-\frac{|x-y|^2}{4t}} \delta_{y_0}(y) \, dy = K(x, y_0, t).$$

In particular, let us take $y_0 = 0$ and define

(20) $$\widetilde{K}(x,t) \equiv K(x,0,t) \equiv \frac{1}{(4\pi t)^{n/2}} e^{-\frac{|x|^2}{4t}}.$$

As $t \to 0$ the temperature distribution should tend to its initial value, so we expect

(21) $$\lim_{t \to 0^+} \widetilde{K}(x,t) = \delta_0(x) = \delta(x)$$

as distributions on $\mathbf{R}^n$ [i.e., for all $v \in C_0^\infty(\mathbf{R}^n)$ we expect $\langle \widetilde{K}(\cdot, t), v \rangle = \int K(x,0,t) v(x) dx \to \langle \delta, v \rangle = v(0)$]. But this is an immediate consequence of Theorem 1 applied to $g = v$. This means that $\widetilde{K}(x,t)$ is the *fundamental solution of the initial value problem* as defined in (62) of Section 2.3; in fact, (19) of Theorem 1 is an instance of (63) in Section 2.3.

Another way of viewing the situation is to extend $\widetilde{K}(x,t)$ by zero to $t \leq 0$:

(22) $$\widetilde{K}(x,t) = \begin{cases} \dfrac{1}{(4\pi t)^{n/2}} e^{-\frac{|x|^2}{4t}} & \text{for } t > 0 \\ 0 & \text{for } t \leq 0. \end{cases}$$

Notice that $\widetilde{K}$ is smooth for $(x,t) \neq (0,0)$ (see Exercise 6). The following shows that $\widetilde{K}$ is the fundamental solution of the "free space" heat equation, as defined in the Remark in Section 2.3d.

Proposition. *As distributions in $\mathbf{R}^{n+1}$ we have*

(23) $$(\partial_t - \Delta)\widetilde{K}(x,t) = \delta(x,t),$$

where $\delta(x,t)$ is the Dirac distribution with singularity at $x=0, t=0$.

Proof. To verify (23) as distributions, we must show that for any $\phi \in C_0^\infty(\mathbf{R}^{n+1})$

(24) $$\int_{\mathbf{R}^{n+1}} \widetilde{K}(x,t)(-\partial_t - \Delta)\phi \, dx \, dt = \int_{\mathbf{R}^{n+1}} \delta(x,t)\,\phi(x,t)\, dx\, dt \equiv \phi(0,0).$$

To do this, let us take $\epsilon > 0$ and define

(25) $$\widetilde{K}_\epsilon(x,t) = \begin{cases} \dfrac{1}{(4\pi t)^{n/2}} e^{-\frac{|x|^2}{4t}} & \text{for } t > \epsilon \\ 0 & \text{for } t \leq \epsilon. \end{cases}$$

Then $\widetilde{K}_\epsilon \to \widetilde{K}$ as distributions (cf. Exercise 9 in Section 2.3), so it suffices to show $(\partial_t - \Delta)\widetilde{K}_\epsilon \to \delta$ as distributions. Now

(26) $$\int \widetilde{K}_\epsilon(-\partial_t - \Delta)\phi \, dx\, dt = \int_\epsilon^\infty \left(\int_{\mathbf{R}^n} \widetilde{K}(x,t)(-\partial_t - \Delta)\phi(x,t)\, dx \right) dt$$
$$= \int_\epsilon^\infty \left(\int_{\mathbf{R}^n} (\partial_t - \Delta)\widetilde{K}(x,t)\phi(x,t)\, dx \right) dt + \int_{\mathbf{R}^n} \widetilde{K}(x,\epsilon)\,\phi(x,\epsilon)\, dx.$$

But for $t > \epsilon$, $(\partial_t - \Delta)\widetilde{K}(x,t) = 0$; moreover, by (21), $\widetilde{K}(x,\epsilon) \to \delta_0(x)$ as $\epsilon \to 0$, so the last integral in (26) tends to $\phi(0,0)$ as desired. ♠

d. The Nonhomogeneous Equation

Let us add a forcing term $f(x,t)$ to the heat equation and attempt to solve the pure initial value problem with homogeneous initial condition:

(27) $$\begin{cases} u_t = \Delta u + f(x,t) & \text{for } x \in \mathbf{R}^n,\ t > 0 \\ u(x,0) = 0 & \text{for } x \in \mathbf{R}^n. \end{cases}$$

Duhamel's principle (cf. Section 2.3 and Section 3.1) suggests we find the solution of (27) in the form

(28) $$u(x,t) = \int_0^t \int_{\mathbf{R}^n} \widetilde{K}(x-y, t-s)\,f(y,s)\, dy\, ds,$$

provided this integral makes sense. In fact, we might at least hope that (28) defines a distribution or, even better, a *weak solution* of (27); this naturally means that $u \in L^1_{loc}(\mathbf{R}^n \times [0,\infty))$ satisfies

(29) $$\int_0^\infty \int_{\mathbf{R}^n} (uv_t + u\Delta v + fv)\, dx\, dt = 0 \quad \text{for all } v \in C_0^\infty(\mathbf{R}^{n+1}).$$

(See Exercise 8.)

Theorem 2. *Suppose $f(x,t)$ is continuous and bounded on $\mathbf{R}^n \times [0,T]$ for every fixed $T > 0$. Then (28) defines a continuous weak solution of (27).*

Proof. Let $\widetilde{f}(x,t)$ be $f(x,t)$ extended by zero for $t < 0$. Then (28) just states $u = \widetilde{K} \star \widetilde{f} = \int_{-\infty}^{\infty} \int_{\mathbf{R}^n} \widetilde{K}(x-y, t-s)\widetilde{f}(y,s)\,dy\,ds$, so $u_t - \Delta u = \widetilde{f}$ as distributions in $\mathbf{R}^{n+1}$. But $u(x,t) \equiv 0$ for $t < 0$, so this implies (29). It only remains to show that u is continuous. For s and t satisfying $0 \leq s < t$, let us define $w(x,t,s) = \int_{\mathbf{R}^n} K(x,y,t-s)f(y,s)\,dy$. We can apply Theorem 1 with $g(y) = f(y,s)$ to conclude that $w(x,t,s)$ is continuous in $x \in \mathbf{R}^n$ and $0 \leq s \leq t$. Therefore,

$$u(x,t) = \int_0^t w(x,t,s)\,ds$$

is continuous in $(x,t) \in \mathbf{R}^{n+1}$. ♠

Increased regularity of $f(x,t)$ guarantees increased regularity of $u(x,t)$. The key step here is to observe that

(30)
$$\begin{aligned}\partial_{x_i}\widetilde{K}(x-y,t-s) &= -\partial_{y_i}\widetilde{K}(x-y,t-s) \\ \partial_t \widetilde{K}(x-y,t-s) &= -\partial_s \widetilde{K}(x-y,t-s).\end{aligned}$$

These formulas, together with integration by parts, yield the following result (see Exercise 9).

Theorem 3. *If f, f_t, f_{x_i}, and $f_{x_i x_j}$ are all continuous and bounded in $\mathbf{R}^n \times [0,T]$ for every fixed $T > 0$, then (28) defines a classical solution $u \in C^{2;1}(\mathbf{R}^n \times (0,\infty)) \cap C(\mathbf{R}^n \times [0,\infty))$ of (27).*

Remark. Theorem 3 remains true if we only assume that f and f_{x_i} are continuous and bounded in $\mathbf{R}^n \times [0,T]$ for every fixed $T > 0$; see Exercise 9.

Combining (19) and (27), we find that the solution of the initial value problem

(31)
$$\begin{cases} u_t = \Delta u + f(x,t) & \text{for } x \in \mathbf{R}^n, t > 0 \\ u(x,0) = g(x) & \text{for } x \in \mathbf{R}^n \end{cases}$$

is given by

(32) $\quad u(x,t) = \displaystyle\int_{\mathbf{R}^n} \widetilde{K}(x-y,t)g(y)\,dy + \int_0^t \int_{\mathbf{R}^n} \widetilde{K}(x-y,t-s)f(y,s)\,dy\,ds.$

Exercises for Section 5.2

1. *Proof of Theorem 1.* Let $K(x,y,t)$ be the Gaussian kernel (18).
   (a) Show that $K(x,y,t)$ is C^∞ and satisfies $(\partial_t - \Delta_x)K(x,y,t) = 0$ for $x,y \in \mathbf{R}^n$ and $t > 0$.
   (b) Prove that for every $x \in \mathbf{R}^n$ and $t > 0$, $\int_{\mathbf{R}^n} K(x,y,t)\,dy = 1$.
   (c) For any $\delta > 0$, show that
   $$\lim_{t \to 0^+} \int_{|x-y|>\delta} K(x,y,t)\,dy = 0$$
   uniformly for $x \in \mathbf{R}^n$.
   (d) Use (a)–(c) and $g \in C_B(\mathbf{R}^n)$ to show $u(x,t) \to g(x)$ as $t \to 0^+$, proving the theorem.

2. Let $g(x)$ be bounded and continuous for $x \in \mathbf{R}^n$ and define u by (19).
   (a) Show $|u(x,t)| \leq \sup\{|g(y)| : y \in \mathbf{R}^n\}$.
   (b) If, in addition, $\int_{\mathbf{R}^n} |g(y)|\,dy < \infty$, show that $\lim_{t \to \infty} u(x,t) = 0$ uniformly in $x \in \mathbf{R}^n$.

3. Let $g(x) \in C^k(\mathbf{R}^n)$ with $D^\alpha g$ uniformly bounded on $\mathbf{R}^n$ for each $|\alpha| \leq k$. Show that u defined by (19) satisfies $u \in C^k(\mathbf{R}^n \times [0,\infty))$.

4. Formally check that
   $$u(x,t) = \sum_{k=0}^\infty \frac{1}{(2k)!} x^{2k} \frac{d^k}{dt^k} e^{-1/t^2}$$
   satisfies $u_t = u_{xx}$ for $t > 0$, $x \in \mathbf{R}$ with $u(x,0) = 0$ for $x \in \mathbf{R}$. (Proving the convergence of the infinite series is not easy; see [John, 1] or [Widder].)

5. (a) Prove the following weak maximum principle for the heat equation in $U = \mathbf{R}^n \times (0,T)$: Let u be bounded and continuous on $\overline{U} = \mathbf{R}^n \times [0,T]$ with $u_t, u_{x_i x_j} \in C(U)$ and $u_t - \Delta u \leq 0$ in U. Then
   $$M \equiv \sup_{(x,t) \in U} u(x,t) = \sup_{x \in \mathbf{R}^n} u(x,0) \equiv m.$$
   (b) Use the maximum principle of (a) to show that the solution (19) is the unique solution that is bounded in $\mathbf{R}^n \times [0,T]$.

6. For $|x| \neq 0$, show that (22) is smooth (C^∞) in $t \in \mathbf{R}$. Conclude that (22) is smooth for $(x,t) \neq (0,0)$.

7. Heat conduction in a semi-infinite rod with initial temperature $g(x)$ leads to the equations
   $$\begin{cases} u_t = u_{xx} & \text{for } x > 0, t > 0 \\ u(x,0) = g(x) & \text{for } x > 0. \end{cases}$$

Assume that g is continuous and bounded for $x \geq 0$.
(a) If $g(0) = 0$ and the rod has its end maintained at zero temperature, then we must include the boundary condition $u(0,t) = 0$ for $t > 0$. Find a formula for the solution $u(x,t)$.
(b) If the rod has its end insulated so that there is no heat flow at $x = 0$, then we must include the boundary condition $u_x(0,t) = 0$ for $t > 0$. Find a formula for the solution $u(x,t)$. Do you need to require $g'(0) = 0$?

8. If $u \in C^{2,1}(\mathbf{R}^n \times [0, \infty))$ and $f \in C(\mathbf{R}^{n+1})$, show that (29) is equivalent to (27).

9. (a) Use (30) to prove Theorem 3.
(b) Verify the following elementary fact: If $0 < \alpha < 1$ and $\beta \geq 0$, then there exists a constant $M = M(\alpha, \beta) > 0$ so that $z^\beta e^{-z} \leq M e^{-\alpha z}$ for all $z \geq 0$.
(c) Use (b) to verify the following estimates:

$$\partial_t \widetilde{K}(x,t) \leq \frac{M_1}{t} \widetilde{K}(x, 2t), \qquad \partial_{x_i} \widetilde{K}(x,t) \leq \frac{M_2}{\sqrt{t}} \widetilde{K}(x, 2t),$$

$$\partial_{x_i x_j} \widetilde{K}(x,t) \leq \frac{M_3}{t} \widetilde{K}(x, 2t).$$

(d) Use (30) and (c) to confirm the Remark following Theorem 3.

10. If $f(x,t)$ is continuous and bounded on $\mathbf{R}^{n+1}$ and vanishes for t sufficiently negative [i.e., $f(x,t) = 0$ for $t \leq \tau < 0$], show that the convolution $u = \widetilde{K} \star f$ defines a continuous weak solution of the nonhomogeneous heat equation $u_t = \Delta u + f(x,t)$ in $\mathbf{R}^{n+1}$.

11. Find a formula for the solution of the initial value problem

$$\begin{cases} u_t = \Delta u - u & \text{for } t > 0, x \in \mathbf{R}^n \\ u(x,0) = g(x) & \text{for } x \in \mathbf{R}^n, \end{cases}$$

where g is continuous and bounded. Is the solution bounded? Is it the only bounded solution?

5.3 Regularity and Similarity

In this section we discuss two important features of the heat equation that are independent of the initial/boundary values. The first is the smoothness of solutions. In the previous section we saw that the solution of the pure initial value problem (12) is C^∞ for $t > 0$, even when the initial value $u(x,0) = g(x)$ is only continuous. This suggests that *any* solution of $u_t = \Delta u$ is C^∞ in x and t; we confirm this. The second feature of the heat equation we consider is a scale invariance property of solutions. This leads to the similarity method for constructing special solutions.

a. Smoothness of Solutions

We would like to show that *any* solution of

(33) $$u_t = \Delta u \quad \text{in} \quad U \subset \mathbf{R}^{n+1}$$

is in fact a smooth solution. We could start with $u \in C^{2;1}(U)$ satisfying (33) in the classical sense, but in fact we need only assume that u is a distribution solution of (33).

Theorem. *If u is a distribution solution of (33), then $u \in C^\infty(U)$.*

Proof. Pick $\xi = (x_0, t_0) \in U$ and $\epsilon > 0$ so small that the ball $B_{4\epsilon}(\xi) \subset U$. Choose $\phi \in C_0^\infty(B_{4\epsilon}(\xi))$ with $\phi \equiv 1$ on $B_{3\epsilon}(\xi)$. Let $w = \phi u$ and $v = \partial_t w - \Delta w$ which are both distributions in U with support in $B_{4\epsilon}(\xi)$, and $v = 0$ on $B_{3\epsilon}(\xi)$. In particular, the convolution

(34) $$\widetilde{K} \star v(x,t) = \int_{\mathbf{R}^{n+1}} \widetilde{K}(x-y, t-s) v(y,s) \, dy \, ds$$
$$= \int_{\mathbf{R}^{n+1}} \widetilde{K}(y,s) v(x-y, t-s) \, dy \, ds$$

is well defined as a distribution. But

(35) $$v(x-y, t-s) = (\partial_t - \Delta_x) w(x-y, t-s) = (-\partial_s - \Delta_y) w(x-y, t-s),$$

so for $g \in C_0^\infty(U)$ we compute

(36) $$\langle \widetilde{K} \star v, g \rangle = \int \widetilde{K}(y,s)(-\partial_s - \Delta_y) w(x-y, t-s) g(x,t) \, dy \, ds \, dx \, dt$$
$$= \int (\partial_s - \Delta_y) \widetilde{K}(y,s) w(x-y, t-s) g(x,t) \, dy \, ds \, dx \, dt$$
$$= \int \delta(y,s) w(x-y, t-s) g(x,t) \, dy \, ds \, dx \, dt$$
$$= \int w(x,t) g(x,t) \, dx \, dt.$$

In other words, $\widetilde{K} \star v = w$ as distributions.

Now choose $\psi \in C^\infty(\mathbf{R}^{n+1})$ with $\psi \equiv 0$ on $B_\epsilon(0)$ and $\psi \equiv 1$ on $\mathbf{R}^{n+1} \backslash B_{2\epsilon}(0)$. Then ψ vanishes in a neighborhood of the singularity of $\widetilde{K}$, so $\psi \widetilde{K}$ is a smooth function, and hence so is $(\psi \widetilde{K}) \star v$ (cf. Exercise 6 in Section 2.3). But for $(x,t) \in B_\epsilon(\xi)$ and $(y,s) \in \operatorname{supp} v \subset B_{4\epsilon}(\xi) \cap (\mathbf{R}^{n+1} \backslash B_{3\epsilon}(\xi))$, we have $|(x,t) - (y,s)| \geq 2\epsilon$, so $\psi(x-y, t-s) = 1$. In other words, $(\psi \widetilde{K}) \star v(x,t) = \widetilde{K} \star v(x,t) = w(x,t)$ for $(x,t) \in B_\epsilon(\xi)$, which shows that w is smooth on $B_\epsilon(\xi)$. But $w = \phi u$, so u is also smooth on $B_\epsilon(\xi)$. Since $\xi \in U$ was arbitrary, we conclude that u is smooth on U. ♠

An analogous result holds for the nonhomogeneous heat equation; see Exercise 1.

Remark. We found in Section 4.2e that harmonic functions are not only smooth, but are real analytic. The corresponding result for the heat equation is *not* true. If it were, then unique continuation would imply that any solution of $u_t = \Delta u$ in $U \subset \mathbf{R}^{n+1}$ that vanishes on an open set $U_1 \subset U$ must vanish on all of U. But the fundamental solution (22) gives a counterexample. However, it can be shown that a solution $u(x,t)$ of the heat equation is real analytic in x when t is fixed. See [John] or [Widder] for a discussion of this fact.

b. Scale Invariance and the Similarity Method

The heat equation has a natural symmetry that enables us to transform the dependent variables by a scaling factor without changing the solutions. This *scale invariance* allows us to "scale out" the t-dependence by introducing a new dependent variable and new spatial variables called *similarity variables*. In fact, the scale invariance and similarity method may be applied to other PDEs, so we first describe the method in general before applying it to the heat equation.

Let us consider the following scaling transformation:

(37a) $$\tilde{u} = \epsilon^\alpha u, \qquad \tilde{x} = \epsilon^\beta x, \qquad \tilde{t} = \epsilon^\gamma t,$$

where α, β, and γ are constants; in other words, we have defined

(37b) $$\tilde{u}(\tilde{x}, \tilde{t}) = \epsilon^\alpha u(\epsilon^{-\beta}\tilde{x}, \epsilon^{-\gamma}\tilde{t}).$$

To "scale out" the t-dependence means that we should take $\epsilon^\gamma = t^{-1}$, so that $\tilde{t} = 1$; that is if we introduce $w(\tilde{x}) = \tilde{u}(\tilde{x}, 1)$ and $z = \tilde{x}$, then

(38a) $$u(x,t) = t^{\alpha/\gamma} w(z), \qquad z = t^{-\beta/\gamma} x$$

defines the *similarity transformation* to the *similarity variables* w and z. Generally, we find that at least one of α, β, and γ may be given an arbitrary value; if we take $\gamma = 1$, then (38a) simplifies to

(38b) $$u(x,t) = t^\alpha w(z), \qquad z = t^{-\beta} x.$$

We now substitute (38b) into the given PDE and choose appropriate values of α and β so that we obtain a PDE for w that is independent of t. This is useful because we may be able to solve explicitly the time-independent PDE for w, or at least conclude properties of w, either of which will translate to conclusions about the solutions u of the original time-dependent PDE.

Let us carry out this program for the heat equation. First let us determine β. From (38b) we find that $u_t = \alpha t^{\alpha-1} w - \beta t^{\alpha-1} z \cdot \nabla_z w$ and $\Delta_x u = t^{\alpha-2\beta} \Delta_z w$, so that

$$u_t - \Delta_x u = t^{\alpha-1}(\alpha w - \beta z \cdot \nabla_z w) - t^{\alpha-2\beta} \Delta_z w.$$

In order to obtain a time-independent PDE for w, the powers of t must agree, which leads us to the conclusion that $\beta = 1/2$; consequently, (38b) for the heat equation becomes

(39) $$u(x,t) = t^\alpha w(z), \qquad z = \frac{x}{\sqrt{t}},$$

and $w(z)$ satisfies

(40) $$\alpha w(z) - \frac{1}{2} z \cdot \nabla w(z) = \Delta_z w(z).$$

Notice that α in (39) and (40) is arbitrary: this is not surprising since the heat equation is linear and homogeneous, so the scaling $\tilde{u} = \epsilon^\alpha u$ does not affect the solutions. But making an appropriate choice of α may be useful in solving the PDE for w, as we shall see.

If $n = 1$, then (40) is an ODE, but otherwise (40) appears difficult to solve. However, if we recall the original symmetry of the heat equation in the x variables, we might try to find w that is a radial function of z [i.e., $w(z) = w(|z|)$]. In this case (40) simplifies to the ODE

(41) $$w'' + \frac{n-1}{r} w' + \frac{1}{2} r w' - \alpha w = 0, \quad \text{where } ' = d/dr \text{ and } r = |z|.$$

Let us multiply (41) by r^{n-1} and take the special value $\alpha = -n/2$ so that (41) may be rewritten

(42) $$(r^{n-1} w')' + \frac{1}{2} (r^n w)' = 0.$$

But (42) implies

(43a) $$r^{n-1} w' + \frac{1}{2} r^n w = A = \text{const.}$$

At this point the constant A is arbitrary, so let us also assume

(43b) $$A = 0.$$

Now (43a,b) implies $w' = -(r/2) w$ which is solved easily by separation of variables to conclude
$$w(r) = C e^{-r^2/4}.$$

The similarity method, with the special choices we have made, has produced the solution

(44) $$u(x,t) = C \frac{1}{t^{n/2}} e^{-\frac{|x|^2}{4t}}.$$

which (with an appropriate choice of C) coincides with the fundamental solution of the heat equation (see also Exercise 2).

Remark. The similarity method is an important tool for constructing special solutions and applies to other linear and nonlinear equations, as we see in the Exercises. However, the method requires the equation to have certain symmetries to be scale invariant. Moreover, even if the equation is scale invariant, the symmetries may be broken by trying to impose initial or boundary conditions. For this reason, the method applies most readily when no side conditions are required; in this case, there may be a free parameter (such as α) that can be chosen to simplify the equation for w.

Exercises for Section 5.3

1. Suppose u is a distribution solution of the nonhomogeneous heat equation $u_t = \Delta u + f(x,t)$ in $U \subset \mathbf{R}^{n+1}$, where $f \in C^\infty(U)$. Show that $u \in C^\infty(U)$.

2. Consider the heat equation $u_t = \Delta u$ in $t > 0, x \in \mathbf{R}^n$ with the initial condition $u(x,0) = \delta(x)$. Since $\int u(x,t)\,dx$ represents the total heat content in $\mathbf{R}^n$ at time t, we encounter the condition $\int u(x,t)\,dx = 1$ for all $t \geq 0$.
   (a) Show that the choice $\alpha = -n/2$, which was made to simplify (41) to (42), is a consequence of the scale invariance of $\int u(x,t)\,dx$ for $t > 0$.
   (b) Show that (43b) is a consequence of $\int u(x,t)\,dx < \infty$.
   (c) Use $\int u(x,t)\,dt = 1$ to evaluate the constant C in (44), and conclude that $u(x,t)$ coincides with the heat kernel.

3. Apply the similarity method to the linear transport equation $u_t + au_x = 0$ to obtain the special solutions $u(x,t) = c(x - at)^\alpha$.

4. Consider the inviscid Burgers equation $u_t + uu_x = 0$.
   (a) Use the similarity method to reduce this equation to the ODE $ww' - (\alpha + 1)zw' + \alpha w = 0$.
   (b) Take $\alpha = -1$ and obtain the rarefaction wave solutions discussed in Section 1.2.

5. (a) For $n = 1$ and $\alpha = 0$, express the solution of (41) in terms of the *error function*
$$\mathrm{erf}(y) = \frac{2}{\sqrt{\pi}} \int_0^y e^{-s^2}\,ds.$$
   (b) Use (a) to solve the heat equation on a half-line: $u_t = u_{xx}$ for $x, t > 0$, with $u(x,0) = 1$ for $x > 0$, and $u(0,t) = 0$ for $t > 0$.

6. Apply the similarity method to the porous medium equation $u_t = \Delta(u^\gamma)$ in $\mathbf{R}^n$, where $\gamma > 1$, to obtain *Barenblatt's solution*:
$$u(x,t) = t^\alpha \left(C - \frac{(\gamma-1)\beta|x|^2}{2\gamma t^{2\beta}} \right)^{1/(\gamma-1)},$$

where C is a positive constant, $\beta = (n(\gamma - 1) + 2)^{-1}$, and $\alpha = n\beta$.
Remark. Here $u(x,t)$ is defined and positive for $(\gamma - 1)\beta|x|^2 < C2\gamma t^{2\beta}$, but may be extended by zero to a weak solution on $\mathbf{R}^n \times (0, \infty)$. In this way, u exhibits finite propagation speed, unlike the solution of the heat equation $\gamma = 1$.

5.4 Applications to Fluid Dynamics

In this section we apply the heat equation to two distinct aspects of fluid dynamics. The first application involves a continuous fluid model of a viscous fluid that is moving slowly enough that we may linearize the Navier-Stokes equations. The second application involves the Brownian motion of microscopic particles suspended in a liquid.

a. Slow Viscous Incompressible Flow

Viscous incompressible fluid flow in a domain $\Omega \subset \mathbf{R}^3$ is governed by the Navier-Stokes equations [see (33) in the Inroduction or (32) in the Appendix]. These are nonlinear equations due to the term $(\vec{u} \cdot \nabla)\vec{u}$. But if the flow is very slow (i.e. $\vec{u}$ and $\nabla \vec{u}$ are small), then it is reasonable to ignore this nonlinear term and use as a model the linear system

$$(45) \qquad \vec{u}_t + \frac{1}{\rho_0}\nabla p = \nu \Delta \vec{u},$$

called the *Stokes equations*. Here the velocity field $\vec{u}(x,t)$ satisfies div $\vec{u} = 0$, ρ_0 is the (constant) density, $p(x,t)$ is the pressure, and ν is the kinematic viscosity. We need to impose an initial condition

$$(46) \qquad \vec{u}(x,0) = \vec{u}_0(x),$$

where $\vec{u}_0$ also satisfies div $\vec{u}_0 = 0$, and if Ω has a boundary $\partial \Omega$ we impose the nonslip boundary condition $\vec{u} = 0$ on $\partial \Omega$.

If we take the curl of (46), we use (14a) in the Appendix to find that the *vorticity*

$$(47) \qquad \vec{v}(x,t) = \operatorname{curl} \vec{u}(x,t)$$

satisfies the simplified system

$$(48) \qquad \vec{v}_t = \nu \Delta \vec{v} \quad \text{for} \quad x \in \Omega, \ t > 0.$$

Notice that this is a decoupled system of three linear heat equations. The initial condition (46) easily transforms to

$$(49) \qquad \vec{v}(x,0) = \vec{g}(x) \equiv \operatorname{curl} \vec{u}_0(x,0) \quad \text{for} \quad x \in \Omega.$$

But boundary conditions are problematic, so we restrict our attention to the case $\Omega = \mathbf{R}^3$; this will also enable us to obtain an explicit formula for the solution. In fact, let us assume that (componentwise) $\vec{u}_0 \in \mathcal{S}(\mathbf{R}^3)$, the Schwartz space. Then each component v_i of $\vec{v}$ may be found using a Gaussian integral:

$$(50) \qquad v_i(x,t) = \frac{1}{(4\pi\nu t)^{3/2}} \int_{\mathbf{R}^3} e^{-|x-y|^2/4\nu t} g_i(y)\, dy;$$

in fact, $\vec{v}$ is smooth in $x \in \mathbf{R}^3$ and $t > 0$. We may then recover $\vec{u}$ from $\vec{v}$ by solving

$$(51) \qquad \operatorname{curl} \vec{u} = \vec{v}, \quad \text{and} \quad \operatorname{div} \vec{u} = 0;$$

notice that we can solve (51) (for each fixed t) using Section 4.5a since $\vec{v}(x)$ decays exponentially as $|x| \to \infty$. Moreover, $\vec{u}$ will inherit the smoothness of $\vec{v}$ and will vanish as $|x| \to \infty$.

Finally, we need to find $p(x,t)$. But from (45) we want p to satisfy

$$(52) \qquad \nabla p = \vec{w} \equiv \rho_0(\nu\Delta\vec{u} - \vec{u}_t),$$

where $\operatorname{curl} \vec{w} = 0$ [since $\vec{v} \equiv \operatorname{curl} \vec{u}$ satisfies the heat equation (48)]. By vector calculus, we can solve (52) to find p (up to an arbitrary additive constant). We have therefore proved the following:

Theorem. *If $\vec{u}_0 \in \mathcal{S}(\mathbf{R}^3)$ satisfies $\operatorname{div} \vec{u}_0 = 0$, then there is a unique smooth divergence-free vector field $\vec{u}(x,t)$ and a unique (up to an additive constant) smooth function $p(x,t)$ that solve (45),(46).*

b. Brownian Motion

Microscopic particles suspended in a liquid are in constant random motion; this is called *Brownian motion*. A probabilistic model for their motion leads to a diffusion equation like (1), which enables us to apply the results of this chapter. For simplicity, we describe the probabilistic model and resultant diffusion equation for a Brownian motion on the real line $(-\infty, \infty)$.

To derive the probabilistic model, let $u(x,t)$ be the *probability density* for the motion of a single given particle; that is, for any finite interval $[a,b] \subset (-\infty, \infty)$, the probability $P_{[a,b]}(t)$ that at time t the particle is in the interval $[a,b]$ is given by

$$(53) \qquad P_{[a,b]}(t) = \int_a^b u(x,t)\, dx.$$

We want to show that $u(x,t)$ satisfies a diffusion equation.

To derive the diffusion equation, assume that, in a small increment of time Δt, the particle will move with equal probability either to the right or to the left a fixed distance $\epsilon = \epsilon(\Delta t)$. Thus

(54) $$u(x, t + \Delta t) = \frac{1}{2}u(x - \epsilon, t) + \frac{1}{2}u(x + \epsilon, t).$$

But if we use the Taylor approximation $u(x \pm \epsilon, t) \approx u(x,t) \pm u_x(x,t)\epsilon + O(\epsilon^2)$ in (54), we find $u(x, t + \Delta t) \approx u(x,t) + \frac{1}{2}u_{xx}(x,t)\epsilon^2$. Rearranging and dividing by Δt, we obtain

$$\frac{u(x, t + \Delta t) - u(x, t)}{\Delta t} \approx \frac{1}{2}u_{xx}(x,t)\frac{\epsilon^2}{\Delta t}.$$

If we now assume that as $\Delta t \to 0$, the quantity $\epsilon^2/\Delta t$ tends to a constant γ called the *diffusion constant*, then

(55) $$u_t = \frac{\gamma}{2}u_{xx}.$$

For example, suppose we know with certainty that the particle is located at $x = 0$ when $t = 0$. This imposes the initial condition $u(x, 0) = \delta(x)$, where $\delta(x)$ is the delta distribution, since $\lim_{t \to 0} \int_{-\epsilon}^{\epsilon} u(x,t)\,dx = 1$ for any $\epsilon > 0$. Consequently, the probability density function is given by

(56) $$u(x,t) = \frac{1}{\sqrt{2\gamma\pi t}} e^{-x^2/2\gamma}.$$

Using (53), we can now compute the probability $P_{[a,b]}(t)$ for any finite interval $[a,b]$ and time $t > 0$.

Exercises for Section 5.4

1. Suppose that the probability that a Brownian motion particle will move to the left a distance ϵ is p and to the right is q, where $p + q = 1$. If $p \neq q$, find the diffusion equation that the probability density satisfies. The constant $p - q$ is called the *drift constant*; can you explain why?

Further References for Chapter 5

The most complete treatment of additional issues associated with the heat equation is [Widder]. The connections between the heat equation and random walks in probability are explored in more detail in [Freedman]. An exhaustive study of potential theory (elliptic and parabolic) and its relation to probability is given in [Doob].

The Fourier transform, which we treated very briefly, is a major tool in mathematics. For more information about Fourier analysis on Euclidean spaces, see [Stein-Weiss]. For extensions of Fourier analysis to harmonic analysis, see [Hewitt-Ross].

The scale invariance and similarity method discussed in Section 5.3 is a special case of *symmetry methods* used to study differential equations; see [Bluman-Kumei] or [Olver].

6 Linear Functional Analysis

In this chapter, we develop some tools from linear functional analysis that we will use to study more complicated partial differential equations and systems than those treated so far. We discuss Banach and Hilbert spaces of functions, including the important Sobolev spaces, and linear operators and functionals; the Riesz representation theorem provides a weak solution of the Dirichlet problem. We also discuss duality, compactness, and imbedding theorems involving the Sobolev spaces, which leads us to the notions of Hölder continuity and smooth approximations of functions. Finally, we consider unbounded linear operators and spectral theory.

6.1 Function Spaces and Linear Operators

We have already encountered various *function spaces*; that is, classes of functions such as $C(\Omega)$ or $C(\overline{\Omega})$ having the property that we remain within the class if we add any two such functions or multiply any such function by a constant. This means that these function spaces are *vector spaces*, although they will be *infinite dimensional* because no finite collection of functions may be used as a basis for all functions in the class.

The function spaces are particularly useful if they come equipped with a means of measuring the size or *norm* of each of its elements. The norm will induce a metric and allow us to consider the distance between functions and the convergence of sequences of functions.

We may also consider linear operators between function spaces. The existence of norms on the spaces will enable us to discuss continuity of the operators.

With this as background motivation, let us now discuss the abstract properties we are interested in and then exhibit them in examples of function spaces that occur in the rest of this text.

a. Banach and Hilbert Spaces

Let X be a vector space over a scalar field (the real numbers $\mathbf{R}$ or the complex numbers $\mathbf{C}$) on which we have defined a *norm* $\|\cdot\|$; that is, for every $x \in X$, $\|x\|$ is a nonnegative real number satisfying the following properties: (i) $\|x\| = 0$ if and only if $x = 0$, (ii) $\|\alpha x\| = |\alpha| \cdot \|x\|$ for all scalars α, and (iii) $\|x+y\| \leq \|x\| + \|y\|$ (the triangle inequality). The norm induces a *distance function* $d(x,y) = \|x - y\|$ so that X is a metric space, called a *normed vector space*. If the normed vector space X is a *complete* metric space (i.e. all Cauchy sequences converge), then we call X a *Banach space*.

It may be that the norm is associated to an *inner product* on X. That is, for every $x, y, z \in X$ and scalars λ and μ, the inner product $\langle x, y \rangle$ is a scalar satisfying (i) $\langle x, x \rangle > 0$ if $x \neq 0$ and $\langle 0, 0 \rangle = 0$, (ii) $\langle x, y \rangle = \overline{\langle y, x \rangle}$ (where the bar indicates complex conjugation in case the scalar field is $\mathbf{C}$), and (iii) $\langle \lambda x + \mu y, z \rangle = \lambda \langle x, z \rangle + \mu \langle y, z \rangle$. Letting $\|x\| = \sqrt{\langle x, x \rangle}$, we see that conditions (i) and (ii) of a norm are satisfied. Moreover, for any $x, y \in X$ we have the *Cauchy-Schwarz inequality* (see Exercise 1)

(1) $$|\langle x, y \rangle| \leq \|x\| \, \|y\|.$$

Using this inequality and $\text{Re}\langle x, y \rangle \leq |\langle x, y \rangle|$, we find $\|x + y\|^2 = \|x\|^2 + 2\text{Re}\langle x, y \rangle + \|y\|^2 \leq (\|x\| + \|y\|)^2$ which establishes the triangle inequality for $\|\cdot\|$, showing that indeed it is a norm. If X is complete in this norm, then we say X is a *Hilbert space*.

A collection $\{x_n\}_{n=1}^N$ (where $N = \infty$ or a positive integer) of vectors in a Hilbert space X is *orthonormal* if $\|x_n\| = 1$ for all n and $\langle x_n, x_m \rangle = 0$ for all $n \neq m$; the collection is *complete* if $x = \sum_{n=1}^N \langle x, x_n \rangle x_n$ (a convergent series if $N = \infty$) for all $x \in X$. A complete, orthonormal set is called an *orthonormal basis* for X, and N is called the *dimension* of X.

Let us now exhibit some of the function spaces that will be important to us.

Example 1. Let K be a compact subset of $\mathbf{R}^n$ and let $C(K)$ denote the space of continuous real-valued functions on K. Since every $u \in C(K)$ achieves its maximum and minimum values on K, we may define

(2) $$\|u\|_\infty = \max_{x \in K} |u(x)|.$$

(The subscript ∞ is chosen for consistency with the norm of L^∞ in Example 2.) It is easy to check that $\|\cdot\|_\infty$ is indeed a norm on $C(K)$ and (since a uniform limit of continuous functions is continuous) that $C(K)$ is a Banach space. However, this norm cannot be derived from an inner product, so $C(K)$ is *not* a Hilbert space.

Notice that, instead of taking real-valued functions, we could just as well have taken complex-valued functions in the definition of $C(K)$. (However, we do not introduce extra notation to distinguish between the space of continuous real-valued functions on K and that of complex-valued functions.) ♣

If Ω is a domain in $\mathbf{R}^n$, then $C(\Omega)$, the space of continuous functions on Ω, is *not* a Banach space since such functions need not be bounded [and (2) need not be finite]. Of course, if Ω is itself a bounded subset of $\mathbf{R}^n$, then the closure $\overline{\Omega}$ is compact and we may take $C(\overline{\Omega})$ as in Example 1, but this requires functions to have continuous extensions to $\overline{\Omega}$. Let us consider another space of functions on Ω that does not have this shortcoming. We assume a familiarity with Lebesgue measure and integration in $\mathbf{R}^n$, so a quick review of these concepts will suffice for our purposes; for a complete treatment, a book on measure theory and real analysis, such as [Royden], should be consulted.

Example 2. Let Ω be a domain in $\mathbf{R}^n$, and let μ denote *Lebesgue measure* in $\mathbf{R}^n$. The sets on which μ is well defined are called *measurable sets*, and the functions $f(x)$, for which $\{x \in \mathbf{R}^n : f(x) < a\}$ is a measurable set for every real number a, are called *measurable functions*. (For purposes of analysis, sets of measure zero are negligible, so measurable functions will be identified if they agree, except possibly on a set of measure zero.) The measurable functions are desirable, because it is for this general class of functions that the Lebesgue integral is defined. For $1 \leq p < \infty$, let $L^p(\Omega)$ denote the measurable real-valued functions u on Ω for which $\int_\Omega |u(x)|^p \, dx < \infty$. In addition, let $L^\infty(\Omega)$ denote the measurable real-valued functions that are bounded, or at least *essentially bounded* in that the function is bounded except on a set of measure zero. For $u \in L^p(\Omega)$ we may define

$$\|u\|_p = \left(\int_\Omega |u(x)|^p \, dx\right)^{1/p} \quad \text{for } 1 \leq p < \infty,$$
$$\|u\|_\infty = \operatorname*{ess\,sup}_{x \in \Omega} |u(x)| \equiv \inf\{M : \mu\{x : u(x) > M\} = 0\}.$$
(3)

Let us just mention two important inequalities involving the norm (3). The first is the *Hölder inequality*

(4) $$\int_\Omega |uv| \, dx \leq \|u\|_p \|v\|_{p'},$$

which holds for $u \in L^p(\Omega)$ and $v \in L^{p'}(\Omega)$, where $\frac{1}{p} + \frac{1}{p'} = 1$; in particular, this shows $uv \in L^1(\Omega)$. The second is the *Minkowski inequality*

(5) $$\|u+v\|_p \leq \|u\|_p + \|v\|_p,$$

which holds for $u, v \in L^p(\Omega)$ and in particular shows $u + v \in L^p(\Omega)$. Using the Minkowski inequality, we find that $\|\cdot\|_p$ is a norm on $L^p(\Omega)$; the *Riesz-Fischer theorem* asserts that $L^p(\Omega)$ is complete in this norm, so $L^p(\Omega)$ is a Banach space under the norm (3). If $p = 2$, then $L^2(\Omega)$ is a Hilbert space with inner product

(6) $$\langle u, v \rangle = \int_\Omega uv \, dx.$$

[Again we may use complex-valued functions provided that in (6) we use $\langle u, v \rangle = \int u\bar{v} \, dx$.] ♣

Section 6.1: Function Spaces and Linear Operators

Note. Unless otherwise stated, we assume that Banach spaces have $\mathbf{R}$ as the scalar field, and we omit all complex conjugate signs.

Example 3. Let Ω be a bounded domain in $\mathbf{R}^n$ and let $C^1(\Omega)$ denote the functions on Ω having continuous first-order derivatives on Ω, and let $C^1(\overline{\Omega})$ denote those functions that, along with their first-order derivatives, extend continuously to the compact set $\overline{\Omega}$. Then $C^1(\overline{\Omega})$ is a Banach space under the norm

$$(7) \qquad \|u\|_{1,\infty} = \max_{x \in \overline{\Omega}}(|\nabla u(x)| + |u(x)|).$$

[Notice that $C^1(\Omega)$ is *not* a Banach space since (7) need not be finite for $u \in C^1(\Omega)$.] ♣

In the next subsection, we encounter function spaces that are equipped with a norm involving the derivatives of the function measured in L^p rather than the supremum norm.

b. Sobolev Spaces

A *Sobolev space* is simply a space of functions whose distributional derivatives (up to some fixed order) exist in an L^p-space. However, there are a couple of different ways of defining these spaces. We only consider first-order derivatives in this section; higher-order derivatives are considered in Section 6.5.

Let Ω be a domain in $\mathbf{R}^n$, and let us introduce

$$(8a) \qquad \langle u, v \rangle_1 = \int_\Omega (\nabla u \cdot \nabla v + uv)\, dx,$$

$$(8b) \qquad \|u\|_{1,2} = \sqrt{\langle u, u \rangle_1} = \left(\int_\Omega (|\nabla u|^2 + |u|^2)\, dx\right)^{1/2},$$

when these expressions are defined and finite. [In the subscripts of (8b), the 1 refers to the appearance of the first-order derivatives, and the 2 refers to the L^2-nature of the norm.] For example, (8) is defined for functions in $C_0^1(\Omega)$. However, $C_0^1(\Omega)$ is *not complete* under the norm (8b), and so does not form a Hilbert space. On the other hand, suppose $\{u_j\} \in C_0^1(\Omega)$ is a sequence that is Cauchy in the norm (8b). Then $\{u_j\}$ is Cauchy in the L^2-norm since $\|u_j - u_k\|_2 \leq \|u_j - u_k\|_{1,2} \to 0$. Since $L^2(\Omega)$ is complete, we must have $u_j \to u$ for some $u \in L^2(\Omega)$. Thus we may obtain a Hilbert space, which we denote by $H_0^{1,2}(\Omega)$, by including all L^2-limits of sequences that are Cauchy in the norm (8b). In other words, $H_0^{1,2}(\Omega)$ is the *completion* of $C_0^1(\Omega)$ in the norm (8b): $H_0^{1,2}(\Omega) = \overline{C_0^1(\Omega)}$ where the bar denotes closure in the norm (8b). Notice that the norm (8b), originally defined only on $C_0^1(\Omega)$, extends by continuity to all of $H_0^{1,2}(\Omega)$ (see Exercise 1b).

Functions in $H_0^{1,2}(\Omega)$ need not be differentiable in the classical sense. Instead, a function $u \in H_0^{1,2}(\Omega)$ will have derivatives $\partial u/\partial x_j \in L^2(\Omega)$

defined by $\partial u/\partial x_j = \lim_{k\to\infty} \partial u_k/\partial x_j$, where $u_k \in C_0^1(\Omega)$ are Cauchy in the norm (8b) and converge to u in $L^2(\Omega)$. This is sometimes expressed as u having L^2-derivatives and is a special case of the weak derivative discussed in Section 2.3. This observation suggests that a more general weak Sobolev space, which we denote by $W^{1,2}(\Omega)$, may be defined as those $u \in L^2(\Omega)$ whose first-order weak derivatives $\partial u/\partial x_j$ are all in $L^2(\Omega)$. Clearly, $W^{1,2}(\Omega)$ is a Hilbert space under the norm (8).

One difference between $H_0^{1,2}(\Omega)$ and $W^{1,2}(\Omega)$ is that the condition "$u = 0$ on $\partial\Omega$," which holds in a generalized sense for $u \in H_0^{1,2}(\Omega)$, need not hold for $u \in W^{1,2}(\Omega)$. We could also avoid the restriction "$u = 0$ on $\partial\Omega$" by defining a Sobolev space in another way. Let us consider $\{u \in C^1(\Omega) : \|u\|_{1,2} < \infty\}$ and take the completion in the norm $\|\cdot\|_{1,2}$ exactly as before to obtain a Hilbert space that we denote by $H^{1,2}(\Omega)$. Notice that $H_0^{1,2}(\Omega) \subset H^{1,2}(\Omega)$, but generally the spaces are not equal. For example, if Ω is a bounded domain, then the constant functions are in $H^{1,2}(\Omega)\backslash H_0^{1,2}(\Omega)$; on the other hand, $H_0^{1,2}(\mathbf{R}^n) = H^{1,2}(\mathbf{R}^n)$ (see Exercise 6). Also notice that $H^{1,2}(\Omega) \subset W^{1,2}(\Omega)$; in fact, we see in Section 6.5 that these spaces are identical!

We can generalize our Sobolev spaces to $p \neq 2$ as follows. For $u \in C_0^1(\Omega)$ and $1 \leq p < \infty$, let us define

$$(9) \qquad \|u\|_{1,p} = \left(\int_\Omega \left(\sum_{i=1}^n |u_{x_i}|^p + |u|^p\right) dx\right)^{1/p}.$$

As before, we may define Banach spaces $H_0^{1,p}(\Omega)$, $W^{1,p}(\Omega)$, and $H^{1,p}(\Omega)$, which all use (9) as a norm.

Definition. For any domain in $\mathbf{R}^n$ and $1 \leq p < \infty$, let $\|\cdot\|_{1,p}$ be defined by (9). Define the Sobolev spaces

(10a) $\quad H_0^{1,p}(\Omega) = \overline{C_0^1(\Omega)},$

(10b) $\quad W^{1,p}(\Omega) = \{u \in L^p(\Omega) : D^\alpha u \in L^p(\Omega) \text{ for all } |\alpha| \leq 1\},$

(10c) $\quad H^{1,p}(\Omega) = \overline{\{u \in C^1(\Omega) : \|u\|_{1,p} < \infty\}},$

where the bar denotes the completion in the norm (9).

Let us consider a simple example.

Example 4. Let $\Omega = (0,1) \subset \mathbf{R}^1$, and

$$u(x) = \begin{cases} x & \text{for } 0 \leq x \leq 1/2, \\ 1-x & \text{for } 1/2 \leq x \leq 1. \end{cases}$$

Notice that $u = 0$ on $\partial\Omega = \{0,1\}$, and u has weak derivative

$$\frac{du}{dx} = \begin{cases} 1 & \text{for } 0 < x < 1/2, \\ -1 & \text{for } 1/2 < x < 1; \end{cases}$$

Section 6.1: Function Spaces and Linear Operators   167

see also Exercise 5a. This shows that $du/dx \in L^p(\Omega)$, and hence $u \in W^{1,p}(\Omega)$. In fact, $u \in H_0^{1,p}(\Omega)$ since there exist $u_k \in C_0^1(\Omega)$ with $du_k/dx \to du/dx$ and $u_k \to u$ in L^p as $k \to \infty$. [Constructing u_k requires smoothing out the corner at $x = 1/2$ as well as pulling the support inside of $(0,1)$; this is most effectively achieved using mollifiers as in Section 6.5b.] ♣

Sobolev spaces have become very important in the study of partial differential equations, especially the theory of *weak solutions*, as we shall see in subsequent chapters. The reader should be aware, however, that notation in the literature is not standardized: the indices 0, 1, and p may appear in almost any combination of superscripts and subscripts, and the H, W distinction used here is not universally accepted.

c. Linear Operators and Functionals

It will also be important for us to consider mappings between function spaces. An important class of such mappings are those that preserve the linear structure of the vector spaces; a mapping $T: X \to Y$ between two vector spaces X and Y is called a *linear operator* if $T(\lambda_1 x_1 + \lambda_2 x_2) = \lambda_1 T(x_1) + \lambda_2 T(x_2)$ for all $\lambda_j \in \mathbf{R}$ and $x_j \in X$. If, moreover, X and Y have norms (respectively, $\|\cdot\|_X$ and $\|\cdot\|_Y$), then we distinguish those linear operators that stretch the norms only a bounded amount in all directions as *bounded* linear operators; that is, there is a constant M such that

$$\|Tx\|_Y \le M \|x\|_X \quad \text{for all } x \in X.$$

The smallest such value M is the *operator norm* of T and is denoted by $\|T\|_{X \to Y}$ or simply $\|T\|$:

(11) $$\|T\|_{X \to Y} \equiv \|T\| \equiv \sup_{x \ne 0} \frac{\|Tx\|_Y}{\|x\|_X}.$$

Notice that linearity prevents $T(X)$ from being bounded in Y; however, the image under T of a *bounded* set in X *is* bounded in Y. Moreover, the following result emphasizes the importance of boundedness for linear operators.

Theorem 1. *A linear operator $T: X \to Y$ between normed vector spaces is bounded if and only if it is continuous.*

Proof. Let $x_j \to x$ in X. Then $\|Tx_j - Tx\|_Y \le M\|x_j - x\|$ implies that $Tx_j \to Tx$ in Y; in other words, T is continuous.
Conversely, if $T: X \to Y$ is continuous, then for any $\epsilon > 0$ there exists $\delta > 0$ such that $\|v\| < \delta$ implies $\|Tv\| < \epsilon$. But for any $u \in X$ let $v = \lambda u$, where $\lambda = \delta(2\|u\|)^{-1}$. So $\|v\| < \delta$ which implies $\|Tv\| < \epsilon$ or $\lambda\|Tu\| < \epsilon$. Hence $\|Tu\| < 2\epsilon\delta^{-1}\|u\|$, which proves that T is bounded. ♠

In the special case $Y = \mathbf{R}$, we call a linear operator $F: X \to \mathbf{R}$ a *linear functional*, and a bounded linear operator $F: X \to \mathbf{R}$ a *bounded linear functional* (or, equivalently, a *continuous linear functional*). Here are some examples corresponding to the previous Examples 1 and 2 of Banach spaces.

Example 1 (Revisited). If K is a compact subset of $\mathbf{R}^n$ and $X = C(K)$ with the norm (1), pick $x_0 \in K$ and define $F(u) = F_{x_0}(u) = u(x_0)$. Clearly $F : X \to \mathbf{R}$ is linear, and the following shows that F is bounded:

$$|F_{x_0}(u)| = |u(x_0)| \leq \max_{x \in K} |u(x)| = \|u\|_X.$$

In fact, if we use $u \in X$, which achieves its maximum at x_0, we see that $\|F_{x_0}\|_{X \to \mathbf{R}} = 1$. ♣

Example 2 (Revisited). If $X = L^p(\Omega)$ for $1 \leq p < \infty$, then point evaluation as in the previous example does not work (since points are of measure zero). However, if $v \in L^{p'}(\Omega)$ where $\frac{1}{p} + \frac{1}{p'} = 1$, then

$$F_v(u) = \int_\Omega u(x) v(x) \, dx$$

is clearly a linear functional, and the Hölder inequality (4) shows that F_v is a bounded linear functional with norm $\|F_v\| \leq \|v\|_{p'}$; in fact, it can be shown (cf. [Royden]) that $\|F_v\| = \|v\|_{p'}$. In the particular case $p = 2$, notice that for $v \in X = L^2(\Omega)$, the bounded linear functional F_v is defined by the inner product:

$$F_v(u) = \int_\Omega uv \, dx = \langle u, v \rangle. \; ♣$$

d. Hahn-Banach and Riesz Representation Theorems

A *subspace* of a normed vector space X is a subset S that is itself a vector space (i.e., S is closed under scalar multiplication and vector addition of its elements); if S contains all limit points of its elements, then S is a *closed* subspace. If $F : X \to \mathbf{R}$ is a bounded linear functional on the normed vector space X and S is a subspace of X (not necessarily closed in the topology of X), then the restriction $F : S \to \mathbf{R}$ is certainly well defined and bounded. The following famous theorem shows that the converse is also true: A bounded linear functional on a subspace may be extended to the entire normed vector space.

Theorem 2 (Hahn-Banach). *Suppose S is a subspace of a normed vector space X and $f : S \to \mathbf{R}$ is a linear functional satisfying $f(s) \leq \|s\|$ for all $s \in S$. Then there is a linear functional F on X such that $F(x) \leq \|x\|$ for all $x \in X$, and $F(s) = f(s)$ for all $s \in S$.*

We do not discuss the proof of this theorem, which requires the axiom of choice and may be found in any book on functional analysis (for example, [Royden] or [Rudin, 2]). The theorem is often stated in a more general form, but the preceding suffices for our purposes. In fact, we are principally interested in the following consequence.

Section 6.1: Function Spaces and Linear Operators

Corollary. Let X be a normed vector space and fix $x_0 \in X$. There is a bounded linear functional $F : X \to \mathbf{R}$ such that $F(x_0) = \|F\| \|x_0\|$.

Proof of Corollary. Let $S = \{\lambda x_0 : \lambda \in \mathbf{R}\}$ and define $f(\lambda x_0) = \lambda \|x_0\|$. By the Hahn-Banach theorem, we may extend f to $F : X \to \mathbf{R}$ so that $F(x) \leq \|x\|$ for $x \in X$. Since $-F(x) = F(-x) \leq \|-x\| = \|x\|$, we find $|F(x)| \leq \|x\|$; in particular, F is bounded. On the other hand, $F(x_0) = f(x_0) = \|x_0\|$, so $\|F\| = 1$ and $F(x_0) = \|x_0\| = \|F\| \|x_0\|$. ♠

When X is a Hilbert space, the result of the corollary may be obtained easily without recourse to the Hahn-Banach theorem: if $x_0 \in X$, then $F(x_0) = \langle x, x_0 \rangle$ defines the desired bounded linear functional. (In fact, for a Hilbert space, the Hahn-Banach theorem may be proved without the axiom of choice; cf. Exercise 13.) Taking the inner product with a fixed vector in a Hilbert space gives an important example of a bounded linear functional, as we saw in Example 2 with $p = 2$. The following result simply states that *all* bounded linear functionals on Hilbert spaces are of this form.

Theorem 3 (Riesz Representation). *For every bounded linear functional F on a Hilbert space X over the scalar field $\mathbf{R}$, there is a unique element $v \in X$ such that $F(u) = \langle u, v \rangle$ for all $u \in X$. Moreover, the norm of F agrees with that of v: $\|F\| = \|v\|$.*

We prove this theorem after some preparations concerning closed subspaces of a Hilbert space.

Lemma 1. *Let Y be a closed subspace of the Hilbert space X. For every $u \in X \backslash Y$ there is a unique $v \in Y$ that is closest to u:*

$$\|u - v\| = \inf_{w \in Y} \|u - w\|.$$

Proof. Let d be the distance from u to Y:

$$d \equiv \mathrm{dist}(u, Y) \equiv \inf_{w \in Y} \|u - w\|.$$

Choose $w_n \in Y$ with $\|u - w_n\| \to d$ as $n \to \infty$. Use the parallelogram law

$$\|u + v\|^2 + \|u - v\|^2 = 2(\|u\|^2 + \|v\|^2)$$

(which is easily verified by expanding the left-hand side) with $v = \frac{1}{2}(w_n - w_m)$ and u replaced by $u - \frac{1}{2}(w_n + w_m)$ to obtain

$$\|u - w_m\|^2 + \|u - w_n\|^2 = 2\|u - \frac{1}{2}(w_n + w_m)\|^2 + \frac{1}{2}\|w_n - w_m\|^2.$$

As $n, m \to \infty$, the sum on the left tends to $2d^2$ and the first term on the right is $\geq 2d^2$; hence $\|w_n - w_m\| \to 0$. Since Y is closed, w_n converges to a unique $v \in Y$. Moreover, $\|u - v\| \leq \|u - w_n\| + \|w_n - v\|$, and letting $n \to \infty$ yields $\|u - v\| \leq d$. But $d = \mathrm{dist}(u, Y)$ then implies $\|u - v\| = d$. ♠

Lemma 2 (The Projection Theorem). *Let Y be a closed subspace of the Hilbert space X. For every $u \in X$ there is a unique $v \in Y$ and $w \in X$ such that $u = v + w$ and $\langle w, Y \rangle = 0$.*

Proof. If $u \in Y$, let $v = u$ and $w = 0$. If $u \notin Y$, use Lemma 1 to find $v \in Y$ such that $\|u - v\| = \operatorname{dist}(u, Y)$. We need only show $\langle u - v, y \rangle = 0$ for all $y \in Y$. But for $y \in Y$ and $\lambda \in \mathbf{R}$ we have

$$\|u - v\|^2 \leq \|u - v - \lambda y\|^2 = \|u - v\|^2 - 2\lambda \langle u - v, y \rangle + \lambda^2 \|y\|^2.$$

In particular, for $\lambda = \langle u - v, y \rangle \|y\|^{-2}$ we find

$$\|u - v\|^2 \leq \|u - v\|^2 - \frac{\langle u - v, y \rangle^2}{\|y\|^2}$$

so $\langle u - v, y \rangle = 0$. Uniqueness is left for Exercise 10. ♠

Proof of Theorem 3. Let N be the nullspace of $F : X \to \mathbf{R}$, which is a closed subspace of X (see Exercise 11). If $N = X$, let $v = 0$; otherwise, by the projection theorem there exists $w \in X$, $w \neq 0$, such that $\langle w, N \rangle = 0$. Notice that $Fw \neq 0$ and for any $u \in X$ we have $u - (Fu/Fw)w \in N$. Hence

$$\langle u - \frac{Fu}{Fw} w, w \rangle = 0 \quad \text{or} \quad \langle u, w \rangle = \frac{Fu}{Fw} \|w\|^2.$$

If we let $v = (Fw/\|w\|^2)w$, we find that $F(u) = \langle u, v \rangle$.

If v' also represents F, then $\|v - v'\|^2 = \langle v - v', v - v' \rangle = \langle v - v', v \rangle - \langle v - v', v' \rangle = F(v - v') - F(v - v') = 0$, so v is unique.

Finally, $F(v) = \|v\|^2$, so

$$\|F\| = \sup_{u \neq 0} \frac{|F(u)|}{\|u\|} \geq \frac{|F(v)|}{\|v\|} = \|v\|.$$

On the other hand, for any $\epsilon > 0$ there exists $u \in X$ such that $\|u\| = 1$ and $\|F\| \leq F(u) + \epsilon$, so $\|F\| - \epsilon \leq F(u) = \langle u, v \rangle \leq \|v\|$. Thus $\|F\| = \|v\|$. ♠

In the next section, we see how to use the Riesz representation theorem to obtain weak solutions for Dirichlet problems.

Exercises for Section 6.1

1. (a) If X is a normed vector space, prove that $|\, \|x\| - \|y\| \,| \leq \|x - y\|$.
   (b) Show that the norm defines a continuous function on X.
   (c) If X is a real inner product space, prove (1).

2. Use Hölder's inequality to prove the *generalized Hölder's inequality*

$$\int_\Omega |uvw|\, dx \leq \|u\|_p \|v\|_q \|w\|_r$$

for $u \in L^p(\Omega)$, $v \in L^q(\Omega)$, and $w \in L^r(\Omega)$, where $p^{-1} + q^{-1} + r^{-1} = 1$.

3. Use Hölder's inequality to prove
$$\|u\|_q \leq \|u\|_p^\lambda \|u\|_r^{1-\lambda} \quad \text{for } u \in L^r(\Omega),$$
where $p \leq q \leq r$ and $q^{-1} = \lambda p^{-1} + (1-\lambda) r^{-1}$.

4. Let ℓ^2 denote the space of all sequences of real numbers $\{a_n\}_{n=1}^\infty$ such that $\sum a_n^2 < \infty$.
   (a) Verify that $\langle \{a_n\}, \{b_n\} \rangle = \sum a_n b_n$ is an inner product on ℓ^2.
   (b) Verify that ℓ^2 is a Hilbert space (i.e., complete in the induced norm).

5. Let $\Omega = (0,1)$.
   (a) Compute the weak derivative of the function u of Example 4.
   (b) For what values of α is $u(x) = |x|^\alpha$ in $H^{1,2}(\Omega)$?

6. Prove that $H_0^{1,p}(\mathbf{R}^n) = H^{1,p}(\mathbf{R}^n)$.

7. Suppose Ω is a bounded domain, and let $S \equiv C(\overline{\Omega}) \subset X \equiv L^p(\Omega)$. Pick $x_0 \in \Omega$. Does the functional $F_{x_0}(u) = u(x_0)$ for $u \in S$ extend to a bounded linear functional on X? If not, why not?

8. If X is a Hilbert space and S is any subset of X, define
$$S^\perp = \{y \in X : \langle x,y \rangle = 0 \quad \text{for all} \quad x \in S\}.$$

   (a) Show that $S^\perp$ is a closed subspace of X.
   (b) Show that $S \cap S^\perp$ can contain only the zero vector.
   (c) If $S \subset T$ are both subsets of X, show that $T^\perp \subset S^\perp$.
   (d) If $\overline{S}$ is the closure of S in X, show that $S^\perp = \overline{S}^\perp$.

9. For vector-valued functions $\vec{u} = (u_1, \ldots, u_N)$ on a domain $\Omega \subset \mathbf{R}^n$, it is natural to define $L^p(\Omega, \mathbf{R}^N)$ as the functions $\vec{u} : \Omega \to \mathbf{R}^N$ for which $|\vec{u}| \in L^p(\Omega)$; that is,
$$\|\vec{u}\|_p \equiv \left(\int_\Omega |\vec{u}|^p \, dx \right)^{1/p} < \infty, \quad \text{where } |\vec{u}|^2 = \sum_{i=1}^N u_i^2.$$

   Similarly, if each $u_i \in C^1(\Omega)$, we may define
$$\|\nabla \vec{u}\|_p \equiv \left(\int_\Omega |\nabla \vec{u}|^p \, dx \right)^{1/p}, \quad \text{where } |\nabla \vec{u}|^2 = \sum_{i=1}^N \sum_{j=1}^n (\partial u_i / \partial x_j)^2.$$

   (a) Define $H_0^{1,p}(\Omega, \mathbf{R}^N)$.
   (b) If $N = n$, show that div: $H_0^{1,p}(\Omega, \mathbf{R}^n) \to L^p(\Omega)$ is a continuous linear operator.
   (c) Show that $\tilde{H}_0^{1,p}(\Omega, \mathbf{R}^n) \equiv \{\vec{u} \in H_0^{1,p}(\Omega, \mathbf{R}^n) : \operatorname{div} \vec{u} = 0\}$ is a closed subspace of $H_0^{1,p}(\Omega, \mathbf{R}^n)$.

10. In Lemma 2 in this section, the vector $v \in Y$ was claimed (but not proved) to be unique. Establish this uniqueness.

11. If $F: X \to Y$ is a bounded linear operator between normed vector spaces X and Y, show that the nullspace $N = \{x \in X : F(x) = 0\}$ is a closed subspace of X.

12. If Y is a subspace of a Hilbert space X. define $Y^\perp$ as in Exercise 8. Show $(Y^\perp)^\perp = \overline{Y}$.

13. Suppose S is a subspace of a Hilbert space X and $f: S \to \mathbf{R}$ is a linear functional with $|f(s)| \leq C\|s\|$ for all $s \in S$. Prove that there is a unique linear functional $F: X \to \mathbf{R}$ extending f (i.e., $F(s) = f(s)$ for $s \in S$) and preserving the norm:

$$\|F\| = \sup_{x \in X, x \neq 0} \frac{|F(x)|}{\|x\|} = \sup_{s \in S, s \neq 0} \frac{|F(s)|}{\|s\|}.$$

(Notice that this special case of the Hahn-Banach theorem does *not* require the axiom of choice.)

14. Suppose X is a Hilbert space and $\{x_n\}_{n=1}^{\infty}$ is a collection of orthonormal vectors. Given $u \in X$, define its *Fourier coefficients* by $\alpha_n = \langle u, x_n \rangle$.
(a) Prove *Bessel's inequality:* $\sum_{n=1}^{\infty} \alpha_n^2 \leq \|u\|^2$.
(b) Let Y be the finite-dimensional subspace of X generated by taking linear combinations of $x_1, \ldots, x_N$. Show that $v = \sum_{n=1}^{N} \alpha_n x_n$ is the element of Y that minimizes $\|u - v\|$ as in Lemma 1 of this section.
(c) Repeat (b) when Y is the infinite dimensional subspace generated by taking linear combinations and limits of all the x_n's.

15. A *bounded bilinear form* on a real Hilbert space X is a map $B: X \times X \to \mathbf{R}$ satisfying
  (i) $B(\alpha x + \beta y, z) = \alpha B(x, z) + \beta B(y, z)$.
  (ii) $B(x, \alpha y + \beta z) = \alpha B(x, y) + \beta B(x, z)$.
  (iii) $|B(x, y)| \leq C\|x\|\|y\|$ for all $x, y, z \in X$ and $\alpha, \beta \in \mathbf{R}$.
Under these assumptions, show that there is a unique bounded linear operator $A: X \to X$ such that $B(x, y) = \langle Ax, y \rangle$ for all $x, y \in X$, where $\langle, \rangle$ denotes the inner product on X.

16. If X is a Hilbert space and $T: X \to X$ is a bounded linear operator, the *adjoint of* T is an operator $T^*: X \to X$ defined as follows:
(a) For $y \in X$, use the Riesz representation theorem to define $T^*y \in X$ satisfying $\langle Tx, y \rangle = \langle x, T^*y \rangle$ for all $x, y \in X$.
(b) Show that $T^*: X \to X$ is a bounded linear operator with $\|T^*\| = \|T\|$.

17. If $T: X \to X$ is a bounded linear operator on a Hilbert space X, let $N = \{x \in X : T(x) = 0\}$ be the nullspace of T and $R = \{Tx : x \in X\}$ be the range of T. Similarly, let N^* and R^* denote the nullspace and range of the adjoint T^* defined in Exercise 16. Prove $(N^*)^\perp = \overline{R}$ (where $S^\perp$ is defined in Exercise 8).

6.2 Application to the Dirichlet Problem

An elementary application of functional analysis to PDEs is the use of the Riesz representation theorem and the Hilbert space $H_0^{1,2}(\Omega)$ to obtain a weak solution of the Dirichlet problem for the Poisson equation. We discuss this application and some natural generalizations.

a. Weak Solutions of the Poisson Equation

Let Ω be a bounded domain, and let us consider the Dirichlet problem for the Poisson equation:

(12) $$\begin{cases} \Delta u = f & \text{in } \Omega \\ u = 0 & \text{on } \partial\Omega. \end{cases}$$

In Chapter 4 we discussed classical solutions, where $u \in C^2(\Omega)$; but now we want to consider *weak solutions* of (12).

Recall from Section 2.3 that a *weak solution* $u \in L^1_{\text{loc}}(\Omega)$ of $\Delta u = f \in L^1_{\text{loc}}(\Omega)$ satisfies $\int_\Omega u \Delta v \, dx = \int_\Omega fv \, dx$ for all $v \in C_0^2(\Omega)$. If $u \in C^1(\Omega)$, then we may write this condition as

(13) $$-\int_\Omega \nabla u \cdot \nabla v \, dx = \int_\Omega fv \, dx$$

for all $v \in C_0^2(\Omega)$. But for $u, v \in H_0^{1,2}(\Omega)$ and $f \in L^2(\Omega)$, (13) still makes sense, and we still have $u = 0 = v$ on $\partial\Omega$, at least in a generalized sense. Therefore, it is natural to say that $u \in H_0^{1,2}(\Omega)$ is a *weak solution* of (12) if $f \in L^2(\Omega)$ and (13) holds for all $v \in H_0^{1,2}(\Omega)$.

Let us define

(14) $$(u, v)_1 = \int \nabla u \cdot \nabla v \, dx,$$

which satisfies the conditions for an inner product on $C_0^1(\Omega)$. [Notice that $(u, u)_1 = 0$ for $u \in C_0^1(\Omega)$ implies $u \equiv 0$.] What happens if we take the completion of $C_0^1(\Omega)$ in the associated norm $|u|_{1,2} = \sqrt{(u,u)_1}$? Do we still get $H_0^{1,2}(\Omega)$? To answer these questions we require the following result.

Theorem 1 (The Poincaré Inequality). *If $\Omega \subset \mathbf{R}^n$ is a bounded domain, then there is a constant $C = C(\Omega)$ such that*

$$\int_\Omega |u(x)|^2 \, dx \leq C \int_\Omega |\nabla u(x)|^2 \, dx$$

for all $u \in C_0^1(\Omega)$, and by completion for all $u \in H_0^{1,2}(\Omega)$.

Proof. For $u \in C_0^1(\Omega)$, we may assume that Ω is contained in the cube $Q = \{x \in \mathbf{R}^n : |x_j| < a\}$. Then an integration by parts in the x_1-direction, followed by the Cauchy-Schwarz inequality yields

$$\|u\|_2^2 = \int_\Omega 1 \cdot |u|^2 \, dx = -\int_\Omega x_1 \frac{\partial}{\partial x_1} |u|^2 \, dx$$

$$= -2\int_\Omega x_1 u \frac{\partial u}{\partial x_1} \, dx \leq 2a \int_\Omega |u| \left|\frac{\partial u}{\partial x_1}\right| \, dx$$

$$\leq 2a \|u\|_2 \left\|\frac{\partial u}{\partial x_1}\right\|_2 \leq 2a \|u\|_2 \|\nabla u\|_2.$$

This yields the desired inequality with $C = 4a^2$.

For $u \in H_0^{1,2}(\Omega)$, let $u_j \in C_0^1(\Omega)$ with $\|u_j - u\|_2 \to 0$ and $\{\nabla u_j\}$ Cauchy in $L^2(\Omega)$. But $\|u_j\|_2 \to \|u\|_2$ (by Exercise 1 in Section 6.1) and $\|\nabla u_j\|_2 \to \|\nabla u\|_2$ (by definition of ∇u), so the Poincaré inequality for u_j implies the same inequality for u. ♠

Remarks. The Poincaré inequality generalizes easily to $p \neq 2$ (see Exercise 2). Also, from the proof, it is clear that the inequality holds for domains that are not bounded but are contained in a "slab" $\{x \in \mathbf{R}^n : a \leq x_1 \leq b\}$ or any rotation of this slab (see Exercise 3). However, the Poincaré inequality, as stated, does *not* hold for $u \in H^{1,2}(\Omega)$, since it fails when $u \equiv \mathrm{const}$. On the other hand, if we add the condition $\int_\Omega u \, dx = 0$, then we may extend the inequality to $u \in H^{1,2}(\Omega)$; this is discussed in Section 6.5.

Now let us compare the two norms $|\cdot|_{1,2}$ and $\|\cdot\|_{1,2}$. By the Poincaré inequality we have

$$|u|_{1,2}^2 \leq \|u\|_{1,2}^2 \leq (1+C)|u|_{1,2}^2$$

for $u \in C_0^1(\Omega)$. This means that the two norms are *equivalent*; in general, two norms $\|\cdot\|_a$ and $\|\cdot\|_b$ on a space X are *equivalent* if there exist constants $c_1, c_2 > 0$ so that $c_1 \|u\|_a \leq \|u\|_b \leq c_2 \|u\|_a$ holds for all $u \in X$. Notice that, for purposes of convergence, equivalent norms are interchangeable. In particular, a sequence $\{u_j\}$ is Cauchy in $|\cdot|_{1,2}$ if and only if it is Cauchy in $\|\cdot\|_{1,2}$. This implies that the completion of $C_0^1(\Omega)$ in the norm $|\cdot|_{1,2}$ is equal to $H_0^{1,2}(\Omega)$, and the inner product (14) extends by continuity to $u, v \in H_0^{1,2}(\Omega)$.

We may now reformulate the Dirichlet problem (12) as follows.

Weak Formulation of the Dirichlet Problem. For $f \in L^2(\Omega)$, find $u \in H_0^{1,2}(\Omega)$ so that

(15) $$(u, v)_1 = -\int_\Omega fv \, dx$$

for all $v \in H_0^{1,2}(\Omega)$.

Solution. We consider $H_0^{1,2}(\Omega)$ with the norm $|\cdot|_{1,2}$ and define $F(v) = -\int fv \, dx$ for $v \in H_0^{1,2}(\Omega)$. Then $F : H_0^{1,2}(\Omega) \to \mathbf{R}$ is a bounded linear functional since

$$|F(v)| = \left|\int fv\,dx\right| \leq \|f\|_2 \|v\|_2 \leq \|f\|_2 \, C^{1/2} \, |v|_{1,2},$$

where we have used the Cauchy-Schwarz inequality and then the Poincaré inequality. The Riesz representation theorem now gives our solution $u \in H_0^{1,2}(\Omega)$:

$$(u,v)_1 = (v,u)_1 = F(v) = -\int fv\,dx$$

for all $v \in H_0^{1,2}(\Omega)$.

Remarks. The Riesz representation theorem also shows that the weak solution $u \in H_0^{1,2}(\Omega)$ is *unique*. In addition, the hypothesis $f \in L^2(\Omega)$ may be relaxed somewhat; the important issue is the continuity of the functional $F(v) = \int fv\,dx$. On the other hand, as we saw in Section 4.3, the regularity of the solution u of (12) depends on that of f; if f is more regular than simply being L^2, then we should expect u to be more regular. We further investigate the regularity of weak solutions in Chapter 8 using elliptic estimates.

b. Weak Solutions of the Stokes Equations

An incompressible viscous fluid in a domain $\Omega \subset \mathbf{R}^3$ satisfies the Navier-Stokes equations [cf. (33) in the Introduction or (32) in the Appendix]. If the motion is slow, then (as in Section 5.4a) it is reasonable to linearize the system by ignoring the term $(\vec{u}\cdot\nabla\vec{u})$. A solution $\vec{u}$ is *stationary* if $\vec{u}_t \equiv 0$ (i.e. $\vec{u}$ does not depend on t). We assume $\rho_0 = 1$, but to make life more interesting, we add an external force $\vec{f}$ (also independent of t), which leads us to the *Stokes equations for stationary (forced) flow*:

(16) $$\begin{cases} \nabla p = \nu\Delta\vec{u} + \vec{f} & \text{in } \Omega \\ \operatorname{div}\vec{u} = 0 & \text{in } \Omega \\ \vec{u} = 0 & \text{on } \partial\Omega, \end{cases}$$

where we have added the natural "nonslip" boundary condition for a viscous fluid. As a linear system of four PDEs in Ω, (16) seems to be much more complicated than the Poisson equation (12). However, we shall see that, by building the condition $\operatorname{div}\vec{u} = 0$ into our Hilbert space, the existence and uniqueness of a solution follows from the Riesz representation theorem.

Let us introduce some notation that applies in all dimensions n. For a domain $\Omega \subset \mathbf{R}^n$, we can easily define $L^2(\Omega, \mathbf{R}^n)$ and $H_0^{1,2}(\Omega, \mathbf{R}^n)$ as Hilbert spaces of vector-valued functions (cf. Exercise 9 of Section 6.1). If $\vec{f} \in L^2(\Omega, \mathbf{R}^n)$, it is natural to define a *weak solution* of (16) to be

176   Chapter 6: Linear Functional Analysis

$\vec{u} \in H_0^{1,2}(\Omega, \mathbf{R}^n)$ satisfying $\operatorname{div} \vec{u} = 0$ and $p \in L^2(\Omega)$ for which

(17)
$$\int_\Omega (-\nu \nabla \vec{u} : \nabla \vec{v} + \vec{f} \cdot \vec{v})\, dx = -\int_\Omega p \operatorname{div} \vec{v}\, dx,$$
$$\text{where } \nabla \vec{u} : \nabla \vec{v} \equiv \sum_{i,j=1}^{n} \frac{\partial u_i}{\partial x_j} \frac{\partial v_i}{\partial x_j},$$

for all $\vec{v} \in H_0^{1,2}(\Omega, \mathbf{R}^n)$. But if we assume $\operatorname{div} \vec{v} = 0$, then (17) becomes

(18)
$$\int_\Omega \nu \nabla \vec{u} : \nabla \vec{v}\, dx = \int_\Omega \vec{f} \cdot \vec{v}\, dx,$$

which is just n copies of the weak formulation of Poisson's equation. This suggests that we try to satisfy (18) for "divergence-free" vector fields $\vec{u}, \vec{v}$:

Definition. Let $\widetilde{H}_0^{1,2}(\Omega, \mathbf{R}^n)$ denote those vector fields $\vec{u} \in H_0^{1,2}(\Omega)$ that are divergence-free; that is, $\operatorname{div} \vec{u} = 0$ (in the sense of distributions).

Now $\widetilde{H}_0^{1,2}(\Omega, \mathbf{R}^n)$ is a Hilbert space under the same norm (Exercise 9c in Section 6.1), and if Ω is *bounded*, then the norms induced by the two inner products

$$\langle \vec{u}, \vec{v} \rangle_1 \equiv \int_\Omega (\nabla \vec{u} : \nabla \vec{v} + \vec{u} \cdot \vec{v})\, dx \quad \text{and} \quad (\vec{u}, \vec{v})_1 \equiv \int_\Omega \nabla \vec{u} : \nabla \vec{v}\, dx$$

are equivalent. We can therefore use the Riesz representation theorem to obtain the following:

Proposition 1. *If Ω is a bounded domain in $\mathbf{R}^n$ and $\vec{f} \in L^2(\Omega, \mathbf{R}^n)$, then there is a unique $\vec{u} \in \widetilde{H}_0^{1,2}(\Omega, \mathbf{R}^n)$ such that (18) holds for every $\vec{v} \in \widetilde{H}_0^{1,2}(\Omega, \mathbf{R}^n)$.*

But for a weak solution of (16), we also need to know the pressure p.

Returning to $n = 3$, we can use classical methods to find p from the $\vec{u}$ of Proposition 1, provided that Ω is simply connected and we assume greater regularity of $\vec{u}$ and $\vec{f}$. To see this, let $\vec{g} = \nu \Delta \vec{u} + \vec{f}$, which by (18) satisfies $\langle \vec{g}, \vec{v} \rangle = 0$ for all $\vec{v} \in \widetilde{H}_0^{1,2}(\Omega, \mathbf{R}^3)$. Let us now assume that $\vec{g} \in C^1(\Omega, \mathbf{R}^3)$ and pick $\vec{\phi} \in C_0^2(\Omega, \mathbf{R}^3)$. By (14b) in the Appendix, we know that $\operatorname{div}(\operatorname{curl} \vec{\phi}) = 0$; but this means that $\vec{v} \equiv \operatorname{curl} \vec{\phi} \in \widetilde{H}_0^{1,2}(\Omega, \mathbf{R}^3)$, and, consequently, $\langle \vec{g}, \vec{v} \rangle = 0$. But we can then integrate by parts to conclude that $\langle \operatorname{curl} \vec{g}, \vec{\phi} \rangle = 0$, and hence $\operatorname{curl} \vec{g} \equiv 0$ in Ω since $\vec{\phi}$ was arbitrary. But $\operatorname{curl} \vec{g} \equiv 0$ in the simply connected domain Ω implies that $\vec{g} = \nabla p$ for some $p \in C^1(\Omega)$, which is the desired pressure p.

We see in the next paragraph that we can weaken the assumption $\vec{g} = \nu \Delta \vec{u} + \vec{f} \in C^1(\Omega, \mathbf{R}^3)$ to $\vec{g} \in L^2(\Omega, \mathbf{R}^3)$, but this last fact is not immediately clear since Proposition 1 only asserts that $\vec{u} \in \widetilde{H}_0^{1,2}(\Omega)$. However, using *elliptic regularity* (discussed in detail in Chapter 8), we conclude that a solution $\vec{u} \in \widetilde{H}_0^{1,2}(\Omega, \mathbf{R}^3)$ of (18), where $\vec{f} \in L^2(\Omega, \mathbf{R}^3)$, in fact satisfies $\Delta \vec{u} \in L^2(\Omega, \mathbf{R}^3)$. (Some regularity of Ω, such as $\partial \Omega \in C^2$, is also required for this step.)

As to why $\vec{g} \in L^2(\Omega, \mathbf{R}^3)$ will be sufficient to conclude $\vec{g} = \nabla p$ for some $p \in H^{1,2}(\Omega)$, we need to appeal to the *Helmholtz-Weyl decomposition*, which may be expressed as

$$(19) \qquad L^2(\Omega, \mathbf{R}^3) = \widetilde{L}^2(\Omega, \mathbf{R}^3) \oplus G^2(\Omega, \mathbf{R}^3),$$

where $\widetilde{L}^2(\Omega, \mathbf{R}^3)$ denotes the completion of $\widetilde{H}_0^{1,2}(\Omega, \mathbf{R}^3)$ in the L^2-norm, $G^2(\Omega, \mathbf{R}^3) = \{\vec{w} \in L^2(\Omega, \mathbf{R}^3) : \vec{w} = \nabla p \text{ for some } p \in H^{1,2}(\Omega)\}$, and $\oplus$ indicates that the two subspaces of L^2 are orthogonal to each other in the L^2-inner product. We do not discuss the proof, but let us remark that (19) is a generalization of the Helmholtz decomposition discussed in Section 4.5b and does *not* require Ω to be simply-connected. (For a proof of (19), see [Galdi] or [Ladyzhenskaya, 1]; in fact, the n-dimensional version of (19) is also true, see [Galdi].)

As a consequence of these facts, we can prove the following:

Proposition 2. *Let Ω be a bounded domain in $\mathbf{R}^3$ with C^2-boundary, and let $\vec{f} \in L^2(\Omega, \mathbf{R}^3)$. Then $\vec{u} \in \widetilde{H}_0^{1,2}(\Omega, \mathbf{R}^3)$ satisfies (18) for all $\vec{v} \in \widetilde{H}_0^{1,2}(\Omega, \mathbf{R}^3)$ if and only if there exists $p \in H^{1,2}(\Omega)$ such that $\nabla p = \nu \Delta \vec{u} + \vec{f}$; moreover, p is unique up to an additive constant.*

Proof. If $\vec{u} \in \widetilde{H}_0^{1,2}(\Omega, \mathbf{R}^3)$ satisfies (18) for all $\vec{v} \in \widetilde{H}_0^{1,2}(\Omega, \mathbf{R}^3)$ and $\vec{g} = \nu \Delta \vec{u} + \vec{f} \in L^2(\Omega, \mathbf{R}^3)$, then $\langle \vec{g}, \vec{v} \rangle = 0$ holds for all $\vec{v} \in \widetilde{L}^2(\Omega, \mathbf{R}^3)$, not just for $\vec{v} \in \widetilde{H}_0^{1,2}(\Omega, \mathbf{R}^3)$. But this means $\vec{g} \in \widetilde{L}^2(\Omega, \mathbf{R}^3)^\perp$, which by (19) implies $\vec{g} \in G^2(\Omega, \mathbf{R}^3)$; in other words, $\vec{g} = \nabla p$ for some $p \in H^{1,2}(\Omega)$.

Conversely, if $\nu \Delta \vec{u} + \vec{f} = \nabla p$ for $p \in H^{1,2}(\Omega)$, then $\int \nabla p \cdot \vec{v}\, dx = 0$ holds for all $\vec{v} \in \widetilde{H}_0^{1,2}(\Omega, \mathbf{R}^3)$, and this implies that (18) holds. If we also had $\nu \Delta \vec{u} + \vec{f} = \nabla p'$ for $p' \in H^{1,2}(\Omega)$, then by the uniqueness of $\vec{u}$ we must have $\nabla(p - p') = 0$ in Ω; in other words, $p - p'$ is a constant. ♠

Combining these two propositions, we have now obtained our weak solution $(\vec{u}, p)$ of the Stokes equations (16):

Theorem 2. *If Ω is a bounded domain in $\mathbf{R}^3$ with C^2-boundary and $\vec{f} \in L^2(\Omega, \mathbf{R}^3)$, then (16) admits a weak solution $(\vec{u}, p)$, where $\vec{u} \in \widetilde{H}_0^{1,2}(\Omega, \mathbf{R}^3)$ and $p \in H^{1,2}(\Omega)$, which is unique up to an additive constant to p.*

c. More General Operators in Divergence Form

It is useful to notice that the Hilbert space method for proving the existence of weak solutions for the Dirichlet problem in the bounded domain

178   Chapter 6: Linear Functional Analysis

Ω extends rather easily to more general operators than the Laplacian. For example, we can consider the operator in *divergence form*

$$(20) \qquad L = \sum_{i,j=1}^{n} \frac{\partial}{\partial x_i} a_{ij}(x) \frac{\partial}{\partial x_j} + c(x),$$

where $a_{ij}(x), c(x) \in C(\overline{\Omega})$ are real valued and satisfy

(21a) $\quad a_{ij}(x) = a_{ji}(x)$ for all $i, j = 1, \ldots, n$,

(21b) $\quad \sum_{i,j=1}^{n} a_{ij}(x)\xi_i\xi_j \geq \epsilon|\xi|^2$ for $x \in \Omega$, $\xi \in \mathbf{R}^n$, $\qquad (\epsilon > 0)$

and

(22) $\qquad c(x) \leq 0$ for $x \in \Omega$.

The symmetry condition (21a) implies that

$$(23) \qquad B_L(u,v) = \int_\Omega \left(\sum_{i,j=1}^{n} a_{ij}(x) \frac{\partial u}{\partial x_i} \frac{\partial v}{\partial x_j} - c(x)uv \right) dx$$

is a *symmetric bilinear form*; that is, $B_L(u,v) = B_L(v,u)$ and for all real scalars $\lambda_1, \lambda_2, \mu_1,$ and μ_2, we have $B_L(\lambda_1 u_1 + \lambda_2 u_2, v) = \lambda_1 B_L(u_1, v) + \lambda_2 B_L(u_2, v)$ and $B_L(u, \mu_1 v_1 + \mu_2 v_2) = \mu_1 B_L(u, v_1) + \mu_2 B_L(u, v_2)$. We also claim that B_L is a *bounded* bilinear form on $H_0^{1,2}(\Omega)$:

$$(24) \qquad |B_L(u,v)| \leq C\|u\|_{1,2}\|v\|_{1,2} \quad \text{for} \quad u,v \in H_0^{1,2}(\Omega).$$

To verify (24) we use the boundedness of $a_{ij}(x)$ and $c(x)$ on Ω, and for each i,j the inequality $|\frac{\partial u}{\partial x_i} \frac{\partial v}{\partial x_j}| \leq |\nabla u||\nabla v|$ to conclude

$$|B_L(u,v)| \leq M \int_\Omega (|\nabla u||\nabla v| + |u||v|) \, dx$$
$$\leq M(|u|_{1,2}|v|_{1,2} + \|u\|_2 \|v\|_2) \leq M(\|u\|_{1,2} \|v\|_{1,2}).$$

The condition (21b) is called *uniform ellipticity* for the operator in (20); cf. Exercise 16 in Section 2.3 and Exercise 8 at the end of this section. As a consequence of (21b) and (22), we claim that B_L is *positive* (or *coercive*) on $H_0^{1,2}(\Omega)$:

$$(25) \qquad B_L(u,u) \geq \epsilon_1 \|u\|_{1,2}^2 \quad \text{for } u \in H_0^{1,2}(\Omega),$$

where $\epsilon_1 > 0$. To verify (25), we use (21b) and (22) to obtain

$$B_L(u,u) = \int_\Omega \left(\sum_{i,j=1}^{n} a_{ij}(x) \frac{\partial u}{\partial x_i} \frac{\partial u}{\partial x_j} - c(x)u^2 \right) dx \geq \epsilon \int_\Omega |\nabla u|^2 \, dx,$$

Section 6.2: Application to the Dirichlet Problem   179

and then apply the Poincaré inequality to obtain (25) with $\epsilon_1 = \epsilon(1+C)^{-1}$.

Now (24) and (25) together show that $B_L(u,v)$ defines an inner product on $H_0^{1,2}(\Omega)$ whose associated norm is equivalent to $\|u\|_{1,2}^2$. This means that we can use the Riesz representation theorem to find a weak solution of the Dirichlet problem

(26) $$\begin{cases} Lu = f & \text{in } \Omega \\ u = 0 & \text{on } \partial\Omega, \end{cases}$$

where $f \in L^2(\Omega)$.

Theorem 3. *If $f \in L^2(\Omega)$ and the operator L of (20) satisfies (21a), (21b), and (22), then there exists a unique weak solution $u \in H_0^{1,2}(\Omega)$ of (26).*

Remarks. The symmetry assumption (21a) implies that L and its adjoint L' (as defined in Section 2.3) equal each other, so (20) is *self-adjoint*. Sometimes the adjoint L' is called the *formal adjoint* to distinguish it from the *Hilbert space adjoint* (encountered for bounded operators in Exercise 16 of Section 6.1, and for unbounded operators such as L in Section 6.6).

What happens if the symmetry condition (21a) fails? Given the form (20) of the operator L, we can try to replace a_{ij} by $(a_{ij} + a_{ji})/2$; this will make the leading order term symmetric, but at the cost of introducing first-order terms, $\sum_{k=1}^n b_k \partial/\partial x_k$. In the next subsection we discuss how to handle such nonsymmetric operators, provided they are still positive.

d. The Lax-Milgram Theorem

Clearly, the divergence form of the operator (20) was essential in our obtaining a weak solution of (26). For more general operators of the form

(27) $$L = \sum_{i,j=1}^n \frac{\partial}{\partial x_i} a_{ij}(x) \frac{\partial}{\partial x_j} + \sum_{k=1}^n b_k(x) \frac{\partial}{\partial x_k} + c(x),$$

the associated bilinear form $B_L(u,v)$ may *not* be symmetric in u,v. However, provided the form is still positive and bounded, the Riesz representation theorem may be generalized to the following.

Theorem 4 (Lax-Milgram). *If B is a positive, bounded bilinear form on a Hilbert space X and F is a bounded linear functional, then there exists a unique $u \in X$ such that*

$$F(v) = B(u,v) \quad \text{for} \quad v \in X.$$

Proof. Let Y be the subspace of $y \in X$ for which there corresponds $w_y \in X$ such that

(28) $$B(w_y, v) = \langle y, v \rangle \quad \text{for all } v \in X.$$

First observe that w_y is uniquely determined by y. Indeed, if w'_y also satisfies (28), then $B(w_y, v) = B(w'_y, v)$ implies $B(w_y - w'_y, v) = 0$ for all $v \in X$. In particular, with $v = w_y - w'_y$, the positivity $B(v,v) \geq \epsilon \|v\|^2$ of B implies $\|w_y - w'_y\| = 0$; in other words, $w_y = w'_y$.

In fact, positivity also yields $\epsilon \|w_y\|^2 \leq B(w_y, w_y) = \langle y, w_y \rangle \leq \|y\| \cdot \|w_y\|$, so that

(29) $$\epsilon \|w_y\| \leq \|y\| \qquad \text{for all } y \in Y.$$

From (29), it is easy to see that Y is a *closed* subspace of X (see Exercise 9). If Y is not all of X, then by the projection theorem there exists $z \in X$, $z \neq 0$, with $\langle z, Y \rangle = 0$. Now $B(z,x)$ defines a bounded linear functional on $x \in X$, so by the Riesz representation theorem, there exists $y \in X$ such that $B(z,x) = \langle y, x \rangle$ for all $x \in X$. But by the definition of Y, this implies $y \in Y$. Taking $x = z$, we have $B(z,z) = \langle y, z \rangle$, but $\langle y, z \rangle = 0$, so the positivity of B implies $z = 0$, a contradiction. Thus Y must be all of X. ♠

Application to Solving (26). If we want to solve (26) for an operator of the form (27), we can use the Lax-Milgram theorem provided we can show that the associated bilinear form

(30) $$B_L(u,v) = \int_\Omega \left(\sum_{i,j=1}^n a_{ij}(x) \frac{\partial u}{\partial x_i} \frac{\partial v}{\partial x_j} - \sum_{k=1}^n b_k v \frac{\partial u}{\partial x_k} - c(x) uv \right) dx$$

is bounded and positive on $X = H_0^{1,2}(\Omega)$. The boundedness is readily proved if we assume $a_{ij}(x), b_k(x), c(x) \in C(\overline{\Omega})$. To check positivity, let us assume the uniform ellipticity condition (21b) and the sign condition (22); then from (30) we find

$$B_L(u,u) \geq \int_\Omega \left(\epsilon |\nabla u|^2 - \sum_{k=1}^n b_k u \frac{\partial u}{\partial x_k} - c(x) u^2 \right) dx.$$

Now conditions on ϵ, b_k, and $c(x)$ may be obtained (cf. Exercise 8) under which the bilinear form will be positive, and the Lax-Milgram theorem applies to prove the existence of a weak solution of (26).

Remarks. (i) For second-order uniformly elliptic operators of the general form (27), the existence and uniqueness of solutions may need to be studied by other methods when the Lax-Milgram theorem does not apply; for example, in Section 8.4 the *method of continuity* is used to show existence and uniqueness of solutions of (26) whenever (22) holds. (ii) The Lax-Milgram theorem holds for mth-order elliptic operators in divergence form (see [John, 1] or [Bers-John-Schechter]).

Exercises for Section 6.2

1. Assume that $u \in H_0^{1,2}(\Omega)$ and $f \in L^2(\Omega)$, and (13) holds for all $v \in C_0^1(\Omega)$. Show that (13) holds for all $v \in H_0^{1,2}(\Omega)$.

2. Show that the Poincaré inequality generalizes to $p \neq 2$; that is, if Ω is a bounded domain and $1 \leq p < \infty$, then
$$\|u\|_p \leq C \|\nabla u\|_p, \qquad \text{where} \quad C = C(\Omega, p)$$
for all $u \in C_0^1(\Omega)$, and by completion for all $u \in H_0^{1,p}(\Omega)$.

3. Suppose Ω is unbounded but lies between two parallel hyperplanes P_1 and P_2.
   (a) Show that the Poincaré inequality holds for Ω.
   (b) Show that there is a weak solution of the Dirichlet problem $\Delta u = f$ in Ω and $u = 0$ on $\partial \Omega$, where $f \in L^2(\Omega)$.

4. For a bounded domain $\Omega \subset \mathbf{R}^n$, let
$$\lambda_1 = \inf_{u \in C_0^1(\Omega)} \frac{\int_\Omega |\nabla u|^2 \, dx}{\int_\Omega |u|^2 \, dx}.$$

   As we see in Section 7.2, λ_1 is indeed the first eigenvalue of the Laplacian as defined in Section 4.4.
   (a) Prove that $\lambda_1 > 0$.
   (b) Prove that, for every $f \in L^2(\Omega)$ and constant $c < \lambda_1$, the Dirichlet problem $\Delta u + cu = f$ in Ω, $u = 0$ on $\partial \Omega$ has a weak solution $u \in H_0^{1,2}(\Omega)$.

5. Consider the Dirichlet problem for the Laplace equation as studied in Chapter 4: $\Delta u = 0$ in Ω, $u = g$ on $\partial \Omega$, where Ω is a bounded domain. Suppose that $g \in C(\partial \Omega)$ satisfies $g = G|_{\partial \Omega}$ for some $G \in C^2(\overline{\Omega})$. How can you use the results in this section to find a *weak solution* for this problem?

6. For $n = 2$, compare the definition (21) of uniform ellipticity with the definition of ellipticity given in Section 2.2, and explain why they agree.

7. Use (29) to show that the space Y defined by the condition (28) is closed.

8. Assume that the operator in (27) satisfies (21b), $|b_k(x)| \leq \eta$ for $x \in \Omega$ in (27), and $c(x) \equiv 0$. Find a smallness condition on $\eta > 0$ (in terms of ϵ and the Poincaré constant) that guarantees that the bilinear form (30) is positive, so that the Lax-Milgram theorem applies to (26).

6.3 Duality and Compactness

In this section, we discuss dual spaces of Banach spaces, weak convergence, and various notions of compactness. Throughout this section X denotes a Banach space.

a. Dual Spaces

If X is a Banach space, let X^* denote the collection of all bounded linear functionals on X. X^* is called the *dual space of* X and is itself a Banach space under the operator norm (see Exercise 1). If X is a Hilbert space, then the Riesz representation theorem shows that X^* is *isometrically isomorphic* to X (i.e., the mapping $v \to F_v$ is one to one, onto, and satisfies $\|F_v\| = \|v\|$).

For a general Banach space X, we may define an injection of X into $X^{**} = (X^*)^*$ (the *dual of the dual space* or the *double dual*) as follows: if $u \in X$, then $\iota u \in X^{**}$ is defined by $\iota u(F) = F(u)$ for $F \in X^*$. Notice that

$$\|\iota u\|_{X^{**}} = \sup_{F \in X^*} \frac{|F(u)|}{\|F\|_{X^*}} \leq \sup_{F \in X^*} \frac{\|F\|_{X^*} \|u\|_X}{\|F\|_{X^*}} = \|u\|_X$$

so $\|\iota u\|_{X^{**}} \leq \|u\|_X$. On the other hand, by the corollary to the Hahn-Banach theorem, there exists $F \in X^*$ such that $\|F\|_{X^*} = 1$ and $F(u) = \|u\|_X$. Thus $\|\iota u\|_{X^{**}} = \|u\|_X$ and we discover that $\iota: X \to X^{**}$ is an isometry. In case $\iota: X \to X^{**}$ is also surjective (i.e., onto all of X^{**}), we say X is *reflexive*, and we may identify X with X^{**}.

Example 1. Every Hilbert space is reflexive.

Example 2. Let Ω be a domain in $\mathbf{R}^n$ and $X = L^p(\Omega)$, where $1 \leq p < \infty$. As we saw in Example 2 of Section 6.1, for $p' = p/(p-1)$ the map $v \to F_v$ defines an isometry $L^{p'}(\Omega) \to X^*$ (since $\|F_v\| = \|v\|_{p'}$). For $p = 2 = p'$, the Riesz representation theorem implies that this isometry is also a surjection, so that $(L^2(\Omega))^*$ may be identified with $L^2(\Omega)$. In fact, this surjectivity is also true for general $1 < p < \infty$ and is sometimes again called the *Riesz representation theorem* (cf. [Royden]): Every bounded linear functional F on $L^p(\Omega)$ is of the form $F(u) = \int uv\, dx$ for some $v \in L^{p'}(\Omega)$ satisfying $\|F\| = \|v\|_{p'}$. Hence $(L^p(\Omega))^*$ may be identified with $L^{p'}(\Omega)$. If we apply this identification to $L^{p'}(\Omega)$, we find that $(L^{p'}(\Omega))^*$ may be identified with $L^p(\Omega)$. In other words, $(L^p(\Omega))^{**}$ may be identified with $L^p(\Omega)$, so $L^p(\Omega)$ is reflexive for all $1 < p < \infty$. In fact, $(L^1(\Omega))^* = L^\infty(\Omega)$, but there are linear functionals on $L^\infty(\Omega)$ that are not given by integration with respect to an $L^1(\Omega)$ function; thus $L^1(\Omega)$ is *not* reflexive.

Example 3. Let Ω be a domain in $\mathbf{R}^n$ and $X = H_0^{1,2}(\Omega)$. As a Hilbert space, X is reflexive; but let us further investigate its dual space. For $F \in X^*$, the Riesz representation theorem produces a unique $v \in X$ with the property that $F(u) = (u, v)_1 = \int_\Omega \nabla u \cdot \nabla v\, dx$ for all $u \in X = H_0^{1,2}(\Omega)$. In particular, we may consider u as a test function and conclude $F(u) = \int_\Omega (-\Delta v) u\, dx$ as a distributional integral (i.e., $F = -\Delta v$ as distributions). We conclude that

$$\left(H_0^{1,2}(\Omega)\right)^* = \{\Delta v : v \in H_0^{1,2}(\Omega)\}.$$

Since differentiation reduces the order of the Sobolev space $[v \in H_0^{1,2}(\Omega)$ implies $\partial v/\partial x_i \in L^2(\Omega)]$, it is natural to write "$\Delta v \in H^{-1,2}(\Omega)$." For this reason, the notation $H^{-1,2}(\Omega)$ is sometimes used for the dual space of $H_0^{1,2}(\Omega)$, and $\|\cdot\|_{-1,2}$ as its norm. We can therefore express the unique weak solvability of Poisson's equation as follows.

Theorem 1. *If Ω is a bounded domain, then $\Delta : H_0^{1,2}(\Omega) \to H^{-1,2}(\Omega)$ is an isomorphism (one to one and onto), where $H^{-1,2}(\Omega)$ denotes the dual space of $H_0^{1,2}(\Omega)$.*

b. Weak Convergence

We know by continuity that $F(u_n) \to F(u)$ as $n \to \infty$, whenever $F \in X^*$ and $\|u_n - u\|_X \to 0$ as $n \to \infty$. But it may be that $F(u_n) \to F(u)$ occurs for all $F \in X^*$ without assuming the strong condition of norm convergence $u_n \to u$; this is called weak convergence. More precisely, a sequence $\{u_n\} \subset X$ *converges weakly* to $u \in X$ if $F(u_n) \to F(u)$ for all $F \in X^*$. Weak convergence is sometimes denoted $u_n \rightharpoonup u$ to distinguish it from norm convergence $u_n \to u$; however, we usually write "$u_n \rightharpoonup u$ weakly in X" to avoid any possible misunderstanding.

Let us consider a simple example that shows that weak convergence is indeed strictly weaker than norm convergence.

Example 4. Let $X = \ell^2 = \{\{a_i\}_{i=1}^\infty : \sum_{i=1}^\infty a_i^2 < \infty, a_i \in \mathbf{R}\}$, which is a Hilbert space under

$$\langle \{a_i\}, \{b_i\} \rangle = \sum_{i=1}^\infty a_i b_i \quad \text{and} \quad \|\{a_i\}\|^2 = \sum_{i=1}^\infty a_i^2.$$

Let $A_1 = \{1, 0, \ldots\}, A_2 = \{0, 1, 0, \ldots\}$, and so on, so that $\|A_i\| = 1$. Let us now consider the sequence $\{A_i\}$. Notice that $\|A_i - A_j\| = \sqrt{2}$ for all i, j, so $\{A_i\}$ is not Cauchy, hence does not converge in the norm. On the other hand, by the Riesz representation theorm, any $F \in X^*$ may be represented by some $B = \{b_i\} \in X$; that is, $F(A) = \langle A, B \rangle$ for every $A \in X$. If we apply F to our sequence $\{A_i\}$, we find $F(A_i) = \langle A_i, B \rangle = b_i \to 0$ as $i \to \infty$. We conclude that $\{A_i\}$ converges weakly to zero: $\{A_i\} \rightharpoonup 0$. ♣

If $\{F_n\}$ is a sequence in X^*, then the weak convergence corresponding to functionals in $\iota(X)$ (instead of all of X^{**}) is called *weak* convergence;* in other words, $F_n \xrightarrow{*} F$ if $F_n(u) \to F(u)$ for all $u \in X$. Of course, if X is reflexive, then weak* convergence = weak convergence in X^*.

c. Compactness

If S denotes any topological space, recall that S is *compact* if every open cover $\{U_\alpha\}$ of S has a finite subcover. Also recall that, if S is a metric space, then S is compact if and only if S is complete and *totally bounded* (i.e., S is covered by a finite number of ϵ-balls, for any $\epsilon > 0$).

Let us introduce two additional types of compactness.

Definitions. (i) A topological space S is *sequentially compact* if every infinite sequence contains a convergent subsequence.

(ii) $S \subset X$ is *weakly sequentially compact* if every infinite sequence in S contains a subsequence that converges weakly to a point in S.

It is obvious that, for subsets S of X, *sequentially compact* $\Rightarrow$ *weakly sequentially compact*. Somewhat less obvious is that *compact* $\Leftrightarrow$ *sequentially compact* (see Exercise 5).

A familiar example of compactness in an infinite-dimensional setting is provided by the Arzela-Ascoli theorem:

Example 5. Let $X = C(K)$ as in Example 1 of Section 6.1, and let $S \subset X$. Suppose S is *equicontinuous*; that is, for any $\epsilon > 0$ there exists $\delta > 0$ such that

(31) $\qquad |u(x) - u(y)| < \epsilon \quad$ for all $\quad |x - y| < \delta \quad$ and $\quad u \in S$.

Also, suppose S is *uniformly bounded*; that is,

(32) $\qquad |u(x)| \leq M \quad$ for all $\quad x \in K \quad$ and $\quad u \in S$.

Then the *Arzela-Ascoli theorem* concludes that S is precompact (i.e., the closure $\overline{S}$ is compact in X). This is equivalent to saying that every sequence of functions $\{u_n\}$ in S contains a subsequence $\{u_{n_k}\}$ that converges uniformly on K to a function $u \in X$.

In a finite-dimensional Banach space X (e.g., $\mathbf{R}^n$), a subset S is compact if and only if it is closed and bounded. This certainly fails for infinite-dimensional Banach spaces. For example, in Example 4 the unit ball $S = \{\{a_i\} : \|\{a_i\}\| \leq 1\}$ is closed and bounded but not compact because the collection $\{A_i\}$ does not contain a convergent subsequence. On the other hand, $\{A_i\}$ is weakly convergent (in fact to zero), so we might expect that a closed bounded set is at least *weakly* sequentially compact. As the following result shows, this expectation is equivalent to X being reflexive.

Theorem 2. *A Banach space X is reflexive if and only if every closed bounded set is weakly sequentially compact.*

For a proof of this theorem, see [Conway]. As an immediate consequence,

Corollary. *Every bounded sequence in a Hilbert space contains a weakly convergent subsequence.*

If the Hilbert space is *separable* (i.e., if it contains a countable dense set), then Exercise 7 shows that the corollary of Theorem 2 may be obtained independently from the theorem. We use the existence of weakly convergent subsequences in Chapter 7 when studying the critical points of continuous functionals on Banach spaces.

When X is not reflexive, the following result is sometimes useful.

Theorem 3 (Alaoglu's Theorem). *If X is a separable Banach space and $\{F_j\}_{j=1}^{\infty}$ is a bounded sequence in X^*, then there is a weak* convergent subsequence.*

For a proof, see Exercise 9.

The assumption of separability is not too restrictive in applications, since it is usually satisfied by function spaces. For example, as a consequence of the *Weierstrass approximation theorem*, the collection of all polynomials is a dense set in $C(\overline{\Omega})$, where Ω is a bounded domain; hence the Banach space $C(\overline{\Omega})$ is separable. This fact may also be used to show that the L^p and Sobolev spaces are separable (cf. Exercise 8); the following well-known result is useful (cf. [Royden]):

Lemma. *If Ω is an arbitrary domain in $\mathbf{R}^n$, then $C_0(\Omega)$, the continuous functions with compact support, is dense in $L^p(\Omega)$ for $1 \leq p < \infty$.*

Exercises for Section 6.3

1. If X is a Banach space, show that the dual space X^* is itself a Banach space under the operator norm:
   (a) Show that X^* is a vector space.
   (b) Show that the operator norm gives a norm on X^*.
   (c) Show that X^* is complete.

2. A Banach space X is reflexive if and only if X^* is reflexive.

3. Suppose X is a Hilbert space and the sequence $\{x_n\}$ converges weakly to x. Show that $\|x_n\| \to \|x\|$ implies $x_n \to x$.

4. If X and Y are Banach spaces, and $T : X \to Y$ is a bounded linear operator, show that $x_n \rightharpoonup x$ weakly in X implies $Tx_n \rightharpoonup Tx$ weakly in Y.

5. If a metric space S is compact, then it is sequentially compact. *Note:* The converse is also true but is much more difficult to prove; see books on topology.

6. Prove that the unit ball in a Banach space X is compact if and only if X is finite-dimensional.

7. If X is a separable Hilbert space, show that every bounded sequence contains a weakly convergent subsequence.

8. Suppose Ω is an arbitrary domain in $\mathbf{R}^n$ and $1 \leq p < \infty$.
   (a) Using the lemma of this section, prove that $L^p(\Omega)$ is separable.
   (b) If N is a positive integer, show that the N-fold direct product $(L^p(\Omega))^N = L^p(\Omega) \times \cdots \times L^p(\Omega)$ with the norm $\|(u_1, u_2, \ldots, u_N)\| = (\|u_1\|_p^p + \cdots + \|u_N\|_p^p)^{1/p}$ is a separable Banach space.
   (c) Let $N = n + 1$ and show that the map $T : W^{1,p}(\Omega) \to (L^p(\Omega))^N$, defined by $Tu = (u, u_{x_1}, \ldots, u_{x_n})$, is linear, continuous, one to one, and its image $R = T(W^{1,p}(\Omega))$ is a closed subspace of $(L^p(\Omega))^N$.
   (d) Conclude that $H_0^{1,p}(\Omega)$, $H^{1,p}(\Omega)$, and $W^{1,p}(\Omega)$ are all separable.

9. If X is a separable Banach space and $\{F_j\}_{j=1}^{\infty}$ is a bounded sequence in X^*, show that there is a weak* convergent subsequence.

6.4 Sobolev Imbedding Theorems

In this section, we consider some relationships between some of the Banach spaces introduced in Section 6.1. Throughout this section, Ω denotes a domain of $\mathbf{R}^n$ and C denotes a generic constant (whose value may change with different occurrences).

If X and Y are Banach spaces such that $X \subset Y$ (as sets) and

(33) $$\|u\|_Y \leq C \|u\|_X \quad \text{for all} \quad u \in X,$$

then we say that X is (continuously) imbedded in Y; equivalently, the identity map $\iota: X \to Y$ is a bounded linear map. Whenever we write $X \subset Y$ for Banach spaces, we intend that (33) shall hold. For example, we know that $H_0^{1,p}(\Omega) \subset L^p(\Omega)$, but we might wonder when $H_0^{1,p}(\Omega) \subset L^q(\Omega)$ for $q \neq p$. This is one of the questions we wish to answer in this section. [Imbeddings of $H^{1,p}(\Omega)$ require some regularity conditions on the boundary of Ω and will be discussed in Section 6.5.]

Incidentally, to prove $X \subset Y$ we need only verify the inequality (33) for u in a dense subspace of X: if $u_j \to u$ in X, then we may apply (33) to $u_j - u_k$ to conclude that $u_j - u_k$ is Cauchy in Y and hence u_j converges to some $w \in Y$; we must have $w = u$, and continuity of the norms implies (33) holds for u. (This is the same "by completion" argument used in proving the Poincaré inequality.)

a. The Sobolev Inequality for $p < n$

We begin our investigation of imbedding theorems with an important inequality:

Sobolev Inequality. *If Ω is a domain in $\mathbf{R}^n$, $1 \leq p < n$, and $q = np/(n-p)$, then there exists a constant $C = C(n,p)$ such that*

(34) $$\|u\|_q \leq C \|\nabla u\|_p$$

for all $u \in C_0^1(\Omega)$.

In this inequality, $\|\nabla u\|_p$ is an abbreviation for the L^p-norm of the function $|\nabla u|$; namely, $\|\nabla u\|_p = (\int |\nabla u|^p \, dx)^{1/p}$.

The proof of (34) is elementary (using only the fundamental theorem of calculus and the Hölder inequality), but rather lengthy, and so has been relegated to the end of this section. It is also possible to obtain the "best constant" C, which is sometimes important in applications but does not concern us here. (For a further discussion of the best constant, see [Aubin] or [Talenti].)

Section 6.4: Sobolev Imbedding Theorems

The Sobolev inequality is easily generalized to obtain the following:

(35) $$\|u\|_q \leq C \|\nabla u\|_p^\lambda \|u\|_r^{1-\lambda} \qquad C = C(n,p,r,q)$$

for all $u \in C_0^1(\Omega)$, provided $0 \leq \lambda \leq 1$ satisfies

(36) $$\frac{1}{q} = \lambda \left(\frac{1}{p} - \frac{1}{n} \right) + (1-\lambda)\frac{1}{r}$$

(see Exercise 3). Taking $\lambda = 0$ and $\lambda = 1$, we see that (35) *interpolates* between the equality $\|u\|_q = \|u\|_q$ and the Sobolev inequality. Moreover, it will be useful in determining when we have an imbedding $H_0^{1,p}(\Omega) \subset L^q(\Omega)$, especially when we observe (see Exercise 5) that $\|u\|_{1,p}$ is equivalent to the norm

(37) $$\|u\|_{1,p}^* \equiv \left(\|\nabla u\|_p^p + \|u\|_p^p \right)^{1/p}.$$

b. The Sobolev Imbedding Theorem for $p < n$

Let us now use the Sobolev inequality to establish imbedding theorems when $p < n$.

Theorem 1. *If Ω is a domain in $\mathbf{R}^n$, then $H_0^{1,p}(\Omega) \subset L^q(\Omega)$ is a continuous imbedding provided $p < n$ and $p \leq q \leq \frac{np}{n-p}$; in particular, the inequality*

(38) $$\|u\|_q \leq C\|u\|_{1,p} \qquad C = C(n,p,q)$$

holds for all $u \in C_0^1(\Omega)$, and by completion for all $u \in H_0^{1,p}(\Omega)$.

Proof. Take $r = p$ in (36) to obtain

$$\frac{1}{q} = \lambda \left(\frac{1}{p} - \frac{1}{n} \right) + (1-\lambda)\frac{1}{p} = \frac{1}{p} - \frac{\lambda}{n}$$

which can be satisfied with $0 \leq \lambda \leq 1$ since $p \leq q \leq \frac{np}{n-p}$. Then (35) becomes

(39) $$\|u\|_q \leq C \|\nabla u\|_p^\lambda \|u\|_p^{1-\lambda}.$$

We next invoke *Young's inequality*

(40) $$ab \leq \frac{a^s}{s} + \frac{b^t}{t}$$

which holds for $a,b > 0$ provided $s,t > 1$ satisfy $s^{-1} + t^{-1} = 1$. [See Exercise 4 for a proof of (40).] With $s = 1/\lambda$ and $t = 1/(1-\lambda)$ we find

$$\|\nabla u\|_p^\lambda \|u\|_p^{1-\lambda} = \left(\|\nabla u\|_p^{2\lambda} \|u\|_p^{2(1-\lambda)} \right)^{1/2}$$

$$\leq \left(\frac{\|\nabla u\|_p^2}{1/\lambda} + \frac{\|u\|_p^2}{1/(1-\lambda)} \right)^{1/2}$$

so that (39) implies (38) for $u \in C_0^1(\Omega)$ (cf. Exercise 5). Since $C_0^1(\Omega)$ is dense in $H_0^{1,p}(\Omega)$, this completes the proof. ♠

It is important to emphasize that (34) and (38) hold for an *arbitrary* domain Ω; if we assume that Ω is a *bounded* domain, then we may consider more general q, but at the cost of allowing the constant in (34) to depend on Ω.

Theorem 2. *If Ω is a bounded domain in $\mathbf{R}^n$, $p < n$, and $1 \leq q \leq \frac{np}{n-p}$, then*

$$(41) \qquad \|u\|_q \leq C \, \|\nabla u\|_p \qquad C = C(n,p,q,\Omega)$$

holds for all $u \in C_0^1(\Omega)$, and by completion for $u \in H_0^{1,p}(\Omega)$; consequently, $H_0^{1,p}(\Omega) \subset L^q(\Omega)$ is a continuous imbedding.

Proof. For $r > 1$, the Hölder inequality implies

$$\|u\|_q^q = \int_\Omega 1 \cdot |u|^q \, dx \leq \left(\int_\Omega 1^{r'} \, dx \right)^{1/r'} \cdot \left(\int_\Omega |u|^{rq} \, dx \right)^{1/r} = (\text{vol}\,\Omega)^{1/r'} \|u\|_{rq}^q$$

where $r' = r/(r-1)$. Assuming $1 \leq q < q_0 = \frac{np}{n-p}$ and taking $r = q_0/q > 1$, this implies

$$(42) \qquad \|u\|_q \leq (\text{vol}\,\Omega)^{\frac{1}{r'q}} \|u\|_{q_0}.$$

Combined with the Sobolev inequality (34), (42) yields the inequality (41) for $u \in C_0^1(\Omega)$. Since $\|\nabla u\|_p \leq C\|u\|_{1,p}$, we obtain the imbedding $H_0^{1,p}(\Omega) \subset L^q(\Omega)$, and the extension of (42) to $u \in H_0^{1,p}(\Omega)$ follows as before. ♠

c. The Sobolev Inequality and Imbedding for $p > n$

In the previous section we found that $\nabla u \in L^p(\Omega)$ for $p < n$ implies $u \in L^q(\Omega)$ for $q = np/(n-p)$. What happens if $p > n$?

Example. Let $u \in C^\infty(\mathbf{R}^n \setminus \{0\})$ such that

$$u(x) = \begin{cases} |x|^a & \text{for } |x| < 1 \\ 0 & \text{for } |x| > 2. \end{cases}$$

Notice that $\nabla u = O(|x|^{a-1})$ as $|x| \to 0$, so $\nabla u \in L^p \Leftrightarrow p(a-1) > -n \Leftrightarrow a > 1 - \frac{n}{p}$. Thus for $p > n$, $\nabla u \in L^p$ implies u is *continuous* at $|x| = 0$. ♣

This example suggests that $\nabla u \in L^p(\Omega)$ for $p > n$ implies $u \in C(\Omega)$, which is a much stronger conclusion than we obtained for $p < n$. In fact, we have the following result.

Section 6.4: Sobolev Imbedding Theorems

Theorem 3. *If Ω is a bounded domain in $\mathbf{R}^n$, then*

(43) $$H_0^{1,p}(\Omega) \subset C(\overline{\Omega})$$

is a continuous imbedding provided $p > n$; in particular, the inequality

(44) $$\|u\|_\infty \leq C\|\nabla u\|_p \qquad C = C(n,p,\Omega)$$

holds for all $u \in H_0^{1,p}(\Omega)$.

Proof. Extend $u \in C_0^1(\Omega)$ by zero outside Ω, so $u \in C_0^1(\mathbf{R}^n)$. For $x, y \in \Omega$, let $|x - y| = \sigma$, and let B_σ denote the ball of radius σ centered at $(x+y)/2$. If we let $\bar{\omega}_n$ denote the volume of the unit ball in $\mathbf{R}^n$, then B_σ has volume $|B_\sigma| = \bar{\omega}_n \sigma^n$. Now consider $z \in B_\sigma$ and parameterize the line between x and z by $x + t(z-x)$ for $0 \leq t \leq 1$. Then

$$u(z) - u(x) = \int_0^1 \left[\frac{d}{dt} u(x + t(z-x))\right] dt$$

$$= \int_0^1 \nabla u(x + t(z-x)) \cdot (z-x)\, dt.$$

Let us integrate z over B_σ to obtain

$$\left|\int_{B_\sigma} u(z)\,dz - \bar{\omega}_n \sigma^n u(x)\right| \leq \int_{B_\sigma} \int_0^1 |\nabla u(x + t(z-x))|\,|z - x|\,dt\,dz.$$

But $|z-x| \leq 2\sigma$, and if we change the integration variable to $\bar{z} = x + t(z-x)$, we obtain

$$\left|\int_{B_\sigma} u(z)\,dz - \bar{\omega}_n \sigma^n u(x)\right| \leq 2\sigma \int_0^1 t^{-n} \left(\int_{\bar{B}_{t\sigma}} |\nabla u(\bar{z})|\,d\bar{z}\right) dt,$$

where $\bar{B}_{t\sigma}$ is a certain ball of radius $t\sigma$. Using Hölder's inequality

$$\int_{\bar{B}_{t\sigma}} |\nabla u(\bar{z})|\,d\bar{z} \leq \left(\int_{\bar{B}_{t\sigma}} d\bar{z}\right)^{1/p'} \left(\int_{\bar{B}_{t\sigma}} |\nabla u(\bar{z})|^p\,d\bar{z}\right)^{1/p}$$

$$= \bar{\omega}_n^{1/p'} (t\sigma)^{n/p'} \left(\int_{\bar{B}_{t\sigma}} |\nabla u(\bar{z})|^p\,d\bar{z}\right)^{1/p},$$

and dominating the last term by $\|\nabla u\|_p$, we obtain

$$\left|\frac{1}{\bar{\omega}\sigma^n} \int_{B_\sigma} u(z)\,dz - u(x)\right| \leq 2\sigma^{1-\frac{n}{p}} \bar{\omega}_n^{-\frac{1}{p}} \left(\int_0^1 t^{-\frac{n}{p}}\,dt\right) \|\nabla u\|_p.$$

Combining this with a similar inequality with y in place of x and recalling $\sigma = |x - y|$, we obtain *Morrey's inequality*

(45) $$|u(x) - u(y)| \leq C|x - y|^{1-\frac{n}{p}} \|\nabla u\|_p,$$

where $C = 4\bar{\omega}_n^{-\frac{1}{p}} \int_0^1 t^{-n/p} dt < \infty$ since $n/p < 1$. Fixing $y \in \partial\Omega$ and taking the maximum of (45) for $x \in \Omega$ yields (44). Since $C_0^1(\Omega)$ is dense in $H_0^{1,p}(\Omega)$, this establishes the theorem. ♠

Remarks. (i) The estimate (45) is much sharper than (44) and will enable us to improve the imbedding (43) when we consider Hölder continuity and compactness properties in the next section. (ii) For $p = n$ we might suppose from (41) and (44) that $H_0^{1,p}(\Omega)$ is contained in $L^\infty(\Omega)$, but this is *not* the case (see Exercise 6). In fact, for $p = n$ we have $H_0^{1,p}(\Omega) \subset L^q(\Omega)$ if $p \leq q < \infty$ (see Exercise 7).

The boundedness of the domain in Theorem 3 is only required in order to ensure that $C(\overline{\Omega})$ is a Banach space; indeed, if Ω is not bounded, then $\|u\|_\infty$ need not be finite for functions u that are continuous on $\overline{\Omega}$. However, we may introduce the space

(46) $$C_B(\overline{\Omega}) = \{u \in C(\overline{\Omega}) : u \text{ is uniformly bounded on } \overline{\Omega}\},$$

which is a Banach space, even when Ω is not bounded. Now we may generalize (43) to find that, for an arbitrary domain Ω,

(47) $$H_0^{1,p}(\Omega) \subset C_B(\overline{\Omega})$$

is a continuous imbedding whenever $p > n$.

d. Proof of the Sobolev Inequality for $p < n$

The proof of the Sobolev inequality involves clever manipulations with the Hölder inequality (4) and uses the elementary inequality

(48) $$\left(\prod_{i=1}^n b_i\right)^{\frac{1}{n}} \leq \frac{1}{n}\sum_{i=1}^n b_i \quad \text{for} \quad b_i \geq 0,$$

between the geometric and arithmetics means.

To illustrate the proof, let us take $n = 3$; all steps extend immediately to general n (cf. [Gilbarg-Trudinger]). Let $u \in C_0^1(\Omega)$, but consider $u \in C_0^1(\mathbf{R}^3)$ by extending u to be zero outside of Ω. We proceed in two steps: (i) $p = 1$ and (ii) $1 < p < 3$.

(i) $p = 1$. By the fundamental theorem of calculus,

$$|u(x)| = \left|\int_{-\infty}^{x_i} u_{x_i} dx_i\right| \leq \int_{-\infty}^{\infty} |u_{x_i}| dx_i.$$

Taking the product over $i = 1, 2, 3$, we find

$$|u(x)|^3 \leq \prod_{i=1}^{3} \int_{-\infty}^{\infty} |u_{x_i}| \, dx_i.$$

If we take the square root of both sides and integrate over $\mathbf{R}^3$, we obtain

(49) $$\int_{\mathbf{R}^3} |u(x)|^{3/2} \, dx \leq \int_{\mathbf{R}^3} \prod_{i=1}^{3} \left(\int_{-\infty}^{\infty} |u_{x_i}| \, dx_i \right)^{1/2} dx.$$

Writing all one-dimensional integrations simply as $\int$, and $dx_j dx_k$ as $dx_{j,k}$, the right-hand side of (49) may be written as

(50) $$\int_{\mathbf{R}^2} \left(\int |u_{x_1}| \, dx_1 \right)^{1/2} \left[\int \left(\int |u_{x_2}| \, dx_2 \int |u_{x_3}| \, dx_3 \right)^{1/2} dx_1 \right] dx_{2,3}$$

since $(\int |u_{x_1}| \, dx_1)^{1/2}$ is independent of x_1. Treating $\int |u_{x_i}| \, dx_i$ as a function of x_1, we may apply the Hölder inequality (with $p = 2$) to the x_1-integration to bound the bracketed term in (50) by

$$\left(\int \int |u_{x_2}| \, dx_{1,2} \right)^{1/2} \left(\int \int |u_{x_3}| \, dx_{1,3} \right)^{1/2}.$$

Thus (50) is bounded by

(51) $$\int \left(\int_{\mathbf{R}^2} |u_{x_2}| \, dx_{1,2} \right)^{1/2} \left[\int \left(\int |u_{x_1}| dx_1 \int_{\mathbf{R}^2} |u_{x_3}| dx_{1,3} \right)^{1/2} dx_2 \right] dx_3$$

where we have used the fact that $\int_{\mathbf{R}^2} |u_{x_2}| \, dx_1 dx_2$ is independent of x_2 to pull it outside that integration. Again we may apply the Hölder inequality to bound the bracketed term in (51) by

$$\left(\int_{\mathbf{R}^2} |u_{x_1}| \, dx_{1,2} \right)^{1/2} \left(\int_{\mathbf{R}^3} |u_{x_3}| \, dx_{1,2,3} \right)^{1/2},$$

so that (51) is bounded by

(52) $$\left[\int \left(\int_{\mathbf{R}^2} |u_{x_2}| \, dx_{1,2} \right)^{1/2} \left(\int_{\mathbf{R}^2} |u_{x_1}| \, dx_{1,2} \right)^{1/2} dx_3 \right] \left(\int_{\mathbf{R}^3} |u_{x_3}| \, dx \right).$$

One more application of the Hölder inequality bounds (52) by

$$\left(\int_{\mathbf{R}^3} |u_{x_1}| \, dx_{1,2,3} \right)^{1/2} \left(\int_{\mathbf{R}^3} |u_{x_2}| \, dx_{1,2,3} \right)^{1/2} \left(\int_{\mathbf{R}^3} |u_{x_3}| \, dx_{1,2,3} \right)^{1/2},$$

which has established

$$(53) \qquad \int_{\mathbf{R}^3} |u|^{3/2}\, dx \le \prod_{i=1}^{3}\left(\int_{\mathbf{R}^3} |u_{x_i}|\, dx\right)^{1/2}.$$

Taking 2/3-power in (53) and applying (48) yields

$$\|u\|_{3/2} \le \frac{1}{3}\sum_{i=1}^{3} \|u_{x_i}\|_1 \le \frac{1}{\sqrt{3}} \|\nabla u\|_1,$$

which is the Sobolev inequality for $p = 1$; moreover, we have explicitly found the constant $C(3,1) = 1/\sqrt{3}$.

(ii) $1 < p < 3$. If we replace u by $|u|^r$, where $r > 1$, and use

$$(|u|^r)_{x_i} = \pm r|u|^{r-1} u_{x_i},$$

then the previous case yields

$$\| |u|^r \|_{3/2} \le \frac{r}{\sqrt{3}} \| |u|^{r-1} \nabla u \|_1.$$

Applying the Hölder inequality to the right-hand side, we get

$$(54) \qquad \| |u|^r \|_{3/2} \le \frac{r}{\sqrt{3}} \| |u|^{r-1} \|_{p'} \|\nabla u\|_p,$$

where $p' = p/(p-1)$. Now we choose r in order to match the powers of $|u|$ being integrated on both sides:

$$\frac{3}{2}r = (r-1)p' \quad \Leftrightarrow \quad r = \frac{2p}{3-p}.$$

Notice that indeed $r > 1$, since $1 < p < 3$; moreover, (54) becomes

$$\left(\int |u|^{\frac{3p}{3-p}}\, dx\right)^{2/3} \le \frac{2p}{\sqrt{3}(3-p)} \left(\int |u|^{\frac{3p}{3-p}}\, dx\right)^{\frac{p-1}{p}} \|\nabla u\|_p.$$

Using

$$\frac{2}{3} - \frac{p-1}{p} = \frac{3-p}{3p}$$

this simplifies to

$$\|u\|_q \le \frac{2p}{\sqrt{3}(3-p)} \|\nabla u\|_p, \qquad \text{where}\quad q = \frac{3p}{3-p},$$

which is the Sobolev inequality with $C(3,p) = \frac{2p}{\sqrt{3}(3-p)}$.

Remark. For general n, the preceding proof establishes the Sobolev inequality with the value $C(n,p) = \frac{p(n-1)}{\sqrt{n}(n-p)}$. For the best possible value for the constant C, see [Aubin] or [Talenti].

Exercises for Section 6.4

1. If $X \subset Y$ is a continuous imbedding of Banach spaces such that X is dense in Y, prove that $Y^* \subset X^*$ is a continuous imbedding of their dual spaces.

2. If Ω is a domain in $\mathbf{R}^n$, and $X = H_0^{1,p}(\Omega)$ for some $p < n$, show that $F(f) = \int_\Omega fg\,dx$ defines an element $F \in X^*$ if $g \in L^r(\Omega)$, where $1/r = 1 - (1/p) + (1/n)$.

3. Prove (35) under the condition (36).

4. Prove Young's inequality (40).

5. (a) Prove that $\|\cdot\|_{1,p}$ as in (10) and $\|\cdot\|_{1,p}^*$ as in (37) are equivalent norms by showing

   $$C_1(a_1^p + \cdots + a_n^p) \leq (a_1^2 + \cdots + a_n^2)^{p/2} \leq C_2(a_1^p + \cdots + a_n^p) \quad \text{for all } a_i \geq 0.$$

   (b) In the proof of (38), the inequality

   $$(c_1\|\nabla u\|_p^2 + \|u\|_p^2)^{1/2} \leq C\|u\|_{1,p}$$

   is used near the end of the argument. Why is this true?

6. If $p = n$, then we might suppose that $H_0^{1,p}(\Omega)$ is contained in $L^\infty(\Omega)$. However, consider $u(x) = \log(\log(2/|x|)) - \log(\log 2)$ on $\Omega = B_1(0)$ for $n \geq 2$. Show that $\|u\|_{1,n} < \infty$ but $u \notin L^\infty$. [Using mollification as in the next section, it is easy to show that $u \in H_0^{1,p}(\Omega)$.]

7. If Ω is a bounded domain in $\mathbf{R}^n$ and $p = n$, show that $H_0^{1,p}(\Omega) \subset L^q(\Omega)$ is a continuous imbedding for all $p \leq q < \infty$. (The boundedness of the domain may be removed by a patching argument, cf. [Adams].)

8. If $X \subset Y$ is a continuous imbedding of Banach spaces and $S \subset X$ is totally bounded, then S is totally bounded in Y.

6.5 Refinements and Generalizations

In this section we discuss several refinements of the functional analysis of the previous sections in this chapter. We begin with Hölder continuity, which provides a refinement of the C^k spaces and is essential for sharp regularity results. We next introduce mollifiers and show how they may be used to approximate continuous or L^p functions by smooth functions. These approximations not only provide density results, including the identification of $H^{1,p}$ and $W^{1,p}$, but they also yield an L^p-compactness criterion. This criterion enables us to treat the issue of compactness in the Sobolev imbedding theorems. Finally, we discuss the generalization of many of these results to higher-order Sobolev spaces.

a. Hölder Continuity

It is necessary at times to refine the collection of function spaces $C^k(\Omega)$. For example, notice that there is an "optimal" value of $q = (np)/(n-p)$ in the Sobolev imbedding for $p < n$. In order to find a similar optimal imbedding when $p > n$, we must introduce the notion of *Hölder continuity*. If $0 \leq \alpha < 1$ and u is defined and continuous in a neighborhood U of x_0, then we say that u is *Hölder continuous at x_0 with exponent α* if

$$[u]_{\alpha;x_0} \equiv \sup_{x \in U} \frac{|u(x) - u(x_0)|}{|x - x_0|^\alpha} < \infty. \tag{55}$$

Notice that $[u]_{\alpha;x_0}$ depends on the choice of U, but the condition that u be Hölder continuous at x_0 only requires (55) for some neighborhood U. If (55) is finite for every $x_0 \in \Omega$, then we say u is *(locally) Hölder continuous with exponent α in Ω* and denote the collection of such functions by $C^\alpha(\Omega)$. Notice that $C^0(\Omega) = C(\Omega)$.

Observe that $C^\alpha(\Omega)$ is not itself a Banach space since the supremum in (55) need not be uniform in $x_0 \in \Omega$. On the other hand, suppose Ω is a bounded domain and u extends continuously to the closure $\overline{\Omega}$. Moreover, if

$$[u]_\alpha \equiv \sup_{x,y \in \overline{\Omega}} \frac{|u(x) - u(y)|}{|x - y|^\alpha} < \infty, \tag{56}$$

then we say u is *Hölder continuous with exponent α in Ω* and denote the collection of such functions by $C^\alpha(\overline{\Omega})$. Notice that $C^\alpha(\overline{\Omega})$ is a Banach space under the norm

$$|u|_\alpha = \|u\|_\infty + [u]_\alpha = \sup_{x \in \overline{\Omega}} |u(x)| + \sup_{x,y \in \overline{\Omega}} \frac{|u(x) - u(y)|}{|x - y|^\alpha} \tag{57}$$

and that we have continuous imbeddings

$$C^\beta(\overline{\Omega}) \subset C^\alpha(\overline{\Omega}) \subset C(\overline{\Omega}) \quad \text{for } 0 \leq \alpha \leq \beta < 1. \tag{58a}$$

In fact, by extending functions to a neighborhood of Ω, it is not difficult to see that

$$C^1(\overline{\Omega}) \subset C^\alpha(\overline{\Omega}) \quad \text{for } 0 \leq \alpha < 1, \text{ provided } \partial\Omega \text{ is } C^1. \tag{58b}$$

(The continuous imbedding (58b) also holds for $\partial\Omega \notin C^1$, under additional convexity assumptions on Ω; cf. [Gilbarg-Trudinger] and [Adams].)

Using these Hölder spaces, we can replace (43) by an optimal imbedding:

Section 6.5: Refinements and Generalizations

Theorem 1 (Sharp Sobolev Imbedding Theorem for p > n). *If $p > n$ and Ω is a bounded domain in $\mathbf{R}^n$, then*

(59) $$H_0^{1,p}(\Omega) \subset C^\alpha(\overline{\Omega})$$

is a continuous imbedding for $\alpha = 1 - (n/p)$.

Proof. The proof of this theorem easily follows from Morrey's inequality (45); see Exercise 1. ♠

We can also introduce higher-order Hölder spaces. Namely, for a nonnegative integer m and $0 \leq \alpha < 1$, we may define $C^{m,\alpha}(\Omega)$ to be those functions in $C^m(\Omega)$ whose mth-order derivatives are all in $C^\alpha(\Omega)$, and define $C^{m,\alpha}(\overline{\Omega})$ to be those functions in $C^{m,\alpha}(\Omega)$ whose mth-order derivatives extend continuously to $\overline{\Omega}$ and for which

(60)
$$|u|_{m,\alpha} = \|u\|_{m,\infty} + \max_{|\beta|=m}[D^\beta u]_\alpha$$
$$= \max_{0\leq|\beta|\leq m}\sup_{x\in\overline{\Omega}}|D^\beta u(x)| + \max_{|\beta|=m}\sup_{x,y\in\overline{\Omega}}\frac{|D^\beta u(x) - D^\beta u(y)|}{|x-y|^\alpha}.$$

Then $C^{m,\alpha}(\overline{\Omega})$ is a Banach space under the norm (60). These spaces will be useful later in this section when we have defined higher-order Sobolev spaces and wish to investigate their imbedding properties.

Another important application of Hölder spaces occurs in the study of regularity for elliptic operators. For example, let

(61) $$u(x) = \frac{1}{\omega_n}\int_\Omega \frac{f(y)}{|x-y|^{n-2}}\,dy$$

denote the domain potential discussed in Section 4.2b. According to the Proposition of Section 4.2, $f \in C^0(\overline{\Omega})$ implies $u \in C^1(\Omega)$; but we cannot conclude $u \in C^2(\Omega)$, since differentiating under the integral sign in (61) creates a singularity of order $|x-y|^{-n}$ that is not integrable at $y = x$. However, if $f \in C^\alpha(\overline{\Omega})$ for $\alpha > 0$, then we can use (56) to weaken the singularity to order $|x-y|^{\alpha-n}$, which *is* integrable.

Theorem 2 (C^2-Regularity of Domain Potentials). *If Ω is a bounded domain and $f \in C^\alpha(\overline{\Omega})$, then (61) defines $u \in C^2(\Omega)$, which satisfies $\Delta u = f$ in Ω.*

The proof of this theorem is outlined in Exercise 3. As an immediate consequence, we are able to solve the nonhomogeneous Dirichlet problem for bounded domains that satisfy the boundary regularity of Section 4.3.

Corllary. *If Ω is a bounded domain with regular boundary, $f \in C^\alpha(\overline{\Omega})$, and $g \in C(\partial\Omega)$, then there is a unique solution $u \in C^2(\Omega) \cap C(\overline{\Omega})$ of $\Delta u = f$ in Ω with $u = g$ on $\partial\Omega$.*

Proof. Let $u = u_1 + u_2$, where $u_1 \in C^2(\Omega) \cap C(\overline{\Omega})$ is given by Theorem 2 and u_2 is harmonic (hence smooth) in Ω with $u_2 = g - u_1$ on $\partial\Omega$. ♠

Remark. With additional estimates, we may show that if $f \in C^\alpha(\Omega)$ is bounded on Ω, then (61) defines $u \in C^{2,\alpha}(\Omega)$; moreover, if $\partial\Omega$ is sufficiently smooth, then $f \in C^\alpha(\overline{\Omega})$ implies $u \in C^{2,\alpha}(\overline{\Omega})$. For details of these assertions, see [Gilbarg-Trudinger].

b. Mollifiers and Smooth Approximations

Let $\rho \in C_0^\infty(\mathbf{R}^n)$ satisfy $\rho \geq 0$, $\int \rho(x)\,dx = 1$, and supp $\rho \subset \overline{B_1(0)}$ [i.e., $\rho(x) = 0$ for $|x| \geq 1$]. For example, we could take $\rho(x) = c\exp(-(1-|x|^2)^{-1})$ for $|x| < 1$, with c chosen so that $\int_{|x|<1} \rho\,dx = 1$, and extend ρ by zero to $\mathbf{R}^n$. Such a function ρ is called a *mollifier*. Let us define $\rho_h(x) \equiv h^{-n}\rho(x/h)$ for $h > 0$; the family $\{\rho_h\}$ is called an *approximation to the identity* for reasons we shall now see.

Let Ω be an open set in $\mathbf{R}^n$ and $u \in L^1_{\text{loc}}(\Omega)$. For $x \in \Omega$ and $0 < h < \text{dist}(x, \partial\Omega)$, define the *regularization* u_h by convolution of u and $\rho_h(x)$:

$$(62) \quad u_h(x) = \rho_h \star u(x) = \frac{1}{h^n}\int_\Omega \rho\left(\frac{x-y}{h}\right)u(y)\,dy = \int_{|z|\leq 1} \rho(z)u(x-hz)\,dz.$$

Since u is integrable over every closed ball $\overline{B_h}(x) \subset \Omega$, the function u_h is well defined. In fact, since x-derivatives of u_h fall on the smooth function $\rho((x-y)/h)$, we see that $u_h \in C^\infty(\Omega_h)$ where $\Omega_h = \{x \in \Omega : \text{dist}(x,\partial\Omega) > h\}$ (see Figure 1). We now show that u_h provides a smooth approximation of u, in either the sup- or L^p-norm.

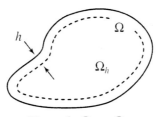

Figure 1. $\Omega_h \subset \Omega$.

Theorem 3 (Smooth Approximations). *If Ω is a domain in $\mathbf{R}^n$, for $h > 0$, let $\Omega_h = \{x \in \Omega : \text{dist}(x, \partial\Omega) > h\}$. For $u \in L^1_{\text{loc}}(\Omega)$, define u_h by (62). Then*
(i) $u_h \in C^\infty(\Omega_h)$.
(ii) *If $u \in C(\Omega)$, then $u_h \to u$ uniformly in $C(K)$ for every compact $K \subset \Omega$.*
(iii) *If $u \in L^p(\Omega)$ for $1 \leq p < \infty$, then $u_h \to u$ in $L^p(\Omega)$.*

Proof. We have already oberved that (i) holds, so let us consider (ii). For $0 < h < \text{dist}(K, \partial\Omega)$, let us use $\int \rho(z)dz = 1$ to write

$$u_h(x) - u(x) = \int_{|z|\leq 1} \rho(z)(u(x-hz) - u(x))\,dz.$$

Then, again using $\int \rho(x)\,dx = 1$, we obtain

(63)
$$\sup_{x \in K} |u_h(x) - u(x)| \le \sup_{x \in K} \sup_{|z| \le 1} |u(x - hz) - u(x)|$$
$$\le \sup_{x,y \in U_h, |x-y| < h} |u(y) - u(x)|,$$

where $U_h = \{y \in \Omega: \operatorname{dist}(y, K) < h\}$. But u is uniformly continuous on $\overline{U}_h$, so (63) establishes the desired convergence.

Now let us prove (iii). If $u \in L^p(\Omega)$, we may extend u by zero to $\mathbf{R}^n$ by defining $u(x) \equiv 0$ for $x \in \mathbf{R}^n \setminus \Omega$; in this way we may consider $u \in L^p(\mathbf{R}^n)$. Writing $\rho = \rho^{1/p'} \rho^{1/p}$ in (62) and applying the Hölder inequality, we obtain

$$|u_h(x)| \le \left(\int_{|z| \le 1} \rho(z)|u(x - hz)|^p \, dz \right)^{1/p}.$$

Taking the power p of both sides, and integrating over $\mathbf{R}^n$, yields

$$\int_{\mathbf{R}^n} |u_h(x)|^p \, dx \le \int_{\mathbf{R}^n} \int_{|z| \le 1} \rho(z)|u(x-hz)|^p \, dz \, dx$$
$$= \int_{|z| \le 1} \rho(z) \int_{\mathbf{R}^n} |u(y)|^p \, dy \, dz = \int_{\mathbf{R}^n} |u(y)|^p \, dy,$$

where we have changed the order of integration and then changed the integration variable: $y = x - hz$ replaces x. This shows that regularization does not increase the L^p-norm:

(64)
$$\|u_h\|_p \le \|u\|_p \quad \text{for} \quad h > 0.$$

In particular, we have $u_h \in L^p(\Omega)$.

To show $u_h \to u$ in $L^p(\Omega)$, let us use the lemma at the end of Section 6.3 to find $\phi \in C_0(\Omega)$ satisfying $\|u - \phi\|_p < \epsilon/3$. By (64) we also have $\|u_h - \phi_h\|_p < \epsilon/3$. Now Exercise 5 shows that there is a compact set $K \subset \Omega$ that contains $\operatorname{supp} \phi$ and $\operatorname{supp} \phi_h$. Applying (i) and $\|\phi - \phi_h\|_p \le (\operatorname{vol} K)^{1/p} \|\phi - \phi_h\|_{\infty;K}$, we conclude $\phi_h \to \phi$ in $L^p(\Omega)$, so we may take h small enough that $\|\phi_h - \phi\|_p < \epsilon/3$. We have shown $\|u_h - u\|_p \le \|u_h - \phi_h\|_p + \|\phi_h - \phi\|_p + \|\phi - u\|_p < \epsilon$. Therefore, $u_h \to u$ in $L^p(\Omega)$. ♠

As a consequence of Theorem 3, we obtain the following [where $C^\infty(\overline{\Omega})$ means the restriction to $\overline{\Omega}$ of functions that are defined on $\mathbf{R}^n$, and are smooth in a neighborhood of Ω].

Corollary A (Density Results). Let Ω be a domain in $\mathbf{R}^n$.
(i) If Ω is a bounded domain, then $C^\infty(\overline{\Omega})$ is dense in $C(\overline{\Omega})$.
(ii) If Ω is an arbitrary domain, then $C_0^\infty(\Omega)$ is dense in $L^p(\Omega)$ and also in $H_0^{1,p}(\Omega)$, for $1 \leq p < \infty$.

Proof. For bounded Ω and $u \in C(\overline{\Omega})$, extend to $\tilde{u} \in C(\tilde{\Omega})$, where $\overline{\Omega} \subset \tilde{\Omega}$ with $\text{dist}(\partial\tilde{\Omega}, \overline{\Omega}) = \epsilon > 0$. Then $\tilde{u}_h \in C^\infty(\overline{\Omega})$ for $0 < h < \epsilon$, and $\tilde{u}_h \to u$ uniformly on $\overline{\Omega}$, proving (i). Part (ii) is proved in Exercise 4. ♠

In (ii) of Corollary A, the following relationship between mollification and differentiation is used for $u \in C^1(\Omega)$:

$$(65) \qquad \frac{\partial}{\partial x_j}(u_h)(x) = \left(\frac{\partial u}{\partial x_j}\right)_h (x), \qquad \text{if } \text{dist}(x, \partial\Omega) > h.$$

In fact, (65) does not require u to have continuous derivatives: if $u \in L^1_{\text{loc}}(\Omega)$ has its weak derivatives $D_j u$ in $L^1_{\text{loc}}(\Omega)$, then (65) is a simple consequence of (52) in Chapter 2. This enables us to prove density results for the first-order Sobolev spaces $W^{1,p}(\Omega)$ and identify weak and strong derivatives.

Corollary B (Weak = Strong). If Ω is an arbitrary domain and $1 \leq p < \infty$, then $C^\infty(\Omega) \cap W^{1,p}(\Omega)$ is dense in $W^{1,p}(\Omega)$. Consequently, $H^{1,p}(\Omega) = W^{1,p}(\Omega)$.

Proof. Given $u \in W^{1,p}(\Omega)$, we want to find $\phi \in C^\infty(\Omega)$ such that $\|u - \phi\|_{1,p} < \epsilon$. For $k = 1, 2, \ldots$, let $U_k = \{x \in \Omega : |x| < k \text{ and } \text{dist}(x, \partial\Omega) > 1/k\}$, so that $\overline{U_k}$ is compact, $\overline{U_k} \subset U_{k+1}$, and $\Omega = \cup_{k=1}^\infty U_k$. Now let us define $\Omega_1 = U_2$ and $\Omega_k = U_{k+1} \cap (\overline{U_{k-1}})^c = \{x \in U_{k+1} : x \notin \overline{U_{k-1}}\}$ for $k \geq 2$. Then $\{\Omega_k\}_{k=1}^\infty$ forms a countable open cover of Ω. Let $\{\phi_k\}_{k=1}^\infty$ be a partition of unity that is subordinate to this open cover [i.e., $\sum_k \phi_k = 1$ on Ω and $\phi_k \in C_0^\infty(\Omega_k)$]. For each k, choose $h_k > 0$ sufficiently small so that $\|(\phi_k u)_{h_k} - \phi_k u\|_{1,p} < \epsilon 2^{-k}$. For h_k sufficiently small, we may assume by Exercise 4(a) that $(\phi_k u)_{h_k} \in C_0^\infty(V_k)$, where $V_k = U_{k+2} \cap (\mathbf{R}^n \setminus \overline{U_{k-2}})$; this means that the infinite sum $\phi = \sum_{k=1}^\infty (\phi_k u)_{h_k}$ has only a finite number of nonzero terms at each $x \in \Omega$, and hence $\phi \in C^\infty(\Omega)$. Moreover, we may compute

$$\|u - \phi\|_{1,p} = \left\|\sum_{k=1}^\infty \phi_k u - \sum_{k=1}^\infty (\phi_k u)_{h_k}\right\|_{1,p}$$
$$\leq \sum_{k=1}^\infty \|\phi_k u - (\phi_k u)_{h_k}\|_{1,p} < \sum_{k=1}^\infty \epsilon 2^{-k} = \epsilon. \spadesuit$$

Theorem 3 also enables us to formulate conditions, analogous to (31) and (32) in the Arzela-Ascoli theorem, under which a set in $L^p(\Omega)$ is precompact.

Section 6.5: Refinements and Generalizations

Corollary C (L^p-Compactness Criterion). Let $1 \leq p < \infty$ and Ω be a bounded domain in $\mathbf{R}^n$. Suppose S is a bounded set in $L^p(\Omega)$; that is,

(66) $$\|u\|_p \leq M \qquad \text{for all} \quad u \in S \subset L^p(\Omega).$$

Also suppose that for every $\epsilon > 0$ there exists $\delta > 0$ such that

(67) $$\int_\Omega |u(y+z) - u(y)|^p \, dy < \epsilon \quad \text{whenever} \quad |z| < \delta \quad \text{and} \quad u \in S$$

(where u has been extended by zero outside Ω). Then S is precompact in $L^p(\Omega)$.

Proof. We shall show that S is totally bounded in $L^p(\Omega)$. Since the closure $\overline{S}$ in $L^p(\Omega)$ is a complete metric space, this shows that $\overline{S}$ is compact; in other words, S is precompact in $L^p(\Omega)$.

Fix $h > 0$ and let $S_h = \{u_h : u \in S\}$. We use the L^p-conditions (66) and (67) to show that S_h satisfies the uniform conditions (31) and (32) for the Arzela-Ascoli theorem. For $u \in S$, extend u by zero to $\mathbf{R}^n$. Using (62), $\rho = \rho^{1/p'} \rho^{1/p}$, and $\int \rho(z)\,dz = 1$, we obtain

(68)
$$|u_h(x)| \leq \int_\Omega \rho_h(x-y)|u(y)|\,dy \leq \left(\int_\Omega \rho_h(x-y)\,|u(y)|^p\,dy\right)^{1/p}$$
$$\leq \left(\sup_{z \in \mathbf{R}^n} \rho_h(z)\right)^{1/p} \|u\|_p.$$

Combined with (66), this establishes (32); in other words, S_h is uniformly bounded in $C(\overline{\Omega})$ (for this fixed h). Moreover, we may write

$$u_h(x+z) - u_h(x) = \int_\Omega (\rho_h(x+z-y)\,u(y) - \rho_h(x-y)\,u(y))\,dy$$
$$= \int_\Omega \rho_h(x-y)\,(u(y+z) - u(y))\,dy,$$

and, similar to (68), obtain

$$|u_h(x+z) - u_h(x)| \leq \left(\sup_{|x|\leq 1} \rho_h(z)\right)^{1/p} \left(\int_\Omega |u(y+z) - u(y)|^p\,dy\right)^{1/p}.$$

Combined with (67), this establishes (31), so that S_h is equicontinuous.

By the Arzela-Ascoli theorem, S_h is precompact in $C(\overline{\Omega})$; in particular, S_h is totally bounded in $C(\overline{\Omega})$, and hence in $L^p(\Omega)$ (see Exercise 8 in Section 6.4). On the other hand, S_h is uniformly close to S in $L^p(\Omega)$, so S is also totally bounded in $L^p(\Omega)$ (see Exercise 6). ♠

We use this L^p-compactness criterion in the next section.

c. Compact Imbeddings of $H_0^{1,p}(\Omega)$

A continuous map $T: X \to Y$ between Banach spaces X and Y is *compact* (or *completely continuous*) if the image of a bounded set in X is precompact in Y [i.e., $\overline{T(A)}$ is compact in Y for every bounded set $A \subset X$]. We are generally concerned with compact maps that are *linear;* we call these *compact linear operators*. In fact, in this subsection we are concerned with some continuous imbeddings $X \subset Y$ that are also compact (i.e., every bounded set $A \subset X$ has compact closure in Y).

As an example, suppose $\Omega \subset \mathbf{R}^n$ is a bounded domain, and consider the imbedding

$$(69) \qquad C^1(\overline{\Omega}) \subset C^0(\overline{\Omega}).$$

The Arzela-Ascoli theorem shows that (69) is a compact imbedding (see Exercise 7a). Similarly, we can show that

$$(70) \qquad C^\beta(\overline{\Omega}) \subset C^\alpha(\overline{\Omega}) \qquad \begin{cases} 0 \leq \alpha < \beta < 1 \\ 0 \leq \alpha < \beta \leq 1 \text{ if } \partial\Omega \in C^1 \end{cases}$$

is a compact imbedding (see Exercise 7c).

Let us now investigate when the Sobolev imbeddings are compact.

Theorem 4 (Kondrachov Compactness). *Suppose Ω is a bounded domain in $\mathbf{R}^n$.*
(i) If $p \leq n$ and $1 \leq q < np/(n-p)$, then

$$(71) \qquad H_0^{1,p}(\Omega) \subset L^q(\Omega)$$

is a compact imbedding.
(ii) If $p > n$ and $0 \leq \alpha < 1 - (n/p)$, then

$$(72) \qquad H_0^{1,p}(\Omega) \subset C^\alpha(\overline{\Omega})$$

is a compact imbedding.

Proof. The case (ii) follows from (70) and Theorem 1, so we need only consider (i). First let us assume $q = 1$. Suppose $S \subset H_0^{1,p}(\Omega)$ is bounded; without loss of generality we may assume $S \subset C_0^1(\Omega)$ and $\|u\|_{1,p} \leq 1$. To show that S is precompact in $L^1(\Omega)$, we use Corollary C to Theorem 3. Now, by Hölder's inequality,

$$\|u\|_1 = \int_\Omega |u(x)|\, dx \leq (\operatorname{vol}\Omega)^{1/p'} \|u\|_p \leq (\operatorname{vol}\Omega)^{1/p'},$$

so (66) holds. To establish (67), we write

$$u(y+z) - u(y) = \int_0^1 \frac{du}{dt}(y+tz)\, dt = \int_0^1 \nabla u(y+tz) \cdot z\, dt.$$

Section 6.5: Refinements and Generalizations

Thus

$$\int_\Omega |u(y+z) - u(y)|\, dy \le |z| \int_\Omega |\nabla u(x)|\, dx$$

$$\le |z|\,(\operatorname{vol}\Omega)^{1/p'} \|\nabla u\|_p \le |z|\,(\operatorname{vol}\Omega)^{1/p'},$$

which shows that (67) holds. Therefore, we conclude that S is precompact in $L^1(\Omega)$.

To extend this compactness to $1 \le q < np/(n-p)$ when $p < n$, let us use Exercise 3 of Section 6.1 to write

$$\|u\|_q \le \|u\|_1^\lambda \|u\|_{np/(n-p)}^{1-\lambda}$$

where λ satisfies $q^{-1} = \lambda + (1-\lambda)(1/p - 1/n)$, and then invoke the Sobolev inequality (34) to write

(73) $$\|u\|_q \le \|u\|_1^\lambda (C\|\nabla u\|_p)^{1-\lambda},$$

Now suppose $\{u_j\}_{j=1}^\infty$ is a sequence in $H_0^{1,p}(\Omega)$ that is uniformly bounded: $\|u_j\|_{1,p} \le M$ for all $j \ge 1$. To show that (71) is a compact imbedding, we must show that $\{u_j\}$ contains a subsequence that converges in $L^q(\Omega)$. But the compactness for $q = 1$ shows that a subsequence $\{u_j'\}$ exists such that $\|u_j' - u_k'\|_1 \to 0$. Applying (73) to $u = u_j' - u_k'$ shows that this subsequence also converges in $L^q(\Omega)$.

Finally, when $p = n$, we must show that (71) is compact for all $1 \le q < \infty$. But given such a q, pick $r > q$. By Exercise 7 in Section 6.4, we have a continuous imbedding $H_0^{1,p}(\Omega) \subset L^r(\Omega)$. Therefore, we may invoke Exercise 9 of this section to conclude that (71) is compact. This completes the proof of the theorem. ♠

Let us distinguish the special case $p = 2$; historically, this was established by F. Rellich before the more general result of V. Kondrachov.

Corollary A (Rellich Compactness). *If Ω is a bounded domain, then $H_0^{1,2}(\Omega) \subset L^2(\Omega)$ is a compact imbedding.*

We shall need these compactness results when we use Sobolev spaces to study linear and nonlinear elliptic equations with Dirichlet boundary conditions. However, for more general boundary conditions, we often need to know compactness of the imbeddings with $H_0^{1,p}(\Omega)$ replaced by $H^{1,p}(\Omega)$. We consider this topic in the next subsection.

d. Imbeddings of $H^{1,p}(\Omega)$

To replace $H_0^{1,p}(\Omega)$ by $H^{1,p}(\Omega)$ in the Sobolev and Kondrachov imbedding theorems requires some subtle regularity conditions on the boundary $\partial\Omega$. A fairly general condition is that $\partial\Omega$ satisfy the following *uniform cone property*: there is a fixed finite cone C such that each $x \in \partial\Omega$ is the vertex of a finite cone $C_x \subset \overline{\Omega}$ such that C_x is congruent to C. For example, a bounded domain Ω with Lipschitz continuous boundary satisfies this condition. However, we are even less general and consider only bounded domains with C^1-boundary.

Theorem 5 (Sobolev-Rellich-Kondrachov Imbeddings). *Suppose Ω is a domain in $\mathbf{R}^n$, with C^1-boundary.*
(i) *If $p < n$ and $p \leq q \leq np/(n-p)$, then*

$$H^{1,p}(\Omega) \subset L^q(\Omega) \tag{74}$$

is continuous. Moreover, if Ω is bounded and $1 \leq q < np/(n-p)$, then (74) is compact.
(ii) *If $p = n$ and $p \leq q < \infty$, then (74) is continuous. Moreover, if Ω is bounded, then (74) is compact for all $1 \leq q < \infty$.*
(iii) *If $p > n$ and Ω is bounded, then*

$$H^{1,p}(\Omega) \subset C^\alpha(\overline{\Omega}) \tag{75}$$

is continuous if $0 \leq \alpha \leq 1 - (n/p)$, and compact if $0 \leq \alpha < 1 - (n/p)$.

The proof of this theorem requires a careful analysis near $\partial\Omega$, which is achieved by means of a partition of unity subordinate to an open cover. For simplicity, assume that Ω is bounded. Since $\overline{\Omega}$ is compact, we may use the C^1 condition to find bounded open sets $U_0, U_1, \ldots, U_N$ in $\mathbf{R}^n$ such that $\overline{U}_0 \subset \Omega$, $\partial\Omega \subset \cup_{i=1}^N U_i$, and for $i = 1, \ldots, N$ a C^1-diffeomorphism $\psi_i : U_i \to B_1(0)$ satisfying $\psi(\partial\Omega \cap U_i) \subset \{x \in \mathbf{R}^n : x_n = 0\}$ and $\psi(U_i \cap B_r(\xi)) \subset \mathbf{R}_+^n$. Moreover, we may choose a partition of unity $\phi_0, \ldots, \phi_N$ subordinate to the open cover $U_0, \ldots, U_N$; that is,

$$\sum_{i=0}^N \phi_i(x) = 1 \quad \text{for } x \in \Omega \quad \text{and} \quad \phi_i \in C_0^\infty(U_i). \tag{76}$$

This partition of unity enables us to reduce the global analysis on Ω to a local analysis in $B_1(0)$, since $\|\cdot\|_{H^{1,p}(U_i)}$ on $u_i \in H^{1,p}(U_i)$ is equivalent to $\|\cdot\|_{H^{1,p}(B_1(0))}$ on $u_i \circ \psi^{-1}$.

One consequence of the C^1 condition on $\partial\Omega$ is the density in $H^{1,p}(\Omega)$ of functions that are smooth up to the boundary.

Lemma 1 (Smooth Approximation). *If Ω is a bounded domain in $\mathbf{R}^n$, with C^1-boundary, then $C^\infty(\overline{\Omega})$ is dense in $H^{1,p}(\Omega)$.*

Proof. Let $u \in C^1(\Omega) \cap H^{1,p}(\Omega)$; it suffices to find $u_h \in C^\infty(\Omega)$ converging to u in $\|\cdot\|_{1,p}$ as $h \to 0$. Use the partition of unity (76) to write $u = \sum_{i=0}^N \phi_i u$, where each $\phi_i u$ has support in U_i. For $i = 0$, we know that $\phi_0 u \in H_0^{1,p}(U_0)$, and $C_0^\infty(U_0)$ is dense in $H_0^{1,p}(U_0)$, so we may find $u_{0,h} \in C_0^\infty(U_0)$ with $u_{0,h} \to \phi_0 u$ in $\|\cdot\|_{1,p}$ as $h \to 0$. For $i = 1, \ldots, N$ we may use the diffeomorphism ψ_i to obtain $v_i = (\phi_i u) \circ \psi_i^{-1} \in C^1(B^+) \cap H^{1,p}(B^+)$, where $B^+ \equiv B_1(0) \cap \mathbf{R}_+^n$. For each i, we must show that $v_i \in C^1(B^+) \cap H^{1,p}(B^+)$ can be approximated in $\|\cdot\|_{1,p}$ by $w_{i,h} \in C^\infty(\overline{B^+})$ as $h \to 0$. The approximation on Ω is then found to be $u_h = u_{0,h} + \sum_{i=1}^N w_{i,h} \circ \psi_i$.

Section 6.5: Refinements and Generalizations

To achieve this approximation of $v \in C^1(B^+) \cap H^{1,p}(B^+)$, we shift and mollify. More specifically, for $x = (x_1, \ldots, x_n) \in B^+$, let $\tilde{x}^h = (x_1, \ldots, x_n + 2h)$ and $V_h = \{x \in \mathbf{R}^n : x_n > -2h\}$; if we define the shifted function $v^h(x) = v(\tilde{x}^h)$, then $v^h \in C^1(V_h) \cap H^{1,p}(V_h)$ (see Figure 2).

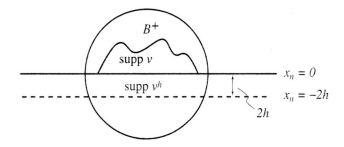

Figure 2.

If ρ denotes a mollifier, let $w_h = \rho_h \star v^h$ [cf. (62)]. Since $\operatorname{dist}(B^+, \partial V_h) = 2h > h$, we have $w_h \in C^\infty(\overline{B^+})$. Since $v^h \to v$ in $L^p(B^+)$ as $h \to 0$, by Theorem 3 (iii) we have $w_h \to v$ in $L^p(B^+)$. Similarly, $D^\alpha w_h = \rho_h \star D^\alpha v^h$ for $|\alpha| = 1$, where $D^\alpha v^h \in L^p(B^+)$ converges to $D^\alpha v$, shows that $D^\alpha w_h \to D^\alpha v$ in $L^p(B^+)$. We conclude that v can be approximated in $\|\cdot\|_{1,p}$ by $w_h \in C^\infty(\overline{B^+})$ as $h \to 0$, as required. ♠

Lemma 2 (Extension to $\mathbf{R}^n$). *If Ω is a bounded domain in $\mathbf{R}^n$, with C^1-boundary, then there is a bounded linear mapping $E \colon H^{1,p}(\Omega) \to H^{1,p}(\mathbf{R}^n)$ and a large bounded set $\tilde{\Omega} \supset \overline{\Omega}$ such that, for every $u \in H^{1,p}(\Omega)$, $E(u)|_\Omega = u$ and the support of $E(u)$ is in $\tilde{\Omega}$.*

Sketch of Proof. Using Lemma 1, it suffices to define the *extension operator* E on $C^\infty(\overline{\Omega})$, and using the partition of unity (76) as in the proof of Lemma 1, it suffices to consider ϕu, where $\phi \in C_0^\infty(U)$ and $U \cap \partial\Omega \neq 0$. Let $\psi \colon \overline{U} \to \overline{B_1(0)}$ be a C^1-diffeomorphism and $v = \phi u \circ \psi^{-1} \in C^\infty(\overline{B^+})$, where $B^+ = B_1(0) \cap \mathbf{R}^n_+$ and v has support as in Figure 2. Extending by zero, we may assume that $v \in C^\infty(\overline{\mathbf{R}^n_+})$.

Let us extend v to $\mathbf{R}^n_-$ by defining

$$(77) \quad v(x) = 4v(x_1, \ldots, x_{n-1}, -\frac{x_n}{2}) - 3v(x_1, \ldots, x_{n-1}, -x_n) \quad \text{for } x_n < 0.$$

It is a straightforward calculation to check (cf. Exercise 11a) that v and $D^\alpha v$ for all $|\alpha| = 1$ are continuous at $x_n = 0$, so $v \in C^1(\mathbf{R}^n)$. In fact, $v(x) \equiv 0$ for all $x_n < -2$, and $\|v\|_{H^{1,p}(\mathbf{R}^n)} \leq C\|v\|_{H^{1,p}(\mathbf{R}^n_+)}$ for some constant C independent of v (cf. Exercise 11b).

By replacing $B_1(0)$ by a larger ball (or, equivalently, by modifying the diffeomorphism ψ), we can arrange that the extension of v to $\mathbf{R}^n_-$ vanishes outside the ball. This enables us to form the composition $v \circ \psi$ to get a function u defined on U.

The extension operator E is defined on $u \in C^\infty(\overline{\Omega})$ by writing $u = \sum_{i=0}^{N} \phi_i u$ and then defining $E(\phi_i u) = v_i \circ \psi_i$, where v_i has been extended to $B_1(0)$ as previously. The required properties of E are easily verified from the construction. ♠

Proof of Theorem 5. Since the support of the extension $E(u)$ defined in Lemma 2 is contained in $\tilde{\Omega}$, we may consider $E\colon H^{1,p}(\Omega) \to H_0^{1,p}(\tilde{\Omega})$. Using Exercise 8, the conclusions for (74) follow from the composition $H^{1,p}(\Omega) \to H_0^{1,p}(\tilde{\Omega}) \to L^q(\tilde{\Omega}) \to L^q(\Omega)$, where the last map is just the restriction to Ω. The conclusions for (75) follow analogously; cf. Exercise 12. ♠

As an application of Theorem 5, let us consider the appropriate version of the Poincaré inequality for functions in $H^{1,p}(\Omega)$ instead $H_0^{1,p}(\Omega)$. For $u \in L^1(\Omega)$, let u_Ω denote the *average value* of u over Ω:

$$(78) \qquad u_\Omega = \frac{1}{|\Omega|} \int_\Omega u(x)\, dx$$

where $|\Omega|$ denotes the volume of Ω.

Corollary (Poincaré Inequality). *If Ω is a connected, bounded domain in $\mathbf{R}^n$, with C^1-boundary, and $1 \leq p < \infty$, then*

$$(79) \qquad \|u - u_\Omega\|_p \leq C \|\nabla u\|_p \qquad \text{where} \quad C = C(n,p,\Omega)$$

for all $u \in H^{1,p}(\Omega)$.

Proof. Replacing u by $v = u - u_\Omega$, it suffices to assume $u_\Omega = 0$. If (79) fails, then (by Lemma 1) we can find $u_k \in C^\infty(\overline{\Omega}) \cap H^{1,p}(\Omega)$ with $(u_k)_\Omega = 0$ and $\|u_k\|_p > k\|\nabla u_k\|_p$ as $k \to \infty$. If we let $w_k = u_k/\|u_k\|_p$, then we have $\|w_k\|_p = 1$ and so $\|\nabla w_k\|_p < 1/k$. In particular, the sequence $\{w_k\}$ is bounded in $H^{1,p}(\Omega)$. Using the compactness of the imbedding $H^{1,p}(\Omega) \to L^p(\Omega)$ (by Theorem 5), there is a subsequence $\{w_{k_j}\}$ that converges in $L^p(\Omega)$ to a function w and by continuity w satisfies

$$(80) \qquad w_\Omega = 0 \qquad \text{and} \qquad \|w\|_p = 1.$$

Now $w \in L^p(\Omega) \subset L^1_{loc}(\Omega)$ has weak derivatives that satisfy $\int \nabla w \phi \, dx = -\int w \nabla \phi \, dx$ for all $\phi \in C_0^\infty(\Omega)$. But $-\int w \nabla \phi \, dx = -\lim_{k_j \to \infty} \int w_{k_j} \nabla \phi \, dx$, and $\lim_{k_j \to \infty} \int \nabla w_{k_j} \phi \, dx = 0$ since $\|\nabla w_k\|_p \to 0$. This means that $\nabla w = 0$, which implies w is a constant (since Ω is connected). But this is impossible by (80). This contradiction proves that (79) must hold. ♠

e. Higher-Order Sobolev Spaces and Strong Solutions

In this subsection we define higher-order Sobolev spaces and investigate their imbedding properties.

Given a domain Ω in $\mathbf{R}^n$, a nonnegative integer k, and $1 \leq p < \infty$, we introduce the norm

$$(81) \qquad \|u\|_{k,p} = \left(\sum_{|\alpha| \leq k} \int_\Omega |D^\alpha u|^p \, dx \right)^{1/p}$$

which generalizes (9). Certainly (81) is finite for any $u \in C_0^k(\Omega)$ but is also finite for some u not having compact support.

Definition. For a domain Ω in $\mathbf{R}^n$, let $H_0^{k,p}(\Omega)$ denote the completion of $C_0^k(\Omega)$ in the norm (81), and $H^{k,p}(\Omega)$ denote the completion of $\{u \in C^k(\Omega) : \|u\|_{k,p} < \infty\}$ in the norm (81).

The following may be proved by iteration of the cases $k = 1$ and $m = 0$ discussed previously (see Exercises 13 and 14).

Theorem 6 (Sobolev-Rellich-Kondrachov Imbeddings). *Suppose Ω is a domain in $\mathbf{R}^n$.*
(i) If $kp < n$ and $p \leq q \leq \frac{np}{n-kp}$, then

$$(82) \qquad H_0^{k,p}(\Omega) \subset L^q(\Omega)$$

is a continuous imbedding; moreover, if Ω is bounded and $1 \leq q < \frac{np}{n-kp}$, then (82) is a compact imbedding.
(ii) If $kp = n$ and $p \leq q < \infty$, then (82) is a continuous imbedding; moreover, if Ω is bounded and $1 \leq q < \infty$, then (82) is a compact imbedding.
(iii) If $kp > n$, Ω is bounded, and $(k-m)p > n \geq (k-m-1)p$, then

$$(83) \qquad H_0^{k,p}(\Omega) \subset C^{m,\alpha}(\overline{\Omega})$$

is a continuous imbedding provided $0 \leq \alpha \leq k - m - (n/p)$ [and provided $\alpha < 1$ in the case $k-m-(n/p) = 1$] and a compact imbedding for $0 \leq \alpha < k - m - (n/p)$.

In case Ω is unbounded, we cannot expect the imbedding (83) to be compact, but it will still be continuous. In fact, analogous to (46), we may define $C_B^{m,\alpha}(\overline{\Omega})$ to be functions for which $D^\beta u$ is uniformly bounded and continuous on $\overline{\Omega}$ for all $|\beta| \leq m$. Then (83) may be replaced by

$$(84) \qquad H_0^{k,p}(\Omega) \subset C_B^{m,\alpha}(\overline{\Omega}),$$

which is continuous for $0 \leq \alpha \leq k - m - (n/p)$.

Analogous to the preceding subsection, it is also possible to replace $H_0^{k,p}(\Omega)$ by $H^{k,p}(\Omega)$ in the imbeddings of the theorem under additional regularity assumptions on Ω. In fact, as in Theorem 5, it suffices to assume that $\partial\Omega$ is C^1, but weaker conditions are also sufficient (see [Adams] for details of these assertions.) It is also possible to replace $L^q(\Omega)$ in (82) by a Sobolev space; see Exercise 15.

Remark. There is a simple mnemonic device for remembering the conditions under which (82) and (83) are continuous or compact imbeddings. If we take $k - (n/p)$ as the measure of regularity of $H_0^{k,p}(\Omega)$, then (82) is a continuous imbedding provided $H_0^{k,p}(\Omega)$ has greater regularity than $L^q(\Omega)$:

$$k - \frac{n}{p} \geq -\frac{n}{q} \quad \Leftrightarrow \quad q \leq \frac{np}{n - kp} \quad \text{(when } kp < n\text{)}.$$

Similarly, if we take $m + \alpha$ as the measure of regularity of $C^{m,\alpha}(\overline{\Omega})$, then the conditions under which (83) is a continuous imbedding may be obtained from

$$k - \frac{n}{p} \geq m + \alpha.$$

In both these inequalities, strict inequality corresponds to compactness of the corresponding imbedding.

The higher-order Sobolev spaces are also very important in the study of regularity properties of solutions of elliptic equations. For example, to say that $u \in H_0^{1,2}(\Omega)$ is a weak solution of $\Delta u = f$ only requires u to have first-order derivatives in $L^2(\Omega)$. On the other hand, since $f \in L^2(\Omega)$ and the equation is second-order, we might expect u to have all *second-order* derivatives also in $L^2(\Omega)$. In this case, $u \in H^{2,2}(\Omega)$ and we say that u is a *strong solution* of $\Delta u = f$. The regularity question of when a weak solution is in fact a strong solution is discussed in Chapter 8.

However, the bounded linear mapping $\Delta : H^{2,2}(\Omega) \to L^2(\Omega)$ is not the best choice for solving PDEs; even if it is onto, it will not be one-to-one, since we have not imposed boundary conditions [specifically, of course, Δ is zero on any constant function in $H^{2,2}(\Omega)$]. For this reason, it is useful to consider the mapping

(85) $$\Delta : H^{2,2}(\Omega) \cap H_0^{1,2}(\Omega) \to L^2(\Omega).$$

As we see in Chapter 8, if Ω is a bounded domain with C^2-boundary, then (85) is indeed an isomorphism (i.e., a bounded linear map that is one to one, onto, and has a bounded inverse map). This fact will also be useful in studying linear and nonlinear evolution equations.

Exercises for Section 6.5

1. Use Morrey's inequality (45) to prove Theorem 1 of this section.
2. If (55) holds with $\alpha = 1$, then we say that u is *Lipschitz continuous* at x_0. With $n = 1$, give an example of a function u that is Lipschitz continuous but is not continuously differentiable at $x_0 = 0$. [The collection of functions $u \in C(\overline{\Omega})$ for which (56) holds with $\alpha = 1$ is sometimes denoted by $C^{0,1}(\overline{\Omega})$ to distinguish it from $C^1(\overline{\Omega})$.]
3. Assume that Ω is a bounded domain and $f \in C^\alpha(\overline{\Omega})$ for some $0 < \alpha < 1$.
   (a) Let $\tilde{\Omega}$ be a smooth bounded domain containing $\overline{\Omega}$, and define
   $$w_{ij}(x) = \int_{\tilde{\Omega}} \partial_i \partial_j K(x-y) f(y) \, dy$$
   $$= \int_{\tilde{\Omega}} \partial_i \partial_j K(x-y) \left(f(y) - f(x)\right) dy - f(x) \int_{\tilde{\Omega}} \partial_i \partial_j K(x-y) \, dy.$$
   Show $w_{ij} \in C(\Omega)$.
   (b) Show that $w_{ij} = \partial_i \partial_j u$, and prove Theorem 2.
4. Assume that Ω is an arbitrary domain in $\mathbf{R}^n$.
   (a) If $u \in C_0(\Omega)$ and $0 < h < \text{dist}(\text{supp}\, u, \partial\Omega)$, show that $u_h \in C_0^\infty(\Omega)$.
   (b) Show that $C_0^\infty(\Omega)$ is dense in $L^p(\Omega)$ for $1 \leq p < \infty$.
   (c) Show that $C_0^\infty(\Omega)$ is dense in $H_0^{1,p}(\Omega)$ for $1 \leq p < \infty$.
   (d) Show that neither $C_0(\Omega)$ nor $C(\overline{\Omega})$ is dense in $L^\infty(\Omega)$.
5. Show that if $u \in L^p(\Omega)$ for $1 \leq p < \infty$ and $z \in \mathbf{R}^n$, then the shift $S_z u(x) = u(x-z)$ converges in $L^p(\Omega)$ to $u(x)$ uniformly as $|z| \to 0$.
6. Suppose X is a Banach space containing the sets S and $\{S_h\}_{h>0}$. If S_h is totally bounded for each $h > 0$ and uniformly close to S, show that S is totally bounded.
7. (a) Prove that (69) is compact.
   (b) Show that for $0 < \alpha < 1$ and $\Omega \subset \mathbf{R}^n$ bounded, $C^\alpha(\overline{\Omega}) \subset C(\overline{\Omega})$ is a compact imbedding.
   (c) Show that for $0 < \alpha < \beta < 1$ and $\Omega \subset \mathbf{R}^n$ bounded, $C^\beta(\overline{\Omega}) \subset C^\alpha(\overline{\Omega})$ is a compact imbedding.
8. If X, Y, and Z are Banach spaces and $T: X \to Y$ and $S: Y \to Z$ are both bounded linear maps, with T or S (or both) compact, then $S \circ T: X \to Z$ is compact.
9. If $\Omega \subset \mathbf{R}^n$ is a bounded domain, and $H_0^{1,p}(\Omega) \subset L^r(\Omega)$ is a continuous imbedding, then $H_0^{1,p}(\Omega) \subset L^q(\Omega)$ is a compact imbedding for $1 \leq q < r$.
10. If $T: X \to X$ is a compact linear operator on a Hilbert space X, show that the adjoint $T^*: X \to X$ is also compact. (The adjoint T^* was defined in Exercise 16 of Section 6.1.)
11. Let $v \in C^\infty(\overline{\mathbf{R}^n_+}) \cap H^{1,p}(\mathbf{R}^n_+)$.
    (a) Show that the extension of v to $\mathbf{R}^n$ defined by (77) is in $C^1(\mathbf{R}^n)$.

(b) Show that $\|v\|_{H^{1,p}(\mathbf{R}^n)} \leq C\|v\|_{H^{1,p}(\mathbf{R}^n_+)}$ for some constant C independent of v.

12. Use Lemmas 1 and 2 to prove the conclusions for (75) in Theorem 5.
13. If $\Omega \subset \mathbf{R}^n$ is a bounded domain, $2p < n$, and $1 \leq q < \frac{np}{n-2p}$, then use Theorem 1 of Section 6.4 and Theorem 4 of Section 6.5 to prove that $H_0^{2,p}(\Omega) \subset L^q(\Omega)$ is a compact imbedding.
14. If $\Omega \subset \mathbf{R}^n$ is a bounded domain, $2p > n$, and $0 \leq m < 2 - \frac{n}{p}$, then use Theorem 3 of Section 6.4 and Theorem 4 of Section 6.5 to prove that $H_0^{2,p}(\Omega) \subset C^m(\overline{\Omega})$ is a compact imbedding.
15. If $\Omega \subset \mathbf{R}^n$ is a bounded domain with C^1-boundary, use Theorem 6 (with $H^{k,p}$ in place of $H_0^{k,p}$) to show that $H^{k+l,p}(\Omega) \subset H^{l,q}(\Omega)$ is a compact imbedding whenever $kp \leq n$ and $1 \leq q < \frac{np}{n-kp}$.

6.6 Unbounded Operators and Spectral Theory

Instead of considering differential operators as linear operators between Banach spaces [such as in (85)], it is sometimes convenient to consider operators as being defined on some subset of a single Banach space. The analogy for (nonlinear) functions of a single variable is a function like $f(x) = 1/x$, which operates on real numbers but is not defined on all of $\mathbf{R}$; however, we may consider f as a function on $\mathbf{R}$ provided that we restrict to an appropriate domain such as $(0, \infty)$ or $\mathbf{R}\backslash\{0\}$. As in the function case, the operators that we consider in this section are unbounded on some portion of the Banach space on which they act.

a. Densely Defined and Closed Operators

Suppose that D is a linear subspace of a Banach space X, and $T\colon D \to X$ is a linear operator (as in Section 6.1c); we say that T *is a linear operator on X with domain D.* For brevity, we may just say that T *is a linear operator on X*, with the understanding that there is a specific domain associated with T; in this case, we write $\mathrm{dom}(T)$ when we want to consider the domain of T. (We could even consider linear operators $T\colon D \to Y$, where D is a linear subspace of a Banach space X that is distinct from the Banach space Y, but we avoid this level of generalization.)

Generally, the domain D is *not closed* in X (otherwise, because D itself would then be a Banach space, we could view $T\colon D \to X$ as an operator between Banach spaces). However, the domain D is generally a *dense* subspace in X (i.e., $\overline{D} = X$); in this case, we say that $T\colon D \to X$ is *densely defined*.

To formulate what we mean by a closed operator, we first introduce $G(T)$, the *graph* of the linear operator $T\colon D \to X$, which is defined by

(86) $$G(T) \equiv \{(x, Tx) : x \in X\} \subset X \times X.$$

Section 6.6: Unbounded Operators and Spectral Theory

Then T is a *closed operator* if $G(T)$ is a closed subset of $X \times X$; in other words, for any $x_i \in D$ such that $x_i \to x$ and $Tx_i \to y$, we must have $x \in D$ and $Tx = y$. Closed operators are naturally better objects for analysis than operators that are not closed.

Suppose that T_1 and T_2 are two linear operators on X with respective domains D_1 and D_2. We say that T_2 is an *extension* of T_1 if $D_1 \subset D_2$ and $T_1 = T_2$ on D_1; in this case, we write $T_1 \subset T_2$. A linear operator T is *closable* if it has an extension that is closed; every closable operator T has a smallest closed extension $\overline{T}$ called its *closure*. A bounded linear operator $T: D \to X$ is closed if and only if D is closed (cf. Exercise 1); more generally, any linear operator T is closable if and only if the closure of its graph, $\overline{G(T)}$, is a graph, and in this case $G(\overline{T}) = \overline{G(T)}$ (cf. Exercise 3).

Example 1. Let $T = d/dx$ and $X = L^2(\Omega)$, where $\Omega = (0,1)$. If we choose $D_0 \equiv C_0^1(\Omega) \subset X$, we obtain a densely defined linear operator $T: D_0 \to X$. But this operator is *not closed*; to see this, pick $u \in H_0^1(\Omega) \backslash C_0^1(\Omega)$ and $u_n \in C_0^1(\Omega)$ for which $u_n \to u$ in the H^1-norm; then $u_n \to u$ and $Tu_n \to v$ in X, where $v = u'$, but $u \notin D_0$. On the other hand, if we let $D = H_0^1(\Omega)$, then $T: D \to X$ is a closed, densely defined operator. In fact, $T: H^1(\Omega) \to L^2(\Omega)$ is also a closed, densely defined operator. ♣

Example 2. Let $\Omega \subset \mathbf{R}^n$ be a bounded domain with smooth boundary, $X = L^2(\Omega)$, and $T = \Delta$. If we let $D_0 = C_0^\infty(\Omega)$, we obtain a densely defined linear operator $\Delta: D_0 \to L^2(\Omega)$, but to obtain a closed operator we should use $D = H_0^1(\Omega) \cap H^2(\Omega)$ as domain. To verify this, we require one of the a priori estimates from Chapter 8: $\|u\|_{H^2} \leq C(\|\Delta u\|_{L^2} + \|u\|_{L^2})$ for all $u \in D$. In fact, if $u_j \to u$ and $\Delta u_j \to v$ in L^2, then we may apply this estimate to the difference $u_j - u_k$ to conclude that u_j is Cauchy in H^2, and hence $u \in H^2(\Omega)$. Since we also knew that $u_j \in H_0^1(\Omega)$, we have $u \in H_0^1(\Omega) \cap H^2(\Omega) = D$, proving that $\Delta: D \to L^2(\Omega)$ is closed. ♣

An operator $T: D \to X$, whether bounded or not, may have a bounded inverse. To state a specific result that is suitable for our purposes, let us introduce the *nullspace* and *range* of T:

$$N_T = \{x \in D : Tx = 0\} \quad \text{and} \quad R_T = \{y \in X : Tx = y \text{ for some } x \in D\}.$$

Propopsition 1. *Let $T: D \to X$ be a linear operator on a Banach space X for which $N_T = \{0\}$, $R_T = X$, and $G(T)$ is closed. Then $T^{-1}: X \to X$ is bounded.*

The proof of this proposition is an immediate consequence of a standard result of functional anlaysis, the *closed graph theorem*, which states that a linear operator between Banach spaces $T: X \to Y$ for which the graph $G(T)$ is closed must be continuous; see [Conway].

b. Adjoints

The *adjoint* of a densely defined linear operator on a Banach space X will be a linear operator on the dual space X^* (cf. Section 6.3a).

Definition. Suppose $T: D \to X$ is a densely defined linear operator on the Banach space X. Let D^* denote those $F \in X^*$ for which there exists $G \in X^*$ such that $F(Tx) = G(x)$ for all $x \in D$. If we set $T^*F = G$, then we have defined a linear operator $T^*: D^* \to X^*$ called the *adjoint* of T:

(87a) $\qquad F(Tx) = (T^*F)(x) \quad$ for all $x \in D$ and $F \in D^*$.

Since D is dense in X, T^* is unique. Moreover, if X is a Hilbert space, then identifying X^* with X allows us to write (87a) as

(87b) $\qquad \langle Tx, y \rangle = \langle x, T^*y \rangle \quad$ for all $x \in D$ and $y \in D^*$.

Note. When T is a differential operator on a Hilbert space of functions and $\langle\,,\,\rangle$ represents integration, T^* is sometimes referred to as the "Hilbert space adjoint" to distinguish it from the adjoint introduced in Section 2.3a.

In a sense, the adjoint interchanges the roles of densely defined operators and closable operators; for simplicity, we restrict our attention to Hilbert spaces.

Proposition 2. *If T is a densely defined linear operator on the Hilbert space X, then T^* is closed. Moreover, T is closable if and only if T^* is densely defined; in this case, $\overline{T} = T^{**}$.*

Proof. Consider the isomorphism $J: X \times X \to X \times X$ defined by $J(y, z) = (-z, y)$. Now we claim that in $X \times X$ we have

(88) $\qquad\qquad [JG(T)]^\perp = G(T^*).$

To see this, observe that $(y, z) \in [JG(T)]^\perp$ implies $\langle (y, z), (-Tx, x) \rangle = 0$ for all $x \in D \equiv \mathrm{dom}(T)$; but this latter condition may be written $\langle y, Tx \rangle = \langle z, x \rangle$ for all $x \in D$, or equivalently as $\langle T^*y, x \rangle = \langle z, x \rangle$ for all $x \in D$. Since D is dense, we conclude that $T^*y = z$, or $(y, z) \in G(T^*)$. All these steps are reversible, so we conclude $[JG(T)]^\perp = G(T^*)$. But $[JG(T)]^\perp$ is closed in $X \times X$, so $G(T^*)$ is closed; in other words, T^* is closed.

Now $G(T)$ is a linear subspace of $X \times X$ and $J^2 = I$, so $\overline{G(T)} = (G(T)^\perp)^\perp = (J^2G(T)^\perp)^\perp = (J(JG(T))^\perp)^\perp = (JG(T^*))^\perp$, where we have also invoked (88). But if T^* is densely defined, then we can apply (88) with T^* in place of T to conclude that $\overline{G(T)}$ is the graph of T^{**}, and hence T is closable (Exercise 2). Conversely, if $D^* \equiv \mathrm{dom}(T^*)$ is not dense, let z be a nonzero vector in $(D^*)^\perp$; then $\langle (z, 0), (x, T^*x) \rangle = \langle z, x \rangle = 0$ for all $x \in D^*$ implies $(z, 0) \in G(T^*)^\perp$. But $G(T^*)^\perp = [JG(T)]^{\perp\perp} = \overline{JG(T)} = J\overline{G(T)}$, so we conclude that $\overline{G(T)}$ contains $(0, z)$. Since $z \neq 0$ and T is linear, this means $\overline{G(T)}$ is not a graph; using Exercise 3, T is not closable. ♠

Example 1 (Revisited). With $T = d/dx$, $X = L^2(\Omega)$, and $D = H_0^1(\Omega)$, let $D^* = \{v \in X : \langle Tu, v \rangle = \langle u, w \rangle \text{ for some } w \in X \text{ and all } u \in D\}$, and define $T^*v = w$. To identify T^* and D^* explicitly, we need to analyze the meaning of $\langle Tu, v \rangle = \langle u, w \rangle$. We know that $\langle u, w \rangle \equiv \int_0^1 u\overline{w}\,dx$, and formally integrating by parts we obtain $\langle Tu, v \rangle \equiv \int_0^1 u'\overline{v}\,dx = -\int_0^1 u\overline{v'}\,dx + u\overline{v}|_0^1$. But $u \in H_0^1(\Omega)$ may be approximated by $u_j \in C_0^1(\Omega)$, so $\langle Tu, v \rangle = -\int_0^1 u\overline{v'}\,dx$; consequently, $\langle Tu, w \rangle = \langle u, w \rangle$ implies $w = -v'$ (distributional derivative). This means that every $v \in D^*$ has its derivative $v' \in L^2(\Omega)$; in other words, $D^* \subset H^1(\Omega)$. Conversely, if $v \in H^1(\Omega)$ we let $w = -v'$ and verify (87b) by approximation. Consequently, $T^*v = -v'$ and $D^* = H^1(\Omega)$. ♣

The main result of this subsection concerns normal solvability; this generalizes a familiar result in finite-dimensional linear algebra (as well as the corresponding result for bounded operators on Hilbert spaces; cf. Exercise 17 in Section 6.1). Recall that $S^\perp$ denotes the collection of vectors that are perpendicular to a given set S in a Hilbert space.

Theorem 1. *If T is a densely defined linear operator on a Hilbert space X, then $(R_T)^\perp = N_{T^*}$.*

Proof. If $z \in (R_T)^\perp$, then $\langle z, Tx \rangle = \langle T^*z, x \rangle = 0$, for all $x \in D$. But D is dense, so $T^*z = 0$, i.e., $z \in N_{T^*}$. We have shown $(R_T)^\perp \subset N_{T^*}$. Reversing the steps shows that $N_{T^*} \subset (R_T)^\perp$, so we conclude $(R_T)^\perp = N_{T^*}$. ♠

Corollary. *If T is a densely-defined and closed linear operator on a Hilbert space X, then $(R_T)^\perp = N_{T^*}$ and $(R_{T^*})^\perp = N_T$.*

Remark. Unlike the finite-dimensional case, we cannot in general conclude that $R_T = (N_{T^*})^\perp$; this latter result also requires R_T to be a closed subspace of X, a condition that is *not* implied by the assumption that T is a closed operator.

c. Resolvents and Spectra

When X is a finite-dimensional vector space, a linear operator T is always bounded and may be represented by a square matrix. Such a matrix has eigenvalues and eigenvectors; to accomodate complex eigenvalues, we generally assume that X is a complex vector space. Let us see how to generalize this to the Banach space setting.

To begin with, assume that X is a complex Banach space and $T: D \to X$ is a linear operator. For any $\lambda \in \mathbf{C}$ we may consider the linear operator $T - \lambda : D \to X$. Provided that $T - \lambda$ is one to one, we may consider its inverse, at least as an unbounded operator $(T - \lambda)^{-1} : (T - \lambda)(D) \to X$, where $(T - \lambda)(D)$ is the image of D under $T - \lambda$. Of course, it may be that $(T - \lambda)^{-1}$ actually extends to a bounded operator on all of X.

Definitions. Suppose $T: D \to X$ is a closed linear operator on a complex Banach space X. The *regular* or *resolvent set* of T is the set

$$\rho(T) \equiv \{\lambda \in \mathbf{C} : T - \lambda \text{ is one to one and } (T - \lambda)^{-1} : X \to X \text{ is bounded}\}.$$

and the operator $(T - \lambda)^{-1}$ is called the *resolvent* for T. The *spectrum* of T, denoted by $\sigma(T)$, is the complement of $\rho(T)$ in $\mathbf{C}$. Of particular interest is the *point spectrum* $\sigma_p(T) \subset \sigma(T)$, consisting of the *eigenvalues* of T, namely $\lambda \in \sigma_p(T)$ if $Tx = \lambda x$ for some $x \in X$; x is called an *eigenvector* (or *eigenfunction* if X is a function space) corresponding to the eigenvalue λ, and the dimension of the eigenspace is the *multiplicity* of the eigenvalue.

Many results in spectral theory rely on the following:

Theorem 2. *If $T: D \to X$ is a closed linear operator on a complex Banach space X, then $\rho(T)$ is an open set in $\mathbf{C}$ and $\lambda \to (T - \lambda)^{-1}$ is an analytic function on each connected component of $\rho(T)$. Moreover, if X is a Hilbert space and T is densely defined, then $\sigma(T^*) = \{\bar{\lambda} : \lambda \in \sigma(T)\}$, and for any $\lambda \in \rho(T)$ we have $[(T - \lambda)^*]^{-1} = [(T - \lambda)^{-1}]^*$.*

Sketch of Proof. Let us assume that $T: X \to X$ is in fact bounded; the generalization is not substantially different (cf. [Conway] or [Kato]). We must also explain that analyticity for the function $f(\lambda) = (T - \lambda)^{-1}$, whose values are bounded linear operators on X, is analogous to the classical case, namely $f(\lambda)$ is *analytic at* λ_0 if it admits a derivative at all λ near λ_0 [i.e. $f'(\lambda) = \lim_{h \to 0}(f(\lambda + h) - f(\lambda))/h$ exists]. Many results from classical function theory, such as Liouville's theorem, continue to hold in this setting.

Now we appeal to a geometric series involving powers of T. For instance,

$$(89) \quad (T - \lambda)^{-1} = -\lambda^{-1}(I - (T/\lambda))^{-1} = -\lambda^{-1}(I + \sum_{n=1}^{\infty}(T/\lambda)^n))$$

is well defined and converges if $|\lambda| > \|T\|$. Analogously, for $\lambda, \mu \in \rho(T)$,

$$(90) \quad \begin{aligned}(T - \lambda)^{-1} &= (T - \mu)^{-1}(I - (\lambda - \mu)(T - \mu)^{-1})^{-1} \\ &= (T - \mu)^{-1}[I + \sum_{n=1}^{\infty}((\lambda - \mu)(T - \mu)^{-1})^n],\end{aligned}$$

which converges if $\|(\lambda - \mu)(T - \mu)^{-1}\| < 1$. Thus, for fixed $\mu \in \rho(T)$, we conclude that $\lambda \in \rho(T)$ provided that λ is sufficiently close to μ, in fact provided that $|\lambda - \mu| < \|(T - \mu)^{-1}\|^{-1}$. This shows that $\rho(T)$ is open; and because $(T - \lambda)^{-1}$ has a uniformly convergent power series expansion in λ, it admits a derivative and hence is analytic.

The remainder of the proof is given in Exercise 4. ♠

One immediate consequence of this theorem is that the spectrum $\sigma(T)$ is always a closed subset of $\mathbf{C}$. Moreover, for a bounded operator $T: X \to X$ it can be shown that $\sigma(T)$ is a nonempty compact set (Exercise 5); in particular, the *spectral radius* of T, $\sigma^{\mathrm{rad}}(T) \equiv \sup\{|\lambda| : \lambda \in \sigma(T)\}$, is finite.

For compact operators, much more is true; the following is called the *Riesz-Schauder theorem*:

Section 6.6: Unbounded Operators and Spectral Theory

Theorem 3. *If $T: X \to X$ is a compact operator on a complex Hilbert space X, then $\sigma(T)$ is a discrete set having no nonzero limit points; moreover, every nonzero $\lambda \in \sigma(T)$ is an eigenvalue of finite multiplicity.*

Proof. See Exercises 12–14. ♠

For unbounded operators, the spectrum is generally unbounded; but an important case is when T has *compact resolvent* [i.e., $(T-\lambda)^{-1}: X \to X$ is compact for every $\lambda \in \rho(T)$] because the spectrum of such operators consists entirely of isolated eigenvalues of finite multiplicity (Exercise 15).

In the next subsection, we investigate operators for which $\sigma(T) \subset \mathbf{R}$.

d. Symmetric and Self-Adjoint Operators

When X is a Hilbert space, T and T^* are both linear operators on X, so we may ask whether they are equal or at least one extends the other.

Definitions. Suppose that $T: D \to X$ is a densely defined operator on a (real or complex) Hilbert space X. Then T is *symmetric* if $T \subset T^*$; that is, if

(89) $\qquad \langle Tx, y \rangle = \langle x, Ty \rangle \quad \text{for all } x, y \in D.$

Moreover, T is *self-adjoint* if $T = T^*$ (meaning $D^* = D$ and (89) holds).

The operator in Example 1 is not symmetric, but multiplication by i will make it so (cf. Exercise 7). The Laplacian with proper choice of domain as in Example 2 is self-adjoint (cf. Exercise 10). Symmetry and self-adjointness coincide for bounded linear operators on X (since $D^* = D = X$) but not in general (cf. Exercise 7).

For a symmetric operator T, all eigenvalues must be real [i.e. $\sigma_p(T) \subset \mathbf{R}$]; this is true since $\langle Tx, x \rangle \in \mathbf{R}$ for all $x \in D$ (by Exercise 8) and $Tx = \lambda x$ together imply $\lambda \in \mathbf{R}$. But, as we shall see, the stronger conclusion $\sigma(T) \subset \mathbf{R}$ requires T to be self-adjoint.

We first establish the following result:

Lemma. *Let $T: D \to X$ be a closed symmetric linear operator on a complex Hilbert space X. If $\operatorname{Im} \lambda \neq 0$, then $R_{(T-\lambda)}$ is a closed subspace of X.*

Proof. Write $\lambda = \alpha + i\beta$, where $\alpha, \beta \in \mathbf{R}$. For any $x \in D$, a simple calculation shows that

$$\|(T-\lambda)x\|^2 = \|(T-\alpha)x\|^2 + 2\operatorname{Re} i\langle (T-\alpha)x, \beta x\rangle + \beta^2 \|x\|^2$$
$$= \|(T-\alpha)x\|^2 + \beta^2 \|x\|^2,$$

since $\langle (T-\alpha)x, \beta x \rangle = \beta \langle Tx, x \rangle - \alpha\beta \|x\|^2 \in \mathbf{R}$ by the symmetry of T. In particular, we have the inequality $\|(T-\lambda)x\|^2 \geq \beta^2 \|x\|^2$. Now if we consider $y \in \overline{R_{(T-\lambda)}}$, there must exist $x_n \in D$ with $(T-\lambda)x_n \to y$, so this inequality implies that $\{x_n\}$ is a Cauchy sequence and must converge to some $x \in X$. But $T - \lambda$ is a closed operator, so we must have $x \in D$ and $(T-\lambda)x = y$; i.e., $y \in R_{(T-\lambda)}$. We conclude that $R_{(T-\lambda)}$ is closed. ♠

Theorem 4. Suppose $T: D \to X$ is a closed symmetric linear operator on a complex Hilbert space X. Then the following are equivalent: (i) T is self-adjoint, (ii) $N_{(T^*-\lambda)} = \{0\}$ for all $\lambda \in \mathbf{C}$ with $\operatorname{Im}\lambda \neq 0$, and (iii) $\sigma(T) \subset \mathbf{R}$.

Proof. (i) $\Rightarrow$ (ii): If $\lambda \in \mathbf{C}$ with $\operatorname{Im}\lambda \neq 0$ and $T^*x = Tx = \lambda x$ for some $x \in D$, then $\overline{\lambda}\langle x, x\rangle = \langle x, \lambda x\rangle = \langle x, Tx\rangle = \langle Tx, x\rangle = \lambda\langle x, x\rangle$ shows that $\overline{\lambda}\|x\|^2 = \lambda\|x\|^2$, which implies either $\lambda \in \mathbf{R}$ or $x = 0$.

(ii) $\Rightarrow$ (iii): Let $\lambda \in \mathbf{C}$ with $\operatorname{Im}\lambda \neq 0$, so $N_{(T-\lambda)} = \{0\}$ by the symmetry of T. By hypothesis we have $N_{(T^*-\overline{\lambda})} = \{0\}$, so Theorem 1 implies that $R_{(T-\lambda)}$ is dense in X. But the preceding lemma then implies that $R_{(T-\lambda)} = X$, so we can apply Proposition 1 to conclude that $(T-\lambda)^{-1}: X \to X$ is bounded, and hence $\lambda \in \rho(T)$. We have shown $\sigma(T) \subset \mathbf{R}$.

(iii) $\Rightarrow$ (i): Let $y \in D^* \equiv \operatorname{dom}(T^*)$; we want to show that $y \in D$. Since $\sigma(T) \subset \mathbf{R}$, we know that $T + i: X \to X$ is boundedly invertible, so there exists $x \in D$ such that $(T+i)x = (T^*+i)y$. But $T + i \subset T^* + i$, so we also have $(T^* + i)x = (T^* + i)y$. Since $N_{(T^*+i)} = \{0\}$, we conclude that $x = y$; in particular, $y \in D$. ♦

Again we can say more for compact operators; the following is known as the *Hilbert-Schmidt theorem*:

Theorem 5. If $T: X \to X$ is a compact, self-adjoint operator on a Hilbert space X, then there is an orthonormal basis of eigenvectors $\{x_n\}$.

Proof. For each eigenvalue of T, find an orthonormal basis for its (finite-dimensional) eigenspace. Since eigenvectors for distinct eigenvalues are orthogonal, we obtain an orthonormal collection $\{x_n\}$ of eigenvectors. Let us denote the closed linear span of $\{x_n\}$ by $\tilde{X}$; we would like to show $\tilde{X} = X$. But $T: \tilde{X} \to \tilde{X}$ and $T: \tilde{X}^\perp \to \tilde{X}^\perp$ are both compact, self-adjoint operators; so if $\tilde{X}^\perp \neq \{\emptyset\}$, then we could find an eigenvector for T in $\tilde{X}^\perp$, contradicting the definition of $\tilde{X}$. Consequently, $\tilde{X}^\perp = \{\emptyset\}$, so $\tilde{X} = X$. ♦

For a self-adjoint operator $T: D \to X$ that is unbounded but has compact resolvent, we can also conclude the existence of a basis of eigenvectors (cf. Exercise 15); in Section 7.2, we encounter this phenomenon in the specific instance of an elliptic operator in divergence form.

Exercises for Section 6.6

1. If $T: D \to X$ is a bounded linear operator (i.e. $\|Tx\| \leq C\|x\|$ for all $x \in D$), show that T is closed if and only if D is closed in X.

2. For any linear operator $T: D \to X$, define the *graph norm* on D by $\|x\|_g = \|x\|_X + \|Tx\|_X$. Show that T is closed if and only if D is a Banach space under the graph norm.

3. Show that $T: D \to X$ is closable if and only if $\overline{G(T)}$ is a graph, and in this case $G(\overline{T}) = \overline{G(T)}$.

4. Prove the second sentence in Theorem 2.

5. Suppose $T: X \to X$ is a bounded linear operator on a complex Hilbert space X.
   (a) Show that $\sigma(T) \subset \{\lambda \in \mathbf{C} : |\lambda| \leq \|T\|\}$.
   (b) Let $f(\lambda) = (T - \lambda)^{-1}$ for $\lambda \in \rho(T)$, and show that $f(\lambda) \to 0$ as $\lambda \to \infty$.
   (c) Conclude that $\sigma(T)$ cannot be empty.
6. Suppose $T: X \to X$ is a bounded linear operator on a complex Hilbert space X. Show $\lim_{n \to \infty} \|T^n\|^{1/n}$ exists and equals $\sigma^{\mathrm{rad}}(T)$.
7. Consider $T = i\, d/dx$ on $X = L^2(\Omega)$ (complex valued) with $D = H_0^1(\Omega)$, where $\Omega = (0, 1)$.
   (a) Show that $T: D \to X$ is a closed, densely defined linear operator.
   (b) Show that T is symmetric, but $T \neq T^*$.
   (c) Show that $\sigma_p(T^*) = \mathbf{C}$.
8. If $T: D \to X$ is a densely defined operator on a complex Hilbert space X, then T is symmetric if and only if $D \subset D^*$ and $\langle Tx, x \rangle \in \mathbf{R}$ for all $x \in D$.
9. If $T: X \to X$ is a self-adjoint compact linear operator on a Hilbert space X, then either $\|T\|$ or $-\|T\|$ is an eigenvalue for T.
10. With $D = H_0^1(\Omega) \cap H^2(\Omega)$, show that $\Delta: D \to L^2(\Omega)$ is self-adjoint.
11. If $T: D \to X$ is a self-adjoint linear operator on a Hilbert space X, then eigenfunctions for distinct eigenvalues are orthogonal.
12. If $T: X \to X$ is a compact linear operator on a complex Hilbert space X and $\lambda \neq 0$, then $R_{(T-\lambda)}$ is closed.
13. If $T: X \to X$ is a compact linear operator on a complex Hilbert space X, and $\lambda \notin \sigma_p(T) \cup \{0\}$, then $\lambda \in \rho(T)$.
14. Prove Theorem 3.
15. Suppose $T: D \to X$ is a closed linear operator on a Hilbert space X such that $(T - \lambda_0)^{-1}: X \to X$ is compact for some $\lambda_0 \in \rho(T)$.
   (a) Show that $\sigma(T)$ consists entirely of isolated eigenvalues, $\{\lambda_n\}_{n=1}^{\infty}$, each λ_n has finite multiplicity, and $\{\lambda_n\}_{n=1}^{\infty}$ has no finite accumulation points.
   (b) Show that $(T - \lambda)^{-1}: X \to X$ is compact for every $\lambda \in \rho(T)$.
   (c) If we additionally assume that T is self-adjoint, then conclude that there is an orthonormal basis for X consisting of eigenvectors for T.

Further References for Chapter 6

For a more complete treatment of functional analysis, see [Conway], [Kato], [Reed-Simon], [Rudin, 2], or [Yoshida]. One topic that we have ignored is the study of seminorms and Frechèt spaces, which are essential for a complete study of distribution theory.

We have also given a very brief treatment of Sobolev spaces. For more details and additional results, [Adams] is an excellent reference. The "weak = strong" result of Corollary B of Theorem 3 in Section 6.5 was originally proved in [Meyers-Serrin]. Sobolev spaces with fractional order, such as $H^{1/2,2}(\Omega)$, may also be defined, and are important in considering "traces" (i.e., the restriction of functions in Ω to a submanifold, such as the boundary $\partial\Omega$); see [Adams]. There are many variants of Sobolev spaces, for example, by the incorporation of weight functions in the definition of the norms.

For higher-order Sobolev spaces, we have focused on the imbedding theorems rather than the inequalities. However, it should be observed that there are generalizations of (34) and (35) to higher-order derivatives. See, for example, [Friedman].

There is a large literature on the Stokes equations and fluid dynamics; for example, see [Chorin-Marsden], [Galdi], or [Temam, 1]. In particular, the Helmholtz-Weyl decomposition (19) may also be obtained using *de Rham theory* for differential forms; see [Temam, 1].

7 Differential Calculus Methods

In this chapter, we investigate differential calculus on Banach spaces, which provides an important tool for the study of linear and nonlinear partial differential equations. We begin with the calculus of scalar-valued functions on Banach spaces and their critical points; this is the basis for *variational methods* for solving PDEs. We shall apply this method to study the existence of weak solutions to linear elliptic boundary value problems, as well as the eigenvalues of the Laplacian. Finally, we consider the calculus of maps between Banach spaces, deriving infinite-dimensional versions of the inverse and implicit function theorems on Banach spaces. (Further applications of these differential calculus methods to nonlinear elliptic equations are given in Chapter 13.)

7.1 Calculus of Functionals and Variations

In Section 6.1. we encountered bounded linear functionals on a Banach space. More generally, a *functional* on a Banach space is a scalar-valued mapping that is continuous but not necessarily linear. In this section we develop differential calculus for nonlinear functionals on Banach spaces; in many ways, this is analogous to the finite-dimensional case. We also generalize notions from optimization theory and the calculus of variations, including critical points, absolute extrema, and saddle points.

a. The Derivative and Critical Points of a Functional

Recall the notions of derivative and differentiability for a map $F\colon \mathbf{R}^n \to \mathbf{R}$. We define the *derivative of F at x in the direction y*, represented by various notations $F'(x)\, y = \nabla F(x)\, y = D_y F(x)$, by the formula

$$F'(x)\, y = \lim_{\epsilon \to 0} \frac{F(x + \epsilon y) - F(x)}{\epsilon}.$$

The *differentiability of F at x*, however, requires the stronger condition that

$$F(x + y) = F(x) + F'(x)\, y + o(|y|) \quad \text{as} \quad |y| \to 0.$$

[See also the Introduction for the notation $o(\cdot)$.] If F is differentiable at x, then $F'(x)\, y$ is linear in y: $F'(x)\, (\alpha_1 y_1 + \alpha_2 y_2) = \alpha_1 F(x)\, y_1 + \alpha_2 F(x)\, y_2$. In other words, $F'(x)$ is a linear functional on $\mathbf{R}^n$; i.e., an element of the dual space $(\mathbf{R}^n)^*$. In this way we may consider $F'\colon \mathbf{R}^n \to (\mathbf{R}^n)^*$.

Now let X be a Banach space and let $F\colon X \to \mathbf{R}$ be a continuous, but not necessarily linear, functional. By analogy with $X = \mathbf{R}^n$, we want to define $F'\colon X \to X^*$; that is, for each $x \in X$, $F'(x)$ should be a bounded linear functional. In fact, let us define the *derivative of F at x in the direction y* by the same formula as for $\mathbf{R}^n$,

$$(1) \qquad F'(x)\,y = \lim_{\epsilon \to 0} \frac{F(x + \epsilon y) - F(x)}{\epsilon},$$

and the *(Frechet) differentiability of F at x* by the condition

$$(2) \qquad F(x + y) = F(x) + F'(x)\,y + o(\|y\|_X) \quad \text{as} \quad \|y\|_X \to 0.$$

In addition, if the map $F'\colon X \to X^*$ is continuous, then we say that F is C^1. We also define x to be a *critical point of F* if $F'(x) = 0$; that is,

$$(3) \qquad F'(x)y = 0 \qquad \text{for all } y \in X.$$

The critical point condition (3) is called the *Euler-Lagrange equation* for the functional $F\colon X \to \mathbf{R}$.

Let us illustrate these concepts with an important example.

Example 1. Suppose $\Omega \subset \mathbf{R}^n$ is a bounded domain, and let $X = H_0^{1,2}(\Omega)$. Fix $f \in L^2(\Omega)$ and let us define

$$(4) \qquad F(u) = \int_\Omega \left(\frac{1}{2}|\nabla u|^2 + fu \right) dx \qquad \text{for} \quad u \in X.$$

Since $|\nabla u|^2$ and fu are both integrable on Ω, $F(u) \in \mathbf{R}$ is well defined. To find F', consider $u, v \in X$ and calculate

$$F'(u)\,v = \lim_{\epsilon \to 0} \frac{\int_\Omega \frac{1}{2}(|\nabla(u+\epsilon v)|^2 - |\nabla u|^2) + f(u + \epsilon v) - fu\,dx}{\epsilon}$$

$$= \lim_{\epsilon \to 0} \int_\Omega \nabla u \cdot \nabla v + \frac{\epsilon}{2}|\nabla v|^2 + fv\,dx = \int_\Omega \nabla u \cdot \nabla v + fv\,dx.$$

Notice that $F'(u)\,v$ is indeed linear in v, and $F'(u)$ is bounded on X; that is,

$$|F'(u)\,v| \leq \|\nabla u\|_2 \|\nabla v\|_2 + \|f\|_2 \|v\|_2 \leq C\,\|v\|_X,$$

where C depends on u and f; hence $F'(u) \in X^*$. Let us verify that F is differentiable:

$$|F(u + v) - F(u) - F'(u)\,v| = \frac{1}{2} \int_\Omega |\nabla v|^2\,dx$$

$$\leq \frac{1}{2}\|v\|_X^2 = o(\|v\|_X) \quad \text{as } \|v\|_X \to 0.$$

Finally, let us verify that F is C^1. This requires showing that $u \to F'(u)$ is continuous $X \to X^*$. For $u, v, w \in X$, we find

$$|(F'(u) - F'(v))w| = \left| \int_\Omega (\nabla u - \nabla v) \cdot \nabla w \, dx \right|$$
$$\leq \|\nabla(u-v)\|_2 \|\nabla w\|_2 \leq \|u - v\|_X \|w\|_X.$$

But then we obtain

$$\|F'(u) - F'(v)\|_{X^*} = \sup_{w \neq 0} \frac{|(F'(u) - F'(v))w|}{\|w\|_X} \to 0 \quad \text{as} \quad \|u - v\|_X \to 0,$$

which shows that $F': X \to X^*$ is continuous.

Notice that if u is a critical point, then u satisfies the Euler-Lagrange equation

(5) $$\int_\Omega \nabla u \cdot \nabla v + fv \, dx = 0 \qquad \text{for all } v \in X.$$

As defined in Section 6.2, this means that u is a *weak* solution of Poisson's equation $\Delta u = f$ in Ω with Dirichlet condition $u = 0$ on $\partial\Omega$. Of course, using the Riesz representation theorem, we already know that a weak solution exists. But we shall solve (5) by showing that the functional (4) has a critical point, in fact an absolute minimum. ♣

In general, finding a weak solution as a critical point of a nonlinear functional is called the *variational method* for solving a partial differential equation and is especially important since it will apply to nonlinear PDEs when we cannot use linear techniques such as the Riesz representation theorem. The variational method is closely related to optimization and the classical theory of the *calculus of variations*. As in the finite-dimensional case, if a C^1-functional F achieves its absolute minimum at a point $x \in X$, then x must be a critical point; for any $y \in X$, the function $f(t) = F(x+ty)$ of the real variable t has a critical point at $t = 0$, so $f'(0) = F'(x)y = 0$, proving that x is a critical point of F. [In the calculus of variations, the quantity $f'(0)$ is called the *first variation of F at x in the direction y* and is denoted $\delta_y F(x)$ or $\delta F(x; y)$.] Notice that, in this calculation, the direction y was allowed to vary over the whole space X. In applications, however, it is sometimes necessary to allow y to vary only over a subset $\mathcal{A}$ of X called the *admissible set*. This occurs naturally in problems involving nonhomogeneous boundary conditions, as the next example illustrates.

Example 2. Suppose we wish to use the variational method to solve Laplace's equation $\Delta u = 0$ in a bounded domain Ω, with nonhomogeneous boundary condition $u = g$ on $\partial\Omega$. Since we no longer have "$u = 0$ on $\partial\Omega$," we are tempted to replace $X = H_0^{1,2}(\Omega)$ by $X = H^{1,2}(\Omega)$ and consider the "energy" functional

(6) $$F(u) = \frac{1}{2} \int_\Omega |\nabla u|^2 \, dx \qquad \text{for } u \in H^{1,2}(\Omega),$$

which is C^1 by the calculations in Example 1. But how do we achieve the boundary condition $u = g$ on $\partial\Omega$? If we assume that g is actually defined on all of $\overline{\Omega}$ with $g \in H^{1,2}(\Omega)$, then a natural way to express the boundary condition is to require $u - g \in H_0^{1,2}(\Omega)$. This means that u is not allowed to range freely over X, but only over the admissible set

(7) $\quad \mathcal{A} = \{u \in H^{1,2}(\Omega) : u - g \in H_0^{1,2}(\Omega)\} = \{u = g + v : v \in H_0^{1,2}(\Omega)\}$.

Suppose we can show that F achieves its minimum on $\mathcal{A}$ at a point u. This means that, for any $v \in H_0^{1,2}(\Omega)$, the function $f(t) = F(u + tv)$ achieves its minimum at $t = 0$. Computing $f'(0) = 0$ yields $F'(u)v = 0$. Using (6), we find that the Euler-Lagrange equation for F is

(8) $\quad \displaystyle\int_\Omega \nabla u \cdot \nabla v\, dx = 0 \qquad \text{for all } v \in H_0^{1,2}(\Omega)$.

In other words, u is a weak solution of Laplace's equation and satisfies the boundary condition in the weak sense that $u - g \in H_0^{1,2}(\Omega)$. Later in this section we show the existence of such a critical point. ♣

In the next subsection we investigate conditions under which functionals possess absolute extrema.

b. Coercive Functionals and Absolute Extrema

Let us consider the possibility that a functional on a Banach space has a critical point that occurs as an absolute extremum. First, let us recall the situation in finite dimensions. If $F\colon \mathbf{R}^n \to \mathbf{R}$ is a C^1 map and if F is maximized or minimized at x_0, then x_0 is a critical point for F: $F'(x_0) = 0$. For example, suppose we know that $F(x) \to \infty$ as $|x| \to \infty$. Clearly, F must attain its (absolute) minimum at some point x_0, which must then be a critical point of F. Of course, we know that the minimum x_0 exists because the set $K = \{x \in \mathbf{R}^n : F(x) \leq M\}$ is closed and bounded, hence compact, so that the continuous function F attains its minimum on K. In Banach spaces, continuous functions need not attain their infimum on closed bounded sets, so let us slightly revise the argument proving the existence of x_0. Since the values of F are bounded below, we can find x_j such that $F(x_j)$ monotonically decreases to $I = \inf\{F(x) : x \in \mathbf{R}^n\}$, and we may assume $F(x_j) \leq I + 1$. Now the x_j belong to the closed bounded set $\{x \in \mathbf{R}^n : F(x) \leq I + 1\}$, so by compactness there is a subsequence x_{j_k} that converges to some point x_0. By continuity, $F(x_0) = \lim F(x_{j_k}) = I$, so F attains its infimum at x_0.

This revised argument for the existence of x_0 is roughly what we shall use in infinite dimensions. Of course the revised argument also requires compactness, and it is this issue that makes the search for critical points of nonlinear functionals on Banach spaces somewhat challenging.

Now let us return to the general setting of a C^1 functional F on the Banach space X and try to find critical points of F as extrema. For example, when can we say that F has a minimum value that is attained at a point $x \in X$? Suppose F satisfies the following inequality:

$$(9) \qquad F(x) \geq C_1 \|x\|_X^2 - C_2 \qquad \text{for} \quad x \in X.$$

where C_1 and C_2 are constants with $C_1 > 0$; such a functional is called *coercive*. The inequality (9) certainly implies that F is bounded below, so $I = \inf\{F(x) : x \in X\} > -\infty$, and we can find $x_j \in X$ such that $F(x_j) \to \inf\{F(x) : x \in X\}$. We must show that at least a subsequence of x_j converges to a point $x_0 \in X$. Figure 1, which represents (9) schematically, suggests that this can be done, but proving it requires the use of compactness and weak convergence; we explain this in the next subsection.

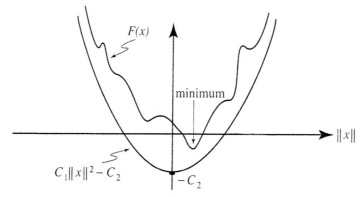

Figure 1. Schematic for a coercive functional

Another important property of functionals is the following: A continuous functional F on a Banach space X is *weakly lower semicontinuous* if

$$(10) \qquad F(u) \leq \liminf_{j \to \infty} F(u_j) \qquad \text{whenever } u_j \rightharpoonup u \text{ weakly in } X.$$

For example, Exercise 3a shows that the norm on a Hilbert space is weakly lower semicontinuous. Notice that weak *upper* semicontinuity need not hold when weak lower semicontinuity does (cf. Exercise 3b).

As in finite dimensions, at a (local) minimum the functional should display certain convexity properties. Let us define the *second variation of F at x in the direction y* to be the value $f''(0)$, where $f(t) = F(x + ty)$; this quantity is sometimes denoted $\delta_y^2 F(x)$ or $\delta^2 F(x; y)$. If x is a minimum point for F on $\mathcal{A}$, then $\delta_y^2 F(x) \geq 0$ for all $y \in \mathcal{A}$. The implications of this fact are explored for certain functionals in Exercise 7.

Later in this section we explore the uniqueness of critical points using global convexity.

c. Weak Existence for Dirichlet & Neumann Problems

Now let us see how to prove the existence of weak solutions for some familiar problems. We assume that Ω is a bounded domain in $\mathbf{R}^n$.

Dirichlet problem for the Poisson equation. As in Example 1, we want to show that the functional F of (4) has a critical point $u \in X = H_0^{1,2}(\Omega)$ so that (5) holds. We must first verify that F is coercive. The Poincaré inequality (Theorem 1 in Section 6.2) shows that it suffices to establish

$$(11) \qquad F(u) \geq C_1 \|\nabla u\|_2^2 - C_2.$$

But, using the Cauchy-Schwarz inequality, the inequality $2ab \leq (a^2/\epsilon^2) + (\epsilon^2 b^2)$, and the Poincaré inequality, we obtain

$$\int_\Omega fu\,dx \leq \|f\|_2 \|u\|_2 \leq \frac{1}{2\epsilon^2}\|f\|_2^2 + \frac{\epsilon^2}{2}\|u\|_2 \leq \frac{1}{2\epsilon^2}\|f\|_2^2 + \frac{\epsilon^2}{2} C \|\nabla u\|_2^2.$$

Therefore,

$$F(u) \geq \frac{1}{2}\|\nabla u\|_2^2 - \frac{1}{2\epsilon^2}\|f\|_2^2 - \frac{\epsilon^2}{2} C \|\nabla u\|_2^2,$$

which yields (11) with $C_1 = (1 - \epsilon^2 C)/2 > 0$ provided that ϵ is small enough.

Next, we show that F attains its minimum on X. As a consequence of (11), we see that F is bounded below, and in fact $F(u) \to \infty$ as $\|u\|_X \to \infty$; we still must show that F attains its minimum on X. Let $I = \inf\{F(u) : u \in X\}$ and pick $u_j \in X$ such that $F(u_j) \leq I + 1$ and $F(u_j) \to I$. By (11) we find that $\{u_j\}$ is bounded in X. But $X = H_0^{1,2}(\Omega) \subset L^2(\Omega)$ is a compact imbedding by Corollary A of Theorem 4 in Section 6.5, so we may find a subsequence $\{u_{j_k}\}$ that converges in $L^2(\Omega)$ to a function $u \in L^2(\Omega)$. Since $f \in L^2(\Omega)$, we also have $\int_\Omega fu_{j_k}\,dx \to \int_\Omega fu\,dx$. Now recall from Theorem 2 (or its corollary) in Section 6.3 that $\{u_{j_k}\}$ must also contain a subsequence (which we continue to denote by $\{u_{j_k}\}$) that converges weakly to some $u' \in X$; recall that we write weak convergence as $u_{j_k} \rightharpoonup u'$. But $u_{j_k} \to u$ in $L^2(\Omega)$ implies that we must have $u = u'$; this means that $u_{j_k} \rightharpoonup u$ weakly in X. A consequence of $u_{j_k} \rightharpoonup u$ is that (cf. Exercise 3a)

$$(12) \qquad \|u\|_X \leq \liminf_{j_k \to \infty} \|u_{j_k}\|_X.$$

Using (12) and $u_{j_k} \to u$ in $L^2(\Omega)$, we obtain

$$\begin{aligned} F(u) &= \frac{1}{2}\|\nabla u\|_2^2 + \int_\Omega fu\,dx \\ &= \frac{1}{2}\|u\|_X^2 - \frac{1}{2}\|u\|_2^2 + \int_\Omega fu\,dx \\ &\leq \liminf_{j_k \to \infty} \left\{ \frac{1}{2}\|u_{j_k}\|_X^2 - \frac{1}{2}\|u_{j_k}\|_2^2 + \int_\Omega fu_{j_k}\,dx \right\} \\ &= \liminf_{j_k \to \infty} F(u_{j_k}) = I. \end{aligned}$$

Hence we see that F attains its minimum at $u \in X$, so u must be a critical point of F, that is, u satisfies (5). Thus we have shown the existence of a weak solution $u \in H_0^{1,2}(\Omega)$ of $\Delta u = f$ in Ω with $u = 0$ on $\partial\Omega$. ♣

In the next example, we use coercivity to establish the existence of an absolute minimum for a functional restricted to an admissible set.

Dirichlet problem for the Laplace equation. As in Example 2, we want to show that the functional F in (6) has a critical point u on the admissible set $\mathcal{A}$ defined in (7), so that (8) holds. Notice that $F(u) \geq 0$, so $I = \inf\{F(u) : u \in \mathcal{A}\} \geq 0$. Pick $u_j \in \mathcal{A}$ with $F(u_j) \leq I + 1$ and $F(u_j) \to I$. Write $u_j = g + v_j$, where $v_j \in H_0^{1,2}(\Omega)$. Using $|\nabla v_j|^2/2 \leq |\nabla g + \nabla v_j|^2 + |\nabla g|^2$ (see Exercise 4a) and the boundedness of $F(u_j)$, we easily conclude that $\int |\nabla v_j|^2\, dx$ is bounded as $j \to \infty$. From here, the existence of a minimizer $u \in \mathcal{A}$ of F follows similarly to that for Example 1 (see Exercise 4b). We conclude that there is a weak solution $u \in H^{1,2}(\Omega)$ of $\Delta u = 0$ in Ω with $u = g$ on $\partial\Omega$ in the sense that $u - g \in H_0^{1,2}(\Omega)$. ♣

The preceding examples have been solved previously by other means. Now let us turn to a Neumann problem for which we have not yet shown existence. We assume that Ω has a smooth boundary.

Neumann problem for the Poisson equation. We want to find a weak solution of $\Delta u = f$ in Ω and $\partial u/\partial \nu = 0$ on $\partial\Omega$; here ν is the exterior unit normal on $\partial\Omega$ and $f \in L^2(\Omega)$ satisfies the compatibility condition $\int_\Omega f(x)\, dx = 0$ [cf. (8) in Chapter 4]. We shall let F be the functional (4) defined on the Banach space $X = \mathcal{A} = H^{1,2}(\Omega)$. The arguments of Example 1 show that F is C^1 on X with $F'(u)v = \int_\Omega (\nabla u \cdot \nabla v + fv)\, dx$, so (5) is the Euler-Lagrange equation. To see why a critical point u of F should be considered as a weak solution of this problem, suppose we have additional regularity; in other words, suppose we know $u \in C^2(\Omega) \cap C^1(\overline{\Omega})$ and we consider an arbitrary $v \in C^1(\overline{\Omega}) \subset X$. Green's identity [cf. (6) in Chapter 4] shows that the Euler-Lagrange equation becomes

$$\int_\Omega (-\Delta u + f)\, v\, dx + \int_{\partial\Omega} v \frac{\partial u}{\partial \nu}\, dS = 0.$$

But since v is arbitrary in $C^1(\overline{\Omega})$, this implies that $\Delta u = f$ in Ω and $\partial u/\partial \nu = 0$ on $\partial\Omega$. [The supposition $u \in C^2(\Omega) \cap C^1(\overline{\Omega})$ is only true under greater regularity of f, but $f \in L^2(\Omega)$ is sufficient to show that a weak solution $u \in H^{1,2}(\Omega)$ satisfies $u \in H^{2,2}(\Omega)$; this is discussed in Section 8.2.]

Now let us consider coercivity. Immediately we encounter a difficulty, because we can let u be any constant function u_0: we find $F(u_0) = 0$, but $\|u_0\|_X \to \infty$ as $u_0 \to \infty$, so (9) fails! However, we may eliminate these constant functions by defining

$$X_0 = \{u \in H^{1,2}(\Omega) : \int_\Omega u\, dx = 0\},$$

which (as the nullspace of the continuous functional $u \to \int u\,dx$) is a closed subspace of $H^{1,2}(\Omega)$ and hence a Banach space under the same norm. In fact, since $u \in X_0$ has zero average over Ω, the Poincaré inequality [cf. (79) in Chapter 4] shows that $\|\nabla u\|_2$ is equivalent to $\|u\|_X$, so it suffices to show (10). Again using the Poincaré inequality, (10) follows exactly as it did for the Dirichlet problem.

Now we must show that F attains its infimum on X_0; again the argument is similar to that for the Dirichlet problem, so we briefly give the details. Let $I = \{F(u) : u \in X_0\}$ and select $u_j \in X_0$ with $F(u_j) \to I$. We know by (10) that $\{u_j\}$ is bounded in X_0 and hence in $X = H^{1,2}(\Omega)$. But $H^{1,2}(\Omega) \subset L^2(\Omega)$ is a compact imbedding by Theorem 5 of Section 6.5, so we may find a subsequence $\{u_{j_k}\}$ that converges in $L^2(\Omega)$ to a function u. We may also assume that $\{u_{j_k}\}$ converges weakly in X, so $u \in X$. In fact, $0 = \int u_{j_k}\,dx \to \int u\,dx$ implies $u \in X_0$. As a consequence of $\int f u_{j_k}\,dx \to \int f u\,dx$ and $\|\nabla u\|_2 \leq \liminf \|\nabla u_{j_k}\|_2$, we conclude that $F(u) = I$; in other words, F attains its infimum on X_0 at u.

We conclude that $u \in X_0$ is a weak solution of $\Delta u = f$ in Ω with $\partial u/\partial \nu = 0$ on $\partial \Omega$. Of course, we can add any constant to u and again get a weak solution, so u is not unique. ♣

This approach may also be used to solve handle a nonhomogeneous Neumann condition $\partial u/\partial \nu = h$ on $\partial \Omega$ (see Exercise 5).

d. Convexity and Uniqueness

We have seen that coercivity of a C^1-functional may be used to show the existence of an absolute minimum, hence a weak solution of the associated Euler-Lagrange equation. As Figure 1 indicates, this absolute extremum is not necessarily the only critical point of the functional. When can we conclude that there is only one critical point? An important factor here is the (global) convexity of the functional.

Suppose F is a C^1-functional on a Banach space X that contains the admissible set $\mathcal{A}$. Then F is *convex on* $\mathcal{A}$ if

(13) $\qquad F(x+y) - F(x) \geq F'(x)y \qquad$ whenever $x, x+y \in \mathcal{A}$.

Moreover, F is *strictly convex on* $\mathcal{A}$ if equality holds in (13) if and only if $y = 0$. For example, it is elementary to show that the functional (4) is strictly convex on $\mathcal{A} = X = H_0^{1,2}(\Omega)$:

$$F(u+v) - F(u) = \int_\Omega \nabla u \cdot \nabla v + \frac{|\nabla v|^2}{2} + fv\,dx \geq \int_\Omega \nabla u \cdot \nabla v + fv\,dx = F'(u)v$$

for all $u, v \in H_0^{1,2}(\Omega)$, with equality if and only if $v \equiv 0$. Similarly, the functional (6) is strictly convex on the admissible set $\mathcal{A}$ as defined in (7).

The significance of convexity is expressed in the following.

Theorem 1. Suppose F is a C^1-functional on a Banach space X, and F is convex on the set $\mathcal{A} \subset X$. If $x_0 \in \mathcal{A}$ satisfies $F'(x_0)y = 0$ whenever $x_0 + y \in \mathcal{A}$, then x_0 minimizes F on $\mathcal{A}$. Moreover, if F is strictly convex on $\mathcal{A}$, then x_0 is the unique minimizer.

Proof. Let $x \in \mathcal{A}$ and $y = x - x_0$. Using (13) and $F'(x_0)y = 0$, we have $F(x) \geq F(x_0)$; in other words, x_0 minimizes F on $\mathcal{A}$. If F is strictly convex, then $F(x) > F(x_0)$, unless $x = x_0$; the minimizer x_0 is unique. ♠

Corollary. The weak solutions for the Laplace and Poisson equations with Dirichlet boundary conditions are unique. The weak solutions for these equations with Neumann boundary conditions are unique up to an additive constant.

e. Mountain Passes and Saddle Points

In the previous subsections, we encountered functionals having critical points that occur as absolute extrema. Of course critical points do not have to be extreme points, so in this subsection we investigate certain functionals having *saddle points:* critical points that are not even local extreme points.

The geometric motivation for this theory is the simple fact that if a mountain range separates two points, then there must be a mountain pass that requires the least climbing to get from one point to the other (see Figure 2).

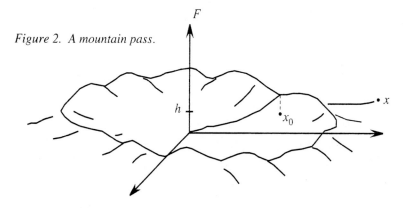

Figure 2. A mountain pass.

To formulate this idea mathematically, suppose X is a real Banach space in which $B_\rho(0)$ denotes the ball of radius ρ at the origin, and suppose F is a C^1 functional on X satisfying the following conditions:
  (i) $F(0) = 0$.
  (ii) For some $\alpha, \rho > 0$, $F|_{\partial B_\rho(0)} \geq \alpha$.
  (iii) For some $\bar{x} \in X \backslash B_\rho(0)$, $F(\bar{x}) \leq 0$.

Consider a path connecting 0 and $\bar{x}$; that is, let $g: [0,1] \to X$ be continuous with $g(0) = 0$ and $g(1) = \bar{x}$. Since g must cross $\partial B_\rho(0)$, we know by (ii) that F attains at least the height α along the path: $\max_{0 \le t \le 1} F(g(t)) \ge \alpha$. Now suppose we let Γ be the collection of all paths connecting 0 and $\bar{x}$ and we try to minimize the height of F along these paths:

(14) $$h \equiv \inf_{g \in \Gamma} \max_{0 \le t \le 1} F(g(t))$$

denotes the height of the "mountain pass"; notice that $h \ge \alpha$. *If* this value h is attained at some $g_0 \in \Gamma$ and $t_0 \in (0,1)$, then let $x_0 = g_0(t_0)$; see Fig. 2.

In a finite-dimensional space X, such a mountain pass has a horizontal tangent plane, so that $F'(x_0) = 0$. Thus it is reasonable to expect that, for a general Banach space X, such a mountain pass x_0 would be a critical point of F. Using a deformation of the curve g, this can be established provided that F satisfies the additional condition

   (iv) If $\{x_m\}$ is a sequence for which $F(x_m)$ is bounded and $F'(x_m) \to 0$ as $m \to \infty$, then $\{x_m\}$ has a convergent subsequence.

The condition (iv) is called the *Palais-Smale condition,* and a sequence satisfying the hypotheses of (iv) is called a *Palais-Smale sequence*. Notice that the coercive functional of Example 1 satisfies this condition (cf. Exercise 10); a more general condition is given in Exercise 11.

We now have all the necessary conditions to conclude the existence of a critical point.

Theorem 2 (The Mountain Pass Theorem). *If F is a C^1 functional on a Banach space X satisfying the conditions (i)–(iv), then F has a critical point x_0 for which $F(x_0) = h$ is given by (14).*

For a proof of this theorem, see [Ambrosetti-Rabinowitz]. We apply this theorem to solve nonlinear elliptic equations in Section 13.3.

Exercises for Section 7.1

1. Let Ω be a bounded domain, $X = L^2(\Omega)$, $Y = H_0^{1,2}(\Omega)$, and $G(u) = \int u^2 \, dx$.
   (a) Show that $G: X \to \mathbf{R}$ and $G: Y \to \mathbf{R}$ are continuous functionals.
   (b) Compute $G'(u)$.
   (c) Show that G is differentiable on both X and Y.
   (d) Show that G is C^1 on both X and Y.

2. Let $X = H_0^{1,2}(\Omega)$, where Ω is a bounded domain, and let $G(u) = \int_\Omega |u(x)|^p \, dx$. Use the Sobolev imbedding theorem to determine the values of $p \ge 1$ for which $G: X \to \mathbf{R}$ is continuous. (The differentiability of G is discussed in Section 13.3.)

3. (a) If X is a Hilbert space and $u_j \rightharpoonup u$ weakly in X, then $\|u\|_X \le \liminf_{j \to \infty} \|u_j\|_X$.
   (b) Show that "weak *upper* semicontinuity" for the norm on a Hilbert space need *not* be true: $u_j \not\rightharpoonup \|u\|_X \ge \limsup \|u_j\|_X$.

Section 7.1: Calculus of Functionals and Variations   227

4. (a) If $a, b \in \mathbf{R}^n$, verify that $|a+b|^2 + |a|^2 \geq |b|^2/2$.
   (b) Complete the argument for the existence of a minimizing function $u \in \mathcal{A}$ for F in Example 2.

5. Suppose we want to solve $\Delta u = 0$ in Ω with $\partial u / \partial \nu = h$ on $\partial \Omega$, where Ω is a smooth bounded domain in $\mathbf{R}^n$ and $h \in C^\infty(\partial\Omega)$ satisfies $\int_{\partial\Omega} h\, dS = 0$. Show how to obtain a weak solution $u \in H^{1,2}(\Omega)$ from the existence theory discussed in this section.

6. Suppose Ω is a bounded domain, $q(x)$ is a bounded function on Ω satisfying $q(x) \leq \eta$, and $f \in L^2(\Omega)$. If $\eta \geq 0$ is sufficiently small, then show that the Dirichlet problem $\Delta u + q(x)u = f$ in Ω, $u = 0$ on $\partial\Omega$ admits a unique weak solution.

7. Many optimization problems involve a functional

$$F(u) = \int_\Omega L(\nabla u, u, x)\, dx$$

for $u \in \mathcal{A}$ [as in (7)], where the function $L = L(p, z, x)$ is called the *Lagrangian* and is smooth in the variables $p \in \mathbf{R}^n$, $z \in \mathbf{R}$, and $x \in \Omega$. In this exercise, let us assume that $L = L(p, x)$ is independent of z.
   (a) If F is minimized at $u \in \mathcal{A}$, then prove that the *Legendre convexity condition* holds:

$$\sum_{i,j}^n L_{p_i p_j}(p, x) \xi_i \xi_j \geq 0 \qquad \text{for all } x \in \Omega, \xi \in \mathbf{R}^n.$$

(This holds even when L depends on z and implies that L is convex in the variable p.)
   (b) Assume the uniform Legendre convexity condition

$$\sum_{i,j}^n L_{p_i p_j}(p, x) \xi_i \xi_j \geq \epsilon |\xi|^2 \qquad \text{for all } x \in \Omega, \xi \in \mathbf{R}^n, \text{ where } \epsilon > 0.$$

Show that F is strictly convex on $\mathcal{A}$.

8. Define the *graph area functional* by

$$F(u) = \int_\Omega \sqrt{1 + u_x^2 + u_y^2}\, dx\, dy \qquad \text{for } u \in H^{1,2}(\Omega),$$

where Ω is a bounded domain in $\mathbf{R}^2$.
   (a) Compute the derivative of F and show that F is C^1 on $H^{1,2}(\Omega)$.
   (b) Given $g \in H^{1,2}(\Omega)$, let $\mathcal{A}$ be defined as in (7). Show that a critical point of F on $\mathcal{A}$ is a weak solution of the *minimal surface equation* $(1 + u_y^2)u_{xx} - 2u_x u_y u_{xy} + (1 + u_x^2)u_{yy} = 0$ in Ω, $u = g$ on $\partial\Omega$.
   (c) Generalize (a) and (b) to dimension n, obtaining the minimal surface equation given by formula (25) in the Introduction.

9. Let Ω be a smooth bounded domain in $\mathbf{R}^n$, $g \in H^{2,2}(\Omega)$, and $h \in H^{1,2}(\Omega)$. Show that a critical point of the functional

$$F(u) = \frac{1}{2} \int_\Omega (\Delta u)^2 \, dx \, dy$$

for $u \in \mathcal{A} \equiv \{u \in H^{2,2}(\Omega) : u - g \in H_0^{2,2}(\Omega) \text{ and } \partial u/\partial \nu - h \in H_0^{1,2}(\Omega)\}$ is a weak solution of the *biharmonic equation* $\Delta^2 u = \Delta(\Delta u) = 0$ in Ω, with Dirichlet boundary conditions $u = g$ and $\partial u/\partial \nu = h$ on $\partial \Omega$.

10. Show that the coercive functional of Example 1 satisfies the Palais-Smale condition (iv) of this section.

11. Suppose F is a C^1 functional on X satisfying (i) $F'(u) = L + K(u)$, where $L : X \to X^*$ is an isomorphism (i.e., a bounded linear operator that is one to one and onto) and $K : X \to X^*$ is a compact map (i.e., maps bounded sets of X to precompact sets of X^* but K is not necessarily linear); and (ii) every Palais-Smale sequence is bounded in X. Show that F satisfies the Palais-Smale condition.

7.2 Optimization with Constraints

In this section we use Lagrange multipliers to optimize a functional on a Banach space that is subject to a constraint in the form of an auxiliary functional having a fixed value. This method is then applied to the study of eigenvalues of the Laplacian.

a. Lagrange Multipliers

An important technique for optimization in finite dimensions is the use of *Lagrange multipliers*. Recall that if we want to minimize the C^1-function $F : \mathbf{R}^n \to \mathbf{R}$ on the hypersurface $\mathcal{C}$ defined by the constraint $G(x) = 0$, where $G : \mathbf{R}^n \to \mathbf{R}$ is also a C^1-function, then at such a minimum point $x_0 \in \mathcal{C}$ we must have $F'(x_0)$ and $G'(x_0)$ colinear; that is, either $F'(x_0) = 0$ or

(15) $\qquad F'(x_0) = \mu G'(x_0) \quad \text{for some } \mu \in \mathbf{R}.$

Similarly, we may have C^1-functionals $F : X \to \mathbf{R}$ and $G : X \to \mathbf{R}$ on a Banach space X and want to minimize $F(x)$ on the constraint set $\mathcal{C} = \{x \in X : G(x) = 0\}$. We would like to know that, at such a minimum point $x_0 \in \mathcal{C}$, the Lagrange multiplier equation (15) must hold. We first observe the following, whose proof reduces to an application of the inverse function theorem in $\mathbf{R}^2$.

Proposition. *Suppose F and G are C^1-functionals on a Banach space X and for $x_0 \in X$ we can find $v, w \in X$ such that*

(16) $$F'(x_0)v \cdot G'(x_0)w \neq F'(x_0)w \cdot G'(x_0)v.$$

Then F cannot have a local extremum at x_0, even when constrained to the level set $\mathcal{C} = \{x \in X : G(x) = G(x_0)\}$.

Proof. Fix $v, w \in X$, and for $s, t \in \mathbf{R}$ consider the real-valued functions

$$f(s,t) = F(x_0 + sv + tw) \quad \text{and} \quad g(s,t) = G(x_0 + sv + tw).$$

Then

$$\frac{\partial f}{\partial s}(0,0) = F'(x_0)v \qquad \frac{\partial f}{\partial t}(0,0) = F'(x_0)w$$
$$\frac{\partial g}{\partial s}(0,0) = G'(x_0)v \qquad \frac{\partial g}{\partial t}(0,0) = G'(x_0)w,$$

so the condition (16) is simply that the Jacobian $|\partial(f,g)/\partial(s,t)|$ is non-vanishing at $(s,t) = (0,0)$. Now F and G being C^1 on X implies that f and g are C^1 on $\mathbf{R}^2$ (see Exercise 1), so we may apply the inverse function theorem to conclude that a local extremum cannot occur at x_0. ♠

We are now in a position to establish (15).

Theorem 1 (Lagrange). *Suppose F and G are C^1-functionals on a Banach space X, $G(x_0) = 0$, and x_0 is a local extremum for F when constrained to the level set $\mathcal{C} = \{x \in X : G(x) = 0\}$. Then either*
(i) *$G'(x_0)v = 0$ for all $v \in X$, or*
(ii) *there exists $\mu \in \mathbf{R}$ such that $F'(x_0)v = \mu G'(x_0)v$ for all $v \in X$.*

Proof. If (i) does not hold, then let us fix $w \in X$ with $G'(x_0)w \neq 0$. By hypothesis and the preceding proposition, we must have $F'(x_0)v \cdot G'(x_0)w = F'(x_0)w \cdot G'(x_0)v$ for every $w \in X$. If we define $\mu = (F'(x_0)w)/(G'(x_0)w)$, then we obtain (ii). ♠

The variational method using Lagrange multipliers is important in studying linear and nonlinear PDEs. We use the method to study a semilinear elliptic equation in Section 13.3, but for now let us use the method to study the eigenvalues of the Laplacian.

b. Application to Eigenvalues of the Laplacian

If Ω is a bounded domain with (sufficiently) smooth boundary, then $T = -\Delta \colon H_0^{1,2}(\Omega) \cap H^{2,2}(\Omega) \to L^2(\Omega)$ is a closed linear operator (as in Section 6.6) whose eigenvalues λ satisfy $Tu = \lambda u$ for some nontrivial u; that is,

(17a) $$\begin{cases} \Delta u + \lambda u = 0 & \text{in } \Omega \\ u = 0 & \text{on } \partial \Omega. \end{cases}$$

In Section 4.4, we listed properties of these eigenvalues and eigenfunctions; we derive some of these properties by studying *weak solutions* of (17a). That is, we try to find $u \in H_0^{1,2}(\Omega)$ satisfying

$$(17\text{b}) \qquad \int_\Omega \nabla u \cdot \nabla v \, dx = \lambda \int_\Omega uv \, dx \qquad \text{for all} \quad v \in H_0^{1,2}(\Omega).$$

It is important to emphasize that, throughout this section, the solutions (λ, u) that we obtain solve (17b) instead of (17a). Using the regularity results of Chapter 8, however, we can show that such weak solutions are in fact smooth functions, and hence (λ, u) will solve (17a).

We use the variational method with $X = H_0^{1,2}(\Omega)$. By linearity, a solution (λ, u) of (17b) determines a one-dimensional eigenspace $\{\alpha u\}$ of solutions, so we must normalize our search, say by requiring $\|u\|_2 = \text{const}$. This condition will act as a constraint on the variational problem, so let us try to use the method of Lagrange multipliers.

In fact, let us define the functionals

$$(18) \qquad F(u) = \int_\Omega |\nabla u|^2 \, dx \qquad \text{and} \qquad G(u) = \int_\Omega u^2 \, dx - 1$$

for $u \in X$ and look for u that minimizes $F(u)$ subject to the constraint $G(u) = 0$ (i.e., $\|u\|_2 = 1$). We find that the derivatives are given by

$$(19) \qquad F'(u)v = 2\int_\Omega \nabla u \cdot \nabla v \, dx \qquad \text{and} \qquad G'(u)v = 2\int_\Omega uv \, dx$$

for $u, v \in X$; moreover, F and G are both C^1 as maps $X \to \mathbf{R}$ (see Example 1 and Exercise 1 of Section 7.1).

Next let us verify that F attains its minimum on the constraint set $\mathcal{C} = \{x \in X : G(x) = 0\}$. Let $I = \inf\{F(u) : u \in \mathcal{C}\} \geq 0$. Select $u_j \in \mathcal{C}$ such that $F(u_j) \to I$ and $F(u_j) \leq I + 1$. Using the Poincaré inequality, $\|u_j\|_X$ is equivalent to $\|\nabla u_j\|_2$ and hence is bounded; using the compactness of the imbedding $H_0^{1,2}(\Omega) \subset L^2(\Omega)$ and the weak sequential compactness of bounded sets in $H_0^{1,2}(\Omega)$, there is a subsequence u_{j_k} and a function $\bar{u} \in H_0^{1,2}(\Omega)$ such that

$$(20) \qquad u_{j_k} \to \bar{u} \quad \text{in } L^2(\Omega) \qquad \text{and} \qquad u_{j_k} \rightharpoonup \bar{u} \quad \text{weakly in } H_0^{1,2}(\Omega).$$

Now $\bar{u} \in \mathcal{C}$ since $\|\bar{u}\|_2 = \lim_{j_k \to \infty} \|u_{j_k}\|_2 \equiv 1$, and $\bar{u}$ minimizes F since [using (12a)]

$$F(\bar{u}) = \|\bar{u}\|_X^2 - \|\bar{u}\|_2^2 \leq \liminf_{j_k \to \infty} \|u_{j_k}\|_X^2 - \lim_{j_k \to \infty} \|u_{j_k}\|_2^2$$

$$= \liminf_{j_k \to \infty}(\|u_{j_k}\|_X^2 - \|u_{j_k}\|_2^2) = \liminf_{j_k \to \infty} F(u_{j_k}) = I.$$

Section 7.2: Optimization with Constraints

Thus $\bar{u}$ satisifes the Euler-Lagrange equation $\int \nabla \bar{u} \cdot \nabla v \, dx = \mu \int \bar{u} v \, dx$ for some Lagrange multiplier μ and all $v \in X$. If we take $\lambda = \mu$, we see that $(\lambda, \bar{u})$ satisfies (17b). In fact, if we take $v = \bar{u}$ in (17b), we find that the eigenvalue λ is just the infimum I:

$$(21) \qquad I = F(\bar{u}) = \int_\Omega |\nabla \bar{u}|^2 \, dx = \lambda \int_\Omega \bar{u}^2 \, dx = \lambda.$$

This means that we could also define λ by the Rayleigh quotient

$$(22) \qquad \lambda = \inf_{u \in X} \frac{\int_\Omega |\nabla u|^2 \, dx}{\int_\Omega u^2 \, dx}.$$

Let us now denote this solution pair by (λ_1, u_1) instead of $(\lambda, \bar{u})$. Clearly $\lambda_1 = F(u_1) \geq 0$. In fact, if $F(u_1) = 0$, then $|\nabla u_1| \equiv 0$, so u_1 is a constant. [There is no danger that u_1 might be nonconstant on a set of measure zero if we invoke the regularity results of Chapter 8 to conclude $u_1 \in C^\infty(\Omega)$.] But the boundary condition $u_1 = 0$ on $\partial\Omega$ would then imply $u_1 \equiv 0$, violating $\|u_1\|_2 = 1$. Thus $\lambda_1 > 0$. A schematic representation of our result appears in Figure 1.

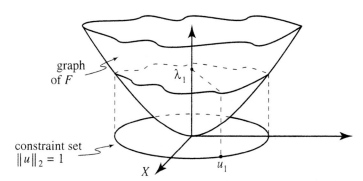

Figure 1. A schematic representation of (λ_1, u_1).

Are there other solutions (λ, u)? Suppose there is another pair (λ_2, u_2) with $\lambda_1 \neq \lambda_2$. Then, invoking (17b) for λ_1 with $v = u_2$ and for λ_2 with $v = u_1$, we obtain

$$(23) \quad \begin{aligned} (\lambda_1 - \lambda_2) \int_\Omega u_1 u_2 \, dx &= \int_\Omega \lambda_1 u_1 u_2 \, dx - \int_\Omega u_1 \lambda_2 u_2 \, dx \\ &= \int_\Omega \nabla u_1 \cdot \nabla u_2 \, dx - \int_\Omega \nabla u_1 \cdot \nabla u_2 \, dx = 0. \end{aligned}$$

Since $(\lambda_1 - \lambda_2) \neq 0$, this implies $\int u_1 u_2 \, dx = \langle u_1, u_2 \rangle = 0$. Thus eigenfunctions corresponding to different eigenvalues are orthogonal in the L^2-inner product $\langle , \rangle$. But this means that we should look for $u_2 \in X_1$, where

$X_1 = \{v \in X : \langle v, u_1 \rangle = 0\}$. Since X_1 is the nullspace of the continuous functional $\langle \cdot, u_1 \rangle$ on X, X_1 is a closed subspace of X and hence a Hilbert space itself under the same inner product.

We can now define

(24) $\quad \lambda_2 \equiv \inf\{F(v) : v \in X_1 \text{ and } \|v\|_2 = 1\} \equiv \inf_{v \in X_1} \dfrac{\int_\Omega |\nabla v|^2 \, dx}{\int_\Omega v^2 \, dx}$

and observe $\lambda_2 \geq \lambda_1$ since $X_1 \subset X$. Moreover, we can repeat the above arguments to show that λ_2 is achieved at some $u_2 \in X_1$.

Proceeding inductively, we let $X_n = \{v \in X : \langle v, u_i \rangle = 0 \text{ for } i = 1, \ldots, n\}$ and

(25) $\quad \lambda_{n+1} = \inf\{F(v) : v \in X_n \text{ and } \|v\|_2 = 1\} = \inf_{v \in X_n} \dfrac{\int_\Omega |\nabla v|^2 \, dx}{\int_\Omega v^2 \, dx}.$

In this way we can generate a sequence of eigenvalues

(26) $\qquad\qquad\qquad 0 < \lambda_1 \leq \lambda_2 \leq \lambda_3 \cdots$

whose associated eigenfunctions u_n are orthonormal in L^2:

(27) $\qquad \begin{cases} \langle u_n, u_n \rangle = \|u_n\|_2^2 = 1 \\ \langle u_n, u_m \rangle = 0 \quad \text{if } n \neq m. \end{cases}$

Let us investigate some further properties of the eigenfunctions u_n. Combining (17b) and (27), we find

(28) $\quad \langle \nabla u_n, \nabla u_m \rangle = \displaystyle\int_\Omega \nabla u_n \cdot \nabla u_m \, dx = \begin{cases} \lambda_n & \text{if } n = m \\ 0 & \text{if } n \neq m. \end{cases}$

Thus the u_n are pairwise orthogonal in both the L^2-inner product, and the natural inner product on X, $\langle u, v \rangle_1 = \int (\nabla u \cdot \nabla v + uv) \, dx$. In fact, if we take any finite linear combination of the u_n we find

(29a) $\qquad\qquad\qquad \left\| \displaystyle\sum_{n=1}^N \alpha_n u_n \right\|_2^2 = \displaystyle\sum_{n=1}^N \alpha_n^2$

(29b) $\qquad\qquad\qquad \left\| \nabla \left(\displaystyle\sum_{n=1}^N \alpha_n u_n \right) \right\|_2^2 = \displaystyle\sum_{n=1}^N \alpha_n^2 \lambda_n.$

We would next like to verify that $\lambda_n \to \infty$. If not, then λ_n is bounded, so (27) and (28) together imply $\|u_n\|_X^2 = \lambda_n + 1 \leq C$. By the Rellich compactness theorem, there is a Cauchy subsequence (which we continue to denote by u_n); that is,

$$\|u_n - u_m\|_2^2 \to 0. \tag{30}$$

But by (27), $\|u_n - u_m\|_2^2 = \|u_n\|_2^2 - 2\langle u_n, u_m\rangle + \|u_m\|_2^2 = 2$ if $n \neq m$, which contradicts (30). We conclude that λ_n must tend to ∞.

A trivial consequence of $\lambda_n \to \infty$ is that each λ_n occurs only finitely many times. This means that the eigenspace associated with a given eigenvalue is finite dimensional.

Finally, we claim that the collection $\{u_n\}_{n=1}^\infty$ forms an orthonormal basis for the infinite-dimensional vector space $L^2(\Omega)$. This means that, for any $f \in L^2(\Omega)$, we can find coefficients $\alpha_n \in \mathbf{R}$ such that f may be represented by the *expansion in the eigenfunctions*

$$f = \sum_{n=1}^\infty \alpha_n u_n; \tag{31a}$$

that is, the remainder ρ_N between f and the Nth partial sum tends to zero in $L^2(\Omega)$:

$$\|\rho_N\|_2 \equiv \left\| f - \sum_{n=1}^N \alpha_n u_n \right\|_2 \to 0 \quad \text{as } N \to \infty. \tag{31b}$$

To prove (31b), we first assume $f \in X = H_0^{1,2}(\Omega)$. For fixed $f \in X$, define $\alpha_n = \langle f, u_n\rangle$. By the orthonormality (27), we have

$$\langle \rho_N, u_n\rangle = 0 \quad \text{for } n = 1, \ldots, N. \tag{32}$$

Moreover, since $f, u_n \in X$, we have $\rho_N \in X$, so we can take $v = \rho_N$ in (17b) with $\lambda = \lambda_n$ and $u = u_n$; using (32), we obtain

$$\langle \nabla \rho_N, \nabla u_n\rangle = 0 \quad \text{for } n = 1, \ldots, N. \tag{33}$$

Thus we may use (29a) and (33) to compute

$$\|\nabla f\|_2^2 = \left\| \nabla \rho_N + \nabla\left(\sum_{n=1}^N \alpha_n u_n\right) \right\|_2^2 = \|\nabla \rho_N\|_2^2 + \sum_{n=1}^N \alpha_n^2 \lambda_n.$$

Since $\lambda_n > 0$ and $\|\nabla f\|_2$ is independent of N, we find $\|\nabla \rho_N\|_2^2 \leq C$ for all N. On the other hand, (32) implies $\rho_N \in X_N$, so by (25) we have $\lambda_{N+1}\|\rho_N\|_2^2 \leq \|\nabla \rho_N\|_2^2$. Combining these two inequalities with the fact that $\lambda_{N+1} \to \infty$, we obtain $\|\rho_N\|_2 \to 0$ as desired.

To establish (31b) for a general $f \in L^2(\Omega)$, recall that $X = H_0^{1,2}(\Omega)$ is dense in $L^2(\Omega)$ (see Section 6.5), so we may find $f_\epsilon \in X$ with $f_\epsilon \to f$ in $L^2(\Omega)$ as $\epsilon \to 0$. Then $\alpha_{n,\epsilon} \equiv \langle f_\epsilon, u_n \rangle \to \alpha_n \equiv \langle f, u_n \rangle$ for each n. Moreover,

$$\left\| f - \sum_{n=1}^{N} \alpha_n u_n \right\|_2 \leq \|f - f_\epsilon\|_2 + \left\| f_\epsilon - \sum_{n=1}^{N} \alpha_{n,\epsilon} u_n \right\|_2 + \left\| \sum_{n=1}^{N} (\alpha_{n,\epsilon} - \alpha_n) u_n \right\|_2. \tag{34}$$

But by (29a) and Bessel's inequality (cf. Exercise 14 in Section 6.1), we have

$$\left\| \sum_{n=1}^{N} (\alpha_{n,\epsilon} - \alpha_n) u_n \right\|_2 = \left(\sum_{n=1}^{N} (\alpha_{n,\epsilon} - \alpha_n)^2 \right)^{1/2} \leq \|f_\epsilon - f\|_2.$$

Therefore, (34) becomes

$$\left\| f - \sum_{n=1}^{N} \alpha_n u_n \right\|_2 \leq 2\|f - f_\epsilon\|_2 + \left\| f_\epsilon - \sum_{n=1}^{N} \alpha_{n,\epsilon} u_n \right\|_2,$$

from which it is clear that (31b) holds.

Putting this all together, we obtain the following (cf. Theorem 5 and Exercise 15 in Section 6.6):

Theorem 2. *The eigenvalues for (17b) form a nondecreasing sequence of positive numbers λ_n tending to infinity. The associated eigenfunctions u_n form an orthonormal basis for $L^2(\Omega)$ (i.e., (31a) holds where $\alpha_n = \langle f, u_n \rangle$). Moreover, if $f \in H_0^{1,2}(\Omega)$, then*

$$\|\nabla f\|_2^2 = \sum_{n=1}^{\infty} \alpha_n^2 \lambda_n. \tag{35}$$

[For the proof of (35), see Exercise 3.]

Remark 1. If f satisfies stronger conditions, then the convergence in (31a) may be stronger. For example, if $f, \Delta f \in C(\overline{\Omega})$ and $f = 0$ on $\partial\Omega$, then (31a) converges absolutely and uniformly in Ω (cf. the expansion theorem in [Courant-Hilbert], vol. I). This stronger convergence may be important in applications, as was seen in Chapter 4.

Remark 2. Some additional properties of the eigenvalues and eigenfunctions should be mentioned. In Section 8.2 we see that the eigenfunctions always satisfy $u_n \in C^\infty(\Omega)$ and, under additional regularity assumptions on $\partial\Omega$, $u_n \in C^\infty(\overline{\Omega})$. Moreover, such additional regularity will allow us to use the strong elliptic maximum principle in Section 8.3 to show that the principal eigenfunction u_1 is never zero in Ω, and the principal eigenvalue λ_1 is a simple eigenvalue (i.e., $\lambda_1 < \lambda_2$).

Remark 3. Theorem 2 may be generalized in several different ways. First, the Laplacian may be replaced by a uniformly elliptic operator in divergence form (cf. Exercise 4). Second, the Dirichlet boundary condition $u = 0$ on $\partial\Omega$ may be replaced by other boundary conditions (such as the Neumann or Robin condition), and the variational method generates the appropriate set of eigenvalues and eigenfunctions (cf. Exercise 5).

c. The Maximin Characterization of Eigenvalues

One disadvantage of the treatment of eigenvalues of the Laplacian in the previous subsection is their recursive definition: λ_1 is used to define λ_2, and so on. We now describe an alternative method that avoids this recursion; this will allow us to conclude additional properties of the eigenvalues and eigenfunctions.

As in the previous subsection, let $X = H_0^{1,2}(\Omega)$ where Ω is a bounded domain and $\langle,\rangle$ denotes the L^2 inner product. Consider the sequence of eigenvalues λ_i and associated eigenfunctions $u_i \in X$ of (17b). For arbitrary $v_1, \ldots, v_{n-1} \in X$, define

(36)
$$m\{v_1, \ldots, v_{n-1}\} = \inf_{u \in X} \left\{ \int_\Omega |\nabla u|^2 dx : \|u\|_2 = 1 \text{ and } \langle u, v_i \rangle = 0 \text{ for } i = 1, \ldots, n-1 \right\}.$$

We now claim that

(37)
$$\lambda_n = \sup_{v_1, \ldots, v_{n-1} \in X} m\{v_1, \ldots, v_{n-1}\}.$$

To verify (37), we may take $v_i = u_i$ for $i = 1, \ldots, n-1$; by definition of λ_n in the previous subsection, $\lambda_n = m\{u_1, \ldots, u_{n-1}\}$, and hence $\lambda_n \leq \sup m\{v_1, \ldots, v_{n-1}\}$. To prove the opposite inequality, let $v_1, \ldots, v_{n-1}$ be arbitrary; we want to determine a function u satisfying $\int |\nabla u|^2 dx \leq \lambda_n$. Let us write $u = \sum_{i=1}^n c_i u_i$ and determine the n constants c_i by satisfying the n conditions: $\sum_i^n c_i^2 = 1$ (i.e., $\|u\|_2 = 1$) and $\langle u, v_i \rangle = 0$ for $i = 1, \ldots, n-1$. Using the orthonormality properties of the u_i, we find $\int |\nabla u|^2 dx = \sum_{i=1}^n c_i^2 \lambda_i$. Combining this with $\lambda_i \leq \lambda_n$ for $i \leq n$ yields

$$m\{v_1, \ldots, v_n\} \leq \int_\Omega |\nabla u|^2 \, dx = \sum_{i=1}^n \lambda_i c_i^2 \leq \lambda_n \sum_{i=1}^n c_i^2 = \lambda_n.$$

Thus (37) is established.

This means that we may describe the nth eigenvalue λ_n of (17b) by using (36) and (37), which do not depend on the determination of the previous eigenvalues $\lambda_1, \ldots, \lambda_{n-1}$. Because (36) and (37) involve taking a supremum (maximum) of a quantity defined as an infimum (minimum), this is called the *maximin characterization* of the eigenvalues.

The maximin characterization sometimes enables us to compare the eigenvalues for different problems. Notice that, if $\widehat{X}$ is a subspace of X, then restricting u and $v_1, \ldots, v_{n-1}$ to lie in $\widehat{X}$ involves minimization over a smaller set and so cannot decrease the maximin. This means that $\lambda_n \leq \widehat{\lambda}_n$ where $\widehat{\lambda}_n$ is defined by (37) with X replaced by $\widehat{X}$. This allows us, for example, to compare $\lambda_n = \lambda_n(\Omega)$ with the eigenvalues $\widehat{\lambda}_n = \lambda_n(\widehat{\Omega})$ of a subdomain $\widehat{\Omega} \subset \Omega$. If we let $\widehat{X}$ denote the closure in the $H_0^{1,2}(\Omega)$-norm of $\{u \in C_0^1(\Omega) : u(x) \equiv 0 \text{ for } x \in \Omega \setminus \widehat{\Omega}\}$, then $\widehat{X} \subset X = H_0^{1,2}(\Omega)$, although we may also identify $\widehat{X}$ with $H_0^{1,2}(\widehat{\Omega})$. In other words, we have proved that the eigenvalues satisfy the following monotonicity with respect to domain.

Theorem 3. *If $\widehat{\Omega} \subset \Omega$ are bounded domains, then $\lambda_n(\widehat{\Omega}) \geq \lambda_n(\Omega)$.*

The monotonicity with respect to the domain does *not* apply to Neumann eigenvalues; see Exercise 6 for a counterexample. However, the maximin method can be used to compare the Dirichlet and Neumann eigenvalues on the same domain (see Exercise 7).

Exercises for Section 7.2

1. If F is a C^1-functional on a Banach space X, fix $x_0, v, w \in X$ and define $f(s,t) = F(x_0 + sv + tw)$. Show that f is C^1 on $\mathbf{R}^2$.

2. In Theorem 2 of this section, we have not actually verified that *every* eigenvalue of (17) is one of the λ_n generated by the functional F. Use the completeness of the eigenfunctions to prove this fact.

3. Prove (35).

4. Consider a uniformly elliptic operator L in divergence form [cf. (20), (21) in Section 6.2], and let us pose the eigenvalue problem

$$\begin{cases} Lu + \lambda u = 0 & \text{in } \Omega \\ u = 0 & \text{on } \partial\Omega. \end{cases}$$

(a) Give a variational formulation for the first eigenvalue λ_1. Observe that if $c(x) \leq c_0$, then $\lambda_1 > -c_0$.
(b) Show that the eigenvalues form a nondecreasing sequence $\lambda_1 \leq \lambda_2 \leq \cdots$ tending to infinity.
(c) Show that each eigenvalue has a finite-dimensional eigenspace, and the collection of all eigenfunctions u_n forms an orthonormal basis for $L^2(\Omega)$.

5. Suppose Ω is a bounded domain with C^1-boundary, and consider the Neumann eigenvalue problem

$$\begin{cases} \Delta u + \mu u = 0 & \text{in } \Omega \\ \dfrac{\partial u}{\partial \nu} = 0 & \text{on } \partial \Omega. \end{cases}$$

A *weak solution* is defined to be $u \in H^{1,2}(\Omega)$ such that $\int_\Omega \nabla u \cdot \nabla v \, dx = \mu \int_\Omega uv \, dx$ for all $v \in H^{1,2}(\Omega)$ (cf. Exercise 5 in Section 7.1).
(a) Clearly one solution is $\mu_1 = 0$ with $u_1 \equiv \text{const}$. Show that there is a weak solution (μ_2, u_2) with μ_2 being the smallest *positive* eigenvalue.
(b) Show that there is a sequence of eigenvalues $0 = \mu_1 < \mu_2 \leq \mu_3 \leq \cdots$ and associated orthonormal eigenfunctions u_n that are weak solutions of the Neumann eigenvalue problem.
(c) Show that the eigenfunctions u_n form a complete set in $H^{1,2}(\Omega)$.

6. Let Ω be the square $\{(x,y) : -1 < x, y < 1\}$ and $\hat{\Omega}$ be the slitted circle $\{(r,\theta) : 0 \leq r < 1, -\pi < \theta < \pi\}$. Although $\hat{\Omega} \subset \Omega$, show that the Neumann eigenvalues do *not* satisfy $\mu_n(\hat{\Omega}) \geq \mu_n(\Omega)$ for all $n \geq 1$.

7. Prove the following relationship between the Dirichlet eigenvalues λ_n and Neumann eigenvalues μ_n of the Laplacian on a bounded domain with C^1-boundary: $\lambda_n \geq \mu_n$ for all $n \geq 1$.

8. Let Ω be a smooth bounded domain in $\mathbf{R}^n$, and consider the eigenvalue problem for the biharmonic operator with Dirichlet boundary conditions

$$\begin{cases} \Delta^2 u - \lambda u = 0 & \text{in } \Omega \\ u = \dfrac{\partial u}{\partial \nu} = 0 & \text{on } \partial \Omega. \end{cases}$$

(See Exercise 9 in Section 7.1.) Show that there is a nontrivial weak solution (λ, u) with $\lambda > 0$.

7.3 Calculus of Maps between Banach Spaces

If X and Z are Banach spaces, let $C(X, Z)$ denote the continuous maps $F \colon X \to Z$. In the previous two sections, we considered differential calculus when $Z = \mathbf{R}$; now we want to consider the more general case. First let us investigate continuous differentiability for such maps. Using (2), it is natural to define $F \colon X \to Z$ to be *(Frechet) differentiable at x_0* if there is a bounded linear map $L \colon X \to Z$ such that

(38) $\quad \|F(x_0 + \xi) - F(x_0) - L\xi\|_Z = o(\|\xi\|_X) \quad$ as $\quad \|\xi\|_X \to 0.$

The linear map L is the *derivative of F at x_0* and is denoted by $DF(x_0)$ or $F'(x_0)$.

If $F'(x)$ exists for all $x \in X$, then we get a mapping $F': X \to \mathcal{B}(X, Z)$, where $\mathcal{B}(X, Z)$ denotes the vector space of all bounded linear operators from X to Z. Since $\mathcal{B}(X, Z)$ is itself a Banach space under the operator norm (for the same reasons that X^* is a Banach space), we may discuss continuity; in particular, let $C^1(X, Z) = \{F \in C(X, Z) : F': X \to \mathcal{B}(X, Z) \text{ is continuous}\}$. Of course, the notion of differentiability is a local one, so we may also consider maps $F: U \to Z$, where U is an open set in X whose derivative F' (if it exists) is a mapping $U \to \mathcal{B}(X, Z)$.

We would like to develop analogues of the inverse and implicit function theorems of finite-dimensional calculus, but we first study mappings that contract distances.

a. The Method of Successive Approximations

Suppose X is a complete metric space with distance function represented by $d(\cdot, \cdot)$. A mapping $T: X \to X$ is a *strict contraction* if there exists $0 < \alpha < 1$ such that

(39) $\qquad d(Tx, Ty) \leq \alpha \, d(x, y) \quad \text{for all} \quad x, y \in X.$

An obvious example on $X = \mathbf{R}^n$ is $Tx = \alpha x$, which shrinks all of $\mathbf{R}^n$, leaving 0 fixed.

Notice that a contraction mapping is necessarily continuous: $x_n \to x \Rightarrow d(x_n, x) \to 0 \Rightarrow d(Tx_n, Tx) \to 0 \Rightarrow Tx_n \to Tx$. Moreover, given an arbitrary $x_0 \in X$, let $x_1 = Tx_0, x_2 = Tx_1 = T(Tx_0) = T^2 x_0, \ldots$. Then we may verify that $\{x_n\}$ is a Cauchy sequence by taking $m \geq n$ and computing

$$\begin{aligned} d(x_n, x_m) &= d(T^n x_0, T^m x_0) \leq \alpha^n d(x_0, x_{m-n}) \\ &\leq \alpha^n [d(x_0, x_1) + d(x_1, x_2) + \cdots + d(x_{m-n-1}, x_{m-n})] \\ &\leq \alpha^n d(x_0, x_1)[1 + \alpha + \alpha^2 + \cdots + \alpha^{m-n-1}] \\ &\leq \alpha^n d(x_0, x_1) \frac{1}{1-\alpha} < \epsilon \quad \text{for } n \text{ large}. \end{aligned}$$

Since X is complete, $\{x_n\}$ must converge to some $x \in X$. By continuity, $Tx = T(\lim x_n) = \lim Tx_n = \lim x_{n+1} = x$. Hence x is a fixed point. In fact, this fixed point is unique; if y is also a fixed point, then $d(x, y) = d(Tx, Ty) \leq \alpha d(x, y)$ and $0 < \alpha < 1$ imply $d(x, y) = 0$, so $x = y$. This argument establishes the following result.

Theorem 1 (The Contraction Mapping Principle). If X is a complete metric space and $T: X \to X$ is a strict contraction, then T has a unique fixed point.

In fact, the proof of the theorem shows that the fixed point x is simply obtained by taking the limit as $n \to \infty$ of $T^n x_0$, where x_0 is arbitrary. For this reason, the result is sometimes referred to as the *method of successive approximations*.

Section 7.3: Calculus of Maps Between Banach Spaces

The contraction mapping principle is used in the next subsection for the proof of the inverse function theorem; but first let us give another application to the existence and uniqueness theorem of Picard-Lindelöf for ODEs.

Example 1. Consider the initial value problem

(40)
$$\begin{cases} y' = f(t,y) \\ y(t_0) = y_0, \end{cases}$$

where $f(t,y)$ is continuous in t, and Lipschitz in y, for (t,y) in a rectangle $R = \{(t,y) : |t - t_0| \leq a, |y - y_0| \leq b\}$. Recall that the Lipschitz condition requires there to be a constant C such that

(41) $\qquad |f(t,y) - f(t,z)| < C|y - z| \qquad$ for $(t,y), (t,z) \in R$.

If we can solve (40), then the solution $y(t)$ must satisfy the integral equation

(42) $$y(t) = y_0 + \int_{t_0}^{t} f(s, y(s))\, ds.$$

This suggests defining a map $y \to \mathcal{T}y$ by $(\mathcal{T}y)(t) = y_0 + \int_{t_0}^{t} f(s, y(s))\, ds$ and then seeking the solution of (40) as a fixed point of $\mathcal{T}$. The contraction mapping principle will yield a unique solution, at least on a small enough time interval (cf. Exercise 2). ♣

The process of replacing a differential equation such as (40) by an integral equation such as (42) also occurs in time-evolution partial differential equations. In fact, the contraction mapping principle is used in Sections 9.2, 10.1, 11.2, and 12.3 to establish the local existence of solutions to various nonlinear evolution equations. It is also used in Chapter 8 in the method of continuity to study linear elliptic equations.

b. The Inverse Function Theorem

Assume that X and Z are Banach spaces. By analogy with finite-dimensional calculus, it is natural to formulate the following *inverse function theorem*.

Theorem 2 (Inverse Function Theorem). *Suppose U is an open neighborhood of 0 in X and $F: U \to Z$ is C^1 such that $F'(0): X \to Z$ is an isomorphism (one to one and onto). Then there is a neighborhood V of $F(0)$ in Z and a unique mapping $G: V \to X$ such that $F(G(z)) = z$ for all $z \in V$.*

Proof. As in the finite-dimensional case, the inverse function theorem states that a nonlinear map is locally invertible near any point where its derivative is invertible. This is not surprising since we can use (38) to write

(43) $\qquad F(x) = F(0) + F'(0)x + R(x),$

where $R(x) = o(\|x\|_X)$ as $\|x\|_X \to 0$, and then use the invertibility of $F'(0)$ to solve for x:

(44) $$x = F'(0)^{-1}[F(x) - F(0) - R(x)].$$

If we let $\mathcal{T}_z x = F'(0)^{-1}[z - F(0) - R(x)]$, then solving $x = \mathcal{T}_z x$ is equivalent to solving $z = F(x)$. Therefore, if we can show that, provided $\|z - F(0)\|_Z$ is sufficiently small, $\mathcal{T}_z$ is a contraction mapping on a small ϵ-ball in X, then we shall be done. Suppose $\|F'(0)^{-1}\|_{Z \to X} = M$ and take ϵ sufficiently small that $\|R(x)\|_Z < \epsilon/2M$ whenever $\|x\|_X \leq \epsilon$. Then $\|\mathcal{T}_z x\|_X \leq M\|z - F(0)\|_Z + M\|R(x)\|_Z$. So if we take $\|z - F(0)\|_Z < \epsilon/2M$, then we find that $\|\mathcal{T}_z x\|_X < \epsilon$, and $\mathcal{T}_z$ maps $\overline{B_\epsilon(0)} \subset X$ into itself. We need only show that $\mathcal{T}_z : \overline{B_\epsilon(0)} \to \overline{B_\epsilon(0)}$ is a strict contraction; this is left for Exercise 4. ♠

Theorem 2 is useful in the *perturbation theory* of PDEs. Namely, if $F(0) = 0$ and the *linear* equation $F'(0)u = f$ is known to admit a unique solution u for every f, then the nonlinear equation $F(u) = f$ will admit a unique solution u, at least for small f. This is illustrated later in this section for the *mean curvature equation*. Of course, this procedure requires that we have Banach spaces X and Y for which (i) $F: X \to Y$ is C^1, and (ii) $F'(0): X \to Y$ is an isomorphism. We return to issue (i) later when X and Y are Sobolev spaces, and issue (ii) for elliptic equations in Chapter 8.

c. The Implicit Function Theorem

The *implicit function theorem* concerns mappings $F: X \times Y \to Z$, where X, Y, Z are all Banach spaces. Differentiability is defined similarly to (38), but we also need to fix $x_0 \in X$ and differentiate $F(x_0, \cdot): Y \to Z$ at a point y_0; we denote the result by $D_y F(x_0, y_0)$.

Theorem 3 (Implicit Function Theorem). *Suppose U is a neighborhood of 0 in X, V is a neighborhood of 0 in Y, and $F: U \times V \to Z$ is C^1. Suppose $F(0,0) = 0$ and $D_y F(0,0): Y \to Z$ is an isomorphism (one to one and onto). Then $F(x, y) = 0$ defines a unique mapping $G: W \to Y$ where W is a neighborhood of 0 in X such that $F(x, G(x)) = 0$ for all $x \in W$.*

Proof. Define $\tilde{F}(x, y) = (x, F(x, y))$ so that $\tilde{F}: U \times V \to X \times Z$, and apply the inverse function theorem to solve $\tilde{F}(x, y) = (x, 0)$. This defines $y = G(x)$ so that $\tilde{F}(x, G(x)) = (x, 0)$; in other words $F(x, G(x)) = 0$ as desired. ♠

We apply the implicit function theorem in the study of a nonlinear eigenvalue problem in Section 13.1.

d. C^1-maps on Sobolev Spaces

For the applications of Theorems 2 and 3 to PDEs, it is necessary to determine when a given map between function spaces is C^1. A simple example of such a map is given by composition with a function

(45a) $$f: \mathbf{R} \to \mathbf{R};$$

Section 7.3: Calculus of Maps Between Banach Spaces

that is, for a given function $u(x)$, we define

(45b) $$F(u)(x) = f(u(x)).$$

If X and Y are function spaces such that $F(u) \in Y$ whenever $u \in X$, then (45b) defines a map

(46) $$F: X \to Y,$$

and we would like to determine when (46) is C^1.

If $f \in C^k(\mathbf{R})$ and $X = Y = C^k(\overline{\Omega})$ where Ω is a smooth, bounded domain, then $F(u) \in Y = C^k(\overline{\Omega})$ for every $u \in X = C^k(\overline{\Omega})$, and it is easy to see that (46) is C^0. Moreover, if $f \in C^{k+1}(\mathbf{R})$, then it is also easy to see that (46) is C^1; see Exercise 6. However, what happens if X and Y are Sobolev spaces?

Theorem 4. *If Ω is a smooth, bounded domain in $\mathbf{R}^n$, and $f \in C^{k+1}(\mathbf{R})$ where $k > n/2$, then (45a,b) defines a C^1-map $F: H^{k,2}(\Omega) \to H^{k,2}(\Omega)$, and $(F'(u)v)(x) = f'(u(x))\,v(x)$.*

Proof. If $u \in C^k(\overline{\Omega})$, then $F(u) \in C^k(\overline{\Omega})$ by the chain rule: $\partial_{x_i} F(u) = f'(u)\partial_{x_i} u$, $\partial_{x_i}\partial_{x_j} F(u) = f''(u)\,\partial_{x_i} u\,\partial_{x_j} u + f'(u)\,\partial_{x_i}\partial_{x_j} u$, and so on. If $u \in H^{k,2}(\Omega)$, let $u_\ell \in C^k(\overline{\Omega})$ with $u_\ell \to u$ in $H^{k,2}(\Omega)$ as $\ell \to \infty$. We claim that $F(u_\ell) \to F(u)$ in $H^{k,2}(\Omega)$. Since $k > n/2$ implies $H^{k,2}(\Omega) \subset C(\overline{\Omega})$ (by the Sobolev imbedding theorem), we know that $\|u\|_\infty \leq C\|u\|_{k,2}$. Let $R = C\|u\|_{k,2}$, so that $\|u\|_\infty \leq R$, and we may assume $\|u - u_\ell\|_\infty \leq 1$ for each ℓ. Now use the fundamental theorem of calculus to write

(47)
$$f(u) - f(v) = \int_0^1 \frac{d}{ds} f(v + s(u-v))\,ds$$
$$= \int_0^1 f'(v + s(u-v))\,(u-v)\,ds.$$

Replace u by $u(x)$, and v by $u_\ell(x)$, and take $\|\cdot\|_{k,2}$ of both sides of (47) to conclude

(48) $$\|F(u) - F(u_\ell)\|_{k,2} \leq \|u - u_\ell\|_{k,2} \max\{|f'(w)| : |w| \leq R + 1\}.$$

This shows that $F(u_\ell) \to F(u)$ in $H^{k,2}(\Omega)$; that is, $F: H^{k,2}(\Omega) \to H^{k,2}(\Omega)$ is a continuous map.

Since $H^{k,2}(\Omega) \subset C(\overline{\Omega})$, we know by Exercise 6 that $F'(u)v$ is given by $f'(u(x))v(x)$. Since $f \in C^2$, we may use an argument similar to (47) to conclude that $F'(u)$ is continuous in u; that is, F is C^1. ♠

e. Application to Small Mean Curvature

If a hypersurface in $\mathbf{R}^{n+1}$ is the graph of a function, namely $x_{n+1} = u(x_1, \ldots, x_n)$, then the *mean curvature* h of the graph at x is defined by

$$(49) \qquad h(x) = \frac{1}{n} \sum_{i=1}^n \partial_i \left(\frac{\partial_i u}{\sqrt{1 + |\nabla u|^2}} \right),$$

where $\partial_i = \partial/\partial x_i$. (The reason for the terminology *mean curvature* is that $h(x)$ represents the *average of the principal curvatures* κ_i, which are defined as the eigenvalues of the Hessian matrix $D^2 u$ at x; cf. [Gilbarg-Trudinger].) This means that, if we are given a function $h(x)$ on a bounded domain Ω and we wish to find a graph $x_{n+1} = u(x)$ having $h(x)$ as its mean curvature function, we must solve the *prescribed mean curvature equation*

$$(50) \qquad \operatorname{div}\left(\frac{\nabla u}{\sqrt{1 + |\nabla u|^2}} \right) = n\, h(x).$$

Notice that the *minimal surface equation* occurs as the special case $h(x) \equiv 0$ in (50); indeed, a minimal surface is simply a surface whose mean curvature is zero.

To solve (50), we must impose boundary conditions; let us take the Dirichlet boundary condition

$$(51) \qquad u(x) = 0 \quad \text{on } \partial\Omega,$$

which geometrically requires the graph of $u(x)$ to be fixed in the hyperplane $x_{n+1} = 0$ along $\partial\Omega$.

To apply the inverse function theorem, let us define for $u \in C^2(\overline{\Omega})$

$$(52) \qquad F(u) = \operatorname{div}\left(\frac{\nabla u}{\sqrt{1 + |\nabla u|^2}} \right).$$

If $\phi \in C_0^1(\Omega)$, then the divergence theorem shows

$$(53) \qquad \langle F(u), \phi \rangle = -\int_\Omega \frac{\nabla u \cdot \nabla \phi}{\sqrt{1 + |\nabla u|^2}}\, dx,$$

where $\langle,\rangle$ denotes the usual L^2-inner product. In this form, it is relatively easy to compute F' and find (cf. Exercise 7)

$$(54) \qquad F'(u)v = \operatorname{div}\left(\frac{\nabla v}{(1 + |\nabla u|^2)^{1/2}} - \frac{(\nabla u \cdot \nabla v)\, \nabla u}{(1 + |\nabla u|^2)^{3/2}} \right).$$

In particular, we obtain

$$(55) \qquad F'(0)v = \Delta v.$$

Since $F(0) = 0$, we seem to be in a good position to use the inverse function theorem to solve $F(u) = f$ when f is "small." However, we must find appropriate Banach spaces X and Y for which (i) $F: X \to Y$ is C^1 and (ii) $\Delta: X \to Y$ is an isomorphism.

If we take $X = \{u \in C^2(\overline{\Omega}) : u = 0 \text{ on } \partial\Omega\}$ and $Z = C(\overline{\Omega})$, then it is easy to see (cf. Exercise 6) that $F: X \to Z$ is C^1; however, $\Delta: X \to Z$ is not an isomorphism. Instead, we should take Hölder spaces $X = \{u \in C^{2,\alpha}(\overline{\Omega}) : u = 0 \text{ on } \partial\Omega\}$ and $Z = C^\alpha(\overline{\Omega})$ for some $0 < \alpha < 1$. In this case, $F: X \to Z$ is still C^1; moreover, the linear elliptic theory of Chapter 8 will show that $\Delta: X \to Y$ is an isomorphism (cf. the corollary to Theorem 5 of Section 8.2). Theorem 2 may then be used to conclude the existence of a unique solution $u \in C^{2,\alpha}(\overline{\Omega})$ provided that $\|h\|_{C^\alpha(\overline{\Omega})}$ is sufficiently small.

An alternative approach would be to use Sobolev spaces for X and Z. In fact, if we take $X = H_0^{1,2}(\Omega)$ and $Z = X^* = H^{-1,2}(\Omega)$, then the Riesz representation theorem guarantees that $\Delta: X \to Y$ is an isomorphism; however, we cannot use Theorem 4 to show that $F: X \to Z$ is C^1 since functions in $H_0^{1,2}(\Omega)$ are not sufficiently smooth. This suggests instead that we take $X = H_0^{1,2}(\Omega) \cap H^{k,2}(\Omega)$ and $Z = H^{k-2,2}(\Omega)$ with k sufficiently large; in fact, writing $F = \text{div} \circ G \circ \nabla$, where $G(\vec{p}) = \vec{p}/\sqrt{1+|\vec{p}|^2}$, we see that Theorem 4 requires $k - 1 > n/2$. Moreover, the linear elliptic theory of Chapter 8 will show that $\Delta: X \to Z$ is an isomorphism. Theorem 2 may then be used to conclude the existence of a unique solution $u \in H^{k,2}(\Omega)$ provided that $\|h\|_{H^{k-2,2}(\Omega)}$ is sufficiently small.

These two results are not exactly the same. For example, when $n = 3$, the Sobolev space method requires $k > 1 + (3/2)$. For example, we could take $k = 3$, so that we are required to take $\|h\|_{H^{1,2}(\Omega)}$ to be small. Now $H^{1,2}(\Omega)$ is not even contained in $C(\overline{\Omega})$, so we get existence for more general functions h; however, the solution $u \in H^{3,2}(\Omega)$ is not a classical (i.e. C^2) solution since $H^{3,2}(\Omega)$ is not contained in $C^2(\overline{\Omega})$.

In both the Hölder and Sobolev space approaches, however, the perturbation result is of limited value, since we have no a priori control on how large $h(x)$ is allowed to become. In fact, if we allow $h(x)$ to become "large," then the solvability of (50)–(51) depends upon curvature properties of the domain Ω (see [Gilbarg-Trudinger] and Section 13.4 of this book).

Exercises for Section 7.3

1. Let X be a complete metric space and $T_1: X \to X$ and $T_2: X \to X$ be two contractions: $d(T_j x, T_j y) \le \alpha_j d(x,y)$ with $0 < \alpha_j < 1$ for $j = 1, 2$. Let $\bar{x}_j$ be the (unique) fixed point of T_j. Assume T_1, T_2 are ϵ-*close*: $d(T_1 x, T_2 x) \le \epsilon$ for all $x \in X$. Show that the fixed points are close: $d(\bar{x}_1, \bar{x}_2) \le \frac{\epsilon}{1-\alpha}$ where $\alpha = \min\{\alpha_1, \alpha_2\}$.

2. Use a contraction mapping principle to prove the Picard-Lindelöf theorem; that is, solve the initial value problem (40) on some small time interval $|t - t_0| < \tau$. What is the space X? Is the solution C^1?

3. Show that the solution $y(t)$ of (40) depends continuously on the initial condition y_0; if $z(t)$ is the solution for the initial condition z_0, then $|y(t) - z(t)| < \epsilon$ for $|t - t_0| < \tau$, provided $|y_0 - z_0| < \delta_\epsilon$.

4. Complete the proof of the inverse function theorem by showing the following:
   (a) Show $T'_z x = F'(0)^{-1}(F'(0) - F'(x))$ and $\|T'_z x\|_{X \to X} < 1/2$ if $\|x\|_X \le \epsilon$ with ϵ sufficiently small.
   (b) Use $T_z x - T_z y = \int_0^1 \frac{d}{dt} T_z(y + t(x-y))\, dt = \int_0^1 T'_z(y + t(x-y))\,(x-y)\, dt$ to show that T_z is a contraction on $\overline{B_\epsilon(0)} \subset X$.

5. In Theorem 2, prove the following additional properties of the inverse function G:
   (a) $G: V \to X$ is continuous.
   (b) $G: V \to X$ is C^1.

6. Let Ω be a bounded domain and $f: \mathbf{R} \to \mathbf{R}$ be C^1. For $u \in C(\overline{\Omega})$, define $F(u)(x) = f(u(x))$.
   (a) Show that $F: C(\overline{\Omega}) \to C(\overline{\Omega})$ is C^1, and that $F'(u)v$ is given by $f'(u(x))v(x)$.
   (b) If f is C^{k+1}, show that $F: C^k(\overline{\Omega}) \to C^k(\overline{\Omega})$ is C^1.

7. Use (53) to establish (54).

8. Suppose the graph $z = u_0(x)$ in a bounded domain $\Omega \subset \mathbf{R}^n$ satisfies $u_0(x) = \phi(x)$ on $\partial\Omega$ and has mean curvature $h_0(x)$ in Ω. Given a prescribed mean curvature function $h(x)$, describe how to show there is a graph $z = u(x)$ in Ω with $u(x) = \phi(x)$ on $\partial\Omega$ and mean curvature h in Ω, provided that $h - h_0$ is sufficiently small.

Further References for Chapter 7

For a more detailed treatment of differential calculus on Banach spaces, see [Dieudonné]. The calculus of variations has many applications to physics and geometry; for example, see [Troutman]. For more about the variational method of solving partial differential equations, consult [Nirenberg, 2] or [Struwe].

8 Linear Elliptic Theory

In this chapter, we study second-order linear elliptic equations and certain inequalities that are called a priori estimates because they must be satisfied by all solutions of the equation (even if we do not know whether a solution exists). The estimates are useful in establishing the regularity, and even existence, of solutions. We begin with Fourier analysis and a priori estimates on a torus. The estimates are then transplanted to the case of elliptic operators on a bounded domain Ω. The question of uniqueness leads us to investigate maximum principles for elliptic operators on Ω. Finally, we use the estimates to establish solvability and Fredholm properties for elliptic operators on Ω with Dirichlet conditions on $\partial\Omega$.

8.1 Elliptic Operators on a Torus

In this section we consider elliptic equations and estimates when no boundary is present. This occurs when we consider functions that are *periodic* on $\mathbf{R}^n$; alternatively we may consider periodic functions as being defined on a torus $\mathbf{T}^n$, which is compact but has no boundary. Fourier analysis enables us to define L^2-Sobolev spaces on the torus and analyze the solvability of elliptic equations with constant coefficients. To extend these results to variable coefficient operators, we first establish a priori estimates; these are used in the next section to establish the regularity of weak solutions.

In Fourier analysis, it is more natural to work with complex-valued functions; consequently, in this section we use as the L^2-inner product $\langle u, v \rangle = \int u \bar{v}\, dx$. Also, throughout this section, and indeed the entire chapter, C denotes a generic constant whose value may change with different occurences, and k always represents a nonnegative integer.

a. Fourier Analysis

Let us consider functions u that are 2π-periodic on $\mathbf{R}^n$; that is, u is defined on the torus $\mathbf{T}^n = \mathbf{S}^1 \times \cdots \times \mathbf{S}^1$, where each $\mathbf{S}^1$ is identified with the interval $[0, 2\pi)$. We can associate with u the Fourier series

(1) $$u(x) \sim \sum_\alpha u_\alpha e^{i\alpha \cdot x},$$

where the sum is taken over all n-tuples $\alpha = (\alpha_1, \ldots, \alpha_n)$ with each α_k an integer (possibly negative), $\alpha \cdot x = \alpha_1 x_1 + \cdots + \alpha_n x_n$, and the u_α are the

246   Chapter 8: Linear Elliptic Theory

Fourier coefficients defined by

(2) $$u_\alpha = (2\pi)^{-n} \int_{\mathbf{T}^n} u(x) e^{-i\alpha \cdot x}\, dx.$$

First, let us consider *finite* sums $u(x) = \sum u_\alpha e^{i\alpha \cdot x}$, which certainly represent C^∞-functions on $\mathbf{T}^n$. Using $\langle e^{i\alpha \cdot x}, e^{i\beta \cdot x}\rangle = \int_{\mathbf{T}^n} e^{i(\alpha-\beta)\cdot x}\, dx = 0$ for $\alpha \ne \beta$ and $\langle e^{i\alpha \cdot x}, e^{i\alpha \cdot x}\rangle = \int_{\mathbf{T}^n} 1\, dx = (2\pi)^n$, we find that two finite sums $u = \sum u_\alpha e^{i\alpha \cdot x}$ and $v = \sum v_\alpha e^{i\alpha \cdot x}$ satisfy

(3) $$\langle u, v\rangle = (2\pi)^n \sum_\alpha u_\alpha \bar{v}_\alpha.$$

In particular, the L^2-norm satisfies

(4) $$\|u\|^2 = \int_{\mathbf{T}^n} |u|^2\, dx = (2\pi)^n \sum_\alpha |u_\alpha|^2.$$

If $u(x)$ is real valued, the Fourier coefficients may be complex valued, but must satisfy $\bar{u}_\alpha = u_{-\alpha}$.

Now, if we take the Hilbert space completion of these finite sums in the norm (4), we obtain (by Riesz-Fischer) $L^2(\mathbf{T}^n)$. In other words, if $u \in L^2(\mathbf{T}^n)$ then its Fourier coefficients u_α satisfy $\sum_\alpha |u_\alpha|^2 < \infty$ and its Fourier series converges to u in $L^2(\mathbf{T}^n)$:

$$u^N(x) \equiv \sum_{|\alpha| \le N} u_\alpha e^{i\alpha \cdot x} \to u(x) \text{ in } L^2(\mathbf{T}^n) \text{ as } N \to \infty.$$

We also want to use L^2-Sobolev spaces: since $p = 2$ is always understood, throughout this section we denote $H^{k,2}$ simply by H^k; in particular, L^2 may be written as H^0, and the norm (4) by $\|\cdot\|_0$. For a finite sum $u = \sum u_\alpha e^{i\alpha \cdot x}$, observe that $\partial u/\partial x_k = \sum i\alpha_k u_\alpha e^{i\alpha \cdot x}$; that is, a derivative affects the Fourier coefficients as a multiplication (a phenomenon common to Fourier transforms, as seen in Section 5.2). Therefore, we obtain

(5)
$$\langle u,v\rangle_1 = \int_{\mathbf{T}^n} (\nabla u \cdot \nabla \bar{v} + u\bar{v})\, dx = (2\pi)^n \sum_\alpha (|\alpha|^2 + 1) u_\alpha \bar{v}_\alpha,$$

$$\|u\|_1^2 = (2\pi)^n \sum_\alpha (|\alpha|^2 + 1)|u_\alpha|^2,$$

$$\langle u,v\rangle_2 = \int_{\mathbf{T}^n} \sum_{|\gamma|=2} D^\gamma u \overline{D^\gamma v}\, dx + \langle u,v\rangle_1$$

$$= (2\pi)^n \sum_\alpha \left(\left(\sum_{i,j=1}^n \alpha_i \alpha_j\right)^2 + |\alpha|^2 + 1\right) u_\alpha \bar{v}_\alpha$$

$$\|u\|_2^2 = (2\pi)^n \sum_\alpha \left(\left(\sum_{i,j=1}^n \alpha_i \alpha_j\right)^2 + |\alpha|^2 + 1\right) |u_\alpha|^2, \text{ and so on.}$$

Section 8.1: Elliptic Operators on a Torus

Let $H^k(\mathbf{T}^n)$ denote the completion of these finite sums $\sum u_\alpha e^{i\alpha\cdot x}$ in the norm $\|\cdot\|_k$. Notice that the condition $u \in H^k(\mathbf{T}^n)$ may be interpreted as a decay condition on the Fourier coefficients u_α (i.e., the larger k, the more rapidly the Fourier coefficients u_α must decay as $|\alpha| \to \infty$). In fact, from (5) it is clear that an equivalent inner product and norm on $H^k(\mathbf{T}^n)$ is

(6) $\quad \langle\langle u, v\rangle\rangle_k = \sum_\alpha (|\alpha|^2 + 1)^k u_\alpha \bar{v}_\alpha, \quad \|\|u\|\|_k^2 = \sum_\alpha (|\alpha|^2 + 1)^k |u_\alpha|^2.$

As an application of Fourier analysis, let us consider the eigenvalue problem for the Laplace operator,

(7) $\qquad\qquad\qquad \Delta u + \lambda u = 0,$

and the associated nonhomogeneous equation

(8) $\qquad\qquad\qquad \Delta u + \lambda u = f,$

where $f \in L^2(\mathbf{T}^n)$. If $u = \sum u_\alpha e^{i\alpha\cdot x} \in H^2(\mathbf{T}^n)$, then approximating by finite sums shows $\Delta u = -\sum |\alpha|^2 u_\alpha e^{i\alpha\cdot x}$; if we write $f(x) = \sum f_\alpha e^{i\alpha\cdot x}$, then the Fourier coefficients u_α and f_α must satisfy

(9) $\qquad\qquad (\lambda - |\alpha|^2) u_\alpha = f_\alpha \quad \text{for all } \alpha.$

This formula not only may be used to try to solve (8) but to *conclude* from $f \in L^2(\mathbf{T}^n)$ that $u \in H^2(\mathbf{T}^n)$.

If $\lambda \neq |\alpha|^2$ for any α, then we can satisfy (8) by letting $u_\alpha = f_\alpha(\lambda - |\alpha|^2)^{-1}$ for all α. It is easy to check (cf. Exercise 2) that these Fourier coefficients guarantee that $u \in H^2(\mathbf{T}^n)$.

On the other hand, suppose $\lambda = |\tilde{\alpha}|^2$ for some specific n-tuple $\tilde{\alpha}$. This means two things. First, we see that $e^{i\tilde{\alpha}\cdot x}$ is a nontrivial solution of (7), so λ is an eigenvalue for the Laplacian (cf. Exercise 3); of course it could be that $\lambda = |\tilde{\alpha}|^2$ for a finite number of multi-indices $\tilde{\alpha}^{(1)}, \ldots, \tilde{\alpha}^{(k)}$, in which case λ will have multiplicity k. Second, we can only satisfy (9) if $f_{\tilde{\alpha}} = \int f(x) e^{-ix\cdot\tilde{\alpha}} dx = 0$, and in that case we may give $u_{\tilde{\alpha}}$ an arbitrary value.

Let us denote the eigenvalues of the Laplacian on $\mathbf{T}^n$ by Λ, which we have identified as the set $\{|\alpha|^2 : \alpha \text{ is an } n\text{-tuple of integers}\}$. Then these observations may be summarized into the following.

Theorem 1. *For $\lambda \in \mathbf{C}$, the following alternative holds for the solvability of (8). Either*
  (i) *$\lambda \in \mathbf{C}\backslash\Lambda$ and (8) admits a unique solution $u \in H^2(\mathbf{T}^n)$ for every $f \in L^2(\mathbf{T}^n)$, or*
  (ii) *$\lambda \in \Lambda$ and (8) is solvable if and only if f is orthogonal to all solutions $v_1, \ldots, v_k$ of (7).*

Remark. If we only consider real-valued functions u, then Theorem 1 still holds if we replace $\mathbf{C}$ by $\mathbf{R}$. However, if we replace Δ by an operator that is not self-adjoint, then we should use complex-valued functions u since Λ may contain complex λ (cf. Exercise 4).

Theorem 1 shows that the solvability of (8) is reminiscent of the solvability of a linear operator A on a finite-dimensional inner product space X: Either A is invertible and $Ax = y$ is solvable for every $y \in X$, or A has nullspace and $Ax = y$ can be solved if and only if y is orthogonal to all elements in the nullspace of the adjoint operator A^*.

It is not difficult to generalize Theorem 1 to any constant coefficient elliptic operator (cf. Exercise 4). To generalize the result to variable coefficient operators on $\mathbf{T}^n$ (i.e., operators with periodic coefficients), we first must establish some a priori estimates, which we consider next.

b. A Priori Estimates and Regularity

Using (6) with $(|\alpha|^2 + 1)^{k+2} = (|\alpha|^2 + 1)^k(|\alpha|^2 + 1)^2$, we find that $|||u|||_{k+2}^2 = ||| - \Delta u + u|||_k^2 \leq (|||\Delta u|||_k + |||u|||_k)^2$; or, equivalently,

(10a) $\qquad \|u\|_{k+2} \leq C\left(\|\Delta u\|_k + \|u\|_k\right)$ for $u \in H^{k+2}(\mathbf{T}^n)$.

In fact, we may improve (10a) by induction on the term $\|u\|_k$ to obtain (see Exercise 5)

(10b) $\qquad \|u\|_{k+2} \leq C\left(\|\Delta u\|_k + \|u\|_0\right)$ for $u \in H^{k+2}(\mathbf{T}^n)$.

Notice that both estimates in (10) *assume* $u \in H^{k+2}$; let us now investigate how to use (10) to *prove* $u \in H^{k+2}$ when $\Delta u, u \in H^k$.

Following Section 6.2, we define $u \in H^1(\mathbf{T}^n)$ to be a *weak solution* of $\Delta u = f \in L^2(\mathbf{T}^n)$ if $\int \nabla u \cdot \nabla \bar{v}\, dx = -\int f\bar{v}\, dx$ for all $v \in H^1(\mathbf{T}^n)$. In terms of the Fourier coefficients of u, v, and f, this means that

(11) $\qquad \sum |\alpha|^2 u_\alpha \bar{v}_\alpha = -\sum f_\alpha \bar{v}_\alpha$ for all $v \in H^1(\mathbf{T}^n)$;

in other words, $-|\alpha|^2 u_\alpha = f_\alpha$ for all α. In turn, this implies that $\Delta u^N = f^N$ for every truncation

(12) $\qquad u^N(x) = \sum_{|\alpha| \leq N} u_\alpha e^{i\alpha \cdot x}$ and $f^N(x) = \sum_{|\alpha| \leq N} f_\alpha e^{i\alpha \cdot x}$.

Of course, u^N and f^N are both C^∞-functions and satisfy as $N \to \infty$:

(13) $\qquad u^N \to u$ in $H^1(\mathbf{T}^n)$, and $f^N \to f$ in $H^k(\mathbf{T}^n)$.

We can apply (10b) to $u^N - u^M$ to obtain

(14) $\qquad \|u^N - u^M\|_{k+2} \leq C\left(\|f^N - f^M\|_k + \|u^N - u^M\|_0\right).$

But (13) with (14) shows that u^N is a Cauchy sequence in $H^{k+2}(\mathbf{T}^n)$. By completeness, we know that u^N converges to an element of $H^{k+2}(\mathbf{T}^n)$, which must agree with u by (13). This means $u \in H^{k+2}(\mathbf{T}^n)$, and we have just proved the following result.

Theorem 2. *If $u \in H^1(\mathbf{T}^n)$ is a weak solution of $\Delta u = f$, where $f \in H^k(\mathbf{T}^n)$ for some $k \in \mathbf{N}$, then $u \in H^{k+2}(\mathbf{T}^n)$.*

Combining this result with the Sobolev imbedding theorems of Chapter 6, we easily obtain the following.

Corollary. *If $u \in H^1(\mathbf{T}^n)$ is a weak solution of $\Delta u = f$, where $f \in C^\infty(\mathbf{T}^n)$, then $u \in C^\infty(\mathbf{T}^n)$.*

Proof. Using Theorem 6 in Section 6.5, we observe that $H^k(\mathbf{T}^n) \subset C^m(\mathbf{T}^n)$ provided that $k > m + \frac{n}{2}$. This means $\cap_{k>0} H^k(\mathbf{T}^n) \subset C^m(\mathbf{T}^n)$ for every m, and hence $\cap_{k>0} H^k(\mathbf{T}^n) \subset C^\infty(\mathbf{T}^n)$. But $C^\infty(\mathbf{T}^n) \subset H^k(\mathbf{T}^n)$ for every k, so

(15) $$\cap_{k>0} H^k(\mathbf{T}^n) = C^\infty(\mathbf{T}^n). \spadesuit$$

Remark. Since a strong solution is also a weak solution, we can apply Theorem 1 and its corollary to $u \in H^2(\mathbf{T}^n)$ to conclude that $\Delta u = f \in H^k(\mathbf{T}^n)$ implies $u \in H^{k+2}(\mathbf{T}^n)$, and $\Delta u = f \in C^\infty(\mathbf{T}^n)$ implies $u \in C^\infty(\mathbf{T}^n)$.

We now want to replace Δ by a second-order operator L of the form

(16) $$L = \sum_{i,j=1}^n a_{ij}(x) \frac{\partial^2}{\partial x_i \partial x_j} + \sum_{i=1}^n b_i(x) \frac{\partial}{\partial x_i} + c(x),$$

where all coefficient functions are C^k functions of $x \in \mathbf{T}^N$ and satisfy the *ellipticity condition*

(17) $$\sum_{i,j=1}^n a_{ij}(x) \xi_i \xi_j \geq \lambda |\xi|^2 \quad \text{for } x \in \mathbf{T}^n, \text{ and } \xi \in \mathbf{R}^n,$$

where $\lambda > 0$ is a constant. We now establish the a priori estimates for L that generalize (10).

Theorem 3. *If L is a second-order elliptic operator with C^k-coefficients on $\mathbf{T}^n$, then the following estimates hold for each $k \geq 0$:*

(18a) $\quad \|u\|_{k+2} \leq C (\|Lu\|_k + \|u\|_k) \quad$ *for all $u \in H^{k+2}(\mathbf{T}^n)$,*

(18b) $\quad \|u\|_{k+2} \leq C (\|Lu\|_k + \|u\|_0) \quad$ *for all $u \in H^{k+2}(\mathbf{T}^n)$.*

Proof. First, let us prove (18a) in the special case that the a_{ij} are all constants and $b_i = 0 = c$. The advantage of this special case is that we can now use Fourier analysis. In fact, by (17) we have $\sum_{ij} a_{ij} \alpha_i \alpha_j \geq \lambda |\alpha|^2$, so squaring, multiplying by $(1+|\alpha|^2)^k|u_\alpha|^2$, and summing over α shows $|\|Lu\||_k \geq \lambda|\|\Delta u\||_k$; thus (18a) follows from (10a).

Now let us try to prove (18a) for the general operator L in (16). The idea is to freeze the coefficients at some point $x_0 \in \mathbf{T}^n$ in order to define the constant coefficient operator

$$(19) \qquad L_0 = \sum_{i,j}^n a_{ij}(x_0)\frac{\partial^2}{\partial x_i \partial x_j}.$$

We know (18a) holds for L_0, so we may write

$$(20) \qquad \begin{aligned} \|u\|_{k+2} &\leq C(\|L_0 u\|_k + \|u\|_k) \\ &\leq C(\|Lu\|_k + \|(L-L_0)u\|_k + \|u\|_k). \end{aligned}$$

We would like to absorb the term $\|(L-L_0)u\|_k$ into the other two terms on that side. Using the continuity of $a_{ij}(x)$ and its derivatives, this can be done provided that u has support in a small ball $B_\epsilon(x_0)$. Indeed, for any $\eta > 0$, we can choose ϵ small enough that $\|(L-L_0)u\|_k \leq \eta\|u\|_{k+2} + C\|u\|_{k+1}$ for all $u \in H^{k+2}(\mathbf{T}^n)$ with supp $u \subset B_\epsilon(x_0)$. Substituting into (20), we obtain

$$(21) \qquad \|u\|_{k+2} \leq C(\|Lu\|_k + \eta\|u\|_{k+2} + \|u\|_{k+1}),$$

where $u \in H_0^{k+2}(B_\epsilon(x_0))$. If we take η sufficiently small, we can arrange $C\eta < 1/2$; this allows us to perform the following absorption trick: Switch the $\|u\|_{k+2}$ term from the right-hand side to the left-hand side of (21), and adjust the constant to obtain

$$(22) \qquad \|u\|_{k+2} \leq C(\|Lu\|_k + \|u\|_{k+1}).$$

The only problem with (22) is the subscript $k+1$, but we can convert it to k by the same absorption trick if we prove the interpolation estimate

$$(23) \qquad \|u\|_{k+1} \leq \eta\|u\|_{k+2} + C_\eta\|u\|_k.$$

This is done in Exercise 6.

We have shown that (18a) holds for $u \in H_0^{k+2}(B_\epsilon(x_0))$; we now patch these results together using a partition of unity. Namely, the balls $B_\epsilon(x)$ form an open cover of $\mathbf{T}^n$, which (by compactness) has a finite subcover: $\mathbf{T}^n \subset B_{\epsilon_1}(x_1) \cup \cdots \cup B_{\epsilon_N}(x_N)$ for some integer N. Then we may find nonnegative functions $\phi_1, \ldots, \phi_N$ with supp $\phi_j \subset B_{\epsilon_j}(x_j)$ and

$$(24) \qquad \sum_{j=1}^N \phi_j^2 = 1.$$

We will be able to apply (18a) to each $\phi_j u$, but first let us state a couple of technical results that we need in the calculation; see Exercise 7 for the proof.

Lemma. Let $\phi, \psi \in C^\infty(\mathbf{T}^n)$ be nonnegative functions.
(i) For any integer $\ell \geq 0$, $|\langle \phi^2 u, u\rangle_{\ell+1} - \langle \phi u, \phi u\rangle_{\ell+1}| \leq C\|u\|_\ell \|u\|_{\ell+1}$, where $u \in H^{\ell+1}(\mathbf{T}^n)$, and C depends on ϕ and its derivatives up to order $\ell + 1$.
(ii) If $u \in H^{k+2}(\mathbf{T}^n)$, then

$$\|\phi Lu\|_k^2 + \|\psi Lu\|_k^2 \leq \int_{\mathbf{T}^n} (\phi^2 + \psi^2) \sum_{|\gamma|=k} |D^\gamma Lu|^2 \, dx + C\|u\|_{k+1}\|u\|_{k+2},$$

where C depends on ϕ and ψ and their derivatives up to order k.

If we use (24) and Lemma (i) with $\ell = k + 1$, we obtain

$$(25) \quad \|u\|_{k+2}^2 = \left\langle \sum_j \phi_j^2 u, u \right\rangle_{k+2} \leq \sum_j \|\phi_j u\|_{k+2}^2 + C\|u\|_{k+1}\|u\|_{k+2}.$$

Now we can apply (18a) and then commute ϕ_j and L to obtain

$$(26) \quad \|\phi_j u\|_{k+2}^2 \leq C(\|L(\phi_j u)\|_k^2 + \|u\|_{k+1}^2) \leq C(\|\phi_j Lu\|_k^2 + \|u\|_{k+1}^2).$$

Inserting (26) in (25) and applying Lemma (ii), we obtain

$$(27) \quad \begin{aligned} \|u\|_{k+2}^2 &\leq C\left(\sum \|\phi_j Lu\|_k^2 + \|u\|_{k+1}^2 + \|u\|_{k+1}\|u\|_{k+2}\right) \\ &\leq C(\|Lu\|_k^2 + \|u\|_{k+1}\|u\|_{k+2}). \end{aligned}$$

If we use the absorption trick twice, once using the elementary inequality $\|u\|_{k+1}\|u\|_{k+2} \leq \eta\|u\|_{k+2}^2 + C_\eta\|u\|_{k+1}^2$, and once using the interpolation inequality (23), we get $\|u\|_{k+2}^2 \leq C(\|Lu\|_k^2 + \|u\|_k^2)$, which implies (18a).

The inequality (18b) is proved by induction upon (18a). ♠

As in Theorem 2, the estimates (18) may be used to show that a weak solution $u \in H^1(\mathbf{T}^n)$ of $Lu = f \in H^k(\mathbf{T}^n)$ is a strong solution with $u \in H^{k+2}(\mathbf{T}^n)$. This is most readily accomplished when all $b_j \equiv 0$ in (16), so that L is in divergence form; a weak solution $u \in H^1(\mathbf{T}^n)$ may then be studied using the symmetric bilinear form B_L of Section 6.2. In this case, the methods of Section 7.2 show that L has real eigenvalues and a complete orthonormal set of eigenfunctions. The resultant eigenfunction expansions for f and u may be used in place of the Fourier series involving the complex exponentials $e^{i\alpha \cdot x}$, and the same truncation method that was applied when $L = \Delta$ [cf. (12)] may be used to show that $u \in H^{k+2}(\mathbf{T}^n)$.

When $b_j \not\equiv 0$, then it is still possible to show regularity of weak solutions by means of difference quotients. The proof is much the same as the case when $\mathbf{T}^n$ is replaced by a bounded domain, which we study in Section 8.2.

c. L^p and Hölder Estimates

We have used L^2-Sobolev spaces to express the a priori estimates (18); however, we could also use general L^p-Sobolev spaces $H^{k,p}$, or Hölder spaces $C^{k,\alpha}$ (cf. Section 6.5). It is clear how to define these spaces on $\mathbf{T}^n$, and we denote their norms respectively by $\|\cdot\|_{k,p}$ and $|\cdot|_{k,\alpha}$. We do not discuss their derivations, but let us record the following results (cf. [Gilbarg-Trudinger], [Bers-John-Schechter]).

Theorem 4. *If L is the second-order elliptic operator (16) with C^k-coefficients on $\mathbf{T}^n$, then the following estimates hold for all $u \in C^\infty(\mathbf{T}^n)$:*

$$\text{(28a)} \quad \|u\|_{k+2,p} \leq C \left(\|Lu\|_{k,p} + \|u\|_{k,p} \right) \quad \text{where } 1 < p < \infty,$$

and, provided the coefficients of L are $C^{k,\alpha}$ with $0 < \alpha < 1$,

$$\text{(28b)} \quad |u|_{k+2,\alpha} \leq C \left(|Lu|_{k,\alpha} + |u|_{k,\alpha} \right).$$

Exercises for Section 8.1

1. Using (6b), the sequence of Sobolev spaces $H^k(\mathbf{T}^n)$ may be imbedded in a *continuous scale* of Sobolev spaces $H^s(\mathbf{T}^n)$ defined for every $s \in \mathbf{R}$ using the norm $\|u\|_s^2 = \sum_\alpha (|\alpha|^2 + 1)^s |u_\alpha|^2$. For $\lambda < 0$ show that $\Delta + \lambda : H^{s+2}(\mathbf{T}^n) \to H^s(\mathbf{T}^n)$ is an isomorphism.

2. Suppose that $f \in L^2(\mathbf{T}^n)$, $\lambda \neq |\alpha|^2$ for any multi-index α, and u has Fourier coefficients $u_\alpha = f_\alpha (\lambda - |\alpha|^2)^{-1}$. Show directly that $\|u\|_2 < \infty$.

3. If $(\Delta + \lambda) e^{\pm i \tilde{\alpha} \cdot x} = 0$ where $\lambda \in \mathbf{R}$ and $\tilde{\alpha}$ an n-tuple, show that there is a *real-valued* function $u(x)$ on $\mathbf{T}^n$ satisfying $(\Delta + \lambda) u = 0$.

4. (a) Let $L = \sum a_{jk} \partial^2 / \partial x_j \partial x_k$ be an elliptic operator [cf. (17)] with constant real coefficients, and consider the equation $Lu + \lambda u = f$ where $f \in L^2(\mathbf{T}^n)$. Determine the eigenvalues λ for which $Lu + \lambda u = 0$ admits a nontrivial solution, and give the version of Theorem 1 that applies in this case.
   (b) Do the same for an elliptic, constant coefficient operator with lower-order terms; that is, $L = \sum a_{jk} \partial^2 / \partial x_j \partial x_k + \sum_{i=1}^n b_j \partial / \partial x_j + c$. Are the eigenvalues λ still all real?

5. Use (10a) to establish (10b).

6. For any $\eta > 0$, show that there exists $C_\eta > 0$ so that (23) holds.

7. Prove the Lemma contained in the proof of Theorem 3.

8. Prove the following:
   (a) If $u \in H^1(\mathbf{T}^n)$ is a weak solution of $\Delta u = 0$, then $u \equiv$ const.
   (b) If $u \in H^1(\mathbf{T}^n)$ is a weak solution of $\Delta u = f$ where $f \in L^2(\mathbf{T}^n)$, then $\int_{\mathbf{T}^n} f \, dx = 0$.

9. If $u \in H^2(\mathbf{T}^n)$ and $\int_{\mathbf{T}^n} u \, dx = 0$, show that $\|u\|_1 \leq C \|\Delta u\|$.

10. Consider an elliptic operator of order $m = 2\ell$ (cf. Exercise 16 in Section 2.3) of the form $L = \sum_{|\alpha|=\ell} D^\alpha a_\alpha(x) D^\alpha + \cdots$, with real-valued coefficients $a_\alpha \in C^\infty(\mathbf{T}^n)$. Prove the following generalization of (18a):

$$\|u\|_{k+m} \leq C\left(\|Lu\|_k + \|u\|_k\right) \quad \text{for all } u \in H^{k+m}(\mathbf{T}^n).$$

11. *Fourier Analysis on $\mathbf{R}^n$.*

(a) Prove the following analogue of (10b):

$$\|u\|_{k+2} \leq C(\|\Delta u\|_k + \|u\|_0), \quad \text{for } u \in H^{k+2,2}(\mathbf{R}^n).$$

(b) If $u \in H^{1,2}(\mathbf{R}^n)$ is a weak solution of $\Delta u = f \in H^{k,2}(\mathbf{R}^n)$, then $u \in H^{k+2,2}(\mathbf{R}^n)$.

8.2 Estimates and Regularity on Domains

In this section, we would like to obtain a priori estimates and regularity for solutions of elliptic equations in a bounded domain $\Omega \subset \mathbf{R}^n$, i.e.,

(29a) $$Lu = f \quad \text{in } \Omega,$$

where L is the uniformly elliptic operator

(29b) $$L = \sum_{i,j=1}^n a_{ij}(x) \frac{\partial^2}{\partial x^i \partial x^j} + \sum_{i=1}^n b_i(x) \frac{\partial}{\partial x^i} + c(x)$$

with bounded coefficients a_{ij}, b_i, and c satisfying *uniform ellipticity:*

(29c) $$\sum_{i,j=1}^n a_{ij}(x) \xi_i \xi_j \geq \lambda |\xi|^2 \quad \text{for all } x \in \Omega \text{ and } \xi \in \mathbf{R}^n \quad (\lambda > 0).$$

Without loss of generality, we assume that $\Omega \subset \{x \in \mathbf{R}^n : 0 < x_i < 2\pi$ for $i = 1, \ldots, n\}$; thus we may consider $\Omega \subset \mathbf{T}^n$ where $\mathbf{T}^n$ is the torus as in the preceding section. Moreover, given $u \in H^k(\Omega)$ and $\phi \in C_0^\infty(\Omega)$, then ϕu vanishes near $\partial \Omega$, so we may consider $\phi u \in H^k(\mathbf{T}^n)$ simply by extending it to be zero on $\mathbf{T}^n \setminus \Omega$. These observations enable us to transfer estimates from $\mathbf{T}^n$ to Ω. Notice that, as in the previous section, we are usually dealing with L^2-Sobolev spaces, and hence we shall omit the index for $p = 2$.

a. Interior Estimates

We use the notation $\Omega' \subset\subset \Omega$ when $\overline{\Omega'}$ is compact and $\overline{\Omega'} \subset \Omega$; we subscript our norms $\|\cdot\|$ with Ω or Ω' to indicate the domain over which the integration is taken. We are interested in estimating $\|u\|_{k+2;\Omega'}$ in terms of $\|Lu\|_{k;\Omega}$ and $\|u\|_{0;\Omega}$; such estimates are used to improve the regularity of weak solutions of (29a), depending on the regularity of f.

First let us prove the interior estimates and regularity for the Laplace operator on Ω.

Theorem 1. If $\Omega \subset \mathbf{R}^n$ is a bounded domain, and $\Omega' \subset\subset \Omega$, then

(30) $\quad \|u\|_{k+2;\Omega'} \leq C \left(\|\Delta u\|_{k;\Omega} + \|u\|_{0;\Omega} \right) \quad$ for all $u \in H^{k+2}(\Omega)$.

Moreover, if $u \in H^1(\Omega)$ is a weak solution of $\Delta u = f$, where $f \in H^k(\Omega)$, then $u \in H^{k+2}(\Omega')$.

Proof. We first verify the estimate (30). Let us introduce a cutoff function $\phi \in C_0^\infty(\Omega)$ such that $\phi \equiv 1$ on Ω'. Then $\phi u \in H^{k+2}(\mathbf{T}^n)$, so we may apply (10b) to obtain

(31)
$$\begin{aligned}\|u\|_{k+2;\Omega'} &\leq \|\phi u\|_{k+2;\Omega} = \|\phi u\|_{k+2;\mathbf{T}^n} \\ &\leq C \left(\|\Delta(\phi u)\|_{k;\mathbf{T}^n} + \|\phi u\|_{0;\mathbf{T}^n} \right) \\ &\leq C \left(\|\Delta(\phi u)\|_{k;\Omega} + \|u\|_{0;\Omega} \right).\end{aligned}$$

Now $\Delta(\phi u) = \phi(\Delta u) + 2\nabla\phi \cdot \nabla u + (\Delta\phi)u$, so we obtain

$$\|\Delta(\phi u)\|_{k;\Omega} \leq \|\phi(\Delta u)\|_{k;\Omega} + 2\|\nabla\phi \cdot \nabla u\|_{k;\Omega} + \|(\Delta\phi)u\|_{k;\Omega}.$$

But ϕ and all its derivatives up to order $k+2$ are bounded on Ω, so we obtain
$$\begin{aligned}\|\Delta(\phi u)\|_{k;\Omega} &\leq C \left(\|\Delta u\|_{k;\Omega} + \|\nabla u\|_{k;\Omega} + \|u\|_{k;\Omega} \right) \\ &\leq C \left(\|\Delta u\|_{k;\Omega} + \|u\|_{k+1;\Omega} \right).\end{aligned}$$

If we insert this into (31), we obtain

(32) $\quad \|u\|_{k+2;\Omega'} \leq C \left(\|\Delta u\|_{k;\Omega} + \|u\|_{k+1;\Omega} \right).$

To reduce the order $k+1$ on the right-hand side in (32), let us take a sequence of sets $\Omega' = \Omega_{k+2} \subset\subset \Omega_{k+1} \subset\subset \cdots \subset\subset \Omega_1 \subset\subset \Omega_0 = \Omega$ and rewrite (32) with Ω_{k+1} in place of Ω:

$$\|u\|_{k+2;\Omega_{k+2}} \leq C \left(\|\Delta u\|_{k;\Omega_{k+1}} + \|u\|_{k+1;\Omega_{k+1}} \right).$$

Replacing k by $k-1$, we also have

$$\|u\|_{k+1;\Omega_{k+1}} \leq C \left(\|\Delta u\|_{k-1;\Omega_k} + \|u\|_{k;\Omega_k} \right).$$

Combining these, we obtain $\|u\|_{k+2;\Omega_{k+2}} \leq C \left(\|\Delta u\|_{k;\Omega_k} + \|u\|_{k;\Omega_k} \right)$ in which the order $k+1$ in (32) has been reduced to k. Proceeding by induction, we obtain $\|u\|_{k+2;\Omega'} \leq C \left(\|\Delta u\|_{k;\Omega_1} + \|u\|_{1;\Omega_1} \right)$. Finally, we may use Exercise 1 to obtain (30). Notice that the constant C in (30) depends on the cutoff functions: the closer Ω' is to $\partial\Omega$, the larger the constant C.

Next, suppose $u \in H^1(\Omega)$ is a weak solution of $\Delta u = f \in H^k(\Omega)$. Then $\phi u \in H^1(\mathbf{T}^n)$ is a weak solution of $\Delta(\phi u) = f'$ where $f' \in H^k(\mathbf{T}^n)$ vanishes outside Ω. We apply Theorem 2 of Section 8.1 to conclude that $\phi u \in H^{k+2}(\mathbf{T}^n)$. Since $\phi u = u$ on Ω', we conclude that $u \in H^{k+2}(\Omega')$. ♠

As applications of this result, let us consider the interior regularity of weak solutions of Poisson's equation and the eigenvalue problems discussed in Section 7.2. Notice that no boundary conditions are assumed, so these regularity results apply equally to weak solutions satisfying Dirichlet, Neumann, or other boundary conditions.

Corollary A. *If Ω is a bounded domain in $\mathbf{R}^n$, $f \in C^\infty(\Omega)$, and $u \in H^1(\Omega)$ is a weak solution of $\Delta u = f$, then $u \in C^\infty(\Omega)$.*

Corollary B. *If Ω is a bounded domain in $\mathbf{R}^n$ and $u \in H^1(\Omega)$ is a weak solution of $\Delta u + \lambda u = 0$, then $u \in C^\infty(\Omega)$.*

The derivation of these results from Theorem 1 is given in Exercise 2.

Now let us turn to the more general elliptic operator (29b,c). We assume that all coefficients are in $C^k(\overline{\Omega})$.

Theorem 2. *Suppose L is a uniformly elliptic second-order operator on a bounded domain Ω, with coefficients in $C^k(\overline{\Omega})$. If $\Omega' \subset\subset \Omega$, then*

$$(33) \qquad \|u\|_{k+2;\Omega'} \le C \left(\|Lu\|_{k;\Omega} + \|u\|_{0;\Omega} \right) \qquad \text{for all } u \in H^{k+2}(\Omega).$$

The inequality (33) is easily derived from (18b) exactly as the proof of Theorem 1 uses (10b); see Exercise 3. We also remark that there are L^p and C^α estimates analogous to (33), which are similarly obtained by transplanting (28) from the torus to Ω.

Of course, we would like to use (33) to obtain regularity results for weak solutions of $Lu = f$. This is done using difference quotients, which we discuss in the next subsection.

b. Difference Quotients

It is often advantageous to study weak derivatives using the limit of difference quotients. For a function u defined in a domain Ω, let us define the *difference quotient*

$$(34) \qquad \delta_j^h u(x) = \frac{u(x + he_j) - u(x)}{h},$$

where $h \ne 0$, e_j denotes the unit vector in the x_j-direction and h is small enough that $0 < |h| < \text{dist}(x, \partial\Omega)$. For $u \in C^1(\Omega)$, $\delta_j^h u(x) \to \partial u/\partial x_j(x)$ as $h \to 0$, but for studying weak derivatives, the following two results are often useful. Recall that we are using $\|\cdot\|_0$ for the L^2-norm.

Proposition 1. *If $u \in H^1(\Omega)$, then $\delta_j^h u \in L^2(\Omega')$ whenever $\Omega' \subset\subset \Omega$ and $0 < |h| < \text{dist}(\Omega', \partial\Omega)$; moreover, we have $\|\delta_j^h u\|_{0;\Omega'} \le \|\partial u/\partial x_j\|_{0;\Omega}$.*

Proof. By density, it suffices to consider $u \in C^1(\Omega)$ with $\|u\|_1 < \infty$. By the fundamental theorem of calculus,

$$\delta_j^h u(x) = \frac{1}{h} \int_0^h \frac{\partial u}{\partial x_j}(x_1, \ldots, x_j + \xi, \ldots, x_n) \, d\xi.$$

By Cauchy-Schwarz, $\int_0^h |\partial u/\partial x_j| d\xi \le (\int_0^h |\partial u/\partial x_j|^2 d\xi)^{1/2} h^{1/2}$, and so

$$|\delta_j^h u(x)|^2 \le \frac{1}{h} \int_0^h \left|\frac{\partial u}{\partial x_j}\right|^2 d\xi.$$

Integrating over $x \in \Omega'$, we let $B_h(\Omega') = \{x \in \Omega : \text{dist}(x, \Omega') < h\}$ to obtain

$$\|\delta_j^h u\|_{0;\Omega'}^2 \leq \frac{1}{h} \int_0^h \int_{B_h(\Omega')} \left|\frac{\partial u}{\partial x_j}\right|^2 dx\, d\xi \leq \int_\Omega \left|\frac{\partial u}{\partial x_j}\right|^2 dx. \spadesuit$$

The next result provides us with a means of determining when a weak derivative is in fact in $L^2(\Omega)$.

Proposition 2. *Suppose that $u \in L^2(\Omega)$ and there is a constant K such that $\|\delta_j^h u\|_{0;\Omega'} \leq K$ whenever $\Omega' \subset\subset \Omega$ satisfies $0 < |h| < \text{dist}(\Omega', \partial\Omega)$. Then the weak derivative $\partial u/\partial x_j \in L^2(\Omega)$ and satisfies $\|\partial u/\partial x_j\|_{0;\Omega} \leq K$.*

Proof. Let $\Omega_h = \{x \in \Omega : \text{dist}(x, \partial\Omega) > 2h\}$, so that $\|\delta_j^h u\|_{0;\Omega_h}$ is uniformly bounded as $h \to 0$. Since a bounded sequence in a Hilbert space contains a weakly convergent subsequence (cf. Section 6.3), there is a sequence $h_k \to 0$ and a function $v \in L^2(\Omega)$, such that $\|v\|_0 \leq K$ and

$$(35a) \qquad \int_\Omega \phi\, \delta_j^{h_k} u\, dx \to \int_\Omega \phi v\, dx \quad \text{as } k \to \infty,$$

whenever $\phi \in C_0^1(\Omega)$. But for $0 < |h| < \text{dist}(x, \partial\Omega)$, we have

$$(35b) \qquad \int_\Omega \phi\, \delta_j^{h_k} u\, dx = -\int_\Omega (\delta_j^{h_k} \phi)\, u\, dx \to -\int_\Omega \frac{\partial \phi}{\partial x_j} u\, dx.$$

Notice that (35a) and (35b) imply that $v = \partial u/\partial x_j$ as weak derivatives. Since $v \in L^2(\Omega)$ and $\|v\|_0 \leq K$, we are done. $\spadesuit$

c. Interior Regularity of Weak Solutions

We would like to use (33) to study the regularity of weak solutions of (29a,b,c). Notice that a weak solution $u \in H^1(\Omega)$ satisfies

$$(36a) \qquad \int_\Omega \sum_{i,j=1}^n a_{ij} \frac{\partial u}{\partial x_i} \frac{\partial v}{\partial x_j}\, dx = \int_\Omega \tilde{f} v\, dx$$

for all $v \in C_0^1(\Omega)$, where $\tilde{f} \in L^2(\Omega)$ is given by

$$(36b) \qquad \tilde{f} = \sum_{i=1}^n \left(b_i - \sum_{j=1}^n \partial_j a_{ij}\right) \frac{\partial u}{\partial x_i} + cu - f.$$

If we let $u = v \in C_0^1(\Omega)$ in (36a) and use ellipticity, we obtain

$$(37) \qquad \int_\Omega |\nabla u|^2\, dx \leq \lambda^{-1} \int_\Omega \sum_{i,j}^n a_{ij} \frac{\partial u}{\partial x_i} \frac{\partial u}{\partial x_j}\, dx = \lambda^{-1} \int_\Omega \tilde{f} u\, dx.$$

From (36b) and the boundedness of $b_i, \partial_j a_{ij}$, and c, we find

$$\left| \int_\Omega \tilde{f} u \, dx \right| \leq C(\|\nabla u\|_0 \|u\|_0 + \|u\|_0^2 + \|f\|_{-1} \|u\|_1),$$

where we have used the Cauchy-Schwarz inequality and the dual space norm $\|\cdot\|_{-1}$ for the linear functionals on $H_0^1(\Omega)$. Combined with (37) and the Poincaré inequality $\|u\|_0 \leq C\|\nabla u\|_0$, we obtain

(38) $$\|u\|_1 \leq C(\|f\|_{-1} + \|u\|_0).$$

By completion, (38) holds whenever $u \in H_0^1(\Omega)$ is a weak solution of $Lu = f$ in Ω.

Using (38) and the higher-order interior estimates (33), we are now able to establish the interior regularity of weak solutions.

Theorem 3. *Suppose L is a uniformly elliptic second-order operator on a bounded domain Ω, with coefficients in $C^{k+1}(\overline{\Omega})$, and $u \in H^1(\Omega)$ is a weak solution of $Lu = f$. Then $f \in H^k(\Omega)$ implies $u \in H^{k+2}(\Omega')$ for every $\Omega' \subset\subset \Omega$.*

Proof. Let δ^h denote the difference quotient operator for some direction x_j, let u_h denote the translation of u by h in the x_j direction [i.e. $u_h(x) = u(x + he_j)$], and let $\delta^h L$ denote the differential operator obtained by taking the difference quotient of each coefficient function. A straightforward calculation (cf. Exercise 6) shows that

(39) $$\delta^h(Lu) = (\delta^h L)(u_h) + L(\delta^h u) \qquad \text{on } \Omega',$$

provided that $0 < |h| < \text{dist}(\Omega', \partial\Omega)$.

It suffices to show that $u \in H^{\ell+1}(\Omega')$ whenever $u \in H^\ell(\Omega)$, for any $1 \leq \ell \leq k+1$. First let us take $\ell = 1$; that is, we want to show that a weak solution $u \in H^1(\Omega)$ is at least a strong solution $u \in H^2(\Omega')$ on every $\Omega' \subset\subset \Omega$. But $\phi \in C_0^\infty(\Omega)$ with $\phi \equiv 1$ on Ω' implies $\phi u \in H_0^1(\Omega)$, so we may apply (38) to ϕu. Since $\|u\|_{1;\Omega'} \leq \|\phi u\|_{1;\Omega}$, we obtain

(40) $$\|u\|_{1;\Omega'} \leq C(\|Lu\|_{-1;\Omega} + \|u\|_\Omega),$$

which is just (33) with $k = -1$.

Take $0 < 2|h| < \text{dist}(\Omega', \partial\Omega)$, and apply (40) to $\delta^h u \in H^1(\Omega')$:

(41) $$\|\delta^h u\|_{1;\Omega'} \leq C(\|L(\delta^h u)\|_{-1;\Omega_h} + \|\delta^h u\|_{\Omega_h}),$$

where $\Omega_h = \{x \in \Omega : \text{dist}(x, \partial\Omega) > |h|\}$. Now we can use (39) in (41) to obtain

(42) $$\|\delta^h u\|_{1;\Omega'} \leq C(\|\delta^h(Lu)\|_{-1;\Omega_h} + \|(\delta^h L)(u_h)\|_{-1;\Omega_h} + \|\delta^h u\|_{\Omega_h}).$$

We claim that the right-hand side of (42) is uniformly bounded as $|h| \to 0$. In fact, $\|\delta^h(Lu)\|_{-1;\Omega_h}$ is bounded uniformly in h, since $Lu = f \in L^2(\Omega)$ implies $\partial f/\partial x_j \in H^{-1}(\Omega) = (H_0^1(\Omega))^*$. Similarly, $\delta^h L$ is a second-order operator with coefficients bounded uniformly in $h > 0$ (since the coefficients of L are at least C^1), so $\|(\delta^h L)(u_h)\|_{-1;\Omega_h} \leq C\|u_h\|_{1;\Omega_h} = C\|u\|_{1;\Omega_h}$ is uniformly bounded. Finally, $\|\delta^h u\|_{0;\Omega_h}$ is uniformly bounded by Proposition 1 since $u \in H^1(\Omega)$. Therefore (42) shows that $\|\delta^h u\|_{1;\Omega'}$ is bounded uniformly as $|h| \to 0$; a natural generalization of Proposition 2 (see Exercise 5) shows that $u \in H^2(\Omega')$.

The proof for $2 \leq \ell \leq k+1$ follows by induction; the details are left for Exercise 7. ♠

d. Global Estimates and Regularity

If $u \in H^1(\Omega)$ is a weak solution of $Lu = f \in L^2(\Omega)$, we may ask whether it is a strong solution on all of Ω; that is, whether $u \in H^2(\Omega)$. More generally, we might hope that $f \in H^k(\Omega)$ implies $u \in H^{k+2}(\Omega)$. These global regularity results would follow if the a priori inequality (33) were to hold with Ω' replaced by Ω. However, this can only be done when the boundary of Ω is sufficiently regular, *and* we add a boundary term to the right-hand side of the inequality.

The regularity of $\partial\Omega$ is expressed by the regularity of the homeomorphism $\psi : B_r(\xi) \to B_1(0)$, which straightens out the boundary near $\xi \in \partial\Omega$: $\psi(\xi) = 0$, $\psi(B_r(\xi) \cap \partial\Omega) \subset \{x \in \mathbf{R}^n : x_n = 0\}$, and $\psi(B_r(\xi) \cap \Omega) \subset \mathbf{R}_+^n$. The boundary $\partial\Omega$ is C^k if, at each $\xi \in \partial\Omega$, the homeomorphism ψ is C^k.

The boundary term to be introduced in the a priori inequality may be defined as follows. For $u \in H^k(\Omega)$ let us define

(43) $\quad \|u\|_{k;\partial\Omega} = \inf\{\,\|\phi\|_{k;\Omega} : \phi \in H^k(\Omega)$ and $u - \phi \in H_0^1(\Omega)\,\}$.

In particular, if $u \in H_0^1(\Omega)$, then we may take $\phi \equiv 0$, so $\|u\|_{k;\partial\Omega} = 0$.

We may now state and prove our global estimates for a general operator L, as in (16), that is uniformly elliptic on Ω and has coefficients extending smoothly to a neighborhood of Ω.

Theorem 4. *Suppose Ω is a bounded domain with C^{k+2}-boundary and L is a uniformly elliptic second-order operator with coefficients in $C^{k+1}(\overline{\Omega})$. Then*

(44) $\quad \|u\|_{k+2;\Omega} \leq C\,(\,\|Lu\|_{k;\Omega} + \|u\|_{0;\Omega} + \|u\|_{k+2;\partial\Omega}\,),$

for all $u \in H^{k+2}(\Omega)$. Moreover, if $u \in H^1(\Omega)$ is a weak solution of $Lu = f$ where $f \in H^k(\Omega)$ and $u - \phi \in H_0^{1,2}(\Omega)$ for some $\phi \in H^{k+2}(\Omega)$, then $u \in H^{k+2}(\Omega)$.

Sketch of Proof. Let us consider only the case $k = 0$. Moreover, replacing u by $u - \phi$, we may assume that $u \in H_0^1(\Omega)$.

Section 8.2: Estimates and Regularity on Domains

Now, cover the compact set $\overline{\Omega}$ by open sets $U_0, \ldots, U_N$ where $\overline{U_0} \subset \Omega$ and $U_1, \ldots, U_N$ are balls of the form $B_\epsilon(\xi)$ with $\xi \in \partial\Omega$ that cover $\partial\Omega$. Next, find a partition of unity $\phi_0, \ldots, \phi_N$ subordinate to this cover [i.e., $\phi_0(x) + \cdots + \phi_N(x) = 1$ for $x \in \Omega$ and $\phi_j \in C_0^\infty(U_j)$ for $j = 0, \ldots, N$]. Using $\|u\| \leq \|\phi_0 u\| + \cdots + \|\phi_N u\|$, it suffices to consider each $\phi_j u$. For $j = 0$ we may use (33). For $j \neq 0$, let $U = U_j$, $\phi = \phi_j$, and use the regularity of $\partial\Omega$ to transform the equation $Lu = f$ in $U^+ = \Omega \cap U$ to an equation $\tilde{L}v = g$ in $V^+ = V = \mathbf{R}_+^n$, where $V = B_1(0)$ and $\tilde{L}$ is the transformed operator that continues to satisfy the hypotheses on L.

In this way, we are led to consider $v \in H_0^1(V^+)$, which vanishes in a neighborhood of ∂V and satisfies $\tilde{L}v = g$ weakly in V^+, with $g \in L^2(V^+)$; see Figure 1.

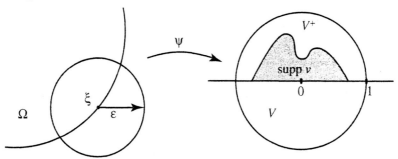

Figure 1. Transformation near the boundary.

We would like to show that $v \in H^2(V^+)$. By (38), we have

(45) $$\|v\|_{1;V^+} \leq C(\|g\|_{-1;V^+} + \|v\|_{0;V^+}).$$

Since the difference quotients $\delta_j^h v$ for $j = 1, \ldots, n-1$ are well defined for h sufficiently small, we may use (45) exactly as in the proof of Theorem 3 to show that all second-order derivatives of v, *except $\partial^2 v / \partial x_n^2$, are in $L^2(\Omega)$*. But we may then use the equation $\tilde{L}v = g$ to solve for $\partial^2 v / \partial x_n^2$ in terms of the other derivatives. This shows that $v \in H^2(V^+)$, and $\|v\|_{2;V^+} \leq C(\|g\|_{0;V^+} + \|v\|_{1;V^+})$. Using (45) we can lower the subscript in the last term from 1 to 0, as required in (44).

Transforming back to Ω, we obtain (44), and the required regularity of a weak solution u follows by using difference quotients. ♠

Corollary A. *Suppose L has C^∞-coefficients and is uniformly elliptic in a smooth, bounded domain $\Omega \subset \mathbf{R}^n$. If $f \in C^\infty(\overline{\Omega})$ and $u \in H^1(\Omega)$ is a weak solution of $Lu = f$, then $u \in C^\infty(\overline{\Omega})$.*

Corollary B. *Suppose L has C^∞-coefficients and is uniformly elliptic in a smooth, bounded domain $\Omega \subset \mathbf{R}^n$. If $u \in H^1(\Omega)$ is a weak solution of $Lu + \lambda u = 0$, then $u \in C^\infty(\overline{\Omega})$.*

Let us record some global estimates in the L^p-Sobolev and the Hölder norms (with $k = 0$ for simplicity); for a proof, see [Gilbarg-Trudinger].

The $C^{2,\alpha}$-regularity of the boundary and the boundary term $|u|_{2,\alpha;\partial\Omega}$ are defined in the obvious way.

Theorem 5. *Suppose Ω is a bounded domain and L is a uniformly elliptic second-order operator with coefficients in $C(\overline{\Omega})$.*
(i) *If $1 < p < \infty$ and $\partial\Omega \in C^2$, then*

(46a) $\quad \|u\|_{2,p;\Omega} \leq C(\|Lu\|_{p;\Omega} + \|u\|_{p;\Omega} + \|u\|_{2,p;\partial\Omega}) \quad$ *for all $u \in H^{2,p}(\Omega)$.*

(ii) *If $0 < \alpha < 1$, $\partial\Omega \in C^{2,\alpha}$, and L has coefficients in $C^\alpha(\overline{\Omega})$, then*

(46b) $\quad |u|_{2,\alpha;\overline{\Omega}} \leq C\left(|Lu|_{\alpha;\overline{\Omega}} + |u|_{\alpha;\overline{\Omega}} + |u|_{2,\alpha;\partial\Omega}\right) \quad$ *for all $u \in C^{2,\alpha}(\overline{\Omega})$.*

As in Theorems 3 and 4, the estimates may be used to obtain regularity results. An example is the following; see also [Gilbarg-Trudinger].

Corollary. *If $\Omega \subset \mathbf{R}^n$ has a $C^{2,\alpha}$-boundary, $f \in C^\alpha(\overline{\Omega})$, and $u \in C^2(\Omega) \cap C(\overline{\Omega})$ satisfies $Lu = f$ in Ω and $u = 0$ on $\partial\Omega$, then $u \in C^{2,\alpha}(\overline{\Omega})$.*

It is important to emphasize that the theorems in this section do not assert the *existence* of a solution to boundary value problems for $Lu = f$ or $Lu + \lambda u = 0$. Of course, when $L = \Delta$, then the existence of a weak solution to the Dirichlet problem was studied in Sections 6.2 and 7.2, respectively. In the next two sections we study the issues of existence and uniqueness for solutions for the more general operator L.

Exercises for Section 8.2

1. If $\Omega_1 \subset\subset \Omega$, prove that $\|\nabla u\|_{0;\Omega_1}^2 \leq C\left(\|\Delta u\|_{0;\Omega}^2 + \|u\|_{0;\Omega}^2\right)$ for all $u \in H^2(\Omega)$.

2. (a) Prove Corollary A of Theorem 1. [Note that we have *not* assumed $f \in L^2(\Omega)$.]
   (b) Prove Corollary B of Theorem 1.

3. Use (18b) to prove (33).

4. Use the corollary to Theorem 5 to prove the following: If $\Omega \subset \mathbf{R}^n$ has a $C^{2,\alpha}$ boundary, $u \in C^2(\Omega) \cap C(\overline{\Omega})$ satisfies $\Delta u = f$ in Ω, where $f \in C^\alpha(\overline{\Omega})$, and $u = g$ on $\partial\Omega$, where $g \in C^{2,\alpha}(\overline{\Omega})$, then $u \in C^{2,\alpha}(\overline{\Omega})$.

5. Generalize Proposition 2 to show that $u \in H^k(\Omega)$ and $\|\delta_j^h u\|_{k;\Omega'} \leq K$, whenever $\Omega' \subset\subset \Omega$ satisfies $0 < |h| < \text{dist}(\Omega', \partial\Omega)$, together imply $\partial u/\partial x_j \in H^k(\Omega)$.

6. Prove (39).

7. Complete the proof of Theorem 3 by considering $2 \leq \ell \leq k+1$.

8. Show by example that the boundary term (43) is necessary in (44); for example, there is no estimate

$$\|u\|_{2;\Omega}^2 \leq C\left(\|\Delta u\|_{0;\Omega}^2 + \|u\|_{1;\Omega}^2\right) \quad \text{valid for all } u \in H^2(\Omega).$$

8.3 Maximum Principles

As we saw in Section 4.1, the Laplace equation satisfies a maximum principle that may be used to show the uniqueness of solutions to boundary value problems. In this section we discuss maximum principles for more general elliptic operators and investigate some of their consequences. Throughout this section, we assume that $\Omega \subset \mathbf{R}^n$ is a bounded domain (recall from the Introduction that this requires Ω to be *connected*).

a. The Weak Elliptic Maximum Principle

Let us consider a function $u \in C^2(\Omega) \cap C(\overline{\Omega})$ that satisfies $\Delta u \geq 0$ in Ω. Recall from Section 4.1 that u must achieve its maximum value on $\overline{\Omega}$ at a point on $\partial\Omega$. We would like to generalize this property by replacing Δ by a general elliptic operator L as in Section 8.2; that is, we consider solutions of the inequality

$$Lu \geq 0, \tag{47}$$

where L is the uniformly elliptic operator (29b). Our first result is the following.

Theorem 1 (Weak Elliptic Maximum Principle). *If $u \in C^2(\Omega) \cap C(\overline{\Omega})$ satisfies (47) with $c(x) \equiv 0$ for $x \in \Omega$, then*

$$\max_{x \in \overline{\Omega}} u(x) \leq \max_{x \in \partial\Omega} u(x). \tag{48}$$

Proof. Let us first establish (48) under the assumption $Lu > 0$ in Ω. If (48) fails, then u achieves its maximum at some $\hat{x} \in \Omega$. At such a local maximum, $\partial u(\hat{x})/\partial x_i = 0$ for $i = 1, \ldots, n$, and the matrix $\partial^2 u(\hat{x})/\partial x_i \partial x_j$ is nonpositive. But this implies $Lu(\hat{x}) \leq 0$ (cf. Exercise 1), contradicting $Lu > 0$ in Ω.

Now consider the general case $Lu \geq 0$ in Ω, and introduce the auxiliary function $u_\eta = u + \eta \exp(Ax_1)$, where $A, \eta > 0$, and A is chosen so large that $A^2 a_{11} + A b_1 > 0$ in Ω. A direct calculation shows that

$$Lu_\eta = Lu + \eta(A^2 a_{11} + A b_1) \exp(Ax_1) > 0 \quad \text{in } \Omega,$$

so u_η satisfies (48). Letting $\eta \to 0$, we see that (48) must hold for u as well.
♠

Remark. It is interesting to note that the proof of Theorem 1 does not actually require the uniform ellipticity (29b) with $\lambda > 0$, but only (29b) with $\lambda = 0$ *and* the positivity $a_{ii}(x) \geq \eta > 0$ for a single direction x_i. This will be significant when deriving a weak maximum principle for parabolic equations in Section 11.1.

If we replace the condition $c(x) \equiv 0$ by $c(x) \leq 0$ in Theorem 1, then we can still obtain conclusions about the maximum of u. To do so, introduce the operator $L_0 = \sum a_{ij} \partial^2/\partial x_i \partial x_j + \sum b_i \partial/\partial x_i$ and apply the theorem to the inequality $L_0 u \geq -cu \geq 0$, which is satisfied in the domain $\Omega^+ = \{x \in \Omega : u(x) > 0\}$. Since $u = 0$ on $\partial\Omega^+ \cap \Omega$, we obtain the following.

Corollary. If $u \in C^2(\Omega) \cap C(\overline{\Omega})$ satisfies (47) with $c(x) \leq 0$ for $x \in \Omega$, then u satisfies

$$(49) \qquad \max_{x \in \overline{\Omega}} u(x) \leq \max_{x \in \partial\Omega} u^+(x), \qquad \text{where } u^+(x) = \max(u(x), 0).$$

These results have immediate application to uniqueness theorems for the Dirichlet problem (see Exercise 2), but they do not rule out the possibility that u satisfying (47) may have maximum points inside Ω as well as on $\partial\Omega$; this is why Theorem 1 is called the weak maximum principle. We soon encounter a strong maximum principle, but first let us apply the weak maximum principle to obtain bounds for the solutions of the nonhomogeneous equation $Lu = f$.

b. Application to a Priori Estimates

The weak maximum principle may be used to obtain pointwise a priori estimates for solutions of the nonhomogeneous equation

$$(50a) \qquad \begin{cases} Lu = f & \text{in } \Omega \\ u = g & \text{on } \partial\Omega, \end{cases}$$

where L is the uniformly elliptic operator (29b). For simplicity, we shall consider a special case of (50a), namely

$$(50b) \qquad \begin{cases} \Delta u + c(x)u = f & \text{in } \Omega \\ u = g & \text{on } \partial\Omega, \end{cases}$$

where $c(x) \leq 0$ is bounded in Ω, $f \in C(\overline{\Omega})$, and $g \in C(\partial\Omega)$. [The more general case (50a) is discussed in Exercise 3.] Consider a solution $u \in C^2(\Omega) \cap C(\overline{\Omega})$ of (50b). If $f(x) \geq 0$ in Ω, then the maximum principle shows that $u(x)$ attains a nonnegative maximum on $\partial\Omega$:

$$f(x) \geq 0 \quad \Rightarrow \quad \max_{x \in \overline{\Omega}} u(x) \leq \max_{x \in \partial\Omega} \{g(x), 0\}.$$

Similarly, if $f(x) \leq 0$, then we may apply the maximum principle to $-u$:

$$f(x) \leq 0 \quad \Rightarrow \quad \max_{x \in \overline{\Omega}} -u(x) \leq \max_{x \in \partial\Omega} \{-g(x), 0\}.$$

Combining these, we obtain

$$f(x) \equiv 0 \quad \Rightarrow \quad \max_{x \in \overline{\Omega}} |u(x)| \leq \max_{x \in \partial\Omega} |g(x)|.$$

This is our a priori bound when $f \equiv 0$.

More generally, if $f(x) \not\equiv 0$, we proceed by introducing a function $w(x)$ satisfying

(51)
$$\begin{cases} \Delta w + c(x)w \leq -M & \text{in } \Omega \\ w \geq N & \text{on } \partial\Omega, \end{cases}$$

where $M = \max_{x \in \overline{\Omega}} |f(x)|$ and $N = \max_{x \in \partial\Omega} |g(x)|$; we explain in a moment how to construct $w(x)$. Now let us define $v(x)$ by $v = u - w$, which satisfies

$$\begin{cases} (\Delta + c(x))v = f - (\Delta + c(x))w \geq f + M \geq 0 & \text{in } \Omega \\ v \leq g - N \leq 0 & \text{on } \partial\Omega. \end{cases}$$

Then (49) implies that $v \leq 0$ in Ω; that is, $u \leq w$ in Ω. Replacing $u - w$ by $u + w$ shows $-u \leq w$ in Ω. Therefore, we obtain

(52)
$$\max_{x \in \overline{\Omega}} |u(x)| \leq \max_{x \in \overline{\Omega}} w(x).$$

This gives us an a priori bound on our solution u, provided that we can construct w explicitly.

To construct w, let us assume that Ω lies in the strip $0 \leq x_1 \leq d$ in $\mathbf{R}^n$. Let
$$w(x) = w(x_1, \ldots, x_n) = M(e^d - e^{x_1}) + N,$$
where the constants M and N are nonnegative. Then

$$\Delta w + c(x)w = -Me^{x_1} + c(x)M(e^d - e^{x_1}) + c(x)N \leq -M \quad \text{in } \Omega,$$

and $w(x) \geq N$ in Ω so (51) is satisfied. Not only have we constructed $w(x)$, but we may formulate the a priori bound (52) as follows.

Theorem 2. If $u \in C^2(\Omega) \cap C(\overline{\Omega})$ satisifes (50b), where $c(x) \leq 0$, $f \in C(\overline{\Omega})$, and $g \in C(\partial\Omega)$, then

(53)
$$\max_{x \in \overline{\Omega}} |u(x)| \leq \max_{x \in \partial\Omega} |g(x)| + C \max_{x \in \overline{\Omega}} |f(x)|,$$

where the constant C depends only on the domain Ω.

One immediate consequence of (53) is the uniqueness of solutions of (50b), although this could also be obtained directly from Theorem 1 (see Exercise 2). We shall also find (53), as well as the a priori L^2-estimates of Section 8.2, useful when studying nonlinear elliptic equations in Chapter 13. Another tool for nonlinear elliptic equations is foreshadowed in the proof of Theorem 2, which involved the construction of a function w satisfying $Lw = (\Delta + c(x))w \geq 0$. Such a function is called a *subsolution* (or *lower function*); these play an analogous role to that of the subharmonic functions for the Laplacian (see Section 4.3).

c. The Strong Elliptic Maximum Principle

We now add the assumption that the bounded domain Ω has a C^2-boundary, $\partial\Omega$. This C^2 assumption implies that for each point $x_0 \in \partial\Omega$, we can find a small ball $B_\epsilon \subset \Omega$ such that $\partial\Omega \cap \overline{B_\epsilon} = \{x_0\}$. (Similarly, we could find an exterior ball $B'_\epsilon \subset \mathbf{R}^n \setminus \overline{\Omega}$ such that $\partial\Omega \cap \overline{B'_\epsilon} = \{x_0\}$; cf. Exercise 3 in Section 4.3.) This condition is required for the validity of statments in this section regarding the positivity of the normal derivative at x_0. Without assuming that $\partial\Omega$ is C^2, the statements regarding interior maximum points are still valid, and the statements regarding the normal derivative are also valid if we require the interior ball condition at x_0 (see Exercise 6).

Recall from Section 4.1 that if a function $u \in C^2(\Omega)$ satisfies $\Delta u \geq 0$ in Ω and achieves its supremum at $x_0 \in \Omega$, then u is a constant; moreover, if u is not constant and achieves its supremum at $x_0 \in \partial\Omega$, then

$$(54) \qquad \frac{\partial u}{\partial \nu}(x_0) > 0,$$

where ν is the unit exterior normal vector on $\partial\Omega$ (cf. Exercise 8 in Section 4.1). Therefore, u satisfies the following.

Strong Elliptic Maximum Principle. *Suppose* $M = \sup\{u(x) : x \in \Omega\} < \infty$.
(a) *If* $u(x_0) = M$ *for some* $x_0 \in \Omega$, *then* u *is constant in* Ω $[u(x) \equiv M$ *for all* $x \in \Omega]$.
(b) *If* $u(x)$ *is not constant in* Ω, *but* $u(x_0) = M$ *for some* $x_0 \in \partial\Omega$ *and* $\partial u/\partial \nu$ *exists at* x_0, *then (54) holds.*

This result is sometimes referred to as the *Hopf maximum principle*, and part (b) as the *boundary point lemma*.

We have stated the strong elliptic maximum principle as a set of properties of u; we must now investigate conditions under which we may conclude that the strong elliptic maximum principle holds for all functions $u \in C^2(\Omega)$ satisfying (47).

Theorem 3. *If the bounded domain Ω has C^2-boundary, L is the uniformly elliptic operator of (29b), and $u \in C^2(\Omega)$ satisfies $Lu \geq 0$, then the strong elliptic maximum principle holds in the following cases:*
  (i) $c(x) \equiv 0$;
  (ii) $c(x) \leq 0$ *and* $M \geq 0$;
  (iii) $c(x)$ *arbitrary and* $M = 0$.

Proof. We consider case (ii); see Exercise 4 for cases (i) and (iii). Let $\Omega^- = \{x \in \Omega : u(x) < M\}$. If $\Omega^- = \emptyset$, we have $u(x) \equiv M$ and the maximum principle certainly holds; otherwise let us suppose the following:

(55) $\begin{cases} \text{there exists } x_0 \in \partial\Omega^- \text{ with } u(x_0) = M \text{ and} \\ \partial\Omega^- \text{ satisfies the interior ball condition at } x_0. \end{cases}$

We now show that (54) holds (where ν is the unit exterior normal vector on $\partial\Omega^-$).

Let us suppose that the ball B_ϵ is centered at 0 (for convenience) and introduce the auxiliary function $v(x) = v(r) = \exp(-Ar^2) - \exp(-A\epsilon^2)$ for $r = |x| < \epsilon$; notice that $v > 0$ for $r = |x| < \epsilon$, but $v(\epsilon) = 0$ and $v'(\epsilon) < 0$.

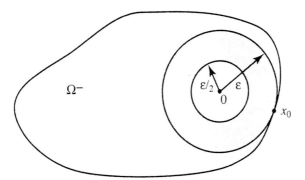

Figure 1. Construction of a barrier.

We may compute

(56)
$$Lv = e^{-Ar^2}\left(4A^2 \sum_{i,j=1}^n a_{ij}x_ix_j - 2A\sum_{i=1}^n (a_{ii} + b_ix_i)\right) + cv$$
$$\geq e^{-Ar^2}\left(4A^2\lambda r^2 - 2A\sum_{i=1}^n (a_{ii} + b_ix_i) + c\right),$$

where we have used (29c) and $c(x)\exp(-A\epsilon^2) \leq 0$. Taking A sufficiently large, we can arrange to have $Lv > 0$ in the annular region $U = B_\epsilon \setminus B_{\epsilon/2}$ (see Figure 1). Since $u(x) - M < 0$ for $x \in \partial B_{\epsilon/2} \subset \Omega^-$, there is a small $\eta > 0$ so that $u(x) - M + \eta v < 0$ for $x \in \partial B_{\epsilon/2}$. But we also have $u(x) - M + \eta v \leq 0$ for $x \in \partial B_\epsilon$ (since $v = 0$). This means that $L(u - M + \eta v) \geq -cM \geq 0$ in U, and $u - M + \eta v \leq 0$ on ∂U; using (49), we conclude that $u - M + \eta v \leq 0$ in U. But the function $w = u - M + \eta v$ being nonpositive in U and vanishing at $x_0 \in \partial U$ implies $\partial w/\partial \nu \geq 0$ at x_0. Therefore, we obtain (54):

$$\frac{\partial u}{\partial \nu}(x_0) \geq -\eta \frac{\partial v}{\partial \nu}(x_0) = -\eta v'(\epsilon) > 0.$$

Now let us prove (a) and (b) of the maximum principle. If $\Omega \neq \Omega^-$, then $\Omega \cap \partial\Omega^- \neq \emptyset$, and u being C^2 enables us to choose $x_0 \in \Omega \cap \partial\Omega^-$

satisfying (55). But x_0 is then an interior maximum point for u, which contradicts the conclusion (54). This proves (a) and shows that we must have $\Omega = \Omega^-$. To prove (b), we assume that $u(x_0) = M$ for some $x_0 \in \partial\Omega$ and use $\partial\Omega \in C^2$ to satisfy (55); we therefore obtain (54). ♠

Remark. The function v introduced in the proof of Theorem 3 is called a *(local) barrier*. The use of barriers is an important method to gain a priori control on the solutions of partial differential equations and inequalities. We already encountered barriers for Laplace's equation in Section 4.3; we encounter local barriers for linear parabolic equations in Section 11.1, and for nonlinear elliptic equations in Chapter 13.

The strong elliptic maximum principle has many applications to linear and nonlinear elliptic equations. In the next subsection we explore one of these, namely the simplicity of the first eigenvalue and the positivity of the associated eigenfunction for the Laplacian.

d. Application to the Principal Eigenvalue

Recall that in Section 7.2 we investigated eigenvalues and eigenfunctions for the Laplacian; that is, values λ for which

$$(57) \quad \begin{cases} \Delta u + \lambda u = 0 & \text{in } \Omega \\ u = 0 & \text{on } \partial\Omega \end{cases}$$

admits a nontrivial solution u. Using variational methods, we found that there is a sequence of eigenvalues $0 < \lambda_1 \leq \lambda_2 \leq \ldots$ and associated nontrivial eigenfunctions $u_1, u_2, \ldots$ that are weak solutions of (57). In fact, u_1 was found by minimizing $F(u) = \int_\Omega |\nabla u|^2 dx$ over $u \in H_0^{1,2}(\Omega)$ satisfying $\|u\|_2^2 = \int_\Omega |u|^2 dx = 1$; moreover, λ_1 is this minimum value: $\lambda_1 = F(u_1)$.

Now suppose that $\partial\Omega$ is C^∞ (actually, $\partial\Omega \in C^{2,\alpha}$ would suffice). This assumption enables us to use the elliptic regularity of Section 8.2 to conclude that each eigenfunction $u_j \in C^2(\overline{\Omega})$. With this regularity, we can apply the strong elliptic maximum principle to prove the following properties of the principal eigenvalue λ_1 and its eigenfunction u_1, which were stated in Section 4.4.

Theorem 4. *If Ω is a smooth, bounded domain, then the principal eigenvalue λ_1 is simple, and its associated eigenfunction u_1 is strictly positive (or strictly negative) in Ω.*

Proof. We first show that u_1 is strictly positive (or strictly negative) in Ω. Assume that $\Omega^+ = \{x \in \Omega : u_1(x) > 0\}$ is nonempty and let $u_1^+ = u_1|_{\Omega^+}$. By the strong maximum principle (applied to $-u_1$), $\partial u_1^+/\partial \nu < 0$ on $\partial\Omega^+ \cap \Omega$. If we let $\Gamma = \{x \in \Omega : u_1(x) = 0\}$, we see that $\Gamma \subset \Omega^+ \cap \Omega$, and so Γ cannot contain any critical points of u_1. Consequently, Γ is an $(n-1)$-dimensional hypersurface; in particular, Γ has measure zero. As a consequence, if we let $\Omega^- = \{x \in \Omega : u_1(x) < 0\}$ and $u_1^- = u_1|_{\Omega^-}$, we find

$$\int_\Omega |\nabla u_1|^2\, dx = \int_{\Omega^+} |\nabla u_1^+|^2\, dx + \int_{\Omega^-} |\nabla u_1^-|^2\, dx.$$

But on Ω^+ we have $|u_1| = u_1$, and so $|\nabla(|u_1|)| = |\nabla u_1|$; similarly, we find $|\nabla(|u_1|)| = |\nabla u_1|$ on Ω^-. This means that $|u_1|$ also minimizes F, so $|u_1|$ is a weak solution of (57) with $\lambda = \lambda_1$. Applying elliptic regularity, we find $|u_1| \in C^2(\overline{\Omega})$. We can therefore apply the strong maximum principle to conclude that $|u_1|$ cannot achieve a nonpositive minimum in Ω; consequently, $|u_1| > 0$ in Ω, as desired.

To show that λ_1 has a one-dimensional eigenspace, suppose otherwise that $\lambda_2 = \lambda_1$; in other words, there exists $u_2 \in H_0^{1,2}(\Omega)$ with $F(u_2) = \lambda_1$ and $\int u_1 u_2\, dx = 0$. But this last condition is impossible since the preceding argument shows that u_2 also has the property $|u_2| > 0$. ♠

Exercises for Section 8.3

1. Recall from linear algebra that if A and B are symmetric $n \times n$ matrices and $A, B \geq 0$, then $C = AB \geq 0$; in particular, the *trace* of $C = (c_{ij})$ is nonnegative: $\mathrm{tr}(C) \equiv c_{11} + c_{22} + \cdots c_{nn} \geq 0$. Use this to show that $Lu(\hat{x}) \leq 0$ at an interior maximum point $\hat{x}$ if $u \in C^2$ and $c(\hat{x}) = 0$.

2. If L is the uniformly elliptic operator (29b) with $c(x) \leq 0$, and Ω is a bounded domain, use Theorem 1 to prove the following.
   (a) There is at most one solution $u \in C^2(\Omega) \cap C(\overline{\Omega})$ of the Dirichlet problem $Lu = f$ in Ω, $u = g$ on $\partial\Omega$.
   (b) If u and v are C^2-functions satisfying $Lu \geq Lv$ in Ω, and $u \leq v$ on $\partial\Omega$, prove that $u \leq v$ in Ω. (This is an example of a *comparison theorem* for an elliptic equation; cf. Section 11.1c for comparison theorems for nonlinear diffusion equations.)
   (c) Show that conclusion (b) need not be true if Ω is unbounded.

3. Prove that a solution $u \in C^2(\Omega) \cap C(\overline{\Omega})$ of (50a) satisfies

   $$\max_{x \in \overline{\Omega}} |u(x)| \leq \max_{x \in \partial\Omega} |g(x)| + C \max_{x \in \overline{\Omega}} |f(x)|,$$

   where the constant C depends on Ω, λ [in (29c)], and $\sup_{x \in \Omega} |b_i(x)|$.

4. Prove that (ii) implies both (i) and (iii) in Theorem 3. That is, prove that both conclusions (a) and (b) of the strong elliptic maximum principle hold, under either of the assumptions (i) or (iii).

5. Let Ω be a smooth, bounded domain, and let $u \in C^2(\Omega) \cap C^1(\overline{\Omega})$ be a solution of $Lu = 0$ in Ω with $\frac{\partial u}{\partial \nu} = 0$ on $\partial\Omega$, where L is the uniformly elliptic operator (29b) with $c(x) \leq 0$.
   (a) Conclude that u is a constant in Ω.
   (b) If $c(x_0) < 0$ for some $x_0 \in \Omega$, conclude that $u \equiv 0$ in Ω.
   (c) Conclude appropriate uniqueness theorems for the Neumann problem associated with the nonhomogeneous equation $Lu = f$.

6. Suppose $\Omega \subset \mathbf{R}^2$ is a bounded domain whose boundary is C^2 *except* at one point, say $(0,0)$, and suppose that $\Omega \cap U = (\mathbf{R}^2 \setminus \overline{Q_1}) \cap U$, where U is a neighborhood of $(0,0)$, and Q_1 is the first quadrant. State and prove appropriate conclusions of a strong elliptic maximum principle for solutions of (47). Discuss the application of this maximum principle to the Neumann problem in Ω.

7. Let Ω be a smooth, bounded domain in $\mathbf{R}^n$.
   (a) Prove that a C^2-solution of $\Delta u = u^2$ in Ω cannot achieve its supremum on Ω unless $u \equiv 0$.
   (b) If $u \in C^2(\Omega) \cap C(\overline{\Omega})$ satisfies $\Delta u = u^3 - u$ in Ω and $u = 0$ on $\partial\Omega$, show that $-1 \leq u(x) \leq 1$ for all $x \in \Omega$. Is it possible to have $u(x_0) = \pm 1$ for $x_0 \in \Omega$?

8. *Generalized weak maximum principle for divergence form operators.* Suppose that $u \in C^1(\Omega) \cap C(\overline{\Omega})$ is a *weak solution* of
$$\sum_{i,j=1}^n \partial_{x_i} a_{ij}(x) \partial_{x_j} u + c(x) u \geq 0 \quad \text{in } \Omega;$$
that is, for any $\phi \in C_0^1(\Omega)$ with $\phi \geq 0$ in Ω,
$$\int_\Omega \left(-\sum a_{ij}(x) \partial_{x_j} u \, \partial_{x_i} \phi + c(x) u \phi \right) dx \geq 0.$$
If the a_{ij} satisfy (29c) and $c(x) \leq 0$ in Ω, then prove the following:
(a) $\max_{\overline{\Omega}} u(x) \leq \max_{\partial\Omega} u(x)$, and (b) $\max_{\overline{\Omega}} |u(x)| \leq \max_{\partial\Omega} |u(x)|$.

8.4 Solvability

In this section we consider the solvability of the Dirichlet problem for the equations $Lu = f$ and $Lu + \lambda u = f$ in a bounded domain Ω, where $f \in L^2(\Omega)$ and L is the uniformly elliptic operator (29b). For convenience of exposition, we assume that *the coefficients of L are smooth (C^∞)*; it will generally be clear how to weaken this hypothesis. As in earlier sections of this chapter, we restrict our attention to L^2-Sobolev spaces, so we omit the index for $p = 2$; we also use $\|\cdot\|_0$ to denote the norm in $H^0(\Omega) = L^2(\Omega)$.

a. Uniqueness and Solvability

When L is in divergence form with $c(x) \leq 0$ (for example, $L = \Delta$), then we know from Section 6.2 that there exists a weak solution $u \in H_0^1(\Omega)$ of $Lu = f \in L^2(\Omega)$ and that the solution is unique. More generally, we say that L [as in (29b)] satisfies *weak uniqueness in Ω* if every weak solution $u \in H_0^{1,2}(\Omega)$ of $Lu = 0$ is identically zero: $u \equiv 0$. In this case, the right-hand side of the a priori estimates discussed in Section 8.2 may be significantly improved. Let us consider the global estimates.

Theorem 1. *Suppose Ω is a bounded domain with C^{k+2}-boundary, and and the operator L of (29b,c) has smooth coefficients on $\overline{\Omega}$ and satisifes weak uniqueness in Ω. Then*

(58) $\qquad \|u\|_{k+2} \leq C\|Lu\|_k \qquad$ for all $u \in H^{k+2}(\Omega) \cap H_0^1(\Omega)$.

Proof. Assume $k = 0$; the more general case is analogous. Appealing to (44), it suffices to show that

(59) $\qquad \|u\|_0 \leq C\|Lu\|_0 \qquad$ for all $u \in H^2(\Omega) \cap H_0^1(\Omega)$.

But if this fails, then there must be a sequence $u_m \in H^2(\Omega) \cap H_0^1(\Omega)$ for which $\|u_m\|_0 = 1$ and $\|Lu_m\|_0 \to 0$. By (44) we have $\|u_m\|_2 \leq C$. Using the weak sequential compactness of bounded sets in $H^2(\Omega)$ (see Section 6.3) together with the compactness of $H^2(\Omega) \to L^2(\Omega)$ (see Section 6.5), there is a subsequence u_{m_j} and $u \in H^2(\Omega)$ such that

(60a) $\qquad u_{m_j} \rightharpoonup u \quad$ weakly in $H^2(\Omega)$,

(60b) $\qquad u_{m_j} \to u \quad$ in $L^2(\Omega)$.

But $\|Lu_{m_j}\|_0 \to 0$ and (60a) together show that u is a weak solution of $Lu = 0$, so $u \equiv 0$ by weak uniqueness; however, this contradicts $\|u\|_0 = \lim \|u_{m_j}\|_0 = 1$, which follows from (60b). Thus (59), and hence (58), must be true. ♠

The inequality (58) enables us to bound the inverse of operators satisfying weak uniqueness. To see this, let $X = H^{k+2}(\Omega) \cap H_0^1(\Omega)$ and $Y = H^k(\Omega)$, which are both Banach spaces under the norms $\|u\|_X = \|u\|_{k+2}$ and $\|u\|_Y = \|u\|_k$. Observe that $\Delta: X \to Y$ is a bounded linear operator. Given $f \in Y$, we know from Section 6.2 that there is a unique weak solution $u \in H_0^1(\Omega)$ of $\Delta u = f$, and Theorem 4 of Section 8.2 shows $u \in X$. Let us write this as $u = \Delta^{-1} f$. The inequality (58) states $\|\Delta^{-1} f\|_X \leq C\|f\|_Y$, so $\Delta^{-1}: Y \to X$ is bounded. Thus $\Delta: X \to Y$ is an isomorphism.

Corollary. *If Ω is a bounded domain with C^{k+2}-boundary, then the map $\Delta: H^{k+2}(\Omega) \cap H_0^1(\Omega) \to H^k(\Omega)$ is an isomorphism.*

Can we do the same for the operator L? Certainly

(61) $\qquad\qquad L: H^{k+2}(\Omega) \cap H_0^1(\Omega) \to H^k(\Omega)$

is continuous. Is the map one to one? Is the map onto? As we see in the next section, there is a connection between these two questions that is the subject of Fredholm theory. In this section, however, we content ourselves with considering a case where (61) being one to one implies that it is also onto. The surjectivity will be established by the *method of continuity*, which continuously deforms L to the operator Δ for which surjectivity is already known.

To insure that L satisfies uniqueness, we assume in (29b) that

(62) $$c(x) \leq 0.$$

Since the elliptic regularity of Section 8.2 shows that a weak solution of $Lu = 0$ is in $C^2(\overline{\Omega})$, we may appeal to the maximum principles of Section 8.3 to conclude that $u \equiv 0$. Now define the continuous one-parameter family of operators

(63) $$L_t = tL + (1-t)\Delta \quad \text{for} \quad 0 \leq t \leq 1$$

so that $L_0 = \Delta$ and $L_1 = L$. Notice that uniqueness holds for each L_t, so we may apply Theorem 1 to conclude

(64) $$\|u\|_X \leq C\|L_t u\|_2 \quad \text{for} \quad 0 \leq t \leq 1 \quad \text{and} \quad u \in X,$$

where we may assume that C is independent of $0 \leq t \leq 1$.

To show that L_t is onto, we must solve $L_t u = f$ for each $f \in L^2(\Omega)$. But we can write this equation as $\Delta u = f + (\Delta - L_t)u = f + t\Delta u - tLu$ or

(65) $$u = \Delta^{-1} f + tu - t\Delta^{-1} Lu \equiv T_t u.$$

In other words, we want a fixed point of the bounded linear map $T_t: X \to X$. But for small t, we find that T_t is a contraction:

$$\|T_t u - T_t v\|_X = \|tu - t\Delta^{-1} Lu - tv + t\Delta^{-1} Lv\|_X$$
$$\leq |t| \|u - v\|_X + |t| \|\Delta^{-1} L\|_{X \to X} \|u - v\|_X$$
$$\leq |t| (1 + \|\Delta^{-1} L\|_{X \to X}) \|u - v\|_X,$$

which shows that T_t is a contraction, provided $\alpha = |t|(1 + \|\Delta^{-1} L\|_{X \to X}) < 1$. We may apply the contraction mapping principle of Section 7.3 to conclude that L_t is onto for $0 \leq t < \delta = (1 + \|\Delta^{-1} L\|_{X \to X})^{-1}$.

We can now try to repeat this procedure. Pick $t_0 \in (0, \delta)$ and write $L_t u = f$ as $L_{t_0} u = f + (L_{t_0} - L_t)u = f + (t - t_0)\Delta u - (t - t_0)Lu$, so

(66) $$u = L_{t_0}^{-1} f + (t - t_0) L_{t_0}^{-1}(\Delta u - Lu) \equiv T_t u.$$

Thus

(67) $$\|T_t u - T_t v\|_X \leq |t - t_0| \, \|L_{t_0}^{-1}(\Delta - L)\|_{X \to X} \, \|u - v\|_X$$

shows that T_t is a contraction, provided $|t - t_0|$ is small. But to use this argument to cover the whole interval $[0, 1]$, we must bound $\|L_{t_0}^{-1}(\Delta - L)\|$ independent of t_0. This can be achieved using (64), namely $\|L_{t_0}^{-1}\|_{L^2 \to X} \leq C$ implies

(68) $$\|T_t u - T_t v\|_X \leq |t - t_0| \, C \, \|\Delta - L\|_{X \to L^2} \, \|u - v\|_X.$$

Using (68) we can extend the surjectivity of L_t step by step to cover $[0, 1]$. Thus $L = L_1: X \to L^2(\Omega)$ is onto, as claimed.

Thus (61) is one to one and onto, so it has an inverse map. But (58) shows that the inverse is bounded, so we have proved the following.

Theorem 2. Suppose Ω is a bounded domain with C^{k+2}-boundary, and the operator L of (29b,c) has smooth coefficients on $\overline{\Omega}$ with $c(x) \leq 0$. Then (61) is an isomorphism. In particular, for every $f \in H^k(\Omega)$ the Dirichlet problem for $Lu = f$ has a unique solution $u \in H^{k+2}(\Omega) \cap H_0^1(\Omega)$.

Remark. If we replace the condition $f \in H^k(\Omega)$ by $f \in C^\alpha(\overline{\Omega})$, where $0 < \alpha < 1$, then replacing (58) by the a priori estimate

(69) $$|u|_{2,\alpha;\overline{\Omega}} \leq C\,|Lu|_{\alpha;\overline{\Omega}} \qquad \text{for } u \in X,$$

where $X = \{u \in C^{2,\alpha}(\overline{\Omega}) : u = 0 \text{ on } \partial\Omega\}$, we could repeat the method of continuity argument to find a unique solution $u \in C^{2,\alpha}(\overline{\Omega})$ of the Dirichlet problem for $Lu = f$.

b. Fredholm Solvability

In this subsection we investigate the solvability of the Dirichlet problem for $Lu = f$ in a bounded domain Ω with C^2-boundary when the operator L of (29b,c) has coefficients in $C^2(\overline{\Omega})$, but we no longer assume weak uniqueness; that is, $c(x)$ is now allowed to take positive values at some points x. This means that the bounded operator

(70) $$L: X \equiv H^2(\Omega) \cap H_0^1(\Omega) \to L^2(\Omega)$$

is not necessarily an isomorphism, so the solvability of the Dirichlet problem for $Lu = f$ depends on whether f is in the image of (70).

As we saw in Section 8.1, the failure of uniqueness for $(\Delta + \lambda)u = f$ on $\mathbf{T}^n$ was associated with a finite number of orthogonality conditions on f in order for existence to hold. This means that $\Delta + \lambda: H^2(\mathbf{T}^n) \to L^2(\mathbf{T}^n)$ is an example of a Fredholm operator: a bounded linear operator $A: X \to Y$ between Hilbert spaces X and Y is *Fredholm* if A has finite-dimensional nullspace N_A and the range $R_A = A(X)$ has finite-dimensional orthogonal complement $R_A^\perp = \{y \in Y : \langle Ax, y \rangle = 0 \text{ for all } x \in X\}$. Furthermore, if we define the *nullity of* A to be $\text{nul}(A) = \dim(N_A)$ and the *deficiency of* A to be $\text{def}(A) = \dim(R_A^\perp)$, then we define the *Fredholm index of* A to be $\text{ind}(A) = \text{nul}(A) - \text{def}(A)$. We find that, not only is $\Delta + \lambda : H^2(\mathbf{T}^n) \to L^2(\mathbf{T}^n)$ a Fredholm operator for each $\lambda \in \mathbf{R}$, but also $\text{ind}(\Delta + \lambda) = 0$.

Is (70) also a Fredholm operator? What is its index? To investigate these questions, let us first suppose $L = \Delta + \lambda$:

(71) $$\Delta + \lambda: X \to L^2(\Omega).$$

For $\lambda = 0$ we know that $\Delta: X \to L^2(\Omega)$ and $\Delta^{-1}: L^2(\Omega) \to X$ are isomorphisms. Thus the composition $T = (\Delta + \lambda)\Delta^{-1} = I + \lambda \Delta^{-1}$ is a bounded operator $L^2(\Omega) \to L^2(\Omega)$. Moreover, showing (71) is a Fredholm operator with index zero is equivalent to showing that this is true for $I + \lambda \Delta^{-1}: L^2(\Omega) \to L^2(\Omega)$. But $H^2(\Omega) \subset L^2(\Omega)$ is a compact injection by Section 6.5, so $\Delta^{-1}: L^2(\Omega) \to L^2(\Omega)$ is a compact linear operator. Therefore, we become interested in the Fredholm theory of

(72) $$I + \lambda K: L^2(\Omega) \to L^2(\Omega),$$

where $K: L^2(\Omega) \to L^2(\Omega)$ is a compact linear operator.

More generally, let $\mathcal{H}$ be a Hilbert space with inner product $\langle,\rangle$ and suppose that $K: \mathcal{H} \to \mathcal{H}$ is a compact linear operator. Fixing the value λ in (72) means we wish to show that

(73) $$A = I + K: \mathcal{H} \to \mathcal{H}$$

is a Fredholm operator of index zero. Let N_A be the nullspace of (73).

Lemma 1. *N_A is finite dimensional.*

Proof. Were N_A not finite dimensional, then (by a Gram-Schmidt procedure) we could find an infinite sequence of orthonormal vectors $u_m \in N_A$: $\|u_m\| = 1$ and $\langle u_m, u_\ell \rangle = 0$ if $m \neq \ell$. However, the compactness of K then implies that $\{Ku_m\}$ contains a convergent subsequence $\{Ku_{m_j}\}$. Since $Au_{m_j} = u_{m_j} + Ku_{m_j} = 0$, we see that $\{u_{m_j}\}$ must converge. But this violates $\|u_{m_j} - u_{m_i}\|^2 = \|u_{m_j}\|^2 - 2\langle u_{m_j}, u_{m_i}\rangle + \|u_{m_i}\|^2 = 2$ for $m_j \neq m_i$. So N_A must be finite dimensional. ♠

The inequality $\|Au\| \leq C_1 \|u\|$ holds by the boundedness of the operator A, whereas the opposite inequality

(74) $$\|u\| \leq C_2 \|Au\|$$

certainly fails for $u \in N_A$; but the following result shows that (74) holds for vectors orthogonal to N_A.

Lemma 2. *There is a constant C_2 so that (74) holds for all $u \in N_A^\perp$.*

Proof. If (74) were not true, then there would exist $u_m \in N_A^\perp$ such that $\|u_m\| = 1$ and $\|Au_m\| \to 0$. But the compactness of K implies that $\{Ku_m\}$ has a convergent subsequence $\{Ku_{m_i}\}$, and writing $u_{m_i} = Au_{m_i} - Ku_{m_i}$ shows that $\{u_{m_i}\}$ itself converges to some $u \in \mathcal{H}$. Since $Au_{m_i} \to 0$, we have $u \in N_A$. On the other hand, since $\langle u, v \rangle = \lim_{m_i \to \infty} \langle u_{m_i}, v \rangle \equiv 0$ for all $v \in N_A$, we see that $u \in N_A^\perp$. But $N_A \cap N_A^\perp = \{0\}$ (see Exercise 8 in Section 6.1), so $u \equiv 0$, which contradicts $\|u\| = \lim_{m_i \to \infty} \|u_{m_i}\| \equiv 1$. ♠

In general, the range of a bounded linear operator on a Hilbert space need not be a closed set; however, the range of a Fredholm operator *is* closed. Let us use (74) to establish that the range $R_A = A(\mathcal{H})$ of the operator (73) is closed.

Lemma 3. *The range R_A of (73) is a closed subspace of $\mathcal{H}$.*

Proof. Suppose $u_m \in \mathcal{H}$ satisfy $v_m = Au_m \to v \in \mathcal{H}$. To show that R_A is closed, we must find $u \in \mathcal{H}$ such that $Au = v$. Use the projection theorem of Section 6.2 to write $u_m = u_m^0 + w_m$, where $u_m^0 \in N_A$ and $w_m \in N_A^\perp$. But $\|v_m\| \to \|v\|$ shows that $\{v_m\} = \{Au_m\} = \{Aw_m\}$ is bounded, and so (74) implies that $\{w_m\}$ is bounded. By the compactness of K, there is a subsequence $\{w_{m_i}\}$ such that $Kw_{m_i} \to w_0 \in \mathcal{H}$. Thus $w_{m_i} = Aw_{m_i} - Kw_{m_i} \to v - w_0$. By continuity of A, $A(v - w_0) = \lim_{m_i \to \infty} A(w_{m_i}) = v$. Letting $u = v - w_0$ shows that R_A is closed. ♠

We have not yet shown that $R_A^\perp$ is finite dimensional, but at this point it is useful to introduce a generalization of the Fredholm property. A bounded linear operator $A: X \to Y$ is *semi-Fredholm* if its range R_A is a closed subspace of Y, and at least one of $\operatorname{nul}(A)$ or $\operatorname{def}(A)$ is finite; in this case, $\operatorname{ind}(A)$ is still well defined, although it could have the values $\pm\infty$.

To exploit the semi-Fredholm property for operators on a Hilbert space, it is effective to use the adjoint of A,

$$(75) \qquad A^*: \mathcal{H} \to \mathcal{H},$$

which was defined in Section 6.6 (cf. also Exercise 16 in Section 6.1).

Lemma 4. *If $A: \mathcal{H} \to \mathcal{H}$ is a semi-Fredholm operator, then so is its adjoint $A^*: \mathcal{H} \to \mathcal{H}$. Moreover, the following relations hold:*

$$(76) \qquad \begin{aligned} \operatorname{nul}(A^*) &= \operatorname{def}(A), \qquad \operatorname{def}(A^*) = \operatorname{nul}(A), \\ \operatorname{ind}(A^*) &= -\operatorname{ind}(A). \end{aligned}$$

Proof. This result follows from the relationships

$$(77) \qquad (N_{A^*})^\perp = R_A \quad \text{and} \quad N_{A^*} = R_A^\perp,$$

which are consequences of Lemma 3, together with Exercises 8 and 17 in Section 6.1. ♠

Lemma 4 has immediate consequences for $A = I + K$ as in (73), which we know to be semi-Fredholm. Since $A^* = I + K^*$, where K^* is also a compact linear operator on $\mathcal{H}$ (see Exercise 10 in Section 6.5), we may apply Lemma 1 to A^* to see that the nullspace N_{A^*} is also finite dimensional, so A is Fredholm. We can summarize these observations in the following.

Proposition 1. *(73) and (75) are both Fredholm operators. Moreover,*
(i) $Au = f$ can be solved if and only if $\langle f, v \rangle = 0$ for all $v \in N_{A^*}$, and
(ii) $A^*v = g$ can be solved if and only if $\langle g, u \rangle = 0$ for all $u \in N_A$.

When A is a self-adjoint Fredholm operator, then $\operatorname{ind}(A) = 0$ by (76). But for $A = I + K$, this is true without self-adjointness:

$$(78) \qquad \operatorname{ind}(I + K) = \operatorname{ind}(I) = 0;$$

see also Exercise 6. In other words, $\dim(N_A) = \dim(N_{A^*})$ so the number of orthogonality conditions on f needed to solve $Au = f$ always equals the dimension of the nullspace of A. (Although the Fredholm operators that we encounter in this section all have index zero, it is possible for a Fredholm operator to have nonzero index: see Exercise 7.)

Now let us see how to use Proposition 1 to study (70). Recall the adjoint operator $L' = \sum_{i,j} \partial_i \partial_j a_{ij}(x) - \sum_i \partial_i b_i(x) + c(x)$, which was introduced in Section 2.3. As in (70), we have $L': X \to L^2(\Omega)$ bounded.

Theorem 3. *Suppose L as in (29b) is uniformly elliptic with coefficients in $C^2(\overline{\Omega})$. The equations $Lu = 0$ and $L'v = 0$ have the same finite number of linearly independent solutions in $X = H^2(\Omega) \cap H_0^1(\Omega)$. Moreover,*
(i) *For $f \in L^2(\Omega)$, the equation $Lu = f$ can be solved if and only if $\langle f, v \rangle = 0$ for every solution $v \in X$ of $L'v = 0$.*
(ii) *For $g \in L^2(\Omega)$, the equation $L'v = g$ can be solved if and only if $\langle g, u \rangle = 0$ for every solution $u \in X$ of $Lu = 0$.*
In other words, (70) is a Fredholm operator of index zero.

Proof. Pick $\mu > 0$ so that $L_\mu = L - \mu$ satisfies uniqueness; that is, $c(x) - \mu \leq 0$ where $c(x)$ is the zero-order term in L [c.f. (29b)]. Then $L_\mu: X \to L^2(\Omega)$ is an isomorphism by Theorem 2, and so we can define $Q \equiv L_\mu^{-1}: L^2(\Omega) \to X$ as a bounded operator, so $Q: L^2(\Omega) \to L^2(\Omega)$ is a compact operator. Now we may define $A = LL_\mu^{-1}$ as an operator on $L^2(\Omega)$. But $A = (L_\mu + \mu)L_\mu^{-1} = I + \mu Q$, so Proposition 1 and (78) show that $A: L^2(\Omega) \to L^2(\Omega)$ is a Fredholm operator of index zero. In fact, for $f \in L^2(\Omega)$, let $u = L_\mu^{-1} f \in X$ to find $Af = 0$ if and only if $Lu = 0$. This shows that $A: L^2(\Omega) \to L^2(\Omega)$ and $L: X \to L^2(\Omega)$ have isomorphic nullspaces: $\mathrm{nul}(A) = \mathrm{nul}(L)$. In particular, the nullspace of (70) is finite-dimensional.

Similarly, we find that $L'v = 0$ has a finite number of linearly independent solutions $v \in X$. In fact, by taking μ sufficiently large, we may assume that *both* $L_\mu = L - \mu$ and $L'_\mu = L' - \mu$ satisfy uniqueness [notice that the zero-order term in L' may involve derivatives of $a_{ij}(x)$ and $b_i(x)$, but these are still bounded]. Thus $Q = L_\mu^{-1}: L^2(\Omega) \to L^2(\Omega)$ and $Q' = L'^{-1}_\mu: L^2(\Omega) \to L^2(\Omega)$ are both compact operators. If we let $A' = I + \mu Q'$, then we conclude that $\mathrm{nul}(L') = \mathrm{nul}(A')$ is finite dimensional.

But A and A' are bounded linear operators on the Hilbert space $L^2(\Omega)$, so we can investigate their adjoints. Indeed, $\langle L_\mu u, v \rangle = \langle u, L'_\mu v \rangle$ for $u, v \in C_0^2(\Omega)$ implies the same for $u, v \in X$. So, with $L'_\mu v = g$, we find $\langle Qf, g \rangle = \langle L_\mu^{-1} f, L'_\mu v \rangle = \langle f, v \rangle = \langle f, L'^{-1}_\mu g \rangle$, which shows that the L^2-adjoint of Q is $Q^* = L'^{-1}_\mu = Q'$, and hence $A^* = A'$. Therefore,

$$L(X) = R_A = (N_{A^*})^\perp = \{g \in L^2(\Omega) : A^*g = 0\}^\perp$$
$$= \{g \in L^2(\Omega) : A'g = 0\}^\perp = \{v \in X : L'v = 0\}^\perp,$$

which establishes (i). If we apply the same arguments to L' we obtain (ii) and $L'(X) = \{u \in X: Lu = 0\}^\perp$. Finally, $0 = \text{ind}(A) = \text{nul}(A) - \text{nul}(A^*) = \text{nul}(L) - \text{nul}(L')$ shows that L and L' have the same number of linearly independent solutions. ♠

c. Spectral Theory

At this point, we should make connection with Section 6.6 by observing that (70) also defines L as an unbounded linear operator on $L^2(\Omega)$ with domain $D = H^2(\Omega) \cap H_0^{1,2}(\Omega)$. In fact, we then realize that the compact operator $L_\mu = (L - \mu)^{-1}$ that occurs in the proof of Theorem 3 is just the resolvent of L. This has significance for the spectral theory of L; for this reason, we now consider $L^2(\Omega)$ to be the complex-valued measurable functions with finite L^2-norm.

Since $(L - \mu)^{-1}$ is a compact operator for this choice of $\mu > 0$, we find that $(L - \lambda)^{-1}$ is defined (and compact) for all λ in the resolvent set of L. In fact, invoking Theorem 5 (and Exercise 15) in Section 6.6, we immediately obtain the following:

Theorem 4. *Suppose L as in (29a,b) is uniformly elliptic with coefficients in $C^2(\overline{\Omega})$. Then the spectrum of the closed, densely defined linear operator $L: D \to L^2(\Omega)$ consists of isolated eigenvalues $\{\lambda_n\}_{n=1}^\infty$, each λ_n has finite multiplicity, and $\{\lambda_n\}_{n=1}^\infty$ has no finite accumulation points.*

Remark. If L is in divergence form [cf. (20), (21) in Section 6.2], then $L: D \to L^2(\Omega)$ is self-adjoint, so $\{\lambda_n\} \subset \mathbf{R}$; moreover, the associated eigenfunctions form an orthonormal basis for $L^2(\Omega)$ (cf. Exercise 15 in Section 6.6 and Exercise 4 in Section 7.2).

Exercises for Section 8.4

Throughout these Exercises, $\mathcal{H}$ denotes a Hilbert space.

1. If $A: \mathcal{H} \to \mathcal{H}$ is Fredholm, show that $A: N_A^\perp \to R_A$ is an isomorphism between Hilbert spaces.

2. (a) If $A: \mathcal{H} \to \mathcal{H}$ is Fredholm, show that there is a bounded linear operator $B: \mathcal{H} \to \mathcal{H}$, called the *Fredholm inverse of A*, such that $AB = I + K_1$ and $BA = I + K_2$, where $K_i: \mathcal{H} \to \mathcal{H}$ are compact operators (in fact, *finite rank* operators, i.e., with finite-dimensional range).
   (b) Conversely, if $A: \mathcal{H} \to \mathcal{H}$ is bounded and there exist $B_i: \mathcal{H} \to \mathcal{H}$ for $i = 1, 2$ such that $B_1 A - I$ and $A B_2 - I$ are compact, then A is Fredholm.

3. If $A: \mathcal{H} \to \mathcal{H}$ is a Fredholm operator and $K: \mathcal{H} \to \mathcal{H}$ is a compact linear operator, then $A + K: \mathcal{H} \to \mathcal{H}$ is a Fredholm operator.

4. If $A: \mathcal{H} \to \mathcal{H}$ with $\|A\| < 1$, then $I + A: \mathcal{H} \to \mathcal{H}$ is an isomorphism.

5. Suppose $A: \mathcal{H} \to \mathcal{H}$ is a Fredholm operator and $B: \mathcal{H} \to \mathcal{H}$ a bounded linear operator with $\|A - B\| < \epsilon$ sufficiently small.

(a) Prove that $B : \mathcal{H} \to \mathcal{H}$ is also Fredholm (i.e., the Fredholm operators, $\Phi(\mathcal{H})$, form an open set in the space of bounded operators, $\mathcal{B}(\mathcal{H})$).
(b) Prove that $\operatorname{ind}(A) = \operatorname{ind}(B)$ [i.e., the Fredholm index is constant on connected components of $\Phi(\mathcal{H})$].

6. If $A: \mathcal{H} \to \mathcal{H}$ is a Fredholm operator and $K: \mathcal{H} \to \mathcal{H}$ is a compact linear operator, then $\operatorname{ind}(A + K) = \operatorname{ind}(A)$.

7. Let $\mathcal{H} = \ell^2$ be square-summable sequences as in Exercise 4 of Section 6.1, and define the *left-shift operator* A as follows: $A(\{a_1, a_2, \ldots\}) = \{a_2, a_3, \ldots\}$.
(a) Verify that $A: \ell^2 \to \ell^2$ is a bounded linear operator.
(b) Verify that $A: \ell^2 \to \ell^2$ is Fredholm, and compute $\operatorname{ind}(A)$.
(c) Describe the adjoint $A^*: \ell^2 \to \ell^2$ and check that $\operatorname{ind}(A) = \operatorname{nul}(A) - \operatorname{nul}(A^*)$.

Further References for Chapter 8

An excellent reference on elliptic estimates for second-order operators is [Gilbarg-Trudinger]. Many of the results of this chapter may be generalized to higher-order elliptic operators and systems of elliptic operators; see [ADN], [BJS], and [Folland]. Elliptic estimates for more general boundary conditions may be found in [ADN] and lead naturally to the notion of boundary "trace" spaces and fractional-order Sobolev spaces; see [Adams]. For details on solving elliptic problems with nonhomogeneous boundary conditions, see [Lions-Magenes].

Maximum principles for elliptic operators have many other applications in analysis and applied mathematics; see [Protter-Weinberger].

Much of the the theory of Fredholm operators, which we have presented in the text and Exercises as pertaining to bounded operators on a Hilbert space, may be performed on operators between Banach spaces; see [Kato] or [Palais].

The study of elliptic operators on a torus is the simplest example of the more general study of elliptic operators on compact manifolds; see [Aubin], [Palais], or [Roe]. In addition, the Fredholm theory of elliptic operators on compact manifolds has strong connections with topology: the Atiyah-Singer index theorem is one example; the de Rham theory of elliptic complexes is another.

9 Two Additional Methods

In this chapter, we collect two additional methods used to study more complicated linear and nonlinear partial differential equations and systems. First we discuss Schauder fixed point theory, which is a powerful tool in establishing existence of solutions to nonlinear elliptic PDEs. Then we discuss linear and nonlinear semigroups of operators, which are useful in studying the dynamics of evolution equations.

9.1 Schauder Fixed Point Theory

In this section we explore some ways of showing that a map on a Banach space has a fixed point. In Section 7.3 we encountered a fixed point theorem for maps that contract distances. *Schauder fixed point theory*, on the other hand, involves compactness (cf. Sections 6.3 and 6.5) and convexity. Applications to the study of the stationary Navier-Stokes equations are given in Section 9.1d and to other nonlinear elliptic equations in Section 13.4.

a. The Brouwer Fixed Point Theorem

Let us recall the following result from finite-dimensional theory.

Theorem 1 (Brouwer Fixed Point Theorem). *If $B \subset \mathbf{R}^n$ is a closed ball and $f \colon B \to B$ is continuous, then f has a fixed point $x \in B$.*

This result is trivial to prove when $n = 1$; for $n > 1$, however, the proof is surprisingly tricky. Many different types of proof are available in the literature: for example, see [Massey] for a topological proof or [Gilbarg-Trudinger] for an analytic proof.

b. The Schauder Fixed Point Theorem

Notice that only continuity is assumed of the mapping f in Theorem 1, but the compactness and convexity of the unit ball B are essential (cf. Exercise 1). If we wish to generalize this to an infinite-dimensional Banach space X, we encounter the difficulty that a ball is not compact (cf. Section 6.3). For this reason, we must consider compact, convex subsets of X instead of a ball. Recall that a subset $A \subset X$ is *convex* if it contains the straight line between any two points of A. One example of a compact, convex subset of X is the *closed convex hull* of finitely many points $x_1, \ldots, x_N \in X$:

$$\mathrm{co}\{x_1, \ldots, x_N\} \equiv \left\{ \sum_{i=1}^N \lambda_i x_i : 0 \leq \lambda_i \leq 1, \sum_{i=1}^N \lambda_i = 1 \right\},$$

which is the smallest convex set containing $x_1, \ldots, x_N$. We can now state and prove the infinite-dimensional generalization of the Brouwer fixed point theorem.

Theorem 2 (Schauder Fixed Point Theorem). *Let A be a compact, convex set in a Banach space X and $T: A \to A$ be continuous. Then T has a fixed point $x \in A$.*

Proof. The proof involves approximating T by finite-dimensional maps and then applying the Brouwer fixed point theorem. If $A = \{x_1\}$, the result is trivial; otherwise, for $\epsilon > 0$ sufficiently small, use the compactness of A to find a finite number of points $x_1, \ldots, x_N \in A$, where $N = N(\epsilon) > 1$, such that A is covered by the balls $B_i = B_\epsilon(x_i)$, but not by any one B_i. Let A_ϵ denote the closed convex hull of $x_1, \ldots, x_N$, and define the mapping $P_\epsilon: A \to A_\epsilon$ by

$$P_\epsilon x = \frac{\sum_{i=1}^{N} \text{dist}(x, A - B_i)\, x_i}{\sum_{i=1}^{N} \text{dist}(x, A - B_i)} \quad \text{for } x \in A.$$

Now P_ϵ is continuous, and for every $x \in A$ we have

$$\text{(1)} \qquad \|P_\epsilon x - x\| \leq \frac{\sum_{i=1}^{N} \text{dist}(x, A - B_i)\, \|x_i - x\|}{\sum_{i=1}^{N} \text{dist}(x, A - B_i)} < \epsilon.$$

Restriction of $P_\epsilon \circ T: A \to A_\epsilon$ to A_ϵ defines a continuous map $A_\epsilon \to A_\epsilon$. But A_ϵ is homeomorphic to the closed unit ball in $\mathbf{R}^M$ for some $M \leq N$, so we may apply the Brouwer fixed point theorem to obtain a fixed point $x_\epsilon \in A_\epsilon$; that is, $(P_\epsilon \circ T)x_\epsilon = x_\epsilon$.

Using the compactness of A, we conclude that there is a subsequence x_{ϵ_j} and a point $x \in A$ such that $x_{\epsilon_j} \to x$ as $\epsilon_j \to 0$. To show that x is our desired fixed point of T, we apply (1) to Tx_{ϵ_j}: $\|x_{\epsilon_j} - Tx_{\epsilon_j}\| = \|(P_{\epsilon_j} \circ T)x_{\epsilon_j} - Tx_{\epsilon_j}\| < \epsilon_j$. By continuity we conclude that $Tx = x$. ♠

In applications, it is often more convenient to replace the compactness of A by that of image set, $T(A)$. Thus we are led to the following formulation.

Theorem 3. *Let A be a closed convex set in a Banach space X, and $T: A \to A$ be continuous such that $\overline{T(A)}$ is compact in X. Then T has a fixed point.*

The derivation of Theorem 3 from Theorem 2 is discussed in Exercise 2.

c. *The Leray-Schauder Fixed Point Theorem*

Let us consider a further modification of the Schauder fixed point theory in which compactness of sets is replaced by compactness of the map $T: X \to X$, and convexity is replaced by the existence of a one-parameter family of maps T_t ($0 \leq t \leq 1$) where $T_1 = T$ and T_0 is a compact map that has a fixed point (such as the trivial map $T_0 = 0$ for which 0 is a fixed point). Provided that we can assure that fixed points of T_t do not leave some large ball B, then we may conclude that $T = T_1$ has a fixed point in B. The following statement (in which $T_t = tT$) shall suffice for our purposes.

Theorem 4 (Leray-Schauder Fixed Point Theorem). *Suppose that $T: X \to X$ is a compact map such that*

(2) $$\|x\|_X < M$$

for every solution (x,t) of $x = tTx$, where $0 \leq t \leq 1$. Then T has a fixed point.

This result may be proved using Theorem 3 (see Exercise 3).

A key step in applying this theorem is establishing the bound (2). Often this is achieved by means of a priori estimates, such as those obtained for linear elliptic equations in Chapter 8. On the other hand, the bound (2) sometimes is achieved by more direct methods, such as the Sobolev inequality. In the next subsection, we see this explicitly in the application of Theorem 4 to the Navier-Stokes equations.

Remark. It should be observed that, unlike the contraction mapping principle of Section 7.3, the fixed points obtained in the Schauder theory need not be unique.

d. Application to Stationary Navier-Stokes

If the motion of an incompressible viscous fluid in a domain $\Omega \subset \mathbf{R}^3$ is subject to an external force $\vec{f}$, then we encounter inhomogeneous Navier-Stokes equations [cf. (32) in the Appendix]

(3) $$\vec{u}_t + (\vec{u} \cdot \nabla)\vec{u} + \rho^{-1}\nabla p = \nu \Delta \vec{u} + \vec{f},$$

where $\vec{u}$ is the velocity of the fluid, p is the pressure, ρ is the (constant) density, and ν is the kinematic viscosity. If the motion is allowed to reach equilibrium (time independence), then we encounter the *stationary (forced) Navier-Stokes equations*; if we normalize $\rho = 1$, the equations become

(4a) $$(\vec{u} \cdot \nabla)\vec{u} + \nabla p = \nu \Delta \vec{u} + \vec{f}.$$

Recall that the incompressibility may be expressed by

(4b) $$\operatorname{div} \vec{u} = 0 \quad \text{in } \Omega,$$

and the viscosity suggests a "nonslip" boundary condition

(4c) $$\vec{u}(x) = 0 \quad \text{for } x \in \partial \Omega.$$

We studied a linearized version of these equations in all dimensions n in Section 6.2b and reduced the problem to finding a weak solution of Poisson's equation on the divergence-free vector fields $\tilde{H}_0^{1,2}(\Omega, \mathbf{R}^n)$. By analogy, we define a *weak solution* of (4abc) to be $\vec{u} \in \tilde{H}_0^{1,2}(\Omega, \mathbf{R}^n)$ such that

$$(5) \qquad \nu \int_\Omega \nabla \vec{u} : \nabla \vec{v} \, dx + \int_\Omega (\vec{u} \cdot \nabla) \vec{u} \cdot \vec{v} \, dx = \int_\Omega \vec{f} \cdot \vec{v} \, dx$$

for all $\vec{v} \in \tilde{H}_0^{1,2}(\Omega, \mathbf{R}^n)$; once $\vec{u}$ is found, then $p \in L^2(\Omega)$ is found using the proposition in Section 6.2b. We now use the Leray-Schauder fixed point theorem to solve (5) when $n = 2$ or 3.

We begin by reducing the solution of (5) to the search for a fixed point of a nonlinear map on a Banach space. Let us rewrite (5) as

$$(6) \qquad \nu \int_\Omega \nabla \vec{u} : \nabla \vec{v} \, dx = \langle \vec{f}, \vec{v} \rangle - \{\vec{u}, \vec{u}, \vec{v}\}$$

where the trilinear form $\{\cdot, \cdot, \cdot\}$ is defined by

$$(7a) \qquad \{\vec{u}, \vec{w}, \vec{v}\} = \int_\Omega (\vec{u} \cdot \nabla) \vec{w} \cdot \vec{v} \, dx \qquad \text{for } \vec{u}, \vec{w}, \vec{v} \in \tilde{H}_0^{1,2}(\Omega, \mathbf{R}^n).$$

It is elementary to verify (see Exercise 5a) that for $\vec{u}, \vec{w}, \vec{v} \in \tilde{H}_0^{1,2}(\Omega, \mathbf{R}^n)$ we have

$$(7b) \qquad \{\vec{u}, \vec{w}, \vec{v}\} = -\{\vec{u}, \vec{v}, \vec{w}\}.$$

In order to apply the Riesz representation theorem, we must confirm that the right-hand side of the equation (6) defines a bounded linear functional on $\vec{v} \in \tilde{H}_0^{1,2}(\Omega, \mathbf{R}^n)$. For $\vec{f} \in L^2(\Omega, \mathbf{R}^n)$, this is certainly true of $\langle \vec{f}, \vec{v} \rangle$, so we need only check the term $\{\vec{u}, \vec{u}, \vec{v}\}$.

For $\vec{u}, \vec{w}, \vec{v} \in \tilde{H}_0^{1,2}(\Omega, \mathbf{R}^n)$, we can easily check the pointwise estimate (see Exercise 5b)

$$(8) \qquad |(\vec{u} \cdot \nabla) \vec{w} \cdot \vec{v}| \leq |\vec{u}| \cdot |\nabla \vec{w}| \cdot |\vec{v}|,$$

where $|\nabla \vec{w}|$ was defined in Exercise 9 of Section 6.1. We may therefore apply the generalized Hölder inequality (cf. Exercise 2 in Section 6.1) to obtain

$$(9) \qquad |\{\vec{u}, \vec{w}, \vec{v}\}| \leq \int_\Omega |(\vec{u} \cdot \nabla) \vec{w} \cdot \vec{v}| \, dx \leq \|\vec{u}\|_p \cdot \|\nabla \vec{w}\|_2 \cdot \|v\|_q$$
$$\text{whenever} \quad p^{-1} + 2^{-1} + q^{-1} = 1.$$

Let us take $p = 4 = q$ in (9). Now the Sobolev imbedding $H_0^{1,2} \subset L^4$ holds for $n \leq 4$, so we have the estimate $\|\vec{u}\|_4 \leq C \|\vec{u}\|_{1,2}$. Combined with (9), we obtain

$$(10a) \qquad |\{\vec{u}, \vec{w}, \vec{v}\}| \leq C \|\vec{u}\|_4 \|\nabla \vec{w}\|_2 \|\vec{v}\|_4 \leq C \|\vec{u}\|_{1,2} \|\vec{w}\|_{1,2} \|\vec{v}\|_{1,2}.$$

Taking fixed $\vec{w} = \vec{u} \in \tilde{H}_0^{1,2}(\Omega, \mathbf{R}^n)$, (10a) shows that indeed $\{\vec{u}, \vec{u}, \vec{v}\}$ defines a bounded linear functional on $\vec{v} \in \tilde{H}_0^{1,2}(\Omega, \mathbf{R}^n)$. Let us observe for future reference that, by virtue of (7b), the same argument shows that

(10b) $$|\{\vec{u}, \vec{w}, \vec{v}\}| \leq C \|\vec{u}\|_4 \|\vec{w}\|_4 \|\nabla \vec{v}\|_2 \leq C \|\vec{u}\|_{1,2} \|\vec{w}\|_{1,2} \|\vec{v}\|_{1,2},$$

which also implies the desired boundedness of $\{\vec{u}, \vec{u}, \vec{v}\}$.

Now let us pick $\vec{w} \in \tilde{H}_0^{1,2}(\Omega, \mathbf{R}^n)$, and consider the *linear* problem

(11a) $$\nu \int_\Omega \nabla \vec{u} : \nabla \vec{v} \, dx = F_{\vec{w}}(\vec{v}) \quad \text{for all } \vec{v} \in \tilde{H}_0^{1,2}(\Omega, \mathbf{R}^n)),$$

where

(11b) $$F_{\vec{w}}(\vec{v}) = \langle \vec{f}, \vec{v} \rangle - \{\vec{w}, \vec{w}, \vec{v}\}.$$

Since $F_{\vec{w}}$ is a bounded linear functional on $X = \tilde{H}_0^{1,2}(\Omega, \mathbf{R}^n))$, the Riesz representation theorem (see Section 6.2b) provides us with a unique solution $\vec{u} \in X$ of (11a). Letting $\vec{u} = T\vec{w}$ defines a map $T: X \to X$. To show that T is compact, suppose $\{\vec{w}_m\}$ is a bounded sequence in X (i.e., $\|\vec{w}_m\|_{1,2} \leq M$ for all $m = 1, 2, \ldots$). We wish to show that $\{T\vec{w}_m\}$ converges in X. Notice that, by definition,

(12) $$\nu \int_\Omega \nabla (T\vec{w}_m - T\vec{w}_n) : \nabla \vec{v} \, dx = \{\vec{w}_n, \vec{w}_n, \vec{v}\} - \{\vec{w}_m, \vec{w}_m, \vec{v}\}.$$

Using (10b) and (12), it is elementary to conclude (see Exercise 5c)

(13) $$\nu \left| \int_\Omega \nabla (T\vec{w}_m - T\vec{w}_n) : \nabla \vec{v} \, dx \right| \leq C \|\vec{w}_n - \vec{w}_m\|_4 \|\vec{v}\|_{1,2}$$

where C is independent of m, n, and $\vec{v}$ but may depend on the uniform bound M. For $n = 2$ or 3, $X \subset L^4(\Omega, \mathbf{R}^n)$ is a compact imbedding (cf. Section 6.5c), so a subsequence of $\{\vec{w}_m\}$, which we also denote by $\{\vec{w}_m\}$, converges in $L^4(\Omega, \mathbf{R}^n)$ (i.e., $\|\vec{w}_m - \vec{w}_n\|_4 \to 0$ as $m, n \to \infty$). Using (13) with $\vec{v}$ replaced by $T\vec{w}_m - T\vec{w}_n$, we obtain

$$\nu \|\nabla (T\vec{w}_m - T\vec{w}_n)\|_2^2 \leq C \|\vec{w}_m - \vec{w}_n\|_4 \|T\vec{w}_m - T\vec{w}_n\|_{1,2},$$

which shows that $\nabla T\vec{w}_m$ converges in L^2. By the Poincaré inequality, this shows that $T\vec{w}_m$ converges in X, as desired. Therefore, $T: X \to X$ is a compact map.

Before we can apply Theorem 4, however, we must establish the a priori bound (2). So let us assume that $\vec{u} \in X$ is a fixed point of tT for some $0 \leq t \leq 1$. This means that $\nu \int \nabla \vec{u} : \nabla \vec{v} \, dx = t(\langle \vec{f}, \vec{v} \rangle - \{\vec{u}, \vec{u}, \vec{v}\})$ for all $v \in X$. In particular, if we let $\vec{v} = \vec{u}$, then (7b) implies

(14) $$\{\vec{u}, \vec{u}, \vec{u}\} = \int_\Omega (\vec{u} \cdot \nabla \vec{u}) \cdot \vec{u} \, dx = 0,$$

and so we obtain

(15)
$$\nu \int_\Omega |\nabla \vec{u}|^2 \, dx = t \int_\Omega \vec{f} \cdot \vec{u} \, dx.$$

An application of the Cauchy-Schwarz inequality and the Poincaré inequality to (15) establishes the desired inequality $\|\vec{u}\|_{1,2} \leq M$, where M depends on $\vec{f}$ and Ω but not on $\vec{u}$.

We are now able to apply Theorem 4 to our problem and immediately conclude the following.

Theorem 5. If $n = 2$ or 3 and $\vec{f} \in L^2(\Omega, \mathbf{R}^n)$, then (4) admits at least one weak solution $(\vec{u}, p)$, where $\vec{u} \in \tilde{H}_0^{1,2}(\Omega, \mathbf{R}^n))$ and $p \in L^2(\Omega)$.

Remark. Notice that, unlike Theorem 2 in Section 6.2, we cannot conclude in general that the function $\vec{u}$ in (4) is unique; in fact, uniqueness may fail (cf. [Galdi, vol. II]). However, for sufficiently small values of $\vec{f}$, the solution $\vec{u}$ will be unique; this follows from an application of the inverse function theorem to the problem (cf. Exercise 6).

Exercises for Section 9.1

1. Show the compactness and convexity of the finite-dimensional ball B are essential to the Brouwer fixed point theorem by giving examples of continuous maps $f: X \to X$ that have no fixed point when
   (a) $X \subset \mathbf{R}^n$ is convex but not compact.
   (b) $X \subset \mathbf{R}^n$ is compact but not convex.

2. For a general subset $K \subset X$, define the *convex hull* of K to be the smallest convex set containing K; that is,

   $$\text{co}(K) = \cap \{A : K \subset A \text{ and } A \text{ is a convex set in } X\},$$

   and $\overline{\text{co}}(K)$, the *closed convex hull of K*, its closure in X. An important theorem of functional analysis states: K is compact $\Rightarrow \overline{\text{co}}(K)$ is compact. Use this to deduce Theorem 3 from Theorem 2.

3. Prove Theorem 4 as follows. Assume $M = 1$; let $B = \{x \in X : \|x\| \leq 1\}$, and define a continuous map $\hat{T} : B \to B$ by $\hat{T}x = Tx$ if $\|Tx\| \leq 1$, and $\hat{T}x = Tx/\|Tx\|$ if $\|Tx\| > 1$.
   (a) Use Theorem 3 to find a fixed point x_0 of $\hat{T}$.
   (b) Show that x_0 is a fixed point of T.

4. Suppose $\Omega \subset \mathbf{R}^n$ is a bounded domain, and $T : C^0(\overline{\Omega}) \to C^0(\overline{\Omega})$ is a continuous map such that

   $$\|Tu\|_{1,\infty} \leq \max(1, \|u\|_{1,\infty}) \quad \text{for} \quad u \in C^1(\overline{\Omega}).$$

   Prove that T has a fixed point $u \in C^1(\overline{\Omega})$.

5. Provide the following details in the proof of Theorem 5.
   (a) Verify (7b) for all $\vec{u}, \vec{w}, \vec{v} \in \tilde{H}_0^{1,2}(\Omega, \mathbf{R}^n)$.
   (b) Verify the pointwise estimate (8).
   (c) Verify (13).

6. When $n = 2$ or 3 and $\vec{f} \in L^2(\Omega, \mathbf{R}^n)$ is sufficiently small, use the inverse function theorem (Theorem 2 of Section 7.3) to show that (4) admits a weak solution $(\vec{u}, p)$ that is unique up to an additive constant to p.

9.2 Semigroups and Dynamics

In this section we are interested in evolution equations of the form $du/dt = f(u)$ with initial condition $u(0) = u_0$; properly interpreted, this will cover time-evolution PDEs as well as systems of ODEs. If there is a unique solution for each u_0, then we may define a solution operator $S(t)$ by $u(t) = S(t)u_0$. These solution operators form a *semigroup* in that $S(t+s) = S(t)S(s)$ and $S(0) = I$.

If the evolution is linear, then the solution operators are also linear; that is, $S(t)(u_0 + v_0) = S(t)u_0 + S(t)v_0$. In this case, we generally have global existence in $t > 0$, although the asymptotics as $t \to \infty$ may depend on the initial conditions.

If the evolution equation is nonlinear, however, then the solution operators will be nonlinear. In this case, the global behavior of solutions becomes even more delicate; for example, $u(t) = S(t)u_0$ may only exist for for a finite time [i.e. for $t \in (0, T)$ where T depends upon u_0].

We begin this section with a brief review of dynamical systems of ODEs to introduce the relevant semigroup concepts and then proceed to first-order evolutions on Banach spaces, with examples to illustrate their use in solving PDEs. We shall also discuss weak solutions of evolution equations which may exist even when strong solutions break down.

a. Finite-Dimensional Dynamics

Let us first consider a linear system of ODEs

$$\text{(16)} \qquad \frac{d\mathbf{x}}{dt} = A\mathbf{x}$$

with initial condition

$$\text{(17)} \qquad \mathbf{x}(0) = \mathbf{x_0},$$

where A is an $n \times n$ matrix with constant coefficients and $\mathbf{x} \in X \equiv \mathbf{R}^n$. The basic theory of linear ODEs states that there is a unique solution $\mathbf{x}(t)$. In fact, exponentiation of matrices enables us to write the solution as

$$\text{(18)} \qquad \mathbf{x}(t) = e^{tA}\mathbf{x_0}.$$

Thus the *solution* or *evolution operator* $S(t): X \to X$, which maps the initial condition $\mathbf{x}_0$ to the value $\mathbf{x}(t)$, is just given by $S(t) = e^{tA}$; moreover, the solution operator defines a *continuous semigroup of operators* in that

(19) $$\begin{cases} S(t+s) = S(t)S(s) & \text{for all } t, s \geq 0 \\ S(0) = I \\ S(t)\mathbf{x} \to S(t_0)\mathbf{x} & \text{for every } \mathbf{x} \in X \text{ when } t \to t_0. \end{cases}$$

We can view $S(t) = e^{tA}$ as defining a *flow* on X, with $\mathbf{x}(t) = e^{tA}\mathbf{x_0}$ being the *orbit* (or *trajectory*) of $\mathbf{x}_0$. Notice that $\mathbf{x}_0 = 0$ is a *fixed point* (or *critical point*) of this flow, corresponding to the trivial solution $\mathbf{x}(t) \equiv 0$. If A has a trivial nullspace, then this is the only fixed point; in particular, $\mathbf{x}_0 = 0$ is an *isolated* fixed point. The nature of the flow near 0 depends on the (complex) eigenvalues $\lambda_1, \ldots, \lambda_n$ of A. Let M_-, M_+, M_0 denote the spaces of (generalized) eigenvectors associated to the eigenvalues whose real part is negative, positive, and zero respectively. If $\mathbf{x}_0$ is itself an eigenvector with *real* eigenvector λ, then $e^{tA}\mathbf{x}_0 = e^{t\lambda}\mathbf{x}_0$ is a scalar multiple of $\mathbf{x}_0$, so in particular remains in the same eigenspace. By taking real and imaginary parts of complex eigenvectors λ, this argument generalizes to show that each of the subspaces M_-, M_+, M_0 is *invariant* under the flow [i.e., if $\mathbf{x}_0 \in M$ then $S(t)\mathbf{x}_0 \in M$ for all $t \geq 0$]. The space M_- is called the *stable subspace*, and for every $\mathbf{x}_0 \in M_-$ we have $S(t)\mathbf{x}_0 \to 0$ as $t \to \infty$. In fact, for $\mathbf{x}_0 \in M_-$, we have

(20) $$\|S(t)\mathbf{x}_0\| \leq Ce^{\beta t}\|\mathbf{x}_0\|$$

for some constants $C > 0$ and $\beta < 0$ independent of $\mathbf{x}_0$. If (20) holds for all $\mathbf{x}_0 \in X$ (for example if $M_0 = M_+ = \emptyset$), then we say that $S(t)$ is a *β-contraction*. The space M_+ is called the *unstable subspace*, and for every $\mathbf{x}_0 \in M_+$ we have $\|S(t)\mathbf{x}_0\| \to \infty$ as $t \to \infty$. In fact, for reasons similar to those for (20), the flow on M_+ satisfies $\|S(t)\mathbf{x}\| \geq Ce^{\beta t}\|\mathbf{x}_0\|$ for some constants $C > 0$ and $\beta > 0$ independent of $\mathbf{x}_0 \in M_+$. The space M_0 is called the *center subspace*. Periodic orbits in the flow are due to $\mathbf{x}_0 \in M_0$, but for multiple eigenvalues it is also possible to have $\|\mathbf{x}(t)\| \to \infty$ as $t \to \infty$.

Let us consider a familiar example to illustrate these properties.

Example 1. Consider the *damped harmonic oscillator*

(21) $$\ddot{y} + c\dot{y} + ky = 0,$$

where $k > 0$, $c \geq 0$, and the dot represents d/dt. We replace the second-order differential equation by the first-order system of the form (16) on $X = \mathbf{R}^2$:

(22) $$\frac{d\mathbf{x}}{dt} = \begin{pmatrix} 0 & 1 \\ -k & -c \end{pmatrix} \mathbf{x} \quad \text{where } \mathbf{x} = \begin{pmatrix} x_1 \\ x_2 \end{pmatrix} = \begin{pmatrix} y \\ \dot{y} \end{pmatrix}.$$

The matrix has eigenvalues $\lambda = \frac{1}{2}(-c \pm \sqrt{c^2 - 4k})$.

Section 9.2: Semigroups and Dynamics

When $c = 0$ (i.e., there is no damping), the eigenvalues $\lambda = \pm i\sqrt{k}$ are pure imaginary, so that $M_0 = X$. The existence of period orbits may be established from the general solution of (22) (cf. Exercise 1a) *or* by observing that the *energy*

$$(23) \qquad \mathcal{E}(t) \equiv \frac{1}{2}(\dot{y}^2 + ky^2)$$

is a constant (cf. Exercise 1b). Thus each solution lies on a closed elliptical orbit in the phase plane $(y, \dot{y})$ about 0; see Figure 1. When $c = 0$, the system (22) is called *conservative* or *Hamiltonian*.

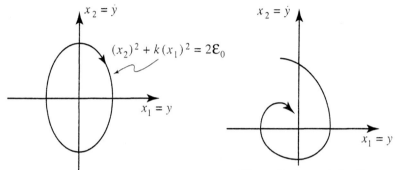

Figure 1. Closed orbit for $c = 0$.

Figure 2. Asymptotically stable for $c > 0$.

When $c > 0$, then both eigenvalues of the matrix in (22) have negative real part, so $M_- = X$. In this case, every solution of (22) is *asymptotically stable* (i.e., approaches 0 as $t \to \infty$); see Figure 2. In this case, we call 0 an *attractor*. The asymptotic decay of solutions can be seen from the general solution (cf. Exercise 1c) or using the energy as a *Liapunov function* (cf. Exercise 1b). Also notice that $S(t)$ is a β-contraction with $\beta < 0$ (cf. Exercise 1c,d). When $c > 0$, the system (22) is called *dissipative*. (Recall that we encountered dissipation and loss of energy in Section 3.4.) ♣

Now let us consider a nonlinear system of ODEs

$$(24) \qquad \frac{d\mathbf{x}}{dt} = f(\mathbf{x}),$$

where $f: \mathbf{R}^n \to \mathbf{R}^n$ is a nonlinear mapping. The system (24) is *autonomous* since f does not depend on t. Provided that f is Lipschitz continuous near $\mathbf{x}_0$, the Picard-Lindelöf theorem (see Section 7.3 for the case $n = 1$) shows that there is always a solution of (24) that satisfies (17) and exists at least for small values of t. If $f(\mathbf{x}_0) = 0$, then $\mathbf{x}_0$ is a *critical point*, and we obtain an *equilibrium solution* $\mathbf{x}(t) \equiv \mathbf{x}_0$: in other words, $\mathbf{x}_0$ is a *fixed point* of the flow defined by the solution operator $S(t)\mathbf{x}_0 \equiv \mathbf{x}(t)$.

Microscopically, the behavior of solutions $\mathbf{x}(t)$ near a critical point $\mathbf{x}_0$ may be determined by *linearization:* Let $A = f'(\mathbf{x}_0)$ denote the derivative of f at $\mathbf{x}_0$, and use the change of variables $\bar{\mathbf{x}} = \mathbf{x} - \mathbf{x}_0$ to replace (24) by a linear system like (16) that is accurate up to order $O(|\bar{\mathbf{x}}|^2)$:

$$(25) \qquad \frac{d\bar{\mathbf{x}}}{dt} = A\bar{\mathbf{x}} + O(|\bar{\mathbf{x}}|^2).$$

The purpose of *linearized stability analysis (LSA)* is to conclude that, in most cases, the behavior near $\mathbf{x}_0$ of trajectories of (24) is qualitatively the same as that for the corresponding linear system. For example, if the linear system has a stable subspace M_- at $\mathbf{x}_0$ (i.e., for $\bar{\mathbf{x}} = 0$), then the nonlinear system admits a set $\tilde{M}_-$ that is invariant under the flow $S(t)$ and consists of those initial conditions $\mathbf{x}$ for which $S(t)\mathbf{x} \to \mathbf{x}_0$; the set $\tilde{M}_-$ is called the *stable manifold* for $\mathbf{x}_0$. Similarly, the *unstable manifold* $\tilde{M}_+$ is a perturbation of the unstable subspace M_+; see Figure 3. On the other hand, LSA fails to predict the behavior of the center subspace M_0 under a nonlinear perturbation, and a more careful analysis must be used to study the existence of a *center manifold* $\tilde{M}_0$.

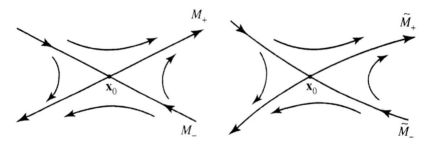

a. Linear system.   b. Nonlinear system.
Figure 3. *Stable and unstable manifolds.*

Macroscopically, the behavior of trajectories for (24) can be quite different from their linearized behavior. To begin with, solutions to a linear system exist for all $t > 0$, but we may only have *local existence* for nonlinear systems. So an important issue to settle for an initial value problem (24),(17) is whether there is *global existence* of the solution. If the solution $\mathbf{x}(t)$ is known to exist globally, then we may inquire about its *asymptotic behavior* as $t \to \infty$. Does $\mathbf{x}(t) \to \infty$? Does $\mathbf{x}(t)$ approach a critical point?

Let us consider these issues for another familiar ODE.

Example 2. Consider the *damped pendulum*

$$(26) \qquad \ddot{y} + c\dot{y} + k\sin y = 0,$$

where $k > 0$, $c \geq 0$. Written as a first-order system, (26) takes the form

$$(27) \qquad \frac{d\mathbf{x}}{dt} = \begin{pmatrix} x_2 \\ -cx_2 - k\sin x_1 \end{pmatrix} \quad \text{where } \mathbf{x} = \begin{pmatrix} x_1 \\ x_2 \end{pmatrix} = \begin{pmatrix} y \\ \dot{y} \end{pmatrix}.$$

The critical points of (27) are $x_1 = \pi\ell$, $x_2 = 0$ for all integers ℓ. The linearization of (27) is just (22) when ℓ is even, but (22) with $-k$ replaced by k if ℓ is odd. For odd values of ℓ, this means that the linear matrix has one positive and one negative eigenvalue ($\dim M_- = \dim M_+ = 1$), so $M_0 = \emptyset$ and we may apply LSA. We can also use LSA when ℓ is even and $c > 0$, since in this case $M_0 = \emptyset$ (and $\dim M_- = 2$). But if ℓ is even and $c = 0$, then we cannot apply LSA, so we appeal to energy methods. For (26), it is natural to define the *energy* to be

$$(28) \qquad \mathcal{E}(t) \equiv \frac{1}{2}\dot{y}^2 + k\cos y,$$

which is constant for every solution of $\ddot{y} + k\sin y = 0$. These considerations produce the phase portraits appearing in Figures 4 and 5 and completely describe the asymptotic behavior of all solutions of (27). ♣

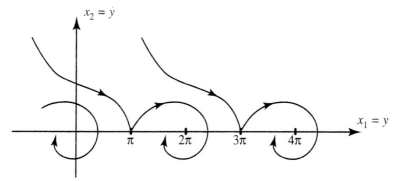

Figure 4. *Asymptotically stable critical points for c > 0.*

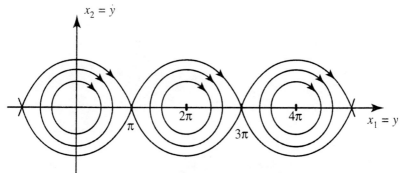

Figure 5. *Periodic orbits for c = 0.*

b. Linear Evolution on Banach Spaces

Let X be a Banach space, and consider the linear evolution equation

(29) $$\begin{cases} \dfrac{du}{dt} = Au & \text{for } 0 < t < \infty, \\ u(0) = g, \end{cases}$$

where $g \in X$ and A is a linear operator that is independent of t. Because A will generally be a differential operator, we need to allow the possibility that A is not a bounded operator on X. In fact, we shall assume that $A \colon D \to X$ is a densely defined, closed linear operator, as discussed in Section 6.6.

In this complicated and abstract setting, what does a "solution" of (29) mean? A *classical solution of (29)* is $u \in C([0,\infty), X) \cap C^1((0,\infty), X)$ such that $u(t) \in D$ for all $0 < t < \infty$, and (29) holds. The solution of (29) is *unique* if, when $g \equiv 0$, the only solution (in the preceding sense) is $u(t) \equiv 0$. As in the finite-dimensional case, solving (29) is closely connected with semigroups. Let us consider an example.

Example 3. Let $X = C_B(\mathbf{R}^n)$ and $A = \Delta$ so that (29) becomes the pure initial value problem for the heat equation that was studied in Section 5.2. Notice that A is not defined on all of X, so let us take $D = C_B^2(\mathbf{R}^n)$ (i.e., functions whose derivatives up to order two are all continuous and bounded on $\mathbf{R}^n$). Then $A \colon D \to X$ is certainly well defined and linear. We know from Section 5.2 that, for any $g \in X$, (29) admits a unique solution given explicitly by

(30) $$u(x,t) = \int_{\mathbf{R}^n} K(x,y,t)\, g(y)\, dy = \frac{1}{(4\pi t)^{n/2}} \int_{\mathbf{R}^n} e^{-\frac{|x-y|^2}{4t}} g(y)\, dy.$$

If we let $(S(t)g)(x) = u(x,t)$, then we have defined a semigroup of operators $S(t) \colon X \to X$ satisfying the properties in (19). Notice that the convergence $S(t)g \to g$ as $t \to 0^+$ occurs in the X-norm and guarantees that the solution $u(x,t)$ satisfies the initial condition in (29). Also, notice that the inequality $\|S(t)g\|_X \le \|g\|_X$ for all $g \in X$, which is a consequence of $\int_{\mathbf{R}^n} K(x,y,t)\, dy = 1$, guarantees that $S(t) \colon X \to X$ for each $t > 0$ is not only a bounded operator, but in fact is also a contraction. ♣

This example illustrates the replacement of the real-valued function $u(x,t)$ by the X-valued function $u(t)$, which allows us to view the time evolution as being governed by a semigroup action on the Banach space X. It also inspires the general definition of *a strongly continuous* (or C^0) *semigroup of operators on a Banach space* X as a family of bounded linear operators $S(t) \colon X \to X$ for $t \ge 0$ satisfying (19); that is, (i) $S(t+s) = S(t)S(s)$, (ii) $S(0) = I$, and (iii) $S(t)u \to S(t_0)u$ for every $u \in X$ as $t \to t_0$. Notice that convergence in (iii) is in the X-norm, and it suffices to have this property at $t = 0$ (cf. Exercise 2). As in the finite-dimensional case, we call the semigroup *contractive* if $\|S(t)u\| \leq \|u\|$ for all $u \in X$ and $t > 0$, and *β-contractive* (or *quasicontractive*) if

(31) $$\|S(t)u\| \leq Ce^{\beta t}\|u\| \quad \text{for all } u \in X,$$

where $C > 0$ and $\beta \in \mathbf{R}$. In fact, every C^0-semigroup of operators is β-contractive for some $\beta \in \mathbf{R}$ (c.f. Exercise 6).

Of course, we are mainly interested in semigroups $S(t)$ that arise as the *solution* or *evolution operator* for (29). In fact, given a C^0-semigroup $S(t)$ on a Banach space X, we may *define* a linear operator $A: D \to X$, so that $S(t)$ is the solution operator for (29). This is achieved by letting

(32a) $$Au = \lim_{t \to 0^+} \frac{S(t)u - u}{t} \quad \text{for } u \in D, \text{ where}$$
(32b) $$D = \{u \in X : \text{the limit in (32a) exists}\}.$$

The operator A (along with its domain D) is called the *infinitesimal generator* of the semigroup; by analogy with the finite-dimensional case, we sometimes write $S(t) = e^{tA}$ in this case.

It is important to confirm that D is nonempty, and $A: D \to X$ has nice properties, including the fact that $S(t)$ is indeed the solution operator for (29). This is achieved by the following.

Theorem 1. Suppose $S(t)$ is a C^0-semigroup on a Banach space X, and $A: D \to X$ is defined by (32a,b). Then the following hold.
(i) The domain D is a dense subset of X.
(ii) $A: D \to X$ is a *closed* operator.
(iii) For $u \in D$, we have $S(t)u \in D$ for all $t \geq 0$, and $AS(t)u = S(t)Au$ for all $t > 0$.
(iv) For $g \in D$, $u(t) = S(t)g$ is a classical solution of (29).

Proof. For $h > 0$, consider the identities

(33a) $$\frac{S(t+h)u - S(t)u}{h} = \frac{S(h) - I}{h}S(t)u = S(t)\frac{S(h) - I}{h}u,$$

which hold for $u \in X$. But if $u \in D$, then we may let $h \to 0$; since $h^{-1}(S(h) - I)u \to Au$, the last identity in (33a) shows that $S(t)u \in D$, and $AS(t)u = S(t)Au$. This establishes (iii). Moreover, if we replace u by g and let $h \to 0$ in (33a), we obtain $D_t^+(S(t)g) = A(S(t)g)$, where D_t^+ denotes the right-hand derivative. For $0 < h < t$, we also have the identity

(33b) $$\frac{S(t)u - S(t-h)u}{h} = S(t-h)\frac{S(h) - I}{h}u,$$

which holds for $u \in X$. If we replace u by $g \in D$ and let $h \to 0$ in (33b), we obtain that $D_t^-(S(t)g) = S(t)Ag = AS(t)g$. Thus $u(t) = S(t)g$ is differentiable and satisfies (29); in fact, this means $u \in C^1([0,\infty), X)$ since $u_t(t) = S(t)Ag$ is continuous in t. This proves (iv).

To show that D is dense in X, let us observe that for all $u \in X$, $\int_0^t S(s)u\,ds$ may be defined as a limit of Riemann sums (cf. Exercise 5a). Moreover,

$$(34) \qquad u = \lim_{h \to 0^+} \frac{1}{h} \int_0^h S(s)u\,ds$$

is a consequence of $S(0) = I$ and the strong continuity of $S(t)$ (cf. Exercise 5b). Now consider

$$(35) \qquad \begin{aligned} \frac{S(h) - I}{h} \int_0^t S(s)u\,ds &= \frac{1}{h} \int_0^t (S(s+h) - S(s))u\,dx \\ &= \frac{1}{h} \int_t^{t+h} S(s)u\,dx - \frac{1}{h} \int_0^h S(s)u\,dx. \end{aligned}$$

If we let $h \to 0^+$, we may use (34) to show that the right-hand side in (35) tends to $S(t)u - u$. In particular, $\int_0^t S(s)u\,ds \in D$ for every $u \in X$ and $t > 0$, so (34) shows that D is dense in X, and (i) is proved.

To show (ii) that $A: D \to X$ is a closed operator, let $u_n \in D$ satisfy $u_n \to u \in X$ and $Au_n \to w \in X$; it must be shown that $u \in D$ and $Au = w$. This is Exercise 4b. ♣

Conversely, there exist general conditions on a densely defined, closed operator $A: D \to X$ that imply the existence of the C^0-semigroup $S(t) = e^{At}$; this is the subject of the *Hille-Yoshida theorem* (cf. [Yoshida]). However, we do not need to invoke this theorem because we always deal with concrete operators A for which we may display $S(t)$ explicitly; this means that we do need to show care in choosing D if we want $A: D \to X$ to be a closed operator. For example, this is not true in Example 3, but let us consider another example.

Example 4. Let $\Omega \subset \mathbf{R}^n$ be a smooth bounded domain, and suppose we want to study the heat equation with Dirichlet boundary conditions (as in Section 5.1):

$$(36) \qquad \begin{cases} u_t = \Delta u & \text{for } x \in \Omega \text{ and } t > 0, \\ u(x, 0) = g(x) & \text{for } x \in \Omega, \\ u(x, t) = 0 & \text{for } x \in \partial\Omega \text{ and } t > 0. \end{cases}$$

Let $X = L^2(\Omega)$ and $A = \Delta$ with $D = H_0^{1,2}(\Omega) \cap H^{2,2}(\Omega)$ so we know that $A: D \to X$ is a closed operator (as in Example 2 in Section 6.6). Furthermore, let $\{\lambda_k, \phi_k(x)\}_{k=1}^\infty$ denote the Dirichlet eigenvalues and eigenfunctions of the Laplacian on Ω (cf. Sections 4.4 and 7.2). Assuming $g \in L^2(\Omega)$, we can use the method of eigenfunction expansions (cf. Section 5.1) to write the solution of (36) as

$$(37) \qquad u(x,t) = \sum_{k=1}^{\infty} a_k e^{-\lambda_k t} \phi_k(x), \qquad \text{where} \qquad g(x) = \sum_{k=1}^{\infty} a_k \phi_k(x).$$

For every $g \in L^2(\Omega)$, the expression (37) satisfies $u \in C([0,\infty), X) \cap C^1((0,\infty), X)$ (cf. Exercise 7a) and $u(t) \in H_0^{1,2}(\Omega) \cap H^{2,2}(\Omega)$ for all $t > 0$ (cf. Exercise 7b). Therefore, (37) defines the solution operator $S(t) = e^{tA}$. Notice that $\{S(t)\}_{t \geq 0}$ is contractive (cf. Exercise 7c), in fact β-contractive for some $\beta < 0$ (cf. Exercise 7d). By analogy with the finite-dimensional case, we might write $M_- = X$ to assert that all solutions decay exponentially.

At times, it is useful to study (36) with additional regularity. For this purpose, we may let $X = H^{k,2}(\Omega)$, and $D = H_0^{1,2}(\Omega) \cap H^{k+2,2}(\Omega)$, for some integer $k \geq 0$. The regularity results of Section 8.2 again show that $A: D \to X$ is the infinitesimal generator of the solution operator $S(t) = e^{tA}$ defined by (37), which again is β-contractive for some $\beta < 0$. ♣

In both Examples 3 and 4, the solution operators $S(t)$ enjoy the property that $S(t)g \in D$ for *every* $g \in X$, not just for $g \in D$; this means that we can solve (29) for every $g \in X$. This is *not* true for all C^0-semigroups, as we see in Chapter 12 when we study the wave equation as a first-order evolution on a Banach space. But Examples 3 and 4 belong to a specialized class called *analytic semigroups*, which are usually defined using certain special properties of A and $S(t)$; see [Henry], [Kato], or [Yoshida]. For our purposes, however, we call a C^0-semigroup $S(t)$ *analytic* if, for every $g \in X$, $u(t) = S(t)g \in C^\infty((0,\infty), X)$ and satisfies (29); in particular, $u(t) = S(t)g \in D$ for every $g \in X$ and $t > 0$.

It is also possible to treat the "nonautonomous" case of (29) when A is replaced by $A(t)$. However, we do not consider this more general case; see [Henry] or [Yoshida].

c. The Nonhomogeneous Equation

We would like to study the nonhomogeneous linear evolution equation on a Banach space X,

$$(38) \qquad \begin{cases} \dfrac{du}{dt} = Au + f(t) & \text{for } 0 < t < T, \\ u(0) = 0, \end{cases}$$

where $f \in C([0,T], X)$ and $T < \infty$. If the associated homogeneous equation $u_t = Au$ has a solution operator $S(t)$, then Duhamel's principle (cf. Sections 2.3d, 3.1d, and 5.2d) suggests that we consider

$$(39) \qquad u(t) = \int_0^t S(t-s) f(s)\, ds.$$

Notice that (39) defines $u \in C([0,T], X)$ with $u(0) = 0$; cf. Exercises 5 and 9. Now let us ask whether u satisfies the PDE in (38). Formally, this is certainly true since $(S(t-s)f(s))_t = AS(t-s)f(s)$ implies $u_t(t) = f(t) + A\int_0^t S(t-s)f(s)\,ds = Au(t) + f(t)$. For this reason, the function u defined by (39) is called a *mild solution* of (38). But establishing that the mild solution is a classical solution [i.e., $u \in C^1((0,T), X)$] depends on the properties of the semigroup as well as the regularity of f.

First let us find conditions under which (39) defines a classical solution when A is the infinitesimal generator of a general C^0-semigroup.

Theorem 2. *If $S(t)$ is a C^0-semigroup with infinitesimal generator A, and $f \in C^1([0,T], X)$, then (39) defines $u \in C^1([0,T], X)$, which is a classical solution of (38).*

Proof. For $h > 0$, consider the identity

$$(40) \quad \frac{u(t+h) - u(t)}{h} = \frac{S(h) - I}{h} u(t) + \frac{1}{h} \int_t^{t+h} S(t+h-s)f(s)\,ds.$$

As $h \to 0$, the continuity of f and the strong continuity of $S(t)$ imply that $h^{-1} \int_t^{t+h} S(t+h-s)f(s)\,ds \to f(t)$. If $u \in C([0,T], D)$, where D is considered as a Banach space under the graph norm (cf. Exercise 2 in Section 6.6), then (40) shows that the right-hand derivative $D^+ u(t)$ exists and is continuous in t; replacing t by $t-h$ in (40) shows that the left-hand derivative $D^- u(t)$ also exists and is continuous [i.e., $u \in C^1([0,T], X)$]. The converse is also true; that is, $u \in C^1([0,T], X)$ if and only if $u \in C([0,T], D)$; moreover, in that case, u is a classical solution of (38). If we rewrite u as

$$(41) \quad u(t) = \int_0^t S(s)f(t-s)\,ds,$$

then we may compute

$$(42) \quad \frac{du}{dt} = S(t)f(0) + \int_0^t S(s)f'(t-s)\,ds.$$

By Exercise 5b, $u \in C^1([0,T], X)$, so u indeed is a classical solution of (38). ♠

For parabolic time evolutions that generate analytic semigroups, Theorem 2 can be improved by relaxing the condition $f \in C^1([0,T], X)$ to Hölder continuity in t; that is, for some $0 < \alpha \leq 1$

$$(43) \quad \|f(t+h) - f(t)\| \leq Ch^\alpha, \quad \text{for all } 0 \leq t < t + h \leq T;$$

we denote the space of functions satisfying (43) by $C^\alpha([0,T], X)$. We need our semigroup to satisfy the following estimate:

$$(44) \quad \|AS(t)g\| \leq Ct^{-1}\|g\|, \quad \text{for all } g \in X \text{ and } 0 < t \leq T.$$

(Under a more stringent definition of analytic semigroup, (44) is automatically true; cf. [Henry], [Kato], or [Yoshida].)

Theorem 2'. If $S(t)$ is an analytic semigroup satisfying (44) and $f \in C^\alpha([0,T], X)$ for some $0 < \alpha \leq 1$, then (39) defines $u \in C^1([0,T], X)$, which is a classical solution of (38).

Proof. We can again obtain from (40) that $u \in C^1([0,T], X)$ if and only if $u \in C([0,T], D)$. Let us rewrite (39) as

$$(45) \quad u(t) = \int_0^t S(t-s)[f(s) - f(t)]\,ds + \int_0^t S(t-s)f(t)\,ds = u_1(t) + u_2(t).$$

Let us consider $u_2(t) = \int_0^t S(t-s)f(t)\,ds$ first. Since $S(t)$ is an analytic semigroup, we have $S(t-s)f(t) \in D$ for all $0 < s < t$. Writing the integral as a limit of Riemann sums, we conclude that $u_2(t) \in D$ for all $t \in [0,T]$, and $Au_2(t) = -\int_0^t \frac{d}{ds}S(t-s)f(t)\,ds = (S(t) - I)f(t)$ (cf. Exercise 10). Clearly, $u_2 \in C([0,T], D)$. Next we want to show that $v(t) = u_1(t) \in C([0,T], D)$. It is again easy to show that $v(t) \in D$ for all $t \in [0,T]$; moreover, for $0 \leq t < t+h \leq T$, let us write

$$(46) \quad \begin{aligned} v(t+h) - v(t) &= I_1(t) + I_2(t) + I_3(t) \quad \text{where} \\ I_1(t) &= \int_0^t (S(h) - I)S(t-s)(f(s) - f(t))\,ds \\ I_2(t) &= \int_0^t S(t+h-s)(f(t) - f(t+h))\,ds \\ I_3(t) &= \int_t^{t+h} S(t+h-s)(f(s) - f(t+h))\,ds. \end{aligned}$$

We claim that $\|AI_k(t)\| \to 0$ as $h \to 0$, for each $k = 1, 2, 3$. In fact, $AI_1(t) = (S(h) - I)\int_0^t AS(t-s)(f(s) - f(t))\,ds$. If we use (43) and (44), we see that the integral is estimated by $\int_0^t (t-s)^{\alpha-1}\,ds$, which converges, so $S(h) \to I$ as $h \to 0$ implies $AI_1(t) \to 0$. Similarly, $AI_2(t)$ is estimated by $\int_0^t (t+h-s)^{-1}h^\alpha\,ds = (\log(t+h) - \log(h))h^\alpha$, so $AI_2(t) \to 0$ as $h \to 0$. Finally, $AI_3(t)$ is estimated by $\int_t^{t+h}(t+h-s)^{\alpha-1}\,ds = h^\alpha$, so $AI_3(t) \to 0$ as $h \to 0$. Thus we see that $Av = Au_1 \in C([0,T], X)$; that is, $u_1 \in C([0,T], D)$. We conclude that $u \in C^1([0,T], X)$. ♠

Remark. In Theorem 2' we can actually show that $u_t \in C^\alpha([\epsilon, T], X)$ for any $\epsilon > 0$ and for $\epsilon = 0$ if $f(0) = 0$ (cf. Exercise 11).

Let us consider an example (cf. also Exercise 12a).

Example 4 (Revisited). For $X = L^2(\Omega)$ and $A = \Delta$ with $D = H_0^{1,2}(\Omega) \cap H^{2,2}(\Omega)$, (39) becomes

$$(47) \quad u(x,t) = \sum_{k=1}^\infty u_k(t)\phi_k(x) = \sum_{k=1}^\infty \left(\int_0^t e^{-\lambda_k(t-s)} b_k(s)\,ds\right)\phi_k(x),$$

where $\sum b_k(t)\phi_k(x)$ is the eigenfunction expansion of $f(x,t)$. If $f(t) = f(\cdot, t) \in C^1([0,T], X)$, that is, if $b'_k(t)$ is continuous and $\sum b'_k(t)\phi_k \in L^2(\Omega)$ for all $t \in [0,T]$, then (47) defines $u(t) = u(\cdot, t) \in C^1([0,T], X)$ by Theorem 2. If $f(t)$ is only assumed to satisfy (43), which in this case takes the form

$$(48) \qquad \|f(t+h) - f(t)\|^2 = \sum_{k=1}^{\infty} |b_k(t+h) - b_k(t)|^2 \le Ch^{2\alpha},$$

then we need the semigroup to satisfy (44) in order to conclude that $u(t) \in C^1([0,T], X)$; this is verified in Exercise 12b. ♣

We can, of course, combine the nonhomogeneous equation with a nonhomogeneous initial condition by superposition; that is, the solution of

$$(49) \qquad \begin{cases} \dfrac{du}{dt} = Au + f(t) & \text{for } 0 < t < T, \\ u(0) = g, \end{cases}$$

is given by

$$(50) \qquad u(t) = S(t)g + \int_0^t S(t-s)f(s)\,ds.$$

For $g \ne 0$, however, we should only expect to have $u \in C^1((0,T], X)$ and not $u \in C^1([0,T], X)$; this is because the term $S(t)g$ has this limitation.

d. Weak Solutions and Energy Methods

Frequently in applications, we encounter evolution equations for which the requirement that u be C^0, let alone C^1, in the variable t is too restrictive. This may be due to insufficient regularity of the coefficients in the equation or the data, or possibly due to mild singularities that may develop in the solution. For such situations, it is preferable to work with some sort of weak solution. Of course, we have encountered weak solutions before, but not specifically for evolution equations of the form (29).

To discuss weak solutions for evolution equations, we may view (29), or the nonhomogeneous case (38), as an equation in the dual space V' of some Banach space V satisfying $V \subset X$. Thus we might define a *weak solution* of (38) to be a function $u: [0,T) \to V$ satisfying $u_t, Au \in V'$ and

$$(51) \qquad \langle u_t(t), v \rangle = \langle Au(t), v \rangle + \langle f(t), v \rangle \quad \text{for every } v \in V,$$

where $\langle u', v \rangle$ denotes the action of $u' \in V'$ on $v \in V$. Notice that (51) holds pointwise in $t \in (0,T)$. In fact, since weak solutions are generally defined through integration, we might only require that (51) holds for *almost every* (a.e.) t in $(0,T)$. In that case, however, it is not clear what u_t or the initial condition $u(0) = g$ mean. We need to impose some additional control on the solution in the t-variable.

For a given Banach space Z and $1 \le p < \infty$, let us define $L^p((0,T), Z)$ to be the Banach space completion of $C([0,T], Z)$ in the norm

$$(52) \qquad \|u\|_{L^p((0,T),Z)} = \left(\int_0^T \|u(t)\|_Z^p \, dt \right)^{1/p}.$$

Similarly, we may define $H^{1,p}((0,T), Z)$ to be the Banach space completion of $C^1([0,T], Z)$ in the norm

$$(53) \qquad \|u\|_{H^{1,p}((0,T),Z)} = \left(\int_0^T \|u(t)\|_Z^p + \|u_t(t)\|_Z^p \, dt \right)^{1/p}.$$

With these definitions, we replace the condition $u_t \in V'$ for a.e. $t \in (0,T)$ by $u \in H^{1,p}((0,T), V')$ for some $1 \le p < \infty$. This also enables us to make sense of the initial condition $u(0) = g$, thanks to the following.

Proposition 1. *If Z is a Banach space and $u \in H^{1,p}((0,T), Z)$ for some $1 \le p < \infty$, then*
(i) $u \in C([0,T], Z)$ (up to a set of measure zero), and
(ii) $u(t) = u(s) + \int_s^t u_\tau(\tau) \, d\tau$ for all $0 \le s \le t$.

The proof of this proposition is Exercise 13.

We generally consider applications where X and V are Hilbert spaces, and we identify X with its dual space to obtain $V \subset X \subset V'$. In addition, we generally have

$$(54) \qquad A \colon V \to V' \quad \text{is a bounded linear operator,}$$

which satisfies the following *coercivity condition:*

$$(55a) \qquad a(u,u) \ge C_1 \|u\|_V^2 - C_2 \|u\|_X^2 \quad \text{for all } u \in V,$$

where C_1 and C_2 are positive constants and $a(u,v)$ is the bilinear form defined by

$$(55b) \qquad a(u,v) = -\langle Au, v \rangle \quad \text{for all } u, v \in V.$$

Moreover, because of (54), we see that $u \in L^p((0,T), V)$ implies $Au \in L^p((0,T), V')$, and to prove (51) it suffices to show that

$$(56) \qquad \int_0^T \langle u_t, w \rangle \, dt + \int_0^T a(u,w) \, dt = \int_0^T \langle f, w \rangle \, dt$$
$$\text{for all } w \in L^{p'}((0,T), V).$$

In fact, by density, it suffices to show (56) holds for all $w \in C_0^1(U_T)$.

The basic approach to finding a solution of (51) consists of several steps. First, approximate the given equation by a sequence of simpler or more regular equations for which solutions are known to exist. Next, obtain a priori estimates, called *energy estimates*, for these approximate solutions. Finally, use compactness results to obtain a subsequence of the approximate solutions that converges to a weak solution [i.e., a solution of (51)]. Let us illustrate the method with a very simple example.

Example 5. Let Ω be a smooth, bounded domain in $\mathbf{R}^n$, and consider

(57a)
$$\begin{cases} u_t = \Delta u + f(x,t) & \text{in } U_T = \Omega \times (0,T) \\ u(x,t) = 0 & \text{for } x \in \partial\Omega \text{ and } 0 < t < T \\ u(x,0) = g(x) & \text{for } x \in \Omega, \end{cases}$$

where we shall assume that

(57b)
$$f \in L^2(U_T) \quad \text{and} \quad g \in L^2(\Omega).$$

Let us take $X = L^2(\Omega)$, $V = H_0^{1,2}(\Omega)$, and $V' = H^{-1,2}(\Omega)$, so that $A = \Delta \colon V \to V'$ is an isomorphism. Moreover, $a(u,v) = \int_\Omega \nabla u \cdot \nabla v \, dx$, and (55a) holds (with $C_2 = 0$) by the Poincaré inequality.

For the moment, suppose u is a sufficiently regular solution of (57a,b). Multiply $u_t = \Delta u + f$ by u and integrate over Ω; using $(u^2)_t = 2uu_t$ and $\int_\Omega u\Delta u \, dx = -\int_\Omega |\nabla u|^2 \, dx$, we find

(58)
$$\frac{d}{dt}\int_\Omega \frac{u^2}{2} \, dx = -\int_\Omega |\nabla u|^2 \, dx + \int_\Omega fu \, dx.$$

Integrating (58) over $t \in (0,T)$, and using $2ab \leq \epsilon^{-1}a^2 + \epsilon b^2$, we obtain

(59)
$$\frac{1}{2}\left(\|u(T)\|_2^2 - \|g\|_2^2\right) + \int_0^T \|\nabla u\|_2^2 \, dt = \int_0^T \langle f, u \rangle \, dt$$
$$\leq \frac{1}{2\epsilon}\int_0^T \|f\|_{-1,2}^2 \, dt + \frac{\epsilon}{2}\int_0^T \|u\|_{1,2}^2 \, dt.$$

We may drop the term $\|u(T)\|_2^2$ in (59), utilize (55a), and take ϵ sufficiently small to obtain the following a priori inequality:

(60)
$$\int_0^T \|u\|_{1,2}^2 \, dt \leq C\left(\|g\|_2^2 + \int_0^T \|f\|_{-1,2}^2 \, dt\right).$$

By (57b), the right hand side in (60) is finite, so this inequality suggests that we seek a weak solution of (57a) with $u \in L^2((0,T), H_0^{1,2}(\Omega))$; the PDE in (57a) then shows $u_t \in L^2((0,T), H^{-1,2}(\Omega))$. Approximating u by smooth functions, we conclude that (60) holds for such u. In particular, if

$u(0) = g \equiv 0$ and $f(x,t) \equiv 0$, then such a weak solution must satisfy $u \equiv 0$. This shows that a weak solution $u \in L^2((0,T), H_0^{1,2}(\Omega))$ of (57a,b) must be *unique*.

Let us now show that (60) in fact implies the *existence* of a weak solution of (57a,b). Of course, if f is sufficiently regular, namely if $f \in C^1([0,T], X)$, then we may use Theorem 2 (or 2') to find a solution. More generally, since $C^1([0,T], X)$ is dense in $L^2((0,T), X)$, we may find $f_\epsilon \in C^1([0,T], X)$ satisfying $f_\epsilon \to f$ in $L^2((0,T), X)$. We then apply Theorem 2 to find $u_\epsilon \in C([0,T], X) \cap C^1((0,T), X)$ satisfying $u_\epsilon(t) \in V$ for all $t \in [0,T]$, and

$$(61) \qquad (u_\epsilon)_t = \Delta u_\epsilon + f_\epsilon(x,t),$$

with the same initial/boundary values as in (57a). Moreover, if we repeat the arguments for (60), we find that u_ϵ satisfies the following a priori estimate:

$$(62) \qquad \int_0^T \|u_\epsilon\|_{1,2}^2 \, dt \leq C \left(\|g\|_2^2 + \int_0^T \|f_\epsilon\|_{-1,2}^2 \, dt \right).$$

Since $\|f_\epsilon\|_{-1,2} \leq \|f_\epsilon\|_2$ and $f_\epsilon \to f$ in $L^2(U_T)$, we see that (62) implies that $\{u_\epsilon\}$ is uniformly bounded in the Hilbert space $\mathcal{H} = L^2((0,T), H_0^{1,2}(\Omega))$. If we apply Theorem 2 of Section 6.3, we may find a subsequence, $\{u_{\epsilon_j}\}_{j=1}^\infty$, that converges weakly in $\mathcal{H}$ to a function $u \in \mathcal{H}$; that is,

$$(63) \qquad \int_0^T \langle \nabla u_{\epsilon_j}, \nabla w \rangle \, dt \to \int_0^T \langle \nabla u, \nabla w \rangle \, dt \quad \text{for all } w \in \mathcal{H}.$$

Since each u_ϵ satisfies (61) and is bounded in $\mathcal{H}$, we find that $(u_\epsilon)_t$ is bounded in the Hilbert space $\mathcal{H}' = L^2((0,T), H^{-1,2}(\Omega))$. If we apply Theorem 2 of Section 6.3 to the subsequence $\{u_{\epsilon_j}\}_{j=1}^\infty$, we obtain another subsequence (which we continue to denote by $\{u_{\epsilon_j}\}_{j=1}^\infty$) that converges weakly in $\mathcal{H}'$ to a function $u' \in \mathcal{H}'$; that is,

$$(64) \qquad \int_0^T \langle (u_{\epsilon_j})_t, w \rangle \, dt \to \int_0^T \langle u', w \rangle \, dt \quad \text{for all } w \in \mathcal{H}.$$

We conclude that there is a unique $u \in \mathcal{H} = L^2((0,T), H_0^{1,2}(\Omega))$ for which $u_t \in L^2((0,T), H^{-1,2}(\Omega))$ and (57a) is satisfied in the weak sense. ♣

We see energy methods applied to the (nonlinear) Navier-Stokes equations in Section 11.4, and to linear hyperbolic systems in Section 12.1.

e. Nonlinear Dynamics

Finally, let us consider a nonlinear evolution equation on a Banach space X

$$(65) \quad \begin{cases} \dfrac{du}{dt} = Au + F(u) & \text{for } 0 < t < T, \\ u(0) = g, \end{cases}$$

where $g \in X$ and $A\colon D \to X$ is a densely defined and closed linear operator. The expression F in (65) is a possibly nonlinear operator that, for now at least, we shall assume is continuous $F\colon X \to X$. There are at least three issues to address:

(i) *Local existence:* Show that a unique solution exists for $0 < t < \tau$ provided $\tau > 0$ is sufficiently small;
(ii) *Global existence:* Does the solution exist for all $0 < t < \infty$, or does it have a finite existence time T?
(iii) *Asymptotic properties:* What happens to the solution as $t \to \infty$?

As we see in the next three chapters, each of these issues requires analysis specially designed for the particular nonlinear dynamics being studied. In this section, however, we want to prove general results for the issues (i) and (ii), which later may be fine-tuned for the specific application.

To study *local existence*, let us observe that a solution of (65) must satisfy (49) with $f(t) = F(u(t))$; by (50) we conclude that u must satisfy

$$(66) \quad u(t) = S(t)g + \int_0^t S(t-s) F(u(s))\, ds \qquad \text{for } 0 < t < \tau.$$

If $u \in C([0, \tau], X)$ satisfies (66), then u is called a *mild solution* of (65). If we let $\psi(t) = S(t)g$ and $\mathcal{T}(u(t)) = \int_0^t S(t-s) F(u(s))\, ds$, then we are reduced to finding a fixed point of the map $u(t) \to \psi(t) + \mathcal{T}(u(t))$. This is generally done by means of the contraction mapping principle; but to show that $\mathcal{T}$ is a contraction, we need to assume that $F\colon X \to X$ is *locally Lipschitz continuous*; that is,

$$(67) \quad \|F(u) - F(v)\| \leq M_R \|u - v\| \qquad \text{for all } \|u\|, \|v\| \leq R.$$

As an example of this method, we prove the following.

Theorem 3 (Local Existence). *If $S(t)$ is a C^0-semigroup with infinitesimal generator A, and $F\colon X \to X$ is continuous and satisfies (67), then for every $g \in D$ there is a positive τ (depending on F and $\|g\|$) such that there exists a unique mild solution $u \in C([0, \tau], X)$ of (65).*

Proof. Let us assume that $S(t)$ is a contraction [i.e., $C = 1$ and $\beta = 0$ in (31); the more general case involves simple modifications. If we introduce the space $Y = C([0, \tau], X)$ with the norm $\|u\|_Y = \max_{0 \leq t \leq \tau} \|u(t)\|$, then we may consider $\psi = S(t)g \in Y$, and $\mathcal{T}: Y \to Y$ is a continuous mapping. We want to show that $u \to \psi + \mathcal{T}(u)$ has a fixed point in Y, provided that τ is sufficiently small.

Let $R = 2\|g\|$ and Y_R be the closed ball of radius R in Y; that is, $Y_R = \{u \in C([0,\tau], X) : \|u(t)\| \leq R \text{ for } 0 \leq t \leq T\}$. Let $\Lambda_R = M_R R + \|F(0)\|$. By (67) with $v = 0$, $\|F(u)\| - \|F(0)\| \leq M_R \|u\|$, which implies $\|F(u)\| \leq \Lambda_R$ when $\|u\| \leq R$. For $u \in Y_R$ this means that

$$\|\mathcal{T}(u(t))\| \leq \int_0^t \|S(t-s)F(u(s))\|\, ds \leq t\Lambda_R \leq \|g\|$$

provided that τ is sufficiently small. Since $\|g\| < R$, we see that $\mathcal{T}: Y_R \to Y_R$. To prove that $\mathcal{T}$ is a contraction, we again use (67):

$$\|\mathcal{T}(u(t)) - \mathcal{T}(v(t))\| \leq M_R \int_0^t \|u(s) - v(s)\|\, ds \leq M_R \tau \|u - v\|_Y.$$

Provided we take τ small enough, this implies that $\mathcal{T}: Y_R \to Y_R$ is a strict contraction.

Finally, we have

$$\|\psi(t) + \mathcal{T}(u(t))\| \leq \|g\| + \|\mathcal{T}(u(t))\| \leq 2\|g\| = R,$$

so $u \to \psi + \mathcal{T}(u)$ maps Y_R to itself and is also a strict contraction. Theorem 1 of Section 7.3 therefore assures the existence of a unique fixed point; that is, (66) admits a unique solution $u \in C([0,\tau], X)$. ♠

Theorem 3 establishes, at least for $0 \leq t \leq \tau$, the existence of a C^0-semigroup of *nonlinear* operators $\mathcal{S}(t)$ on X that may be interpreted as a *continuous flow* on X. Of course, we would like the mild solution of Theorem 3 to be a classical solution [i.e., $u \in C^1((0,\tau), X)$], in which case $\mathcal{S}(t)$ defines a C^1-*flow* on X; but this requires additional special properties of A and f. We encounter these issues in the applications in Chapters 10, 11, and 12.

Next let us consider *global existence* for (65). The simplest approach is to attempt to *iterate* the local existence argument. In other words, let $g_1 = u(\tau)$ be the new initial condition, and find $\tau_1 > 0$ such that a solution u_1 exists for $\tau \leq t \leq \tau + \tau_1$; this means the original problem (65) has a solution for $0 \leq t \leq \tau + \tau_1$. Continuing in this way, we generate a sequence $\tau, \tau_1, \tau_2, \ldots$. If we can show that $\tau + \tau_1 + \tau_2 + \ldots \geq T$, then we will have shown that the solution is globally defined for all $0 < t < T$.

This can certainly be achieved when the solutions are known to satisfy an a priori bound of the form

(68) $$\|u(t)\| \leq K \qquad \text{for } 0 \leq t < T.$$

Taking $R = \max\{2\|g\|, K\}$ in the proof of Theorem 2, we find that the same τ may be used for each iteration (i.e., $\tau_1 = 2\tau$, $\tau_2 = 3\tau$, etc.). As a consequence, we have the following.

Theorem 4 (Global Existence). *If we have local existence for (65), and the solution satisfies the a priori bound (68), then the solution exists for all $0 < t < T$.*

We shall see in applications how to establish a priori bounds like (68).

Exercises for Section 9.2

1. Perform the required ODE analysis of (21):
   (a) Compute the general solution of the system (22) with $c = 0$.
   (b) Show that the energy (23) of a solution of (21) satisfies $\dot{\mathcal{E}}(t) = -c\dot{y}^2$. When $c > 0$, conclude that $\mathcal{E}(t) \to 0$ as $t \to \infty$.
   (c) Find the general solution of (22) when $c^2 < 4k$. Conclude that the solution operator is a β-contraction with $\beta = -c/2$.
   (d) Find the general solution of (22) when $c^2 = 4k$, and conclude that the solution operator is a β-contraction for every $\beta > -c/2$.

2. Show that a semigroup of operators on a Banach space is strongly continuous on $[0,\infty)$, that is, $S(t)g \to S(t_0)g$ as $t \to t_0 \geq 0$, provided this continuity is true at $t_0 = 0$.

3. For a function $u(x)$ defined on $-\infty < x < \infty$, let $(S(t)u)(x) = u(x+t)$, so $\{S(t)\}_{t \geq 0}$ satisfies the first two semigroup properties.
   (a) Is $S(t)$ strongly continuous on $X = L^2(\mathbf{R}^n)$?
   (b) Is $S(t)$ strongly continuous on $X = C_B(\mathbf{R}^n)$?

4. (a) Show that the infinitesimal generator of a C^0-semigroup on X is a closed operator (cf. the proof of Theorem 2).
   (b) Suppose $A: D \to X$ is a closed operator and $u \in C([0,T], X)$ satisfies $u(t) \in D$ for all $t \in [0,T]$. Show that $u \in C([0,T], D)$ if and only if $Au \in C([0,T], X)$.

5. Let X be a Banach space and $f \in C([0,T], X)$.
   (a) Show that the Riemann integral $F(t) = \int_0^t f(s)\,ds$ exists for $0 \leq t \leq T$.
   (b) Show that $F \in C([0,T], X)$ and $f(0) = \lim_{t \to 0+} t^{-1} F(t)$.

6. Let $\{S(t)\}_{t \geq 0}$ be a C^0-semigroup of operators on a Banach space X.
   (a) Let $\phi: [0,\infty) \to \mathbf{R}$ be continuous and *subadditive* [$\phi(s+t) \leq \phi(s) + \phi(t)$]. Show that $\inf_{t > 0} \phi(t)/t = \lim_{t \to \infty} \phi(t)/t$.
   (b) Show that $\phi(t) = \log \|S(t)\|$ is a subadditive function.
   (c) Show that $S(t)$ is a β-contraction for every $\beta > \beta_0$, where $\beta_0 = \lim_{t \to \infty} (\log \|S(t)\|)/t$. (This β_0 is called the *type* of the semigroup.)

7. The following pertain to Example 4. Given $g \in L^2(\Omega)$, define u by (37).
   (a) By direct calculation, show that $u \in C([0,T], X) \cap C^1((0,T), X)$.
   (b) By direct calculation, show that $u(t) \in H_0^{1,2}(\Omega) \cap H^{2,2}(\Omega)$ for all $t > 0$.
   (c) By direct calculation, show that $\|u(t)\|_2 \leq \|g\|_2$.
   (d) Show that $\|u(t)\|_2 \leq Ce^{\beta t} \|g\|_2$ for some constants $C > 0$ and $\beta < 0$.

8. Describe the solution semigroup for the heat equation in a smooth bounded domain with Neumann boundary conditions. Verify that the semigroup is strongly continuous. Is it contractive? Is it β-contractive for some $\beta < 0$?

9. If X is a Banach space, $f \in C([0,T], X)$, and $S(t)$ is a C^0-semigroup, show that $h(s) = S(t-s)f(s) \in C([0,t], X)$ for all $0 \le t \le T$.

10. Let $A \colon D \to X$ be a closed operator on the Banach space X, and $f \in C([0,T], D)$ (cf. Exercise 4c). Let $u(t) = \int_0^t f(s)\,ds$ (cf. Exercise 5). Show that $u \in C([0,T], D)$ and $Au(t) = \int_0^t Af(s)\,ds$.

11. As in Theorem 2', let $f(t)$ satisfy (43), let $S(t)$ be an analytic semigroup satisfying (44), and let $u(t)$ be defined by (39).
    (a) Show that $Au \in C^\alpha([\epsilon, T], X)$ for any $\epsilon > 0$.
    (b) If $f(0) = 0$, show that $Au \in C^\alpha([0,T], X)$.

12. (a) Show that $S(t)$ for Example 3 satisfies (44).
    (b) Show that $S(t)$ for Example 4 satisfies (44).

13. Use a mollifier (cf. Section 6.5b) to prove Proposition 1.

Further References for Chapter 9

Just as the Schauder and Leray-Schauder fixed point theorems are generalizations of the Brouwer fixed point theorem, there is a Leray-Schauder degree for maps on a Banach space that generalizes the Brouwer degree in finite dimensions; for details and applications, see [Nirenberg,1], [Nirenberg,2], or [Smoller].

The application to the Navier-Stokes equations was the original motivation for Leray's work on fixed point theorems; however, we have followed a formulation due to Ladyzhenskaya. Although $n = 2$ or 3 was essential for this method, an alternative proof using *Galerkin's method* works when $n = 4$, and with extra conditions when $n \ge 5$ (cf. [Galdi]). For more on the stationary Navier-Stokes equations, see [Galdi], [Temam, 1], or [Temam, 2].

Our summary of finite-dimensional dynamical systems was only for illustrative purposes and does not do justice to the richness of the subject. For more details, see [Hirsch-Smale] or [Guckenheimer-Holmes].

There are many books dealing with strongly continuous and analytic semigroups, as well as their use in studying evolutionary PDEs. For example, see [Friedman], [Goldstein], [Henry], [Hille-Phillips], [Kato], [Pazy], [Yoshida].

For more information on the use of a priori energy estimates to study linear and nonlinear equations, particularly weak solutions, consult [Lions] or [Lions-Magenes] as well as Sections 11.4 and 12.1 of this book.

10 Systems of Conservation Laws

In this chapter, we study hyperbolic systems of first-order equations in two independent variables, x and t. We first consider the issue of local or short-time existence of solutions. To explore the global behavior of solutions, we consider a special class of quasilinear systems that arise frequently in physical applications: systems of conservation laws. We investigate special weak solutions, including rarefaction waves and shocks; these special solutions are particularly useful in solving Riemann problems, which occur frequently in the physical applications. Finally, we define the Riemann invariants when the system consists of just two equations; these are used to characterize the development of singularities and to define the hodograph transformation. The application to one-dimensional gas flow is discussed in particular detail.

10.1 Local Existence for Hyperbolic Systems

Let us consider the first-order system of PDEs

$$(1) \qquad \vec{u}_t + A\vec{u}_x = \vec{f},$$

where $\vec{u}$ and $\vec{f}$ are N-vectors and A is an $N \times N$ matrix. When A is independent of $\vec{u}$, then (1) is a semilinear system [in fact (1) is linear when $\vec{f}$ depends at most linearly on $\vec{u}$]. In general, however, A may also depend on $\vec{u}$, in which case (1) is quasilinear. Following Section 2.2, we define (1) to be *hyperbolic* when A has N real eigenvalues $\lambda_1, \ldots, \lambda_N$ and a basis of eigenvectors $\vec{V}_1, \ldots, \vec{V}_N$; however, in the quasilinear case, the λ_k and $\vec{V}_k$ may depend on $\vec{u}$ as well as on x and t. Similarly, we define (1) to be *strictly hyperbolic* when the λ_k are distinct.

Given an initial condition for (1), we see that a unique solution exists for at least a short time. When (1) is a linear system, we show that the solution exists for large values of t as well.

a. *Linear Systems*

Let us consider an initial value problem for a linear hyperbolic system

$$(2a) \qquad \begin{cases} \vec{u}_t + A(x,t)\vec{u}_x = D(x,t)\vec{u} + \vec{f}(x,t) & \text{for } x \in \mathbf{R} \text{ and } 0 < t < T, \\ \vec{u}(x,0) = \vec{g}(x) & \text{for } x \in \mathbf{R}, \end{cases}$$

Section 10.1: Local Existence for Hyperbolic Systems

where the $N \times N$ matrices A and D, and the N-vector $\vec{f}$ satisfy

(2b) $\quad A_{jk} \in C_B^1(\mathbf{R} \times [0,T])$ and $D_{jk}, f_k \in C_B^0(\mathbf{R} \times [0,T])$.

We use the method of characteristics to reduce the solution of (2a) to finding a fixed point of a contraction mapping (cf. Section 9.2c).

As in Section 2.2, the substitution $\vec{u} = \Gamma \vec{v}$, where Γ is the matrix of eigenvectors of A, transforms (2a) to the more convenient form

(2') $\quad \begin{cases} \vec{v}_t + \Lambda \vec{v}_x = D'(x,t)\vec{v} + \vec{f}'(x,t) & \text{for } x \in \mathbf{R} \text{ and } t > 0, \\ \vec{v}(x,0) = \vec{g}'(x) & \text{for } x \in \mathbf{R}, \end{cases}$

where $\Lambda = \Lambda(x,t)$ is the diagonal matrix with eigenvalues $\lambda_1, \ldots, \lambda_N$ on the main diagonal, which, along with D' and $\vec{f}'$, are again continuous and uniformly bounded on $\mathbf{R} \times [0,T]$.

The left-hand side of (2') is decoupled; if the right-hand side also decouples, we can solve each equation in (2') separately by the method of characteristics as in Section 1.1. To review this method, we solve $v_t + \lambda(x,t)v_x = c(x,t,v)$ with $v(x,0) = g(x)$ by the following steps to determine the value of v at a given point $\bar{x}, \bar{t}$:

(i) Find the characteristic curve $x(t)$ through $(\bar{x}, \bar{t})$ by solving $dx/dt = \lambda(x,t)$, with initial condition $x(\bar{t}) = \bar{x}$.
(ii) Find the point $\xi = x(0)$ where the characteristic hits the initial curve $t = 0$ (see Figure 1a).
(iii) Solve the ODE $dz/dt = c(x(t), t, z)$ with initial condition $z(0) = g(\xi)$ to find $z(\bar{t})$.
(iv) Let $v(\bar{x}, \bar{t}) = z(\bar{t})$.

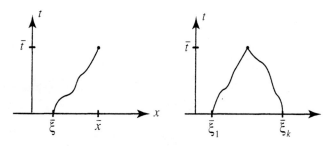

Figures 1a, 1b. The method of characteristics.

Without actually solving the ODE in (iii), however, notice that $z(t)$ satisfies the integral equation $z(\bar{t}) = g(\xi) + \int_0^{\bar{t}} c(x(t), t, z(t))\, dt$. This means that $v(x,t)$ satisfies

(3) $\quad v(\bar{x}, \bar{t}) = g(\xi) + \int_0^{\bar{t}} c(x(t), t, v(x(t), t))\, dt.$

Applying the method of characteristics to (2′) yields a system of integral equations similar to (3) for the vector $\vec{v}$. Namely, each equation of (2′) is of the form

(4)
$$\begin{cases} (v_k)_t + \lambda_k(x,t)(v_k)_x = \sum_{j=1}^{N} D'_{jk}(x,t)v_j + f'_k(x,t) \\ v_k(x,0) = g_k(x) \end{cases}$$

whose characteristic equation is $dx/dt = \lambda_k(x,t)$. In other words, given a point $(\bar{x}, \bar{t})$, we obtain k different characteristics $x_k(t)$ through $(\bar{x}, \bar{t})$, each of which hits the x-axis at some value $\xi_k = x_k(0)$ (see Figure 1b). The values $v_k(\bar{x},\bar{t})$ of the solution satisfy

(5)
$$v_k(\bar{x},\bar{t}) = g_k(\xi_k) + \int_0^{\bar{t}} \left[\sum_{j=1}^{N} D'_{jk}(x_k(t),t) v_j(x_k(t),t)) + f'_k(x_k(t),t) \right] dt.$$

As in Section 9.2c, a solution $\vec{v}$ of (5) is called a *mild solution* of (2′).

Let us write (5) as $\vec{v} = \vec{\phi} + \mathcal{T}(\vec{v})$, where $\phi_k(\bar{x},\bar{t}) = g_k(\xi)$ and $\mathcal{T}$ is the integral operator in (5). Then solving (2′) amounts to finding a fixed point of the mapping $\vec{v} \to \vec{\phi} + \mathcal{T}(\vec{v})$. In order to apply the contraction mapping principle of Section 7.3, we must introduce a complete metric space X on which $\mathcal{T}$ is a contraction mapping. A natural choice is the Banach space $X = C_B(\mathbf{R} \times [0,\tau], \mathbf{R}^N)$ of all N-vector-valued functions $\vec{v}(x,t)$ that are continuous and bounded in $x \in \mathbf{R}$ and $0 \leq t \leq \tau$. If we define

$$\|\vec{v}\| = \sup\{|v_k(x,t)| : x \in \mathbf{R},\ 0 \leq t \leq \tau,\ \text{and}\ k = 1,\ldots,N\},$$

then X is a complete metric space under the metric $d(v,w) = \|v - w\|$.

Notice that for $\vec{v}, \vec{w} \in X$ we have

(6)
$$\|\mathcal{T}(\vec{v}) - \mathcal{T}(\vec{w})\| \leq \tau [\![D']\!] \|\vec{v} - \vec{w}\|, \quad \text{where}$$
$$[\![D']\!] = \sup\{\sum_{j=1}^{N} |D'_{jk}(x,t)| : x \in \mathbf{R},\ 0 \leq t \leq \tau,\ k = 1,\ldots,N\}.$$

If τ is sufficiently small, then $\mathcal{T}$ is a contraction on X. But this means that $\tilde{\mathcal{T}}(\vec{v}) = \vec{\phi} + \mathcal{T}(\vec{v})$ is also a contraction on X, and we obtain the desired fixed point $\vec{v} \in X$. Translating our results for (2′) back to (2), we have shown that (2) admits a unique mild solution $\vec{u} \in C(\mathbf{R} \times [0,\tau], \mathbf{R}^N)$, provided that $\tau > 0$ is sufficiently small.

Can we conclude global existence for (2) on $[0,T]$? As in Section 9.2e, we should try to iterate the local existence argument. This may be done by taking the supremum in (6) over $0 \leq t \leq T$ to find a uniform τ. We can then use the initial condition $\vec{g}_1 = \vec{u}|_{t=\tau}$ to find a unique solution on $[\tau, 2\tau]$ and iterate to cover $[0,T]$.

In this way, we can conclude the following.

Theorem. Suppose $\vec{g} \in C_B(\mathbf{R}, \mathbf{R}^N)$ and (2b) holds. Then (2a) admits a unique mild solution $\vec{u} \in C_B(\mathbf{R} \times [0,T], \mathbf{R}^N)$.

Remark. In order to show that the mild solution of the theorem is a classical solution, we must show that $\vec{v} \in C_B^1(\mathbf{R} \times [0,T], \mathbf{R}^N)$. This may be accomplished by replacing $X = C_B(\mathbf{R} \times [0,\tau], \mathbf{R}^N) = C([0,\tau], C_B(\mathbf{R}, \mathbf{R}^N))$ by $X^1 = C([0,\tau], C_B^1(\mathbf{R}, \mathbf{R}^N))$, with norm $\|\vec{v}\|_1 = \max(\|\vec{v}\|, \|\vec{v}_x\|)$. This requires us to assume $\vec{g} \in C_B^1(\mathbf{R}, \mathbf{R}^N)$ and an additional x-derivative of A, D, and $\vec{f}$ to be uniformly bounded. Notice that we do not need additional assumptions on the t-derivatives, and the continuity of $\vec{u}_t$ follows from $\vec{u} \in X^1$ and (2). By uniqueness, this classical solution agrees with the mild solution of the theorem. (The details of this proof are left for Exercise 1.)

b. Nonlinear Systems

If $A = A(x,t)$, but we allow $\vec{f} = \vec{f}(x,t,\vec{u})$ to depend on $\vec{u}$, then (1) is *semilinear*, and local existence for an initial value problem can again be proved by using the method of characteristics to reduce the problem to an integral equation, and then using a contraction mapping argument. As usual for nonlinear equations, this requires $\vec{f}$ to be locally Lipschitz in $\vec{u}$:

$$(7) \qquad |f_k(x,t,\vec{v}) - f_k(x,t,\vec{w})| \le M_R |\vec{v} - \vec{w}|$$

for all $x \in \mathbf{R}$, $0 \le t \le \tau$, $k = 1, \ldots, N$, and vectors $\vec{v}, \vec{w} \in \mathbf{R}^N$ satisfying $|\vec{v}|, |\vec{w}| \le R$. We leave this as Exercise 5.

If $A = A(x,t,\vec{u})$ is also allowed to depend on $\vec{u}$, then (1) is *quasilinear*, and local existence is somewhat more difficult to prove. In particular, the characteristics now depend on the values of the solution, exactly as they did for a single equation in Section 1.1, making the reduction to an integral equation rather tricky. Nevertheless, this method does work (cf. [Garabedian]). However, there is another method, which involves defining a map $\mathcal{T}: \vec{u} \to \vec{v}$, where $\vec{v}$ solves the *linear* equation

$$(8) \qquad \vec{v}_t + A(x,t,\vec{u}(x,t))\vec{v}_x = \vec{f}(x,t,\vec{u}(x,t)).$$

Again a contraction mapping argument can be used to show that $\mathcal{T}$ admits a fixed point $\vec{u}$ that is the desired solution, but this requires rather delicate estimates of the smoothness of $\vec{v}$ in terms of A, $\vec{f}$, and $\vec{u}$. This procedure is described in [Courant-Hilbert, vol. I] using sup-norm estimates. But the method extends to n-dimensional symmetric hyperbolic systems when energy norms are used for the estimates; we describe this in detail in Section 12.1.

In particular, this local existence argument applies to the special case

$$(9) \qquad \vec{u}_t + A(\vec{u})u_x = 0,$$

which we encounter in the next section in connection with conservation laws.

Exercises for Section 10.1

1. Assuming sufficient regularity of $\vec{g}$, A, D, and $\vec{f}$, show that (2a) admits a unique classical solution $\vec{u} \in C_B^1(\mathbf{R} \times [0,T], \mathbf{R}^N)$.

2. Use Exercise 1 in Section 7.3 to show that the solution $\vec{v}$ of (2) depends continuously on the initial data: If $\vec{v}$ (resp. $\vec{w}$) is the solution on $[0, \tau]$ with initial condition $\vec{g}$ (resp. $\vec{h}$), then $|\vec{v}-\vec{w}|$ can be made small if $|\vec{g}-\vec{h}|$ is small.

3. Suppose the A_{jk}, D_{jk}, and f_k in (2b) are independent of t. Show that the solution of the theorem exists for all $t > 0$ and satisfies $\|\vec{u}(t)\|_{C_B} \leq Ct$.

4. If L, C are positive constants, and $R, G, I_0, V_0 \in C_B^1(\mathbf{R})$, show that the initial value problem for the telegraph system [cf. (24), (25), and (27) of Section 2.2] admits a C^1-solution defined for all $t > 0$.

5. Consider an initial value problem for a semilinear hyperbolic system

$$\begin{cases} \vec{u}_t + A(x,t)\vec{u}_x = \vec{f}(x,t,\vec{u}) & \text{for } x \in \mathbf{R} \text{ and } t > 0, \\ \vec{u}(x,0) = \vec{g}(x) & \text{for } x \in \mathbf{R}, \end{cases}$$

where $\vec{f}$ satisfies (7). If $\vec{g} \in C_B(\mathbf{R}, \mathbf{R}^N)$, show that for $\tau > 0$ sufficiently small there exists a unique solution $\vec{u} \in C_B(\mathbf{R} \times [0,\tau], \mathbf{R}^N)$.

6. Show that the solution of the semilinear initial value problem

$$u_t + u_x - u + w - 1 = 0$$
$$v_t + 2u_x - w_x - 2u + w - v^2 + 2vw - w^2 = 0$$
$$w_t + 2u_x - w_x - 2u + w = 0$$
$$u(x,0) = 1, \quad v(x,0) = 3, \quad w(x,0) = 2$$

has a singularity on the line $t = 1$.

10.2 Quasilinear Systems of Conservation Laws

In this section, we consider a *system of conservation laws*

(10) $$\vec{u}_t + (G(\vec{u}))_x = 0,$$

where $\vec{u}$ is an N-vector and $G: \mathbf{R}^N \to \mathbf{R}^N$ is C^1. If we let $A(\vec{u}) = G'(\vec{u})$, then (10) takes the form (9); as in Section 10.1, we assume that $A(\vec{u})$ is hyperbolic. If we are given smooth initial conditions (and A is sufficiently smooth), then a smooth solution exists for at least for a short time. However, the solution may develop a singularity in a finite time, even if A and $\vec{g}$ are C^∞. Thus we are led naturally to the study of singularities and weak solutions; it is here that the special form (10) is important.

Systems of the form (10) arise frequently in physical applications when the "state" $\vec{u}$ represents quantities that may vary with x and t but are conserved in the sense that the total amount of these quantities in any interval $[x_1, x_2]$ can only change due to the *flux* of the quantities at x_1 and x_2. We encountered the scalar case ($N = 1$) of (10) in Section 1.2. In the application to traffic flow, we found that "conservation of cars" led to a conservation law of the form $\rho_t + q(\rho)_x = 0$. Moreover, weak solutions consisting of constant states separated by rarefaction waves or shocks were of physical significance.

These features generalize to the system (10). Referring to Section 2.3, it is natural to call $\vec{u}(x,t)$ a *weak solution* of (10) in a domain $\Omega \subset \mathbf{R}_+^2$ if

$$\text{(11)} \qquad \int_\Omega \vec{u}(x,t)\, v_t(x,t) + G(\vec{u}(x,t))\, v_x(x,t) \, dx\, dt = 0$$

for every scalar-valued $v \in C_0^1(\Omega)$. As usual, integration by parts shows that if the components of $\vec{u}$ and G are C^1, then (11) implies that $\vec{u}$ is a solution of (10). We investigate weak solutions for (10) formed by patching together constant states $\vec{u}_\ell$ and $\vec{u}_r$ by rarefaction waves or by shocks. In particular, a *Riemann problem*, which is an initial value problem in which the initial condition is a constant state $\vec{u}_\ell$ for $x < 0$ and $\vec{u}_r$ for $x > 0$, may admit such weak solutions. But before exploring these issues, let us begin with some examples that occur in physical applications.

a. Examples and Applications

In Example 8 of Section 2.2, we reduced the wave equation to a first-order system. If we consider a nonlinear wave equation of the form $u_{tt} = a(u_x)u_{xx}$ where the propagation speed depends on u_x, then reduction to a first-order system by letting $w = u_x$ and $v = u_t$ produces the following example.

Example 1. *(The p-system)* Let $\vec{u} = \begin{pmatrix} w \\ v \end{pmatrix}$ and consider the system

$$\text{(12a)} \qquad \begin{cases} w_t = v_x \\ v_t = a(w) w_x, \end{cases}$$

where $a(w) > 0$ for all w. If we let $p(w) = -\int^w a(s)\, ds$, this system can be written

$$\text{(12b)} \qquad \begin{cases} w_t - v_x = 0 \\ v_t + p(w)_x = 0 \end{cases}$$

which is of the form (10) if we let

$$\text{(12c)} \qquad G(\vec{u}) = G\begin{pmatrix} w \\ v \end{pmatrix} = \begin{pmatrix} -v \\ p(w) \end{pmatrix}.$$

The system (12b) is called the *p-system* and arises in the Lagrangian formulation of gas dynamics, as we see in Example 4. ♣

If we consider shallow-water waves, we can normalize the horizontal velocity v and the height of the wave h so that the equations expressing conservation of mass and momentum take a particularly simple form.

Example 2. *(Shallow-water waves)* The following system

(13a)
$$v_t + vv_x + h_x = 0 \quad \text{conservation of mass}$$
$$h_t + hv_x + vh_x = 0 \quad \text{conservation of momentum}$$

provides a model of waves propagating in shallow water (see [Kevorkian]). This is of the form (10) if we let

(13b) $\qquad \vec{u} = \begin{pmatrix} v \\ h \end{pmatrix} \qquad \text{and} \qquad G(\vec{u}) = G\begin{pmatrix} v \\ h \end{pmatrix} = \begin{pmatrix} \frac{1}{2}v^2 + h \\ vh \end{pmatrix}.$ ♣

Consider a gas (or any inviscid, compressible fluid) moving in a one-dimensional pipe. The motion of the gas is governed by conservation laws, such as conservation of mass, conservation of momentum, and conservation of energy. These conservation laws may be expressed as PDEs involving quantities such as the velocity, pressure, density, and energy, but the form of the PDEs depends on the coordinates used to describe the gas. In *Eulerian coordinates*, x represents the spatial coordinate along the length of the pipe and we let $v(x,t)$, $p(x,t)$, $\rho(x,t)$, and $E(x,t)$ denote, respectively, the *velocity, pressure, mass density*, and *energy density* (per unit mass) of the gas at that point x and time t.

Example 3. *(Gas dynamics in Eulerian coordinates)*

(14)
$$\begin{cases} \rho_t + (\rho v)_x = 0 & \text{conservation of mass} \\ (\rho v)_t + (\rho v^2 + p)_x = 0 & \text{conservation of momentum} \\ (\rho E)_t + (\rho E v + pv)_x = 0 & \text{conservation of energy} \end{cases}$$

Notice that there are three equations but four unknown functions, so we need to eliminate one variable. This can be achieved by adding the *equation of state*

(15a) $\qquad p = p(\rho, e) \qquad \text{where} \quad e = E - \dfrac{v^2}{2} \quad \text{is the *internal energy*.}$

If we let $u_1 = \rho$, $u_2 = \rho v$, and $u_3 = \rho E$, then (14) can be realized in the form (10).

A simplified equation of state occurs when the gas is is *isentropic*; that is, the pressure is an increasing function of the density:

(15b) $\qquad\qquad p = p(\rho), \qquad \text{where } dp/d\rho > 0;$

for an isentropic gas, $c = \sqrt{dp/d\rho}$ is called the *speed of sound*. Usually the isentropic dependence (15b) is of the form

(15c) $\qquad p(\rho) = \alpha \rho^\gamma \qquad$ with constants $\alpha > 0$ and $\gamma > 1$,

and the gas is called *polytropic*; in this case, $c^2 = \alpha \gamma \rho^{\gamma-1}$.

Notice that it is possible to ignore conservation of energy and to consider only the first two equations in (14) for ρ and v. We still need, however, an equation of state such as (15c). ♣

Another model of gas dynamics is obtained if we use *Lagrangian coordinates* in which x represents a fixed particle of gas as it moves through the pipe. The advantage of this model is the simplification of the equations expressing the conservation of mass, momentum, and energy. Let $w(x,t)$ denote the *specific volume* of the gas, which is the inverse of the density (i.e., $w = \rho^{-1}$).

Example 4. *(Gas dynamics in Lagrangian coordinates)*

(16) $\qquad \begin{cases} w_t - v_x = 0 & \text{conservation of mass} \\ v_t + p_x = 0 & \text{conservation of momentum} \\ E_t + (wp)_x = 0 & \text{conservation of energy} \end{cases}$

Again we have more variables than equations, so we need the equation of state (15a) to eliminate one variable, even if we consider only the conservation of mass and momentum equations. In fact, under the isentropy condition (15b), the first two equations in (16) coincide with the *p*-system of Example 1. In the polytropic case (15c), for example, $p(w) = \alpha w^{-\gamma}$ so $a(w) = -p'(w) > 0$ and $a'(w) = -p''(w) < 0$. ♣

b. Simple Waves and Rarefaction

Let us consider a special class of solutions of (9) called *simple waves*, which are of the form

(17) $\qquad\qquad\qquad \vec{u}(x,t) = \vec{f}(\theta(x,t)),$

where θ is a *scalar*-valued function; this means that the two-dimensional surface in N-space defined by the graph of $\vec{u}(x,t)$ degenerates to a curve. Using (17) in (9), we obtain $\vec{f}'(\theta)\theta_t + A(\vec{f})\vec{f}'(\theta)\theta_x = 0$, so that $\vec{f}'(\theta)$ is an eigenvector for $A(\vec{f})$ with eigenvalue λ satisfying $\theta_t + \lambda \theta_x = 0$. Thus, if $(\lambda(\vec{u}), \vec{V}(\vec{u}))$ is an eigenpair for $A(\vec{u})$ and we let $\vec{f}(\theta)$ denote a particular solution of the ODE

(18) $\qquad\qquad\qquad \dfrac{d\vec{f}}{d\theta} = \vec{V}(\vec{f}),$

then $\lambda(\theta) = \lambda(\vec{f}(\theta))$ becomes a known function of θ, and we need only take for θ any solution of the PDE

$$\theta_t + \lambda(\theta)\theta_x = 0. \tag{19}$$

If we can solve (18) and (19), we obtain our simple wave (17). Since $A(\vec{u})$ has N eigenvalues, we expect to obtain N simple waves [for each choice of initial condition for (19)]. However, we must emphasize that simple waves are special solutions and the nonlinearity of the system (as well as the distinctness of the eigenvalues) prevents us from taking linear combinations of them.

The ODE (18) should be easy to solve, so let us consider the PDE (19). The characteristic equations for (19) are $dx/dt = \lambda(\theta)$, $d\theta/dt = 0$, so we find $x = \lambda(\theta)t + x_0$ or

$$\lambda(\theta) = \frac{x - x_0}{t}. \tag{20}$$

By the inverse function theorem, (20) determines $\theta(x,t)$, at least provided that $d\lambda/d\theta \neq 0$, and we see that the simple wave $\vec{u}(x,t) = \vec{f}(\theta(x,t))$ is constant along the lines $x = ct + x_0$ where c is any constant. For this reason, simple waves for (9) are also called *rarefaction waves*.

We encountered rarefaction waves in Section 1.2 for the scalar conservation law $u_t + G(u)_x = 0$ and found that they are useful in constructing a continuous weak solution when the initial condition $u_0(x)$ is a step function that has constant values u_ℓ and u_r for $x < 0$ and $x > 0$, respectively. Assuming $G''(u) > 0$ and $u_\ell < u_r$, then the characteristics emanating from the x-axis spread apart, leaving a wedge region that can be filled in by the rarefaction wave.

Let us investigate when rarefaction waves can be used to solve (9) when the initial condition is a step function; that is,

$$\vec{u}(x,0) = \begin{cases} \vec{u}_\ell & \text{for } x < 0, \\ \vec{u}_r & \text{for } x > 0, \end{cases} \tag{21}$$

where $\vec{u}_\ell$ and $\vec{u}_r$ are constant states. The relationship between the equation and the constant states $\vec{u}_\ell$ and $\vec{u}_r$ for which we can find a rarefaction wave can be subtle; we illustrate this construction of weak solutions in the special case of the p-system.

Example 1 (Revisited). Notice that (12a) is in the form (9), where the matrix $A(\vec{u})$ is

$$A\begin{pmatrix} w \\ v \end{pmatrix} = \begin{pmatrix} 0 & -1 \\ -a(w) & 0 \end{pmatrix}. \tag{22}$$

First, notice that A has eigenvalues $\lambda = \pm\sqrt{a(w)}$, so (9) is strictly hyperbolic. With $\lambda_1 = -\sqrt{a(w)}$ we find the eigenvector

$$\vec{V}_1 \begin{pmatrix} w \\ v \end{pmatrix} = \begin{pmatrix} 1 \\ \sqrt{a(w)} \end{pmatrix}. \tag{23}$$

Let us try to find a rarefaction wave using λ_1; we call this a *rarefaction one-wave*. If we write $\vec{f}(\theta) = \begin{pmatrix} w(\theta) \\ v(\theta) \end{pmatrix}$, then (18) becomes

$$\frac{dw}{d\theta}(\theta) = 1, \qquad \frac{dv}{d\theta}(\theta) = \sqrt{a(w(\theta))}. \tag{24}$$

We can solve these ODEs by taking $w = \theta$ and $v = \int^w \sqrt{a(s)}\,ds$, so we may write our one-wave as

$$\vec{u}(x,t) = \begin{pmatrix} w(x,t) \\ \int^{w(x,t)} \sqrt{a(s)}\,ds \end{pmatrix} \tag{25}$$

provided that we can determine w. But w satisfies $w_t - \sqrt{a(w)} = 0$, and we may determine $w(x,t)$ by using (20):

$$-\sqrt{a(w)} = \frac{x}{t}. \tag{26}$$

To ensure that (26) indeed determines $w(x,t)$, we need to assume $a'(w) \neq 0$; in fact, we assume the sign condition

$$a'(w) < 0 \qquad \text{for all } w, \tag{27}$$

which is true in the application to polytropic gas dynamics (cf. Example 4).

This completes our prescription for constructing our rarefaction one-wave, but we have not yet considered what values $\vec{u}_\ell$ and $\vec{u}_r$ can be connected by such a solution. First, since w satisfies $w_t - \sqrt{a(w)}w_x = 0$ where the characteristic speed $dx/dt = -\sqrt{a(w)}$ is increasing in w [by (27)], we must have $w_\ell < w_r$ in order for the characteristics to spread apart and leave a wedge W in the xt-plane (see Figure 1a). Second, we can use (24) to express the direct dependence of v and w as

$$dv/dw = \sqrt{a(w)}. \tag{28}$$

If we integrate (28) between (w_ℓ, v_ℓ) and (w_r, v_r), we get

$$v_r = \int_{w_\ell}^{w_r} \sqrt{a(s)}\,ds + v_\ell. \tag{29}$$

Now (29) implies that $v_r > v_\ell$, and v_r is determined by w_ℓ, w_r, and v_ℓ. In other words, the possible initial states of the form (21) that generate a rarefaction one-wave lie on a curve R_1 in the wv-plane satisfying $w_\ell < w_r$ and $v_\ell < v_r$ (see Figure 1b). (The convexity of the curve in Figure 1b is determined by fixing $\vec{u}_\ell$ and taking the second derivative of v_r with respect to w_r.)

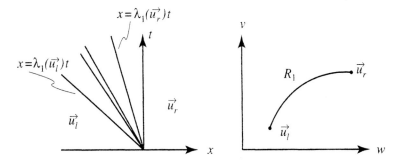

Figures 1a, 1b. A rarefaction one-wave.

Similarly, the rarefaction two-wave for the initial-condition (21) is defined by using the eigenvalue $\lambda_2 = \sqrt{a(w)}$ with eigenvector

(30a) $$\vec{V}_2 \begin{pmatrix} w \\ v \end{pmatrix} = \begin{pmatrix} 1 \\ -\sqrt{a(w)} \end{pmatrix}$$

to generate a simple wave:

(30b) $$\vec{u}(x,t) = \begin{pmatrix} w(x,t) \\ -\int^{w(x,t)} \sqrt{a(s)}\, ds \end{pmatrix}$$

where $w(x,t)$ is a solution of $w_t + \sqrt{a(w)}\, w_x = 0$ and is found from $\sqrt{a(w)} = x/t$ by the inverse function theorem [again assuming (27)].

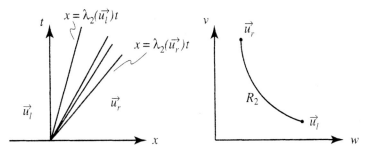

Figures 2a, 2b. A rarefaction two-wave.

Section 10.2: Quasilinear Systems of Conservation Laws   313

To determine what values $\vec{u}_\ell$ and $\vec{u}_r$ can be connected by our rarefaction two-wave, we first observe that the characteristic speed $dx/dt = \sqrt{a(w)}$ now implies that $w_\ell > w_r$ in order for the characteristics to spread (see Figure 2a) and then replace (28) by $dv/dw = -\sqrt{a(w)}$ to conclude

$$(31) \qquad v_r = -\int_{w_\ell}^{w_r} \sqrt{a(s)}\, ds + v_\ell = \int_{w_r}^{w_\ell} \sqrt{a(s)}\, ds + v_\ell.$$

Now (31) again implies $v_\ell < v_r$, and the possible initial states lie on a curve R_2 in the wv-plane (see Figure 2b). ♣

c. Shocks and the Entropy Condition

To explain shocks and the entropy condition, let us first reconsider the *scalar* conservation law $u_t + (G(u))_x = u_t + A(u)u_x = 0$. In Section 1.2 we found that, along a shock discontinuity $(\xi(t), t)$, the speed of the shock satisfies the jump condition

$$(32) \qquad \frac{d\xi}{dt} = \frac{G(u_r) - G(u_l)}{u_r - u_l}.$$

But shocks should only occur when characteristic curves carrying conflicting initial data cross each other. This requires the equation to be *genuinely nonlinear* [i.e., $G''(u) \neq 0$], and the speed of characteristics from the left should exceed the speed of characteristics from the right [i.e., $G'(u_\ell) > G'(u_r)$, since $G'(u) = A(u)$ is the characteristic speed]. Let us assume *strict convexity* $G''(u) > 0$ [as was the case for the example $G(u) = u^2/2$ in Section 1.2], so that

$$(33) \qquad G'(\alpha) < \frac{G(\beta) - G(\alpha)}{\beta - \alpha} < G'(\beta) \qquad \text{whenever} \quad \alpha < \beta.$$

Letting $\alpha = u_r$ and $\beta = u_l$, we obtain from (32) and (33)

$$(34) \qquad G'(u_r) < \frac{d\xi}{dt} < G'(u_l),$$

which states that the shock speed must be intermediate between the characteristic speeds on either side. The condition (34) is called the *entropy condition* and is imposed to eliminate the existence of unnecessary shocks (cf. Exercise 6 in Section 1.2). The use of the term entropy comes from the application to gas dynamics.

Now let us return to the *system* of conservation laws (10) and try to discover when shocks exist and how we might find them. Consider a curve γ, parameterized by $x = \xi(t)$, which divides Ω into two regions Ω_l and Ω_r lying on the left and right of γ respectively. Suppose $\vec{u}$ is a weak solution of (10) such that the restrictions $\vec{u}_l = \vec{u}|_{\Omega_l}$ and $\vec{u}_r = \vec{u}|_{\Omega_r}$ are both C^1 and extend continuously to γ. For a point $(\xi(t), t)$ of γ, let $[\vec{u}] = [\vec{u}](\xi(t), t)$ be the jump discontinuity: $[\vec{u}] = \vec{u}_r - \vec{u}_l$. Similarly, $[G(\vec{u})] = G(\vec{u}_r) - G(\vec{u}_l)$. Then the following *jump condition of Rankine-Hugoniot* generalizes the scalar case: at any point of γ, the jump discontinuity satisfies

$$\frac{d\xi}{dt} \cdot [\vec{u}] = [G(\vec{u})]. \tag{35}$$

But how do we generalize the genuine nonlinearity and the entropy condition (34) for the system (10)? We assume that the system (10) is *strictly hyperbolic* and write the distinct real eigenvalues of $A(\vec{u}) = G'(\vec{u})$ in increasing order: $\lambda_1(\vec{u}) < \lambda_2(\vec{u}) < \cdots < \lambda_N(\vec{u})$. Furthermore, we define the system (10) to be *genuinely nonlinear* when

$$\nabla \lambda_k(\vec{u}) \cdot \vec{V}_k(\vec{u}) \neq 0 \quad \text{for all} \quad \vec{u} \in \mathbf{R}^N \quad \text{and} \quad k = 1, \ldots, N \tag{36}$$

where $\vec{V}_k$ is a basis of eigenvectors associated to the λ_k. When $N = 1$ we have $\lambda(u) = A(u) = G'(u)$ and (36) becomes $G''(u) \neq 0$; so (36) generalizes the nonlinear scalar case. Since $\lambda_k(\vec{u})$ is the speed of the kth characteristic (cf. Section 2.2), it is natural to replace (34) by the condition that for some k we have

$$\lambda_k(\vec{u}_r) < \frac{d\xi}{dt} < \lambda_k(\vec{u}_l) \tag{37a}$$

where $\vec{u}_l$ and $\vec{u}_r$ again denote the limiting values of $\vec{u}$ from the left and right sides of the shock curve $(\xi(t), t)$. However, because every characteristic curve from the left (respectively right) with speed greater than (respectively less than) $d\xi/dt$ will impose a condition on the limiting values of $\vec{u}$ along γ, we must also require

$$\lambda_{k-1}(\vec{u}_l) < \frac{d\xi}{dt} < \lambda_{k+1}(\vec{u}_r) \tag{37b}$$

in order to get the correct number of conditions on the discontinuity (cf. [Lax] or [Smoller]). As a consequence of (37a,b), we see that the shock speed $d\xi/dt$ is intermediate between the characteristic speeds on either side for the index k and for *no other value* of the index. The conditions (37a,b) are called the *entropy* or *shock conditions (of Lax)*, and a shock satisfying them is called a *k-shock*. [If $N = 1$, then (37b) is vacuous and (37a) agrees with (34).]

As in the scalar case, solutions with smooth initial conditions may develop shocks after a finite time, but shocks can also be used to construct weak solutions when the initial condition is a step function as in (21). Let us consider this construction in the special case of the p-system.

Example 1 (Revisited). Let us investigate shock solutions for the system (12b) with initial conditions (21). We shall assume the polytropic condition (15c), so $p(w) = \alpha w^{-\gamma}$ is decreasing in w and $p''(w) > 0$.

Section 10.2: Quasilinear Systems of Conservation Laws

We know from the previous subsection that (12b) is strictly hyperbolic with eigenvalues and eigenvectors $\lambda_1(w) = -\sqrt{a(w)} < 0 < \lambda_2(w) = \sqrt{a(w)}$ and associated eigenvectors

$$(38) \qquad \vec{V}_1(w) = \begin{pmatrix} 1 \\ \sqrt{a(w)} \end{pmatrix} \quad \text{and} \quad \vec{V}_2(w) = \begin{pmatrix} 1 \\ -\sqrt{a(w)} \end{pmatrix}.$$

Using (27), we can show that the system is genuinely nonlinear: $\nabla\lambda_1(w) \cdot \vec{V}_1(w) = -\frac{1}{2}a'(w)/\sqrt{a(w)} > 0$ and $\nabla\lambda_2(w) \cdot \vec{V}_2(w) = \frac{1}{2}a'(w)/\sqrt{a(w)} < 0$. Let us now investigate the two types of shocks that can occur.

For one-shocks, the Lax shock conditions (37a,b) require the shock speed $d\xi/dt$ to satisfy

$$(39) \qquad -\sqrt{a(w_r)} < \frac{d\xi}{dt} < -\sqrt{a(w_\ell)} \quad \text{and} \quad \frac{d\xi}{dt} < \sqrt{a(w_r)}.$$

In particular, (39) shows that a one-shock must satisfy $d\xi/dt < 0$; for this reason it is sometimes called a *back-shock*. Similarly, we find that the speed of a two-shock satisfies

$$(40) \qquad \sqrt{a(w_r)} < \frac{d\xi}{dt} < \sqrt{a(w_\ell)} \quad \text{and} \quad -\sqrt{a(w_\ell)} < \frac{d\xi}{dt},$$

and so it is sometimes called a *front-shock*.

Now let us try to determine what constant states $\vec{u}_\ell$ and $\vec{u}_r$ can be connected by shocks. Let us begin with one-shocks. The Rankine-Hugoniot jump condition (35) states that

$$(41) \qquad (w_r - w_\ell)\frac{d\xi}{dt} = (v_\ell - v_r) \quad \text{and} \quad (v_r - v_\ell)\frac{d\xi}{dt} = (p(w_r) - p(w_\ell)).$$

If we eliminate $d\xi/dt$ from these equations, we get $(v_r - v_\ell)^2 = (w_r - w_\ell)(p(w_\ell) - p(w_r))$ or

$$(42) \qquad v_r - v_\ell = \pm\sqrt{(w_r - w_\ell)(p(w_\ell) - p(w_r))}.$$

To determine the sign in (42), we use $-\sqrt{a(w_r)} < -\sqrt{a(w_\ell)}$ from (39) to conclude that $p'(w_r) < p'(w_\ell)$, and hence $w_\ell > w_r$ by $p''(w) > 0$. Using $w_\ell > w_r$ and $d\xi/dt < 0$ in (41), we find $v_\ell - v_r > 0$, so we must take the minus sign in (42). Thus

$$(43) \qquad v_r = v_\ell - \sqrt{(w_r - w_\ell)(p(w_\ell) - p(w_r))}, \qquad w_\ell > w_r,$$

defines the curve S_1 in the wv-plane corresponding to the constants states $\vec{u}_\ell$ and $\vec{u}_r$ that can be connected by a one-shock; recall that this one-shock has negative speed (see Figure 3).

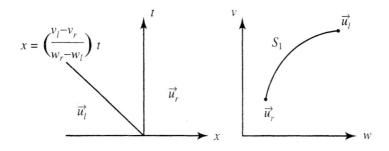

Figures 3a, 3b. Possible initial states $\vec{u}_l$, $\vec{u}_r$ for a one-shock.

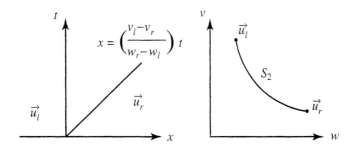

Figure 4a, 4b. Possible initial states, $\vec{u}_l$, $\vec{u}_r$ for a two-shock.

Similarly, the constant states $\vec{u}_\ell$ and $\vec{u}_r$ that can be connected by a two-shock lie on the curve S_2 defined by

(50) $\qquad v_r = v_\ell - \sqrt{(w_r - w_\ell)(p(w_\ell) - p(w_r))}, \qquad w_\ell < w_r.$

Notice that this two-shock has positive speed (see Figure 4.) ♣

Remark. If $\nabla \lambda_k(\vec{u}) \cdot \vec{V}_k(\vec{u}) \equiv 0$ for some k, then we say that λ_k is *linearly degenerate*. The condition (36), on the other hand, states that (10) is nonlinear in each direction $\vec{V}_k$. The existence of a linearly degenerate eigenvalue allows for the possibility of an additional type of discontinuous weak solution called a *contact discontinuity*, in which the curve of discontinuity satisfies $d\xi/dt = \lambda_k(\vec{u}_\ell) = \lambda_k(\vec{u}_r)$. For information on these discontinuities, see [Smoller].

d. Riemann Problems

A *Riemann problem* is an initial value problem for a system of hyperbolic conservation laws (10) in which the initial data $\vec{u}(x,0)$ are step functions as in (21).

This problem occurs naturally in the "shock tube" experiment of gas dynamics. The experiment is performed by separating two gases in a long

Section 10.2: Quasilinear Systems of Conservation Laws

pipe by a membrane. Assume that each gas is initially at rest and has a constant density, pressure, and energy. At $t = 0$, the membrane is removed and the gases are allowed to mix. The resulting movement of the gases will be a solution of a Riemann problem, where the specific system (10) depends on the model chosen to represent the gas (e.g., Eulerian or Lagrangian coordinates).

As we have seen, for certain special initial states $\vec{u}_\ell$ and $\vec{u}_r$, we will get a continuous weak solution consisting of a rarefaction wave connecting these constant states; for other special initial values we will get a weak solution with a single discontinuity connecting the constant states. But what happens in general? The idea is to try to combine rarefaction waves and shocks in order to accomodate *any* choices of $\vec{u}_\ell$ and $\vec{u}_r$ (provided that they are not too far from each other).

We illustrate this method in the simplest case of the p-system (i.e. gas dynamics in Lagrangian coordinates), ignoring energy. This enables us to take advantage of our analysis in the previous two sections of the rarefaction waves and shocks for the p-system.

Example 1 (Revisited). Let us consider the Riemann problem for (12a) with the initial conditions (21).

The method is to introduce an intermediate state $\vec{u}_0$ and try to connect it to $\vec{u}_\ell$ by either a rarefaction one-wave or a one-shock, and to $\vec{u}_r$ by either a rarefaction two-wave or a two-shock. There are four possibilities, depending on the relative positions of $\vec{u}_\ell$ and $\vec{u}_r$. To determine which combination of rarefaction waves and shocks to use, let us fix $\vec{u}_\ell$ and superimpose the four curves R_1, R_2, S_1, and S_2 of Figures 1b, 2b, 3b, and 4b to get a partitioning of state space; see Figure 5.

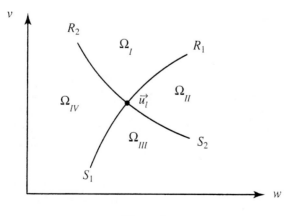

Figure 5

If $\vec{u}_r$ happens to lie on R_1 or R_2, then we know we can solve the Riemann problem with a *single* rarefaction wave. But if $\vec{u}_r$ happens to lie in the region Ω_I between R_1 and R_2, and is sufficiently close to $\vec{u}_\ell$, then there is a point $\vec{u}_0 \in R_1$ which lies on the curve R'_2 of constant states that can be connected to $\vec{u}_r$ by a rarefaction two-wave; see Figure 6a. Since $\vec{u}_0$ can be connected to $\vec{u}_\ell$ ($\vec{u}_r$) by a rarefaction one-wave (two-wave), the resulting continous weak solution is that pictured in Figure 6b.

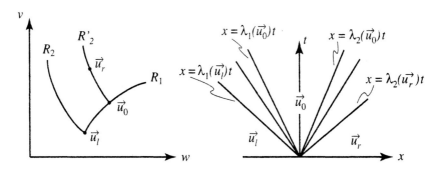

Figures 6a, 6b.

What happens if $\vec{u}_r$ lies in the region Ω_{II} between R_1 and S_2? It appears that there is a point $\vec{u}_0 \in R_1$ that lies on the curve S'_2 of constant states that can be connected to $\vec{u}_r$ by a two-shock; see Figure 7a. Since $\vec{u}_0$ can be connected to $\vec{u}_\ell$ by a rarefaction one-wave and to $\vec{u}_r$ by a two-shock, the resulting weak solution is that pictured in Figure 7b.

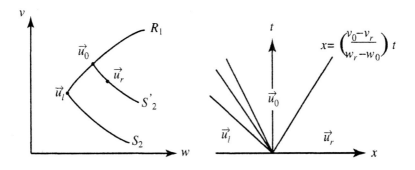

Figures 7a, 7b.

For $\vec{u}_r \in \Omega_{III}$ or $\vec{u}_r \in \Omega_{IV}$, we proceed analogously; in the former case the solution involves of a one-shock and a two-shock, and in the latter case it involves of a one-shock and a rarefaction two-wave.

We conclude that, provided that $\vec{u}_\ell$ and $\vec{u}_r$ are sufficiently close, *the Riemann problem for (12a) and (21) can be solved by a weak solution consisting of three constant states connected by rarefaction waves or shocks satisfying the Lax shock conditions.* ♣

Section 10.2: Quasilinear Systems of Conservation Laws

We should observe that, in this simple example, a refined analysis of the geometry of the curves R_i and S_i in Figure 5 shows that the assumption of $\vec{u}_r$ and $\vec{u}_\ell$ being close is unnecessary (cf. [Smoller]). However, the closeness is necessary for Riemann problems in general. In fact, let us state the basic result which may be proved by the method that we have just illustrated (cf. [Lax] or [Smoller]).

Theorem. *If (10) is strictly hyperbolic and genuinely nonlinear and $\vec{u}_\ell$ and $\vec{u}_r$ in the initial conditions (21) are sufficiently close, then there is a weak solution of the Riemann problem composed of $N+1$ constant states connected by rarefaction waves or shocks satisfying the Lax shock conditions (37a,b).*

Finally, if the hypothesis of genuine nonlinearity is removed, then the theorem is still true if we include the possibility of contact discontinuities between the constant states (cf. [Smoller]).

Exercises for Section 10.2

1. Find the rarefaction wave solution of $u_t - u^2 u_x = 0$ with initial conditions $u(x,0) = 0$ for $x > 0$ and $u(x,0) = u_0 < 0$ for $x < 0$.

2. Consider the quasilinear second-order equation $u_{tt} - (1 + u_x)^2 u_{xx} = 0$.
   (a) Reduce this to a first-order p-system. Is the system hyperbolic? Strictly hyperbolic?
   (b) Find a simple wave solution satisfying the initial condition $u(x,0) = h(x)$, where $h(x)$ is arbitrary.
   (c) Let $h(x) = x^2$ and find the domain of existence of the solution u found in (b).

The p-System

3. For the p-system (12), is it possible to have a shock solution in which only one of the variables u or v suffers a discontinuity? Do you think a similar statement is true for any system of conservation laws (10)?

4. Consider the p-system (12b) with $p(w) = w^{-3}/3$ and $w > 0$. For the initial condition (21) with $w_r = \frac{1}{2} w_\ell$, find the shock path $\xi_1(t)$ for the one-shock; similarly, find $\xi_2(t)$ for the two-shock.

5. Let $p(\rho) = k^2 \rho^2 / 2$ and consider the Riemann problem for (12b). Let $\vec{u}_\ell = (w_\ell, 0)$, where $w_\ell > 0$. Find conditions on $w_r > 0$ and v_r that guarantee that the solution has no shocks.

Shallow-Water Waves

6. Consider the system (13) of Example 2.
   (a) Show that the system is strictly hyperbolic (assuming $h > 0$).
   (b) Describe the simple wave solutions.
   (c) Describe the possible initial constant states $\vec{u}_\ell$ and $\vec{u}_r$ that can be connected by rarefaction one-waves. Do the same for rarefaction two-waves.

(d) Show how two rarefaction waves can be combined to solve the system with initial condition

$$v(x,0) = \begin{cases} v_\ell & \text{for } x < 0 \\ v_r & \text{for } x > 0 \end{cases} \qquad h(x,0) = 1,$$

where $v_\ell < v_r$.

7. Consider the system (13) in Example 2.
   (a) For the constant initial conditions

$$v(x,0) = \begin{cases} v_\ell & \text{for } x < 0 \\ v_r & \text{for } x > 0 \end{cases} \qquad h(x,0) = \begin{cases} h_\ell & \text{for } x < 0 \\ 0 & \text{for } x > 0, \end{cases}$$

   where $h_\ell > 0$, find the relationship between v_ℓ and v_r required to produce a one-shock; do the same for a two-shock.
   (b) Find the shock paths $\xi_i(t)$ in part (a).

Eulerian Gas Dynamics

8. Consider (14) for a polytropic gas (15c).
   (a) Compute $A(\vec{u}) = G'(\vec{u})$, and show that the system is strictly hyperbolic.
   (b) Conclude that along a curve of discontinuity, the following jump condition holds:

$$[\rho v]^2 = [p + \rho v^2][\rho].$$

10.3 Systems of Two Conservation Laws

Let us consider the system of conservation laws (10) when $N = 2$ (i.e., the number of equations equals the number of independent variables). In this case, there is additional structure provided by certain quantities, called Riemann invariants, which are constant along the characteristics and are useful in studying more general solutions than simple waves and shocks. In this section, we introduce these invariants and then apply them to the physical problem of gas dynamics.

a. Riemann Invariants

We again assume strict hyperbolicity so that the 2×2-matrix $A(\vec{u})$ of (9) has distinct real eigenvalues $\lambda_1 < \lambda_2$ and linearly independent eigenvectors $\vec{V}_1, \vec{V}_2$. The eigenvalues and eigenvectors are, of course, functions of $\vec{u}$. We now define Riemannn invariants, which are functions of $\vec{u}$ that will be useful in studying the solutions of (10).

Definition. A k-*Riemann invariant* is a function $r_k \colon \mathbf{R}^2 \to \mathbf{R}$ satisfying

(51) $$\nabla r_k \cdot \vec{V}_k = 0.$$

Notice that (51) simply states that r_k is constant along the trajectories of the vector field $\vec{V}_k$, so it is elementary to find Riemann invariants; in fact, we can choose r_1 and r_2 to have nondegenerate dependence on u_1 and u_2. Indeed, to find the function r_1, simply take any curve $\mathcal{C}$ that is transverse to the trajectories of $\vec{V}_1$, and assign arbitrary, but strictly increasing, values for r_1 along $\mathcal{C}$. In this way, each trajectory of $\vec{V}_1$ is given a distinct constant value for r_1. Similarly, the function r_2 may be defined so that it has a distinct constant value along each trajectory of $\vec{V}_2$.

We may use the Riemann invariants r_1, r_2 in place of u_1, u_2 as new dependent variables in the hyperbolic system. To simplify the notation, let $r = r_1$ and $s = r_2$; similarly, let $\lambda = \lambda_1$ and $\mu = \lambda_2$. If $\vec{u}$ is a solution of (9), consider $r(x,t) \equiv r(\vec{u}(x,t))$. Then (denoting partial derivatives by subscripts)

$$r_t + \mu r_x = \nabla r \cdot \vec{u}_t + \mu \nabla r \cdot \vec{u}_x = \nabla r \cdot (-A(\vec{u}) + \mu)\vec{u}_x = 0,$$

where we have used the fact that ∇r must be a *left* eigenvector for $A(\vec{u})$ with eigenvalue μ (cf. Exercise 1); notice that $r_t + \mu r_x = 0$ simply states that r is constant along the characteristic $dx/dt = \mu$. Similarly, $s_t + \lambda s_x = 0$ implies that s is constant along the characteristic $dx/dt = \lambda$. This means that, when expressed in terms of these new variables, the system (9) becomes diagonalized:

(52) $\quad \vec{v}_t + \Lambda \vec{v}_x = 0, \quad$ where $\vec{v} = \begin{pmatrix} r \\ s \end{pmatrix}$ and $\Lambda = \begin{pmatrix} \mu & 0 \\ 0 & \lambda \end{pmatrix}$.

One application of (52) is to determine when a solution develops singularities. In Exercise 1 of Section 1.2, it was shown that the solution of $u_t + a(u)u_x = 0$ with initial condition $u(x,0) = h(x)$ develops a gradient catastrophe in finite time, unless the function $a(h(x))$ is nondecreasing; if we assume genuine nonlinearity, $a'(u) > 0$, this means that a singularity develops unless $h'(x) \geq 0$ for all x. In order to extend this result to $N = 2$, recall that genuine nonlinearity for (10) is defined by (36); for (52) this condition is simply $\mu_r, \lambda_s \neq 0$.

Theorem 1 (Existence of Singularities). *Suppose $\mu_r, \lambda_s > 0$. The solution of (52) remains continuous if and only if the initial conditions $r(x,0)$ and $s(x,0)$ are nondecreasing functions of x.*

Proof. Let us differentiate $r_t + \mu r_x = 0$ with respect to x; we obtain

$$p_t + \mu p_x + \mu_r p^2 + \mu_s s_x p = 0, \quad \text{where } p = r_x.$$

Notice that $p_t + \mu p_x$ represents differentiation of p in the direction of the characteristic $dx/dt = \mu$. In fact we would like to consider p as the solution of an ODE along this characteristic. If we use ' to denote the directional derivative along $dx/dt = \mu$, then $s' = s_t + \mu s_x = -\lambda s_x + \mu s_x$, and we may write our ODE as

$$(53) \qquad p' + \mu_r p^2 + \frac{\mu_s}{\mu - \lambda} s' p = 0.$$

Let $z(r, s)$ be any function satisfying $z_s = \mu_s/(\mu - \lambda)$; then $r' = 0$ shows that $z' = z_s s' = \mu_s s'/(\mu - \lambda)$ and (53) becomes $p' + \mu_r p^2 + z' p = 0$. This simplifies to

$$(54) \qquad q' + kq^2 = 0, \qquad \text{where } q = e^z p \text{ and } k = e^{-z} \mu_r.$$

It is elementary to solve (54) as an ODE with initial condition $q_0 = q(0)$:

$$(55) \qquad q(t) = \frac{q_0}{1 + q_0 K(t)} \qquad \text{where } K(t) = \int_0^t k(\tau) \, d\tau,$$

and $k(t)$ represents the restriction of k to the characteristic.

Let us analyze the behavior of q from (55). We use $\mu_r > 0$ to conclude that, along any finite length characteristic, μ_r is bounded below by a positive constant; therefore, $k = e^{-z} \mu_r \geq k_0 > 0$, and so $K(t) \geq k_0 t$. If $q_0 \geq 0$, then $q(t)$ remains bounded. On the other hand, if $q_0 < 0$, then $q(t)$ becomes unbounded in a finite time. Since $q = e^z p = e^z r_x$, we conclude that we can avoid a gradient catastrophe for r if and only if $p(x, 0) = r_x(x, 0) \geq 0$ for all $x \in \mathbf{R}$.

We may perform a similar analysis with $p = s_x$ to find that the existence of a gradient catastrophe for s can be avoided exactly when $s_x(x, 0) \geq 0$ for all $x \in \mathbf{R}$. ♠

Example 1. *(The p-system)* For the system (12a), we found $\lambda = \lambda_1 = -\sqrt{a(w)}$ and $\mu = \lambda_2 = \sqrt{a(w)}$, with eigenvectors V_1 and V_2 given respectively by (23) and (30a). By (51), the Riemann invariants $r(w, v) = r_1(w, v)$ and $s(w, v) = r_2(w, v)$ must satisfy

$$(56) \qquad r_w + r_v \sqrt{a(w)} = 0 \quad \text{and} \quad s_w - s_v \sqrt{a(w)} = 0.$$

It is elementary to solve these first-order PDEs to find

$$(57a) \qquad r(w, v) = v - \psi(w) \quad \text{and} \quad s(w, v) = v + \psi(w),$$

where

$$(57b) \qquad \psi'(w) = \sqrt{a(w)}.$$

Recall that the assumption $a'(w) < 0$ was made to ensure nonlinearity; with this assumption and the definitions of r and s, it is elementary to check that the hypothesis $\mu_r, \lambda_s > 0$ of Theorem 1 holds (cf. Exercise 2a).

Given continuous initial conditions $w(x,0)$ and $v(x,0)$, we may therefore use r and s to determine whether the solution remains continuous for all time (cf. Exercise 2b).

b. The Hodograph Transformation

We have seen how to use the Riemann invariants as new dependent variables to diagonalize the system (10). Now we investigate the use of the Riemann invariants as new independent coordinates (i.e., to replace x and t); this is an example of a *hodograph transformation*, which, in general, means interchanging the roles of dependent and independent variables.

Let us assume that, for the given solution $\vec{u}$, the following Jacobian condition holds:

$$(58) \qquad J = \det\left(\frac{\partial(r,s)}{\partial(x,t)}\right) \neq 0,$$

which allows us to consider (r,s) as new independent coordinates. (Incidentally, this condition fails when considering the special case of simple wave solutions; cf. Exercise 3.) In order to convert (10) to the new coordinates, let the characteristic curves $dx/dt = \mu$ be defined implicitly by $\psi(x,t) = \text{const}$. Now r is constant along such a curve, but s is variable, so $\psi_x x_s + \psi_t t_s = 0$. But differentiating $\psi(x,t) = \text{const}$ with respect to t yields $\psi_x x_t + \psi_t = \psi_x \mu + \psi_t = 0$. Combining these two equations yields $x_s = \mu t_s$. Similarly, we obtain $x_r = \lambda t_r$. Therefore, we have obtained a *linear* system

$$(59) \qquad \begin{cases} x_s - \mu t_s = 0 \\ x_r - \lambda t_r = 0, \end{cases}$$

with x,t as dependent variables and r,s as independent variables; notice that λ and μ are computable functions of r and s.

This is a remarkable accomplishment, but there is still additional work to do: we must solve the linear system (59) and then use the answer to find $\vec{u}(x,t)$. In the next subsection, we discuss these issues in the context of the application to gas dynamics.

c. Application to Gas Dynamics

Let us return to one-dimensional gas dynamics in Eulerian coordinates (cf. Example 3 of Section 10.2),

$$(60) \qquad \begin{cases} \rho_t + (\rho v)_x = 0 & \text{conservation of mass} \\ (\rho v)_t + (\rho v^2 + p)_x = 0 & \text{conservation of momentum} \end{cases}$$

We assume that the gas is isentropic (i.e., p is a known function of ρ). It is natural to assume that $p(\rho)$ is strictly increasing in ρ, and this means that $p_x = c^2 \rho_x$ where $c(\rho) = \sqrt{dp/d\rho}$ is defined to be the *speed of sound*. This system is strictly hyperbolic with distinct real eigenvalues $\lambda_1 = v - c$ and $\lambda_2 = v + c$ and characteristics given by

$$\text{(61)} \qquad \frac{dx}{dt} = v + c \quad \text{and} \quad \frac{dx}{dt} = v - c.$$

Notice that, if we interpret the characteristics as sound waves, then their speed dx/dt differs from the flow velocity v by $\pm c$, which leads to the interpretation of c as the speed of sound in the gas.

To simplify calculations, we now assume that the gas flow is polytropic [i.e., $p(\rho) = \alpha \rho^\gamma$, where $\alpha > 0$ and $\gamma > 1$ as in (15c)]. Let us make a change of dependent variables by defining

$$\text{(62)} \qquad h(\rho) \equiv \int_0^\rho \frac{c(r)}{r} \, dr = \frac{2\sqrt{\alpha\gamma}}{\gamma - 1} \rho^{\frac{\gamma-1}{2}} = \frac{2c(\rho)}{\gamma - 1}.$$

Since $dh/d\rho = c(\rho)/\rho > 0$, the correspondence $h \leftrightarrow \rho$ is one to one. We compute $\rho_t = \rho c^{-1} h_t$ and $\rho_x = \rho c^{-1} h_x$, so the equations (62) become

$$\text{(63)} \qquad \begin{cases} h_t + c v_x + v h_x = 0 \\ v_t + v v_x + c h_x = 0, \end{cases}$$

which is of the form (9) with

$$\text{(64)} \qquad A(\vec{u}) = \begin{pmatrix} v & c \\ c & v \end{pmatrix} \quad \text{with} \quad \vec{u} = \begin{pmatrix} h \\ v \end{pmatrix}.$$

In this form, the eigenvalues $\lambda_1 = v - c$ and $\lambda_2 = v + c$ have eigenvectors

$$\text{(65)} \qquad \vec{V}_1 = \begin{pmatrix} 1 \\ -1 \end{pmatrix} \quad \text{and} \quad \vec{V}_2 = \begin{pmatrix} 1 \\ 1 \end{pmatrix}.$$

It is elementary to check that the quantities

$$\text{(66)} \qquad r = \frac{(h+v)}{2} \quad \text{and} \quad s = \frac{(h-v)}{2}$$

are Riemann invariants. Thus r is constant along the characteristic $dx/dt = v + c$, and s is constant along $dx/dt = v - c$.

Now let us assume that, for the solution (ρ, v) we are considering, the Jacobian condition (58) holds. Then the hodograph transformation (59) becomes

$$\text{(67a)} \qquad \begin{cases} x_s = (v + c) t_s \\ x_r = (v - c) t_r, \end{cases}$$

Section 10.3: Systems of Two Conservation Laws

which is a linear system with coefficients

(67b)
$$\begin{cases} v + c = (r-s) + \dfrac{\gamma-1}{2}(r+s) = \dfrac{\gamma+1}{2}r + \dfrac{\gamma-3}{2}s \\ v - c = (r-s) - \dfrac{\gamma-1}{2}(r+s) = \dfrac{3-\gamma}{2}r - \dfrac{\gamma+1}{2}s. \end{cases}$$

We do not address the details of the analysis of the linear system (67a,b), but let us describe the general procedure. The system (67a,b) may be converted to a second-order equation (cf. Exercise 4), and the solution of this second-order equation may be expressed by means of an integral operator involving a *Riemann function* (cf. [Garabedian], [Guenther-Lee], or [Kevorkian]). Of course, this also requires converting initial conditions from the x, t variables to the r, s variables. Finally, we need to express r, s in terms of x, t in order to express ρ and v as functions of x, t. This is clearly a complicated process but can be implemented numerically to yield results; we refer to the aforementioned references for further details.

Exercises for Section 10.3

1. If A is a 2×2-matrix with distinct real eigenvalues λ_1, λ_2 and respective eigenvectors $\vec{v}_1, \vec{v}_2$, and $\vec{v}$ is a vector orthogonal to $\vec{v}_1$, prove that $\vec{v}$ is a left eigenvector for A with eigenvector λ_2 (i.e., $\vec{v} \cdot A\vec{w} = \lambda_2 \vec{v} \cdot \vec{w}$ for all 2-vectors $\vec{w}$).

2. As in Example 1, let $\mu = \sqrt{a(w)}$ and $\lambda = -\sqrt{a(w)}$.
   (a) If $a'(w) < 0$, show that the Riemann invariants r and s defined by (57a,b) satisfy $\mu_r, \lambda_s > 0$.
   (b) Let $a(w) = k^2 w^{-2\beta}$, where $\beta > 1$ as in the application to polytropic gas dynamics. Determine a condition on the initial values under which singularities do not develop.

3. If $\vec{u}$ is a simple wave solution [cf. (17)], show that condition (58) fails (i.e., $J \equiv 0$).

4. Show that the first-order system (67a,b) may be reduced to the second-order equation
$$t_{rs} + \frac{\lambda}{r+s}(t_r + t_s) = 0 \quad \text{where} \quad \lambda = \frac{\gamma+1}{2(\gamma-1)}.$$

5. Consider the shallow-water wave system (13a).
   (a) Find Riemann invariants.
   (b) Determine a condition on the initial values under which singularities do not develop.
   (c) Using the Riemann invariants as independent variables, derive the corresponding system of linear equations.

Further References for Chapter 10

The subject of first-order hyperbolic systems is very rich and has many important applications; we have given only a flavor of it here. More details may be found in [Kevorkian], [Lax], [Smoller], and [Whitham]. The treatment in [Whitham] is not only thorough but also geared toward applications.

For systems of conservation laws in several space dimensions, see [Majda].

11 Linear and Nonlinear Diffusion

In this chapter, we study equations and systems of equations that arise in physical situations involving diffusion. One example is the linear heat equation, which we studied in Chapter 5 using classical methods and as a linear evolution on a Banach space in Section 9.2. A simple way to introduce a nonlinearity is to consider the semilinear heat equation, which we frequently take as a model case. However, we also study the Navier-Stokes equations as an example of a diffusive system of equations.

The heat equation is a special case of a parabolic equation, and we begin the chapter with a study of maximum principles for such equations; these extend the elliptic maximum principles of Section 8.3 and enable us to obtain comparison theorems for semilinear equations. Next we consider local existence and regularity theory for semilinear equations; this uses the contraction mapping principle of Section 7.3 as well as eigenfunction expansions and powers of the Laplacian when boundary conditions are involved. Then we obtain some results on the global behavior of solutions, both by the comparison method and energy methods. Finally, we turn our attention to a specialized study of the Navier-Stokes equations.

11.1 Parabolic Maximum Principles

Let Ω be a bounded domain in $\mathbf{R}^n$. In Section 5.1 we found that solutions of $\Delta u \geq u_t$ in $U_T = \Omega \times (0,T)$ satisfy a weak maximum principle; in this section we generalize and strengthen this result. We consider more general inequalities of the form

(1) $$Lu \geq u_t \quad \text{in} \quad U_T = \Omega \times (0,T),$$

where L is a uniformly elliptic second-order operator; in fact, we can now allow the coefficients of L to depend on t as well as x; that is, we assume that

(2a) $$L = \sum_{i,j=1}^{n} a_{ij}(x,t) \frac{\partial^2}{\partial x^i \partial x^j} + \sum_{i=1}^{n} b_i(x,t) \frac{\partial}{\partial x^i} + c(x,t)$$

satisfies the ellipticity condition

(2b) $$\sum_{i,j=1}^{n} a_{ij}(x,t) \xi_i \xi_j \geq \lambda |\xi|^2 \quad \text{for } (x,t) \in U_T,$$

where $\lambda > 0$ is a constant, and the coefficients a_{ij}, b_i, and c are all bounded functions on U_T. Equations of the form $u_t = Lu + f$ are called *parabolic*, and the strong and weak maximum principles that apply to solutions of (1) are called *parabolic maximum principles*.

a. The Weak Parabolic Maximum Principle

The weak maximum principle of Section 5.1 concludes that

$$(3) \qquad \max_{(x,t) \in U_T} u(x,t) \leq \max_{(x,t) \in \Gamma_T} u(x,t),$$

where $\Gamma_T = (\partial\Omega \times [0,T]) \cup (\Omega \times \{0\})$ is the parabolic boundary and $u \in C^{2;1}(U_T) \cap C(\overline{U_T})$ satisfies (1) for the special case $L = \Delta$. The proof of the following generalization is similar to that in Section 5.1 (Exercise 1).

Theorem 1 (Weak Parabolic Maximum Principle). *Suppose $u \in C^{2;1}(U_T) \cap C(\overline{U_T})$ satisfies (1) with L given by (2) where $c(x,t) \equiv 0$. Then (3) holds.*

Analogous to the elliptic case, this may be generalized to $c(x,t) \leq 0$, and may be used to establish pointwise a priori bounds on solutions of nonhomogeneous parabolic equations (see Exercise 2).

b. The Strong Parabolic Maximum Principle

If we compare with the strong elliptic maximum principle of Section 8.3, we might expect solutions of (1) to satisfy the following.

Strong Parabolic Maximum Principle. *Suppose $M = \sup\{u(x,t) : (x,t) \in U_T\} < \infty$.*
(a) *If $u(x_0, t_0) = M$ for some $(x_0, t_0) \in U_T$, then $u(x,t) \equiv M$ for all $x \in \Omega$ and $0 < t \leq t_0$.*
(b) *If $u(x_0, t_0) = M$ for some $x_0 \in \partial\Omega$ and $0 < t_0 < T$, but $u(x,t) < M$ for all $x \in \Omega, 0 < t < t_0$, then*

$$(4) \qquad \frac{\partial u}{\partial \nu}(x_0, t_0) > 0,$$

provided that the exterior normal derivative $\partial u/\partial \nu$ exists at (x_0, t_0).

We now investigate some conditions on $c(x,t)$ and M under which the parabolic maximum principle holds.

Theorem 2. *If the bounded domain Ω has C^2-boundary, and $u \in C^{2;1}(U_T)$ satisfies (1) with L given by (2), then the strong parabolic maximum principle holds in the following cases:*
   (i) $c(x,t) \equiv 0$;
   (ii) $c(x,t) \leq 0$ and $M \geq 0$;
   (iii) $c(x,t)$ arbitrary and $M = 0$.

Section 11.1: Parabolic Maximum Principles

Proof. We only consider case (ii); see Exercise 3 for cases (i) and (iii). The proof involves the construction of local barriers as in the proof of Theorem 3 in Section 8.3, but is naturally more complicated due to the additional variable t. To illustrate the ideas of the proof, we assume that $n = 1$; then $\Omega = (\alpha, \beta)$ is an interval and the operator (2) may be written $Lu = a(x,t)u_{xx} + b(x,t)u_x + c(x,t)u$; for the general case $n \geq 1$, see [Protter-Weinberger].

We proceed in several steps.

1. *Suppose $B_\epsilon = B_\epsilon(\xi, \tau)$ is a open ball (disk) in U_T on which $u < M$, but $u(x_0, t_0) = M$ for some $(x_0, t_0) \in \partial B_\epsilon$; if $\overline{B_\epsilon} \subset U_T$, then $x_0 = \xi$.*

To prove this statement, we first observe that we may assume that $(x_0, t_0) \in \partial B_\epsilon$ is the *only* point on ∂B_ϵ at which $u = M$; otherwise, we could take a smaller ball in B_ϵ that only meets ∂B_ϵ at (x_0, t_0). We therefore assume $u < M$ on $\overline{B_\epsilon} \setminus \{x_0, t_0\}$. We assume $x_0 \neq \xi$ and derive a contradiction.

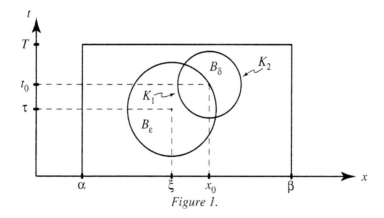

Figure 1.

Let us define the auxiliary function $v(x,t) = \exp(-Ar^2) - \exp(-A\epsilon^2)$, where $A > 0$ is a constant and $r^2 = (x - \xi)^2 + (t - \tau)^2$, which is defined on the ball $B_\epsilon = B_\epsilon(\xi, \tau)$ (see Figure 1). Notice that $v > 0$ on B_ϵ, but $v = 0$ on ∂B_ϵ, and $v < 0$ outside B_ϵ. A calculation shows that $Lv - v_t = \exp(-Ar^2)[4A^2(x-\xi)^2 - 2aA - 2bA(x-\xi) + c + 2A(t-\tau)] - c\exp(-A\epsilon^2)$. Since $x_0 \neq \xi$, we can choose δ sufficiently small that $(x-\xi)^2 \geq \gamma > 0$ for $(x,t) \in B_\delta(x_0, t_0)$; using the positivity of $a(x,t)$ (by ellipticity), we can therefore choose A sufficiently large that $Lv - v_t > 0$ on B_δ. We may also use $\overline{B_\epsilon} \subset U_T$ to pick δ so that $\overline{B_\delta} \subset U_T$, so both u and v are defined on $\overline{B_\delta}$. Let us define the function $w = u - M + \eta v$ on B_δ and observe that $Lw - w_t = Lu - u_t + \eta(Lv - v_t) - cM \geq 0$; by Theorem 1, w achieves its maximum on ∂B_δ. But, since $u < M$ on the closed set $K_1 = \partial B_\delta \cap \overline{B_\epsilon}$, we can choose η sufficiently small that $w < 0$ on K_1; and since $v < 0$ and $u \leq M$ on $K_2 = \partial B_\delta \cap \overline{U_T \setminus B_\epsilon}$, we know $w < 0$ on K_2. Hence $w < 0$ on $K_1 \cup K_2 = \partial B_\delta$, which does not allow w to achieve its maximum on ∂B_δ, since $w(x_0, t_0) = \eta v(x_0, t_0) = 0$. This contradiction shows that we must have $x_0 = \xi$.

2. *If $u(x_0, t_0) < M$ for some $x_0 \in (\alpha, \beta)$ and $t_0 \in (0, T)$, then $u(x, t_0) < M$ for every $x \in (\alpha, \beta)$.*

To verify this statement, suppose otherwise: there exists $x_1 \in (\alpha, \beta)$ at which $u(x_1, t_0) = M$. Without loss of generality, we suppose that $x_0 < x_1$ and $u(x, t_0) < M$ for all $x_0 \leq x < x_1$. For $x \in [x_0, x_1]$, let $d(x)$ denote the distance to the set where $u = M$: $d(x)$ is a continuous function satisfying $d(x_1) = 0$ and $d(x_0) > 0$. Moreover, for every $x \in (x_0, x_1)$, we have by Step 1 either $u(x, t_0 + d(x)) = M$ or $u(x, t_0 - d(x)) = M$. For small $|h| > 0$, the Pythagorean theorem shows that $d(x + h) \leq \sqrt{d(x)^2 + h^2}$; similarly, $d(x) \leq \sqrt{d(x+h)^2 + h^2}$. Combining these yields $\sqrt{d(x)^2 - h^2} - d(x) \leq d(x + h) - d(x) \leq \sqrt{d(x)^2 + h^2} - d(x)$; dividing by h and letting $h \to 0$ shows that $d'(x) = 0$. This means that $d(x)$ is a constant on (x_0, x_1), contradicting $d(x_0) > 0$ and $d(x_1) = 0$.

3. *Suppose $u(x, t) < M$ for $\alpha < x < \beta$ and $0 < t < t_1$, where $t_1 \leq T$. Then $u(x, t_1) < M$.*

On the contrary, suppose $u(x_1, t_1) = M$ for some $x_1 \in (\alpha, \beta)$, and introduce the auxiliary function $v(x, t) = \exp[-(x - x_1)^2 - A(t - t_1)] - 1$ where $A > 0$ is a constant. A computation yields

$$Lv - v_t = e^{-[(x-x_1)^2 + A(t-t_1)]} \left[4a(x - x_1)^2 - 2a - 2b(x - x_1) + A \right] + cv.$$

We can choose $A > 0$ large so that $L(v) > v_t$ on the half-ball $B_\delta^- = \{(x, t) \in B_\delta(x_1, t_1) : t < t_1\}$. Let U denote the points of B_δ^- lying below the parabola Π defined by $(x - x_1)^2 + A(t - t_1) = 0$ (see Figure 2); notice that $\partial U = K_1 \cup K_2$, where $K_1 = \Pi \cap B_\delta$ and $K_2 \subset \partial B_\delta^-$. Since $v = 0$ on K_1 and $u < M$ on K_2, we can choose $\eta > 0$ sufficiently small that $w = u - M + \eta v \leq 0$ on ∂U.

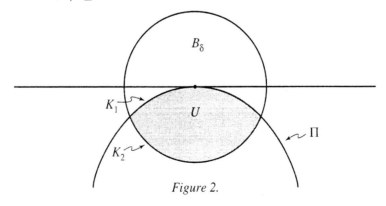

Figure 2.

Since $Lw - w_t \geq -cM \geq 0$ on U, we can apply Theorem 1 to conclude that $w = u - M + \eta v \leq 0$ on U. Therefore, if we take the t-derivative at

(x_1, t_1), we must have $w_t \geq 0$ (i.e., $u_t + \eta v_t = u_t - \eta A \geq 0$); this shows that $u_t > 0$ at (x_1, t_1). But since the maximum of u on $t = t_1$ occurs at $x = x_1$, we must have $u_x = 0$ and $u_{xx} \leq 0$, and hence $Lu = au_{xx} + bu_x + cu \leq cM \leq 0$, at (x_1, t_1). This contradiction with $Lu \geq u_t$ shows that we must have $u(x, t_1) < M$ for all $\alpha < x < \beta$.

4. *Proof of (a).* If $u(x_0, t_0) = M$, consider the line $\ell_{x_0} = \{(x_0, t) : 0 \leq t \leq t_0\}$. Suppose there is a $t_1 \leq t_0$ such that $u(x_0, t) < M$ for $t < t_1$, but $u(x_0, t_1) = M$. By Step 2, $u(x, t) < M$ for all $x \in (\alpha, \beta)$ and $0 < t < t_1$. But then Step 3 shows $u(x_0, t_1) < M$, a contradiction. This shows that $u(x, t) \equiv M$ for all $x \in (\alpha, \beta)$ and $0 < t \leq t_0$, as desired.

5. *Proof of (b).* The proof of (b) requires a barrier construction similar to that in the proof of Theorem 3 in Section 8.3; see Exercise 4. ♠

Let us close this subsection by investigating the issue of uniqueness of solutions to linear parabolic equations. Namely, consider the problem

(5)
$$\begin{cases} u_t = Lu + f(x,t) & \text{in } U_T \\ u(x,0) = g(x) & \text{for } x \in \Omega \\ \alpha(x,t)u(x,t) + \beta(x,t)\dfrac{\partial u}{\partial \nu}(x,t) = h(x,t) & \text{for } x \in \partial\Omega, 0 < t < T, \end{cases}$$

where L is the operator in (2a,b). Then Theorem 2 may be used to prove the following (cf. Exercise 5).

Corollary. *If the functions $\alpha(x,t)$, $\beta(x,t)$ are nonnegative and satisfy $\alpha^2 + \beta^2 \neq 0$ on $\partial\Omega \times (0, T)$, then there is at most one solution $u \in C^{2;1}(U_T) \cap C(\overline{U_T})$ of (5).*

Notice that the boundary condition in (5) covers the special cases of a Dirichlet condition, a Neumann condition, and a Robin condition. Notice also that, unlike uniqueness for elliptic equations, uniqueness for (5) does *not* require $c \leq 0$.

c. Comparison Principles

The strong maximum principle also yields comparison principles that will be useful in the study of global existence in Section 11.3. The statements depend on the boundary condition involved, so we consider Neumann and Dirichlet boundary conditions separately. We state results for $L = \Delta$, but they hold more generally for L in the form (2).

Theorem 3 (Neumann Comparison). *Suppose Ω is a bounded domain with C^2-boundary, $f: \mathbf{R} \to \mathbf{R}$ is C^1, and $u, v \in C^{2;1}(U_T) \cap C^1(\overline{U_T})$ satisfy*

$$u_t - \Delta u - f(u) \geq v_t - \Delta v - f(v) \quad \text{in } U_T = \Omega \times (0, T)$$
$$u(x, 0) > v(x, 0) \quad \text{for } x \in \Omega$$
$$\frac{\partial u}{\partial \nu}(x, t) \geq \frac{\partial v}{\partial \nu}(x, t) \quad \text{for } x \in \partial\Omega, 0 < t < T.$$

Then $u(x, t) > v(x, t)$ for all $(x, t) \in \overline{U_T}$.

Proof. Let $w = v - u$. We must show $w < 0$ in U_T. Since f is C^1 in u, we may use the mean value theorem of calculus to write $f(u) - f(v) = f'(\theta)(u - v)$ for some θ between u and v. This defines a function $\theta(x,t)$, and the hypotheses imply $\Delta w + f'(\theta(x,t))w \geq w_t$. If we let $c(x,t) = f'(\theta(x,t))$, we see that w satisfies the inequality (1), so we may hope to use the parabolic maximum principle to show that $w < 0$. Of course, we do not know if $c(x,t) \leq 0$, so we must try to use condition (iii) of Theorem 2. Now $w(x,0) < 0$ for $x \in \Omega$ and $\partial w/\partial \nu \leq 0$ on $\partial \Omega$ together imply $w(x,0) < 0$ for $x \in \overline{\Omega}$; moreover, continuity implies $w(x,t) < 0$ for $0 < t < t_0$ for some $t_0 > 0$. Suppose $w(x_0, t_0) = 0 = \max\{w(x,t) : x \in \overline{\Omega}, 0 \leq t \leq t_0\}$. We cannot have $x_0 \in \Omega$ by (a) of the maximum principle, and we cannot have $x_0 \in \partial \Omega$ by (b) of the maximum principle since $\partial w/\partial \nu(x_0, t_0) \leq 0$. Thus we must have $w(x,t) < 0$ for all $(x,t) \in U_T$, as desired. ♠

Theorem 4 (Dirichlet Comparison). Suppose Ω is a bounded domain, f is C^1, and $u, v \in C^{2;1}(U_T) \cap C(\overline{U_T})$ satisfy

$$u_t - \Delta u - f(u) \geq v_t - \Delta v - f(v) \quad \text{in} \quad U_T = \Omega \times (0, T)$$
$$u(x, 0) \geq v(x, 0) \quad \text{for} \quad x \in \Omega$$
$$u(x, t) \geq v(x, t) \quad \text{for} \quad x \in \partial\Omega, 0 < t < T.$$

Then $u(x,t) \geq v(x,t)$ for all $(x,t) \in \overline{U_T}$.

Proof. See Exercise 6.

Exercises for Section 11.1

1. Prove Theorem 1.

2. (a) Assume that $u \in C^{2;1}(U_T) \cap C(\overline{U_T})$ satisfies (1) with L given by (2), where $c(x,t) \leq 0$. Prove that $\max_{\overline{U}_T} u(x,t) \leq \max_{\overline{\Gamma}_T} u^+(x,t)$.
   (b) Analogous to (53) in Chapter 8, establish a priori bounds for a solution $u \in C^{2;1}(U_T) \cap C(\overline{U_T})$ of $u_t = \Delta u + c(x,t)u + f(x,t)$ in U_T, with $u(x,0) = \phi(x)$ for $x \in \Omega$, and $u(x,t) = g(x,t)$ for $x \in \partial\Omega$ and $0 < t < T$ [assuming $c(x,t) \leq 0$ and $f, g \in C(\overline{U_T})$].

3. Prove that (ii) implies both (i) and (iii) in Theorem 2.

4. Prove conclusion (b) of the strong parabolic maximum principle under condition (ii) of Theorem 2.

5. Prove the corollary of Theorem 2.

6. Use Theorem 2 to prove Theorem 4.

7. Does Theorem 3 still hold if the hypothesis $u(x,0) > v(x,0)$ is weakened to $u(x,0) \geq v(x,0)$ and the conclusion $u(x,t) > v(x,t)$ is weakened to $u(x,t) \geq v(x,t)$?

11.2 Local Existence and Regularity

In this section, we consider local existence for parabolic equations of the form

(6) $$u_t = \Delta u + f(x, t, u, \nabla u) \quad \text{in } U_T = \Omega \times (0, T),$$

where Ω is a smooth, domain in $\mathbf{R}^n$; it is elementary to generalize (6) by replacing Δ by any uniformly elliptic operator in divergence form (cf. Section 6.2). We often view (6) as a first-order evolution on a Banach space X of functions on Ω and write

(6′) $$u_t = Au + F(t, u) \quad \text{for } 0 < t < T,$$

where $A = \Delta \colon D \to X$ is a linear operator on X as in Section 6.6. The simplest situation involves $F(t, u)$ for which $F \colon \mathbf{R} \times X \to X$ is continuous, but the nonlinearities that we encounter may require us to consider $F \colon \mathbf{R} \times X' \to X$ for some auxiliary Banach space $X' \subset X$.

We impose an initial condition on (6), namely

(7) $$u(x, 0) = g(x) \quad \text{for } x \in \Omega,$$

which for (6′) becomes

(7′) $$u(0) = g \in X.$$

In addition, when $\partial\Omega \neq \emptyset$, we impose a boundary condition such as

(8D) $\quad u(x, t) = 0 \quad$ for $x \in \partial\Omega$ and $t > 0 \quad$ (Dirichlet), or

(8N) $\quad \dfrac{\partial u}{\partial \nu}(x, t) = 0 \quad$ for $x \in \partial\Omega$ and $t > 0 \quad$ (Neumann).

For (6′), the boundary condition (8D) or (8N) must be incorporated into the definition of the Banach space X.

To establish *local existence* and *uniqueness* (i.e., to show that there is a unique solution in U_τ provided that τ is sufficiently small), we use Duhamel's principle for the nonhomogeneous equation and the contraction mapping principle; namely, letting $S(t)$ be the solution operator associated with the linear evolution $u_t = Au$ on X, we find that a solution $u(t)$ of (6′), (7′) satisfies

(9) $$u(t) = S(t)g + \int_0^t S(t - s) F(s, u(s))\, ds,$$

so we want to show that $u(t) \to \psi(t) + \mathcal{T}(u(t))$ has a fixed point, where $\psi(t) = S(t)g$ and $\mathcal{T}(u(t)) = \int_0^t S(t - s) F(s, u(s))\, ds$. To show that $\mathcal{T}$ is a contraction mapping, we need $F: \mathbf{R} \times X' \to X$ to be Hölder continuous in $t \in [0, T]$ and locally Lipschitz in $u \in X'$; that is, for some $0 < \alpha < 1$,

(10) $$\begin{cases} \|F(t,u) - F(s,v)\|_X \leq M_R(|t-s|^\alpha + \|u-v\|_{X'}) \\ \text{for all } 0 \leq s, t \leq T \text{ and } \|u\|_{X'}, \|v\|_{X'} \leq R. \end{cases}$$

As in Section 9.2, the solution $u \in C([0,T], X')$ of (9) is called a *mild solution* of (6'),(7'); a separate regularity argument is required to conclude that $u \in C^1((0,T), X)$; that is, u is a *classical solution* of (6'),(7').

We carry out this program for establishing local existence and uniqueness in some specific cases.

a. Pure Initial Value Problems

Let us take $\Omega = \mathbf{R}^n$ and $U_\tau = \mathbf{R}^n \times (0, \tau)$. For $g \in X = C_B(\mathbf{R}^n)$, we want to show that (6) and (7) admit a unique solution $u \in C_B(\overline{U_\tau})$ provided that $\tau > 0$ is sufficiently small. For simplicity, we assume that $f = f(u)$ does not depend explicitly on x, t, or ∇u; the case $f = f(x, t, u)$ is considered in Exercise 2.

Recall from Section 5.2 and Example 2 in Section 9.2 that the linear evolution $u_t = \Delta u$, $u(0) = g \in X$ is given by the contractive semigroup on X defined by $(S(t)g)(x) = \int K(x,y,t)\,g(y)\,dy$ where $K(x,y,t)$ is the heat kernel on $\mathbf{R}^n$. We now assume that $f: \mathbf{R} \to \mathbf{R}$ is locally Lipschitz continuous [i.e., $|f(u) - f(v)| \leq M_R |u-v|$ whenever $|u|, |v| \leq R$]. This easily implies (10), where $X' = X = C_B(\mathbf{R}^n)$ and $F(u)(x) = f(u(x))$; moreover, (9) takes the form

(11) $$u(x,t) = \int_{\mathbf{R}^n} K(x,y,t)g(y)\,dy + \int_0^t \int_{\mathbf{R}^n} K(x,y,t-s)f(u(y,s))\,dy\,ds.$$

Indeed, if $u(x,t)$ satisfies (11), then u is a weak solution of (6) and (7) in the sense of Section 2.3 (cf. Theorem 2 in Section 5.2).

The proof of Theorem 2 in Section 9.2 may be repeated word for word to prove the following local existence and uniqueness theorem.

Theorem 1. *If $g \in C_B(\mathbf{R}^n)$ and $f(u)$ is locally Lipschitz in $u \in \mathbf{R}$, then there is a positive τ [depending on f and $\sup_{x \in \mathbf{R}^n} |g(x)|$] such that there exists a unique weak solution $u \in C_B(\mathbf{R}^n \times [0, \tau])$ of (6) and (7).*

If $f(u)$ is C^1 in u, then u is a classical solution of (6),(7); that is, $u \in C^{2;1}(\mathbf{R}^n \times [0,\tau])$ (cf. Exercise 1). Moreover, this theorem may be generalized to allow $f = f(x,t,u)$ when f is continuous and bounded in x, locally Hölder continuous in t, and locally Lipschitz in u (cf. Exercise 2).

b. Initial/Boundary Value Problems

Now assume that Ω is a smooth, bounded domain in $\mathbf{R}^n$, and we want to solve (6) and (7) with either (8D) or (8N). In this case, let $X = L^2(\Omega)$ and $Y = C([0,\tau], X)$ with norms $\|\cdot\|$ and $\|\cdot\|_Y$, respectively.

Section 11.2: Local Existence and Regularity

The boundary condition determines the heat kernel used in the definition of the solution operator $S(t)$. For example, let us take (8D) and let (λ_j, ϕ_j) denote the eigenvalues and associated eigenfunctions for the Dirichlet problem on Ω, which form a complete orthonormal set for $L^2(\Omega)$; see Section 7.2. As in Example 4 of Section 9.2, the solution operator $S(t)$ for the linear evolution $u_t = \Delta u$, $u(0) = g \in X$ may be described using eigenfunction expansions: If $g = \sum a_k \phi_k$ then $u(t) = S(t)g = \sum a_k e^{-\lambda_k t} \phi_k$ satisfies $u(t) \in H_0^{1,2}(\Omega) \cap H^{2,2}(\Omega)$ for each $t > 0$, so that $\Delta u \in X$, and the boundary condition (8D) is satisfied in the generalized sense. We would like to find a classical solution of $(6'), (7')$; that is, $u(t)$ continuous for $t \geq 0$, $u(t) \in D$ and $du(t)/dt$ continuous for $t > 0$.

Now let us define $F(t, u)(x) = f(t, x, u(x, t), \nabla u(x, t))$. If F satisfies (10) with $X' = X = L^2(\Omega)$, then (9) admits a unique local solution for every $g \in L^2(\Omega)$; this is essentially Theorem 3 of Section 9.2. However, many nonlinearities that we encounter require taking $X' \subset X$, for example, $X' = H_0^{1,2}(\Omega)$. For this purpose, it is important to take advantage of the smoothing properties of $S(t)$ by observing (cf. Exercise 5a) that

(12a) $\quad \|S(t)g\|_{1,2} \leq Ct^{-1/2}\|g\|_2 \qquad$ for all $g \in X$, and

(12b) $\quad \|(S(t) - I)v\|_2 \leq Ct^{1/2}\|v\|_{1,2} \qquad$ for all $v \in D$.

Let us first note that the mild solution of (9) is in fact a classical solution.

Proposition. Let $X = L^2(\Omega)$ and $X' = H_0^{1,2}(\Omega)$. If $F: \mathbf{R} \times X' \to X$ satisfies (10) and $u \in C([0,T], X')$ satisfies (9), then $u \in C^1((0,T], X)$ satisfies $(6'), (7')$.

This proposition may be derived from Theorem 2' in Section 9.2 if we view $(6')$ as a nonhomogeneous linear equation with $f(t) = F(t, u(t))$ (cf. Exercise 6).

To establish local existence for $(6'), (7')$, we may apply a fixed point argument to (9) that is similar to that used to prove Theorem 3 of Section 9.2, the main difference being the use of (12a) in place of $\|S(t)g\| \leq \|g\|$; this is outlined in Exercise 7 when $F(u)$ is independent of t. We therefore obtain the following local existence result.

Theorem 2. Let $X = L^2(\Omega)$ and $X' = H_0^{1,2}(\Omega)$. If $F: \mathbf{R} \times X' \to X$ satisfies (10), and $g \in X'$, then there is a positive τ (depending on F and $\|g\|_{X'}$) such that there exists a unique solution $u \in C([0,\tau], X') \cap C^1((0,\tau], X)$ of $(6'), (7')$.

An analogous result holds when the boundary condition (8D) is replaced by (8N) (cf. Exercise 8).

Let us consider some examples of nonlinear equations to which Theorem 2 applies.

Example 1. Let $f \in C^1(\mathbf{R})$, and consider

(13)
$$\begin{cases} u_t = u_{xx} + f(u) & \text{for } 0 < x < 1 \text{ and } t > 0, \\ u(x,0) = g(x) & \text{for } 0 < x < 1, \\ u(0,t) = 0 = u(1,t) & \text{for } t > 0. \end{cases}$$

We let $\Omega = (0,1)$, $X = L^2(\Omega)$, $X' = H_0^{1,2}(\Omega)$, and $F(u)(x) = f(u(x))$. To show that (10) is satisfied, we can appeal to Theorem 4 of Section 7.3. More explicitly, let us use the Sobolev imbedding $X' \subset C(\overline{\Omega})$ with $\|u\|_\infty \leq \|u\|_{1,2}$. For $\|u\|_{1,2}, \|v\|_{1,2} \leq R$, we have $|u(x)|, |v(x)| \leq R$, and so

$$\|F(u) - F(v)\|_2^2 = \int_0^1 |f(u(x)) - f(v(x))|^2 \, dx = \int_0^1 \left(\int_{v(x)}^{u(x)} f'(s) \, ds \right)^2 dx$$

$$\leq C_R \int_0^1 |u(x) - v(x)|^2 \, dx \leq C_R \|u - v\|_2^2 \leq C_R \|u - v\|_{1,2}^2,$$

where $C_R = \max\{|f'(s)| : -R \leq s \leq R\}$. We conclude that, for $g \in H_0^{1,2}(\Omega)$, (13) admits a unique solution $u \in C([0,\tau], X') \cap C^1((0,\tau], X)$, provided that $\tau > 0$ is sufficiently small. ♣

The next example shows that we may allow f in (6) to depend on ∇u.

Example 2. Let $f \in C^1(\mathbf{R})$, and consider

(14)
$$\begin{cases} u_t + uu_x = u_{xx} + f(u) & \text{for } 0 < x < 1 \text{ and } t > 0, \\ u(x,0) = g(x) & \text{for } 0 < x < 1, \\ u(0,t) = 0 = u(1,t) & \text{for } t > 0. \end{cases}$$

Take X and X' as in Example 1; we want to verify (10) for $F(u) = f(u) - uu_x$. First observe that $F: X' \to X$, since $u \in X' \subset C([0,1])$ implies $f(u) \in L^2(\Omega)$ (as in Example 1), and $\|uu_x\|_2 \leq \|u\|_\infty \|u_x\|_2 \leq \|u\|_{1,2}^2$. Moreover, we may verify (10) by using Example 1 and

$$\|uu_x - vv_x\|_2 \leq \|u(u_x - v_x)\|_2 + \|(u-v)v_x\|_2$$
$$\leq \|u\|_\infty \|u - v\|_{1,2} + \|u - v\|_\infty \|v\|_{1,2}$$
$$\leq (\|u\|_{1,2} + \|v\|_{1,2}) \|u - v\|_{1,2} \leq 2R \|u - v\|_{1,2}.$$

For $g \in H_0^{1,2}(\Omega)$, we can apply Theorem 2 to obtain a unique solution $u \in C([0,\tau], X') \cap C^1((0,\tau], X)$, provided that $\tau > 0$ is sufficiently small. ♣

c. Additional Smoothness

Examples 1 and 2 benefit from $n = 1$ in that then $H_0^{1,2}(\Omega) \subset C(\overline{\Omega})$. For $n \geq 2$, the Sobolev imbedding changes and restricts the nonlinearities

Section 11.2: Local Existence and Regularity   337

to which Theorem 2 applies. Therefore, it is frequently useful to generalize the spaces X' for which we consider $F: \mathbf{R} \times X' \to X$ satisfying (10).

To define the space X', let $A = -\Delta$ and for $\alpha \geq 0$ let us use eigenfunction expansions to define the operator A^α by

(15a) $$A^\alpha \psi = A^\alpha \left(\sum_{j=1}^\infty a_j \phi_j \right) = \sum_{j=1}^\infty \lambda_j^\alpha a_j \phi_j,$$

provided that ψ belongs to the domain of A^α which we denote X_α:

(15b) $$\operatorname{dom}(A^\alpha) \equiv X_\alpha \equiv \left\{ \psi = \sum_{j=1}^\infty a_j \phi_j \in X : A^\alpha \psi \in X \right\};$$

X_α may be considered as a Banach space under its *graph norm*

(15c) $$\|\psi\|_{X_\alpha} = \|\psi\|_X + \|A^\alpha \psi\|_X.$$

Notice that $X_1 = D$, and for $0 \leq \alpha \leq 1$, we have $X_1 \subset X_\alpha \subset X = X_0$ since the condition $A^\alpha \psi \in X$ imposes extra-stringent decay conditions on the eigenfunction coefficients a_j. For the semigroup $S(t) = e^{-tA} : X \to X$ defined by $S(t)(\sum a_j \phi_j(x)) = \sum e^{-\lambda_j t} a_j \phi_j(x)$, the spaces X_α enable us to generalize (12a,b) to the following [cf. Exercise 5(b)]: for $0 < \alpha \leq 1$

(16a) $\|S(t)g\|_{X_\alpha} \leq C_\alpha t^{-\alpha} \|g\|_X$ for all $g \in X$, and

(16b) $\|(S(t) - I)v\|_X \leq C_\alpha t^\alpha \|v\|_{X_\alpha}$ for all $v \in D$.

We have defined X_α in (15) without specifying either X or D. This means that the method is available for various boundary conditions, such as (8D) or (8N), provided that we have a complete set of eigenfunctions ϕ_j. It also enables us to use for X and X_α spaces of functions with additional regularity. But when $X = L^2(\Omega)$ and the ϕ_j are the Dirichlet eigenfunctions, then $X_1 = H_0^{1,2}(\Omega) \cap H^{2,2}(\Omega)$ and $X_{1/2} = H_0^{1,2}(\Omega)$; this last fact is verified by writing $\langle Au, u \rangle = \|A^{1/2} u\|_2^2 = \int_\Omega |\nabla u|^2 \, dx$. In this case, of course, taking $\alpha = 1/2$ in (16a,b) reproduces (12a,b).

For $0 \leq \alpha < 1$, the integrability of $t^{-\alpha}$ at $t = 0$ enables us to generalize Theorem 2 as follows (cf. Exercise 9)

Theorem 3. *Suppose $0 \leq \alpha < 1$, and $F: \mathbf{R} \times X_\alpha \to X$ is continuous and satisfies (10) with $X' = X_\alpha$. If $g \in X_\alpha$, then there is a positive τ (depending on F and $\|g\|_{X_\alpha}$) such that there exists a unique solution $u \in C([0,\tau], X_\alpha) \cap C^1((0,\tau], X)$ of $(6'),(7')$.*

In applications of the theorem, it is frequently important to know more about the spaces X_α, in particular what Sobolev or Hölder spaces contain them. For example, in the Dirichlet case with $X = L^2(\Omega)$, we have $X_{1/2} = H_0^{1,2}(\Omega)$ and $X_1 \subset H^{2,2}(\Omega)$; we might conjecture that "$X_\alpha \subset H^{2\alpha,2}(\Omega)$," at least if we can make sense of $H^{2\alpha,2}$ when 2α is not an integer. This can actually be done using a continuous scale of Sobolev spaces (cf. Exercise 1 in Section 8.1); however, we content ourselves with obtaining the corresponding Sobolev-type imbeddings for the X_α.

Theorem 4. Suppose $\Omega \subset \mathbf{R}^n$ is a smooth bounded domain, $X = L^2(\Omega)$ and $D = X_1 \subset H^{2,2}(\Omega)$. Then for $0 \leq \alpha \leq 1$ we have the imbeddings

(17a) $\qquad X_\alpha \subset H^{1,q}(\Omega)$ if $2\alpha > 1$ and $2\alpha - \dfrac{n}{2} > 1 - \dfrac{n}{q}$,

(17b) $\qquad X_\alpha \subset C^{m,\nu}(\overline{\Omega})$ if $2\alpha - \dfrac{n}{2} > m + \nu \geq 0$.

Proof. The *Gagliardo-Nirenberg inequality* generalizes (35) of Chapter 6:

(18a)
$$\|u\|_{\ell,q} \leq C \|u\|_{m,p}^\theta \|u\|_r^{1-\theta}, \quad \text{for all } u \in H^{2,2}(\Omega), \text{ provided that}$$
$$\ell < m, \ \ell/m \leq \theta \leq 1, \text{ and } \ell - \frac{n}{q} < \theta\left(m - \frac{n}{p}\right) - (1-\theta)\frac{n}{r};$$

for a proof of (18a) see [Friedman] [where the desired estimate is even proved when $\ell - \frac{n}{q} \leq \theta(m - \frac{n}{p}) - (1-\theta)\frac{n}{r}$, provided that $m - \ell - n/p \notin \mathbf{N}$]. Let $Y = H^{1,q}(\Omega)$, take $m = 2 = p = r$, and use the elliptic estimate $\|u\|_{2,2} \leq C \|Au\|_2$ for $u \in D$, to obtain

(18b) $\qquad \|u\|_Y \leq C \|Au\|_2^\theta \|u\|_2^{1-\theta}, \quad \text{for all } u \in D$.

Using Young's inequality, (18b) implies the existence of K for which

(18c)
$$\|u\|_Y \leq \epsilon \|Au\|_2 + K \, \epsilon^{-\theta/(1-\theta)} \|u\|_2 \quad \text{for all } \epsilon > 0 \text{ and } u \in D,$$
$$\text{provided that } 1/2 \leq \theta \leq 1 \text{ and } 1 - \frac{n}{q} < 2\theta - \frac{n}{2}.$$

Another (more general) way of defining $A^{-\alpha}$ for $\alpha > 0$ is inspired by the Laplace transform:

(19) $\qquad A^{-\alpha} = \dfrac{1}{\Gamma(\alpha)} \displaystyle\int_0^\infty e^{-sA} s^{\alpha-1} \, ds.$

Using (19), we may show that $A^{-\alpha} : Y \to Y$ is bounded when $\theta < \alpha \leq 1$; hence $X_\alpha \subset Y$ as claimed in (17a). In fact,

$$\|A^{-\alpha} u\|_Y \leq \frac{1}{\Gamma(\alpha)} \int_0^\delta \|e^{-sA} s^{\alpha-1} u\|_Y \, ds + \int_\delta^\infty \|e^{-sA} s^{\alpha-1} u\|_Y \, ds.$$

Use (18c) with $\epsilon = s^{1-\theta}$ and (44) of Section 9.2 to estimate $\|e^{-sA} s^{\alpha-1} u\|_Y$ by $C \, s^{\alpha-\theta-1} \|u\|_2$, which is integrable near $s = 0$ since $\alpha - \theta > 0$; so the integral over $(0, \delta)$ is bounded by $C \|u\|_2$, which in turn is bounded by $C \|u\|_Y$ since $Y \subset L^2(\Omega)$. Since e^{-sA} is β-contractive with $\beta < 0$, the integral over (δ, ∞) is bounded by $C \|u\|_Y$. Thus $\|A^{-\alpha} u\|_Y \leq C \|u\|_Y$.

The proof of (17b) is similar; in fact, if $2\alpha > 1$ and $m = 0$, then (17a) may be used to obtain (17b). ♠

Section 11.2: Local Existence and Regularity   339

Theorems 3 and 4 enable us to obtain local existence and additional regularity in situations where Theorem 2 does not apply.

Example 3. Let Ω be a smooth, bounded domain in $\mathbf{R}^n$, $f \in C^\infty(\mathbf{R})$, and consider

(20) $$\begin{cases} u_t = \Delta u + f(u) & \text{for } x \in \Omega \text{ and } t > 0, \\ u(x,0) = g(x) & \text{for } x \in \Omega, \\ u(x,t) = 0 & \text{for } x \in \partial\Omega \text{ and } t > 0. \end{cases}$$

Let $X = L^2(\Omega)$ and $D = X_1 = H_0^{1,2}(\Omega) \cap H^{2,2}(\Omega)$. Applying (17b) with $k = m = \nu = 0$, we see that $X_\alpha \subset C(\overline{\Omega})$ for every $1 \geq \alpha > n/4$; arguing as in Example 1, this means that $F(u)(x) = f(u(x))$ defines $F: X_\alpha \to X$ which is locally Lipschitz. Therefore, provided that $n = 1, 2$, or 3 and we take $g \in X_\alpha$ with $1 \geq \alpha > n/4$, we may use Theorem 3 to conclude local existence of a unique strong solution $u \in C([0,\tau], X_\alpha) \cap C^1((0,\tau], X)$; notice that for $n = 2$ or 3, we cannot use $\alpha = 1/2$ as we did in Example 1.

For $n \geq 4$, a substantial modification needs to be made; we need to use a Sobolev space as the Banach space X on which the evolution takes place, namely $X = H^{k,2}(\Omega)$, where $k + 2\alpha > n/2$ for some $0 \leq \alpha < 1$. Then the assumption $g \in X_\alpha \subset C(\overline{\Omega})$ will enable us to use Theorem 3 to obtain a unique strong solution (cf. Exercise 10 when $k = 1$).

This same idea may be used to impart additional regularity of the initial condition to the solution. For example, suppose that $X = H^{k,2}(\Omega)$ for k large enough to ensure $X_\alpha \subset C^2(\overline{\Omega})$ for some $0 \leq \alpha < 1$. For $g \in X_a$, the unique strong solution guaranteed by Theorem 3 must lie in $C^1((0,\tau), X_\alpha)$, which is a subset of $C^1((0,\tau), C^2(\overline{\Omega})) \equiv C^{2;1}(\overline{\Omega} \times (0,\tau))$; in other words, u is a classical solution of (20). ♣

Exercises for Section 11.2

1. Suppose $f(u)$ is C^1 in $u \in \mathbf{R}$; let $F(u)(x) = f(u(x))$ and $X = C_B(\mathbf{R}^n)$.
   (a) Verify that $F: X \to X$ is locally Lipschitz.
   (b) Show that the solution u in Theorem 1 is in $C^{2;1}(\mathbf{R}^n \times [0,\tau])$.

2. Prove that Theorem 1 still holds when $f(u)$ is replaced by $f(x,t,u)$, which is continuous in x, t, u, and satisfies $|f(x,t,u) - f(x,s,v)| \leq C_R(|t-s|^\alpha + |u-v|)$ for all $x \in \mathbf{R}^n$, $s, t \in [0,\tau]$, and $|u|, |v| \leq R$; here $\alpha \in (0,1)$.

3. If $f(u)$ satisfies the global Lipschitz condition $|f(u) - f(v)| \leq M|u-v|$ for all $u, v \in \mathbf{R}$, show that the solution of Theorem 1 is global in time (i.e., we can take τ arbitrarily large).

4. Suppose $a_i, c, f \in C^\alpha([0,T], C_B(\mathbf{R}^n))$ and $g \in C_B^1(\mathbf{R}^n)$. Discuss the solvability of $u_t = \Delta u + \sum_{i=1}^n a_i(x,t)u_{x_i} + c(x,t)u + f(x,t)$ in $U_T = \mathbf{R}^n \times [0,T]$ with initial condition $u(x,0) = g(x)$ for $x \in \mathbf{R}^n$.

5. (a) Prove (12a,b). (b) Prove (16a,b).

6. Let $X = L^2(\Omega)$, $X' = H_0^{1,2}(\Omega)$ and assume that $F: \mathbf{R} \times X' \to X$ satisfies (10). Suppose $u \in C([0,T], X')$ satisfies (9).
   (a) Let $f_u(t) = F(t, u(t))$ and consider $u(t) = S(t)g + \int_0^t S(t-s)f_u(s)\,ds$ as a linear equation with $f_u: \mathbf{R} \to X$ continuous. Show that $u: \mathbf{R} \to X'$ is locally Hölder continuous for some $\alpha > 0$.
   (b) Use (a) to conclude that $f_u: \mathbf{R} \to X$ is locally Hölder continuous for some $\alpha > 0$.
   (c) Use Theorem 2' from Section 9.2 to conclude that $u \in C^1((0,T), X)$.

7. Let $X = L^2(\Omega)$, $X' = H_0^{1,2}(\Omega)$, and $F: X' \to X$ satisfy (10).
   (a) Let $T(u(t)) = \int_0^t S(t-s)F(u(s))\,ds$ and $Y_R = C([0,\tau], X')$ where $R = 2\|g\|'$. For τ sufficiently small, show that $T: Y_R \to Y_R$.
   (b) For τ sufficiently small, show that $T: Y_R \to Y_R$ is a strict contraction.
   (c) Conclude that $u(t) \to \psi(t) + T(u(t))$ has a fixed point in Y_R.

8. (a) If we replace (8D) by (8N), formulate and prove a local existence theorem analogous to Theorem 2.
   (b) Discuss Examples 1 and 2 if the boundary condition is changed to $u_x(0,t) = 0 = u_x(1,t)$.

9. Prove Theorem 3.

10. Let $X = H^{1,2}(\Omega)$ and define X_α by (15); suppose $X_\alpha \subset C(\overline{\Omega})$ for some $0 \le \alpha < 1$. Let $f \in C^\infty(\mathbf{R})$, and define $F(u)(x) = f(u(x))$. Show that $F: X_\alpha \to X$ is locally Lipschitz. Does this extend to $X = H^{k,2}(\Omega)$?

11. Replace the Dirichlet condition in (20) by the Neumann condition and discuss local existence of solutions.

11.3 Global Behavior

In this section, we study the global behavior of solutions of (6). Does the solution exist for all values of t? Does it blow up in finite time? If it exists for all time, what happens as $t \to \infty$? In particular, does the solution approach an equilibrium $v(x)$ as $t \to \infty$? This last question is often very important in applications. For example, if (6) represents a heat diffusion process, then the asymptotic equilibrium v represents the ultimate temperature distribution in the body.

For simplicity, we consider (6) when $f = f(u)$ is a smooth function of $u \in \mathbf{R}$, independent of x, t, and ∇u:

$$(21) \qquad u_t = \Delta u + f(u) \qquad \text{in } U = \Omega \times (0, \infty),$$

with boundary condition (8D) or (8N) on $\partial\Omega$. Global existence generally holds when $f(u)$ is globally Lipschitz (cf. Exercise 3 of Section 11.2), but this condition is rather restrictive and completely rules out the possibility of blow-up. In fact, for many nonlinear equations the issue of global existence depends on the initial condition. This can easily be seen from some ODE examples.

Example 1. Consider the ODE

(22) $$\begin{cases} \dfrac{du}{dt} = u^2 \\ u(0) = u_0, \end{cases}$$

which may be solved easily to find $u(t) = (u_0^{-1} - t)^{-1}$ provided that $u_0 \neq 0$. [If $u_0 = 0$, then clearly $u(t) \equiv 0$.] Whether this solution exists for all time or blows up at some finite time depends on the initial condition u_0: If $u_0 \leq 0$ then the solution $u(t)$ exists for all time and $u(t) \to 0$ as $t \to \infty$; if $u_0 > 0$, then the solution blows up (tends to positive infinity) as $t \to u_0^{-1}$.
♣

Example 2. Consider the ODE

(23) $$\begin{cases} \dfrac{du}{dt} = -u^3 \\ u(0) = u_0, \end{cases}$$

which also may be solved easily to find $u^2(t) = (u_0^{-2} + 2t)^{-1}$, again provided that $u_0 \neq 0$. In this case we see that the solution exists for all $t \geq 0$, and in fact $u(t) \to 0$ as $t \to \infty$, regardless of the initial conditions. ♣

An a priori bound of the form

(24) $$\|u(t)\| \leq K(t) \qquad \text{for } 0 \leq t < \infty,$$

where $K(t)$ is defined (and finite) for all $t > 0$, can be used to iterate the local existence argument and obtain global existence for all $t \geq 0$; this was described in Theorem 3 of Section 9.2 when $F: X \to X$ is locally Lipschitz continuous, but it extends immediately to the case when (10) is satisfied, provided that we have the X'-norm in (24). But how can we establish (24)? When might it fail? Can we say more as $t \to \infty$?

a. The Comparison Method

For semilinear heat equations of the form $u_t = \Delta u + f(u)$, the issue of global existence or blow-up can often be settled by using the comparison theorems of Section 11.1. Suppose we have a classical solution $u(x,t)$. In addition, suppose we can find a function u_- such that $u_- \leq u$ on $U_T = \Omega \times (0,T)$. If $u_-(x,t)$ blows up to $+\infty$ as $t \to T$, then we know that u itself must blow up *at least* by $t = T$. On the other hand, if we can find two functions u_- and u_+, which are both bounded (above and below) as $t \to T$ and such that $u_- \leq u \leq u_+$ on $U_T = \Omega \times (0,T)$, then we know that $|u(x,t)|$ is bounded as $t \to T$.

This comparison method, however, requires finding functions to use for the comparison; good candidates are often solutions of an ODE. Let us consider some examples.

Example 3. Consider the problem

(25)
$$\begin{cases} u_t = \Delta u - u^3 & \text{in } U = \Omega \times (0, \infty) \\ u(x, 0) = g(x) & \text{for } x \in \Omega \\ \dfrac{\partial u}{\partial \nu} = 0 & \text{on } \partial\Omega \times (0, \infty), \end{cases}$$

where $g \in C^\infty(\overline{\Omega})$ with $\partial g/\partial \nu = 0$ on $\partial\Omega$. Let us choose m such that for all $x \in \Omega$ we have $g(x) > m$, and then solve the ODE (23) with $u_0 = m$ to find $u_-(t)$. We can apply Theorem 3 of Section 11.1 to u and $v = u_-$:

$$u_t - \Delta u + u^3 = 0 = (u_-)_t - \Delta(u_-) + (u_-)^3 \quad \text{in } U_T = \Omega \times (0, T)$$
$$u(x, 0) > u_-(0) \quad \text{for } x \in \Omega$$
$$\frac{\partial u}{\partial \nu}(x, t) = 0 = \frac{\partial u_-}{\partial \nu}(t) \quad \text{for } x \in \partial\Omega \text{ and } 0 < t < T$$

implies that $u(x, t) > u_-(t)$ in any U_T. Similarly, choose M such that $g(x) < M$ for all $x \in \Omega$, and then solve the ordinary differential equation (23) with $u_0 = M$ to find $u_+(t)$. Application of the same comparison principle now shows that $u(x, t) < u_+(t)$ in any U_T. Thus we have shown that

$$u_-(t) < u(x, t) < u_+(t) \quad \text{for } x \in \Omega \text{ and } 0 < t < \infty.$$

But we know from Example 2 that $u_\pm(t)$ not only exist for all $0 < t < \infty$, but tend to 0 as $t \to \infty$. We conclude that classical solutions of (25) remain uniformly bounded and tend to 0 uniformly in x as $t \to \infty$. ♣

Example 4. Consider the problem

(26)
$$\begin{cases} u_t = \Delta u + u^2 & \text{in } U = \Omega \times (0, \infty) \\ u(x, 0) = g(x) & \text{for } x \in \Omega \\ \dfrac{\partial u}{\partial \nu} = 0 & \text{on } \partial\Omega \times (0, \infty), \end{cases}$$

where $g(x) > \epsilon > 0$ for $x \in \Omega$. Suppose $u(x, t)$ is a classical solution. Solve the ODE (22) with $u_0 = \epsilon$ to find $u_-(t)$. Example 1 shows that $u_-(t)$ blows up at $t = 1/\epsilon$. But if we apply the comparison principle to u and $v = u_-$, we find $u(x, t) > u_-(t)$, so $u(x, t)$ cannot exist globally and must itself blow up at least by time $t = 1/\epsilon$. ♣

Examples 3 and 4 both used the Neumann comparison principle. Examples requiring the Dirichlet comparison principle occur in the Exercises.

It must be observed, however, that the comparison method alone does not *prove* global existence of the solution. This is because uniform bounds on $\sup u$ need not produce (24). This is rectified when the X'-norm is dominated by the uniform norm:

Theorem 1. *Suppose $F : \mathbf{R} \times X' \to X$ satisfies (10), where $X' \subset C(\overline{\Omega})$. Then a uniform bound on the solution $u(x,t)$ of $(6'),(7')$ [i.e., $|u(x,t)| \leq K(T)$ for $0 < t < T \leq \infty$], implies existence of the solution for $0 < t < T$.*

Proof. Taking an equivalent norm for $X' \subset C(\overline{\Omega})$, we may assume that $\|u\|_\infty \leq \|u\|_{X'}$ for all $u \in X'$. This means that $F : X' \to X$ satisfies (10) for all $\|u\|_\infty, \|v\|_\infty \leq M$. As in Theorem 2, for $\|g\|_\infty \leq M$, we have local existence on $(0, \tau)$, where τ depends only on M. If we fix $T < \infty$ and let $M = K(T)$, then we can apply local existence on $(\tau, 2\tau)$, $(2\tau, 3\tau)$, and so on until we have covered $(0, T)$. ♠

b. Energy Methods for Large Time Existence

In many cases, it is difficult to construct comparison functions for nonlinear equations. In addition, as we have seen, comparison functions can only be used to prove global existence via the a priori estimate (24) when $X' \subset C(\overline{\Omega})$; for $X' = H_0^{1,2}(\Omega)$, this requires $n = 1$. Therefore, it is frequently more expedient to establish (24) directly as an energy estimate.

For a solution u of the equation (21), let us define the *potential energy* at time t to be

$$(27) \quad V(t) = V_u(t) = \int_\Omega \frac{1}{2} |\nabla u|^2 - \Phi(u)\, dx, \quad \text{where } \Phi(u) = \int_0^u f(s)\, ds.$$

We claim that V is a *nonincreasing function* of t. Assuming that u is sufficiently smooth, this is a straight forward calculation; for $n = 1$, for example,

$$\frac{d}{dt} V(t) = \int (u_x u_{xt} - f(u) u_t)\, dx = -\int (u_{xx} + f(u)) u_t\, dx = -\int u_t^2\, dx \leq 0,$$

where we have used the boundary condition (either $u = 0$ or $u_x = 0$) for the integration by parts. But this calculation generalizes, for example, to $u \in C^1((0,T), X)$, where $X = H_0^{1,2}(\Omega)$. We assume that we may use the regularity results of Section 11.2 to ensure that the appropriate mild solution $u \in C([0,T], X)$ of (21) has the requisite regularity.

Now let us assume the following condition on $f(u)$:

$$(28) \quad \limsup_{|u| \to \infty} \frac{f(u)}{u} \leq 0.$$

For example, this condition is satisfied by $f(u) = -u^3$ but not by $f(u) = \pm u^2$. Indeed, we now prove that (28) is sufficient to conclude global existence in $t > 0$. Let $X' = H_0^{1,2}(\Omega)$ and $X = L^2(\Omega)$.

Theorem 2. Suppose f satisfies (28), and $F(u)(x) = f(u(x))$ defines $F: X' \to X$, which satisfies (10). Then the mild solution $u \in C([0,T], X)$ of (21) with Dirichlet boundary condition and initial condition $g \in X'$ exists for all $t > 0$.

Proof. As a consequence of (28), for any $\epsilon > 0$ we have
$$\Phi(u) - \epsilon u^2 = \int_0^u \left[\frac{f(s)}{s} - 2\epsilon\right] s \, ds \le C_\epsilon$$
for some constant C_ϵ, which holds for both $u \ge 0$ and $u \le 0$. For convenience, let us assume $\int_\Omega dx = 1$ and use the Poincaré inequality to conclude that
$$V(t) = \int_\Omega \frac{1}{2}|\nabla u|^2 - \Phi(u)\, dx \ge k\|u\|_{1,2}^2 - \epsilon\|u\|_2^2 - C_\epsilon \ge (k-\epsilon)\|u\|_{1,2}^2 - C_\epsilon.$$
Taking $\epsilon = k/2$, we obtain
$$(29) \qquad \|u\|_{1,2}^2 \le \frac{2}{k}(V(t) + C).$$

Since $V(t)$ is nonincreasing in t, this provides us with the desired bound (24) with $K(t) \equiv 2k^{-1}(V(0) + C)$, and hence global existence. ♠

c. Energy Methods for Asymptotic Behavior

Energy methods may also be used to reach conclusions about the asymptotic behavior of global solutions as $t \to \infty$. As a simple example, let us consider (21) with Neumann boundary condition, and let us assume that
$$(30) \qquad \sup_{-\infty < u < \infty} |f'(u)| < \mu_1,$$
where μ_1 is the first positive eigenvalue for the Laplacian with Neumann boundary condition (cf. Exercise 5 in Section 7.2). If a global solution $u(x,t)$ approaches a constant equilibrium, then we would expect to have
$$(31) \qquad G(t) \equiv \frac{1}{2}\int_\Omega |\nabla_x u(x,t)|^2 \, dx \to 0 \quad \text{as } t \to \infty.$$

Theorem 3. If $f \in C^\infty(\mathbf{R})$ satisfies (30) and $u \in C^{2;1}(U) \cap C(\overline{U})$ is a global solution of (21) with Neumann boundary condition, then (31) holds.

Proof. For simplicity, let us assume that $n = 1$ and $\Omega = (0,1)$ (cf. Exercise 7). Let us use the equation (21) and integration by parts to compute

$$(32) \qquad \begin{aligned} \frac{dG}{dt} &= \int_0^1 u_x u_{xt} \, dx = \int_0^1 u_x (u_{xx} + f(u))_x \, dx \\ &= \int_0^1 u_x u_{xxx} \, dx + \int_0^1 (u_x)^2 f'(u) \, dx \\ &= -\int_0^1 (u_{xx})^2 \, dx + \int_0^1 (u_x)^2 f'(u) \, dx \\ &\le -\int_0^1 (u_{xx})^2 \, dx + 2MG, \end{aligned}$$

where $M = \sup\{|f'(u)| : -\infty < u < \infty\}$. We now require the following, which is proved in Exercise 6.

Lemma. *The first nonzero eigenvalue μ_1 of*

(33) $$\begin{cases} v_{xx} + \mu v = 0 \\ v'(0) = 0 = v'(1) \end{cases}$$

satisfies

(34) $$\mu_1 = \inf \left\{ \frac{\int_0^1 (v_{xx})^2 \, dx}{\int_0^1 (v_x)^2 \, dx} : v \in C^2[0,1], \; v'(0) = 0 = v'(1) \right\}.$$

Of course, the eigenvalues of (33) are well known to be $\mu_n = n^2 \pi^2$. Returning to our calculation of dG/dt, we find

(35) $$\frac{dG}{dt} \leq 2(M - \pi^2)G.$$

Integrating, we obtain

(36) $$G(t) \leq G(0) \, e^{2(M-\pi^2)t}.$$

Thus we find that $G(t) \to 0$ as $t \to \infty$ provided that $M < \pi^2$, which is exactly condition (30) for this case. ♠

Remark. The conclusion $G(t) \to 0$ can be misleading since it need not imply that $u(x,t)$ approaches a constant as $t \to \infty$ (cf. Exercise 5).

Exercises for Section 11.3

1. Let Ω be a smooth, bounded domain, $g \in C^\infty(\overline{\Omega})$, and consider the equation $u_t = \Delta u + e^{-u}$ in $U = \Omega \times (0,\infty)$, with $u(x,0) = g(x)$ for $x \in \Omega$, and $\partial u / \partial \nu = 0$ on $\partial \Omega \times (0,\infty)$. Determine whether the solution exists globally in time.

2. Let Ω be a smooth, bounded domain, and $g \in C^\infty(\overline{\Omega})$ satisfy $g(x) \geq \epsilon > 0$ for $x \in \Omega$. Consider the equation $u_t = \Delta u + u^{3/2}$ in $U = \Omega \times (0,\infty)$, with $u(x,0) = g(x)$ for $x \in \Omega$ and $\partial u/\partial \nu = 0$ on $\partial \Omega \times (0,\infty)$. Determine a time by which the solution is certain to blow up.

3. Let Ω be a smooth, bounded domain, $g \in C^\infty(\overline{\Omega})$, and consider the equation $u_t = \Delta u + u^2$ in $U = \Omega \times (0,\infty)$, with $u(x,0) = g(x)$ for $x \in \Omega$ and $u(x,t) = 0$ for $x \in \partial \Omega$, $t > 0$. Prove that either (i) no global solutions exist, or (ii) for some function $g(x)$ there are global solutions.

4. Consider the linear equation $u_t = \Delta u + f(t)$ in $\Omega \times (0,\infty)$ with initial condition (7), where $g \in X = L^2(\Omega)$, and boundary condition (8D) or (8N) defines $D(\Delta) \subset H^{2,2}(\Omega)$. Suppose $f : [0,\infty) \to X$ is uniformly

Hölder continuous and satisfies $\|f(t) - f_\infty\| \to 0$ as $t \to \infty$ for some $f_\infty \in L^2(\Omega)$. Show that there exists $u_\infty \in D(\Delta)$ such that $\Delta u_\infty = f_\infty$ and $\|u(t) - u_\infty\| \to 0$, $\|du/dt\| \to 0$ as $t \to \infty$.

5. (a) If Ω is a bounded domain and $f(u) > 0$ for $-\infty < u < \infty$, show that (6) does *not* have a global solution $u(x, t) \to u_\infty =$ const.
   (b) Let $f(u) = 1 + \tanh u$. Verify that (30) may hold *and* $f(u) > 0$ for $-\infty < u < \infty$. What has gone wrong with the existence of a steady state?

6. Prove the lemma in the proof of Theorem 3.

7. Generalize the proof of Theorem 3 to $n \geq 1$.

8. Prove *Gronwall's inequality*: If $y(t) \geq 0$ satisfies $dy/dt \leq \phi y$, where $\phi(t) \geq 0$ and $\phi \in L^1((0, T))$, then $y(t) \leq y(0) \exp(\int_0^t \phi(s)\,ds)$.

9. Let $\Omega = (0,1) \times (0,1) \subset \mathbf{R}^2$ and consider the nonlinear evolution $u_t = \Delta u + u(1-u)$ for $x \in \Omega$, $t > 0$, with $u(x,0) = g(x)$ for $x \in \Omega$ and $\partial u/\partial \nu = 0$ for $x \in \partial\Omega$, $t > 0$, where g is a smooth function satisfying $0 \leq g(x) \leq 1$ for all $x \in \Omega$. Prove that there is a global solution $u(x,t) \to u_\infty =$ const. What are the possible values of u_∞?

11.4 Applications to Navier-Stokes

Let us recall from Section 9.1d the inhomogeneous Navier-Stokes equations for a viscous, incompressible fluid in a smooth, bounded domain $\Omega \subset \mathbf{R}^n$ with nonslip boundary conditions:

(37a) $$\vec{u}_t + (\vec{u} \cdot \nabla)\vec{u} + \nabla p = \nu \Delta \vec{u} + \vec{f} \quad \text{in } \Omega$$
(37b) $$\operatorname{div} \vec{u} = 0 \quad \text{in } \Omega$$
(37c) $$\vec{u} = 0 \quad \text{on } \partial\Omega,$$

where $\vec{u}$ is the velocity field, $\vec{f}$ is a forcing term, p is the pressure, and ν is the viscosity (assumed constant). Of course, the important physical cases are $n = 2$ and 3; in fact, in these cases we found in Section 9.1 that, for $\vec{f} \in L^2(\Omega, \mathbf{R}^n)$, there exists a weak equilibrium solution (i.e., $\vec{u}$ independent of t). This gives us hope that we might be able to find a solution of (37a,b,c) satisfying a sufficiently nice initial condition

(38) $$\vec{u}(x, 0) = \vec{g}(x) \quad \text{for } x \in \Omega.$$

Recall from Sections 6.2b and 9.1d that the incompressibility condition is handled by considering divergence-free vector fields: $\tilde{H}_0^{1,2}(\Omega, \mathbf{R}^n) = \{\vec{u} \in H_0^{1,2}(\Omega, \mathbf{R}^n) : \operatorname{div} \vec{u} = 0\}$. In fact, if we let

(39a) $$X = \tilde{L}^2(\Omega, \mathbf{R}^n) = \text{the closure of } \tilde{H}_0^{1,2}(\Omega, \mathbf{R}^n) \text{ in } L^2(\Omega, \mathbf{R}^n),$$

then we recall from (19) in Section 6.2 that

(39b) $\qquad X^\perp = \{\vec{w} \in L^2(\Omega, \mathbf{R}^n) : \vec{w} = \nabla p \text{ for some } p \in H^{1,2}(\Omega)\}.$

a. Local Existence by the Semigroup Method

Let us define $A = \nu\Delta$, which we consider as an unbounded operator on X with domain $D = \tilde{H}_0^{1,2}(\Omega, \mathbf{R}^n) \cap H^{2,2}(\Omega, \mathbf{R}^n)$. We would like to consider $A: D \to X$ as the infinitesimal generator of a linear semigroup on X. To this end we require the following.

Lemma 1. $A: D \to X$ is a densely defined, closed linear operator on X.

Proof (partial). Clearly D is dense, and $\operatorname{div}(\Delta\vec{u}) = \Delta(\operatorname{div}\vec{u})$ shows that $A: D \to X$. But proving that the operator A is closed is nonelementary. As we saw for the scalar case, closedness requires the use of elliptic estimates, which we have not studied for systems. Elliptic estimates for the Stokes system are discussed in [Ladyzhenskaya, 1]; for general elliptic systems, cf. [ADN]. ♠

Lemma 2. There is an orthonormal basis for X consisting of eigenfunctions for A.

Proof. Notice that A is self-adjoint on X (i.e., $\langle A\vec{u}, \vec{v}\rangle = \langle A\vec{u}, \vec{v}\rangle$ for all $\vec{u}, \vec{v} \in D$). Also, by the results of Section 6.2b, $A: D \to X$ is an isomorphism. Since $D \subset X$ is a compact imbedding, we conclude that A has compact resolvent. The result now follows from the Theorem 5 in Section 6.6 (cf. Exercise 15). ♠

We can write (37) as the nonlinear evolution equation

(40a) $\qquad \vec{u}_t = A\vec{u} + \vec{F}(t, \vec{u}),$

where $\vec{u} \in X$ and

(40b) $\qquad F(t, \vec{u}) = P[\vec{f}(t) - (\vec{u} \cdot \nabla)\vec{u}],$

with P denoting the projection of $L^2(\Omega, \mathbf{R}^n)$ onto the closed subspace X. As a consequence of Lemma 2, we can define the spaces X_α as in (15). In order to conclude local existence of a solution, we wish to invoke Theorem 3 of Section 11.2, which requires that

(41) $\qquad F: \mathbf{R} \times X_\alpha \to X$ satisfies (10).

Given the form of F, this requires two conditions:

(42a) $\qquad P\vec{f} \in C^\nu([0, T], X)$
(42b) $\qquad N(\vec{u}) = P[(\vec{u}\cdot\nabla)\vec{u}]$ is locally Lipschitz in $u \in X_\alpha$.

Notice that (42a) is certainly satisfied when $\vec{f} \in L^2(\Omega, \mathbf{R}^n)$ is independent of t or when $\vec{f} \in C^\nu([0, T], X)$.

Lemma 3. *Condition (42b) is satisfied provided that $\alpha > 1/2$ for $n = 2$, and $\alpha > 3/4$ for $n = 3$.*

Proof. Invoke Theorem 4 of Section 11.2 to conclude that (i) $X_\alpha \subset H^{1,2}(\Omega, \mathbf{R}^n)$ provided that $2\alpha > 1$, and (ii) $X_\alpha \subset L^\infty(\Omega, \mathbf{R}^n)$ provided that $4\alpha > n$. The estimate

$$(43) \qquad \|(\vec{u} \cdot \nabla \vec{u})\|_2 \leq \|\vec{u}\|_\infty \|\nabla \vec{u}\|_2 \leq \|u\|_\alpha$$

shows that $N: X_\alpha \to X$ and may be used to show the map is locally Lipschitz (cf. Exercise 1). ♠

Applying Theorem 3 of Section 11.2, we immediately obtain the following.

Theorem 1. *Assume that $\vec{f}$ satisfies (42a); also suppose either $n = 2$ and $1/2 < \alpha < 1$, or $n = 3$ and $3/4 < \alpha < 1$. For every $\vec{g} \in X_\alpha$ there is a positive τ such that there exists a solution $(\vec{u}, p)$ of (37a,b,c) with $\vec{u} \in C([0, \tau], X_\alpha) \cap C^1((0, \tau], X)$ and $\nabla p \in X^\perp$; in fact, $\vec{u}$ is unique.*

This provides us with local existence; we next consider large time existence.

b. Weak Solutions of Navier-Stokes

To prove large time existence, we might try to obtain a priori estimates such as (24). As we saw in Section 11.3, these are most naturally derived as energy estimates; that is, $X' = H_0^{1,2}(\Omega, \mathbf{R}^n) = X_{1/2}$. The restriction $\alpha > 1/2$ in Theorem 1, however, means that these energy estimates cannot be used to extend the solutions obtained by the semigroup method. On the other hand, the energy estimates are exactly right for obtaining weak solutions.

Let us introduce the following notation:

$$(44) \qquad V = \tilde{H}_0^{1,2}(\Omega, \mathbf{R}^n) \quad \text{and} \quad V' = \text{the dual space of } V.$$

As in Section 9.2d, we have the continuous imbeddings

$$(45) \qquad V \subset X \subset V',$$

and we are led to reformulate (37a,b,c),(38) in the following weak form: If $f \in L^2((0, T), V')$ for some $T > 0$ and $\vec{g} \in V$, then we seek $\vec{u} \in L^2((0, T), V)$ such that $u_t \in L^p((0, T), V')$ for some $p \in [1, \infty]$, and

$$(46a) \qquad \langle \vec{u}_t, \vec{v} \rangle + \nu \langle \nabla \vec{u}, \nabla \vec{v} \rangle + \{\vec{u}, \vec{u}, \vec{v}\} = \langle \vec{f}, \vec{v} \rangle \quad \text{for all } \vec{v} \in V$$
$$(46b) \qquad \vec{u}(x, 0) = \vec{g}(x),$$

where $\{\vec{u}, \vec{w}, \vec{v}\} = \int (\vec{u} \cdot \nabla \vec{w}) \cdot \vec{v}\, dx$ is the trilinear form introduced in Section 9.1d, and $\langle , \rangle$ denotes the spatial L^2-inner product. Notice that

Section 11.4: Applications to Navier-Stokes

$\vec{u} \in L^2((0,T), V)$ and $u_t \in L^p((0,T), V')$ imply, by Proposition 1 in Section 9.2 (or Exercise 3 of this section), that $\vec{u}(x,0)$ in (46b) makes sense.

Let us rewrite (46a) as

$$(47) \qquad \vec{u}_t + B(\vec{u}) = \nu \Delta \vec{u} + \vec{f} \quad \text{in } V',$$

where $B(\vec{u}) \in V'$ is defined by $B(\vec{u})\vec{v} = \{\vec{u}, \vec{u}, \vec{v}\}$. If $\vec{u} \in L^2((0,T), V)$ satisfies (47), let us see what we can say about $\vec{u}_t$. The right-hand side of (47) is in $L^2((0,T), V')$, but this may not be true of $B(\vec{u})$. In fact, if we recall the estimates

$$(48a) \qquad \|\vec{u}\|_4 \le C\|\vec{u}\|_{1,2} \quad (\text{since } H^{1,2} \subset L^4 \text{ for } n \le 4),$$

$$(48b) \qquad |B(\vec{u})\vec{v}| \le C\|\vec{u}\|_4^2 \|\vec{v}\|_{1,2} \quad (\text{cf. (10b) in Section 9.1}),$$

we obtain

$$(49) \qquad \int_0^T \|B(\vec{u})\|_{V'} dt \le C \int_0^T \|\vec{u}\|_4^2 dt \le C \int_0^T \|\vec{u}\|_{1,2}^2 dt < \infty.$$

At least we may conclude that $B(\vec{u}) \in L^1((0,T), V')$. Using (47) and $L^2((0,T)) \subset L^1((0,T))$, we conclude that

$$(50) \qquad \vec{u}_t \in L^1((0,T), V').$$

As a consequence of (50) and Proposition 1 in Section 9.2, we have shown the following.

Proposition 1. *If $n \le 4$ and $\vec{u} \in L^2((0,T), V)$ satisfies (47) for a.e. $t \in (0,T)$, then $\vec{u}$ satisfies (50) and $\vec{u} \in C([0,T], V')$.*

This result is not optimal; we discover in the next subsection that we can improve (50) when $n = 2$ or 3.

c. A Priori Estimates

Now let us discuss certain a priori estimates for solutions. In their derivation, we assume $\vec{u}$ is a sufficiently smooth solution; the weak solutions that we construct, however, will also satisfy these estimates. Let $\vec{v} = \vec{u}(t)$ in (46a), and use $\{\vec{u}, \vec{u}, \vec{u}\} = 0$ to obtain $\langle \vec{u}_t, \vec{u}\rangle + \nu\langle \nabla\vec{u}, \nabla\vec{u}\rangle = \langle \vec{f}, \vec{u}\rangle$. But this means that

$$\frac{1}{2}\frac{d}{dt}\|\vec{u}\|_2^2 + \nu\|\nabla\vec{u}\|_2^2 = \langle \vec{f}, \vec{u}\rangle \le \|\vec{f}\|_{-1,2} \|\nabla\vec{u}\|_2,$$

where we use $\|\nabla\vec{u}\|_2$ for the norm on V, and $\|\cdot\|_{-1,2}$ denotes the norm in $V' = (H_0^{1,2}(\Omega, \mathbf{R}^n))^* = H^{-1,2}(\Omega, \mathbf{R}^n)$. Using $2ab \le \nu a^2 + b^2/\nu$, we have

$$\|\vec{f}\|_{-1,2} \|\nabla\vec{u}\|_2 \le \frac{\nu}{2}\|\nabla\vec{u}\|_2^2 + \frac{1}{2\nu}\|\vec{f}\|_{-1,2}^2,$$

so, after rearrangement, we obtain the inequality

$$\frac{d}{dt}\|\vec{u}(t)\|_2^2 + \nu\|\nabla\vec{u}(t)\|_2^2 \leq \frac{1}{\nu}\|\vec{f}\|_{-1,2}^2. \tag{51}$$

If we integrate (51) from $t = 0$ to T, we obtain

$$\int_0^T \|\nabla\vec{u}(t)\|_2^2\, dt \leq K_1(\vec{g},\vec{f},\nu,T) = \left(\frac{\|\vec{g}\|_2^2}{\nu} + \frac{1}{\nu^2}\int_0^T \|\vec{f}(t)\|_{-1,2}^2\, dt\right). \tag{52}$$

Similarly, we may integrate (51) from $t = 0$ to $\tau \leq T$ to find

$$\|\vec{u}(\tau)\|_2^2 \leq \|\vec{g}\|_2^2 + \frac{1}{\nu}\int_0^\tau \|\vec{f}(t)\|_{-1,2}^2\, dt \leq \|\vec{g}\|_2^2 + \frac{1}{\nu}\int_0^T \|\vec{f}(t)\|_{-1,2}^2\, dt.$$

Therefore, we obtain

$$\max_{0 \leq t \leq T} \|\vec{u}(t)\|_2^2\, dt \leq \nu K_1(\vec{g},\vec{f},\nu,T). \tag{53}$$

The inequalities (52) and (53) are the desired a priori estimates. They state that solutions are bounded in $L^2((0,T),V)$ and in $L^\infty((0,T),X)$, independent of $\vec{u}$.

As an application, we now see that we may improve the result (50) when $n = 2$ or 3. Using the Sobolev inequality [cf. formula (35) in Section 6.4],

$$\|\vec{u}\|_4 \leq C\|\nabla\vec{u}\|_2^\lambda \|\vec{u}\|_2^{(1-\lambda)} \quad \text{for} \quad \frac{1}{4} = \lambda(\frac{1}{2} - \frac{1}{n}) + (1-\lambda)\frac{1}{2} \Leftrightarrow \lambda = \frac{n}{4}.$$

Now (53) implies that $\|\vec{u}\|_2^{(1-\lambda)}$ is bounded for $t \in (0,T)$, so we have $\|\vec{u}\|_4 \leq C\|\vec{u}\|_{1,2}^{n/4}$. But from (48b) we have $\|B(\vec{u})\|_{V'} \leq C\|\vec{u}\|_4^2$, so

$$\begin{aligned}\int_0^T \|B(\vec{u})\|_{V'}^p\, dt &\leq C\int_0^T \|\vec{u}\|_4^{2p}\, dt \\ &\leq C\int_0^T \|\vec{u}\|_{1,2}^{pn/2}\, dt < \infty,\end{aligned} \tag{54}$$

provided that $pn/2 = 2$. In other words,

$$\vec{u} \in L^{2p}((0,T),L^4(\Omega)) \quad \text{and} \quad B(\vec{u}) \in L^p((0,T),V') \quad \text{for } p = 4/n. \tag{55}$$

Using $\Delta\vec{u}, \vec{f} \in L^2((0,T),V') \subset L^p((0,T),V')$ if $p \leq 2$ and (47), we conclude that $\vec{u}_t \in L^p((0,T),V')$ if $p = 4/n \leq 2$. We have just proved the following.

Proposition 2. If $n \leq 4$ and $\vec{u} \in L^2((0,T),V) \cap L^\infty((0,T),X)$ satisfies (46a,b), then $\vec{u} \in L^{2p}((0,T),L^4(\Omega))$ and $\vec{u}_t \in L^p((0,T),V')$ for $p = 4/n$.

Notice that, for $n = 4$, there is no improvement over (50).

d. Existence of a Weak Solution: Galerkin's Method

In order to obtain a weak solution, we invoke the *method of Galerkin* to approximate (46) by a finite-dimensional problem. Let $\{\vec{w}_j\}_{j=1}^\infty$ be a complete orthonormal set of eigenfunctions of $A = \nu\Delta$ on X (as in Lemma 2), and let W_ℓ denote the finite-dimensional space obtained by taking the linear span of $\vec{w}_1, \ldots, \vec{w}_\ell$. We now search for an approximate solution of (46a,b) in the form

$$\vec{u}^\ell(t) = \sum_{j=1}^\ell a_j^\ell(t) \vec{w}_j, \tag{56}$$

where the coefficients $a_j^\ell(t)$ are functions of t only. We want $\vec{u}^\ell$ to satisfy (46a,b) restricted to W_ℓ; that is,

$$\langle \vec{u}_t^\ell, \vec{v} \rangle + \nu \langle \nabla \vec{u}^\ell, \nabla \vec{v} \rangle + \{\vec{u}^\ell, \vec{u}^\ell, \vec{v}\} = \langle \vec{f}, \vec{v} \rangle \quad \text{for all } \vec{v} \in W_\ell \tag{57a}$$

$$\vec{u}^\ell(x, 0) = \vec{g}^\ell(x), \tag{57b}$$

where $\vec{g}^\ell$ denotes the projection of $\vec{g}$ on W_ℓ (i.e., $\vec{g}^\ell = \sum_{j=1}^\ell g_j \vec{w}_j$ for $\vec{g} = \sum_{j=1}^\infty g_j \vec{w}_j$). If we pick $0 \leq k \leq \ell$ and we take $\vec{v} = \vec{w}_k$ in (55a), we obtain

$$\dot{a}_k^\ell(t) + \lambda_k a_k^\ell(t) + \sum_{i,j=1}^\ell c_{ijk} a_i^\ell(t) a_j^\ell(t) = f_k(t) \tag{58a}$$

$$a_k^\ell(0) = g_k, \tag{58b}$$

where $\dot{}$ represents d/dt, and $c_{ijk} = \int_\Omega (\vec{w}_i \cdot \nabla \vec{w}_j : \vec{w}_k)\, dx$. This is a nonlinear system of ODEs for the functions $\{a_k^\ell(t)\}_{k=1}^\ell$; by standard existence theory, there is a unique solution that exists on some time interval $[0, T_\ell)$. We need estimates on these functions $a_k^\ell(t)$ in order to show that we may take $T = T_\ell$ and that we may let $\ell \to \infty$ to obtain the solution $\vec{u}$ of (37). But replacing $\vec{v}$ in (55a) by $\vec{u}^\ell$ and arguing as we did for (48) and (49), we obtain

$$\int_0^T \|\nabla \vec{u}^\ell(t)\|_2^2\, dt \leq \left(\frac{\|\vec{g}^\ell\|_2^2}{\nu} + \frac{1}{\nu^2} \int_0^{T_\ell} \|\vec{f}(t)\|_{-1,2}^2\, dt \right) \tag{59a}$$

$$\max_{0 \leq t \leq T} \|\vec{u}^\ell(t)\|_2^2 \leq \nu \left(\frac{\|\vec{g}^\ell\|_2^2}{\nu} + \frac{1}{\nu^2} \int_0^{T_\ell} \|\vec{f}(t)\|_{-1,2}^2\, dt \right). \tag{59b}$$

Moreover, these estimates may be taken independent of ℓ since $\|\vec{g}^\ell\|_2 \leq \|\vec{g}\|_2$ and $\int_0^{T_\ell} \|\vec{f}(t)\|_{-1,2}^2\, dt \leq \int_0^T \|\vec{f}(t)\|_{-1,2}^2\, dt < \infty$. Thus $T_\ell = T$. If we appeal to Proposition 2, we obtain the following.

Proposition 3. *If $n \leq 4$, $\vec{f} \in L^2((0,T), V')$, $\vec{g} \in X$, and $\vec{u}^\ell(t)$ denotes the Galerkin approximation (56) whose coefficients $a_j^\ell(t)$ satisfy (58a,b), then $\vec{u}^\ell(t)$ is defined for $t \in [0, T]$ and satisfies the a priori estimates*

$$(60) \qquad \max_{0 \leq t \leq T} \|\vec{u}^\ell(t)\|_2^2 \leq \nu \int_0^T \|\nabla \vec{u}^\ell(t)\|_2^2 \, dt \leq \nu K_1(\vec{g}, \vec{f}, \nu, T).$$

In addition, $\{\vec{u}^\ell\}$ is bounded in $L^{2p}((0,T), L^4(\Omega))$ and $\{\vec{u}_t^\ell\}$ is bounded in $L^p((0,T), V')$, where $p = 4/n$.

As a consequence of (60), $\{\vec{u}^\ell\}_{\ell=1}^\infty$ is bounded both in $L^2((0,T), V)$ and in $L^\infty((0,T), X)$, which is the dual space of $L^1((0,T), X)$. By Theorems 2 and 3 of Section 6.3, there is a subsequence $\{\vec{u}^{\ell_i}\}_{i=1}^\infty$ such that

(61a) $\qquad \vec{u}^{\ell_i} \rightharpoonup \vec{u}$ weakly in $L^2((0,T), V)$ as $i \to \infty$

(61b) $\qquad \vec{u}^{\ell_i} \rightharpoonup \vec{u}$ weak* in $L^\infty((0,T), X)$ as $i \to \infty$.

But Proposition 3 also guarantees that $\{\vec{u}_t^\ell\}_{\ell=1}^\infty$ is bounded in $L^p((0,T), V')$ for $p = 4/n \geq 1$. In fact, if $n = 2$ or $n = 3$, then we can take $p > 1$; in this case, $L^p((0,T), V')$ is the dual space of $L^{p'}((0,T), V)$, and we may apply Theorem 2 of Section 6.3 to obtain a subsequence of $\{\vec{u}^{\ell_i}\}$ (which we continue to denote by $\{\vec{u}^{\ell_i}\}$) such that

$$(62) \qquad \vec{u}_t^{\ell_i} \rightharpoonup \vec{u}_t \quad \text{weakly in } L^p((0,T), V') \text{ as } i \to \infty, \text{ if } 1 < p = 4/n$$

(cf. Exercise 4).

Let $\vec{w} \in C([0,T], V)$. Since $C([0,T], V) \subset L^2((0,T), V)$ and $\vec{f} \in L^2((0,T), V')$, we conclude that $\int_0^T \langle \vec{f}, \vec{w} \rangle \, dt$ is well defined; moreover, (61a) implies that

$$(63) \qquad \int_0^T \langle \nabla \vec{u}^{\ell_i}, \nabla \vec{w} \rangle \, dt \to \int_0^T \langle \nabla \vec{u}, \nabla \vec{w} \rangle \, dt \qquad \text{as } i \to \infty.$$

Since $C([0,T], V) \subset L^{p'}((0,T), V)$, (62) implies that

$$(64) \qquad \int_0^T \langle \vec{u}_t^{\ell_i}, \vec{w} \rangle \, dt \to \int_0^T \langle \vec{u}_t, \vec{w} \rangle \, dt \qquad \text{as } i \to \infty.$$

We would also like to show that

$$(65) \qquad \int_0^T \langle B(\vec{u}^{\ell_i}), \vec{w} \rangle \, dt \to \int_0^T \langle B(\vec{u}), \vec{w} \rangle \, dt \qquad \text{as } i \to \infty,$$

But for (65), we need the following compactness result.

Section 11.4: Applications to Navier-Stokes   353

Lemma 4 (Compactness). *Suppose $X_1 \subset X_0 \subset X_{-1}$ are continuously imbedded Bannach spaces, where $X_1 \subset X_0$ is a compact imbedding. Suppose $0 < T < \infty$, $1 < p < \infty$, and $\{u^k\}_{k=1}^\infty$ is a sequence that is bounded in $L^p((0,T), X_1)$, and $\{u_t^k\}_{k=1}^\infty$ is bounded in $L^p((0,T), X_{-1})$. Then $\{u^k\}_{k=1}^\infty$ is contained in a compact set in $L^p((0,T), X_0)$.*

This lemma is not too surprising; for example, with $p = 2$, $X_1 = H^{1,2}(\Omega)$ and $X_0 = X_{-1} = L^2(0)$, the result is simply a consequence of the compact imbedding $H^{1,2}(\Omega \times (0,T)) \subset L^2(\Omega \times (0,T))$. Nevertheless, its proof is rather technical, so we do not give it here; see [Lions].

Using Lemma 4, we may now complete the proof of the following.

Theorem 2. *Suppose $n = 2$ or 3. For $\vec{f} \in L^2((0,T), V')$ and $\vec{g} \in X$, there is a weak solution $\vec{u} \in L^2((0,T), V) \cap L^\infty((0,T), X)$ of (37a,b,c), (38).*

Proof. We only prove the case $n = 2$; for $n = 3$, the proof is analogous but requires slightly different technical details.

By Proposition 3, $\{\vec{u}^\ell\}$ is bounded in $L^2((0,T), V)$ and $\{\vec{u}_t^\ell\}$ is bounded in $L^2((0,T), V')$. Therefore, we can apply Lemma 4 with $X_1 = V$, $X_0 = X$, and $X_{-1} = V'$ to obtain a subsequence that converges to $\vec{u}$ strongly in $L^2((0,T), X)$. We may therefore assume that $\{\vec{u}^{\ell_i}\}$ satisfies (61a,b) *and*

(66) $$\vec{u}^{\ell_i} \to \vec{u} \text{ in } L^2((0,T), X).$$

We want to use (61) and (66) to show (65). Let us write

(67) $$\int_0^T \langle B(\vec{u}^{\ell_i}), \vec{w} \rangle \, dt = \int_0^T \{\vec{u}^{\ell_i}, \vec{u}^{\ell_i}, \vec{w}\} \, dt = -\int_0^T \{\vec{u}^{\ell_i}, \vec{w}, \vec{u}^{\ell_i}\} \, dt$$
$$= -\int_0^T \int_\Omega \sum_{j,k}^n u_j^{\ell_i} (\partial_j w_k) u_k^{\ell_i} \, dx dt.$$

Now observe that $L^2((0,T), V) \cap L^\infty((0,T), X) \subset L^4((0,T), L^4(\Omega)) = L^4(\Omega \times (0,T))$; this follows from the Sobolev inequality as used in proving Proposition 2. But then (61) shows that $\vec{u}^{\ell_i} \rightharpoonup \vec{u}$ weakly in $L^4(\Omega \times (0,T))$; this follows from duality (i.e., $Y \subset Z$ implies $Z^* \subset Y^*$). Since a weakly convergent sequence must be bounded, we conclude that $\|\vec{u}^{\ell_i}\|_4$ must be bounded. But $\|w\|_4 = \|w^2\|_2$, so we conclude that each $\|u_j^{\ell_i} u_k^{\ell_i}\|_2$ is bounded as $i \to \infty$, where u_j denotes the components of the vector $\vec{u}$. We also can use (66) to show that $u_j^{\ell_i} u_k^{\ell_i} \to u_j u_k$ in $L^1(\Omega \times (0,T))$ as $i \to \infty$ (cf. Exercise 5a). But these facts imply that $u_j^{\ell_i} u_k^{\ell_i} \rightharpoonup u_j u_k$ weakly in $L^2(\Omega \times (0,T))$ (cf. Exercise 5b). As a consequence, since $(\partial_j w_k) \in L^2(\Omega \times (0,T))$, we conclude from (67) that (65) holds.

Combining (63), (64), and (65) enables us to conclude that

(68) $$\int_0^T [\langle \vec{u}_t, \vec{w} \rangle + \langle B(\vec{u}), \vec{w} \rangle + \nu \langle \nabla \vec{u}, \nabla \vec{w} \rangle] \, dt = \int_0^T \langle \vec{f}, \vec{w} \rangle \, dt.$$

Taking $\vec{w} = \phi\vec{v}$ with $\phi \in C([0,T])$ and $\vec{v} \in V$, we see that (68) implies (55a) holds for a.e. $t \in (0,T)$. In other words, $\vec{u}$ is a weak solution of (37a,b,c), (38). ♠

e. Further Remarks

There are many important distinctions between the dimensions $n = 2$ and $n = 3$ for Navier-Stokes. This was already seen to some extent in Theorem 1 for local existence, but it pertains even more clearly to the large time existence of solutions. Let us simply mention some of these issues; some proofs are given in the Exercises; for further details see [Temam, 1], [Temam, 2], or [Constantin-Foias-Temam].

For $n = 2$, the weak solution of Theorem 2 in fact satisfies $\vec{u} \in C([0,T], X)$ (cf. Exercise 3). However, for $n = 3$, the weak solution is only weakly continuous $[0,T] \to X$. Moreover, the weak solution of Theorem 2 is *unique* when $n = 2$ (cf. Exercise 7). However, the necessary estimates to guarantee uniqueness for $n = 3$ have not yet been proved.

In addition to the weak solutions, it is possible to consider *strong solutions* of Navier-Stokes that satisfy $\vec{u} \in L^2((0,T), D) \cap C([0,T], V)$ and $\vec{u}_t \in L^2((0,T), X)$. For $n = 2$, it is known that there is a unique strong solution for any finite time $T > 0$; however, for $n = 3$, the strong solution is only known to exist for a short time.

The possible breakdown of the strong solution for $n = 3$ was conjectured in the 1930s by J. Leray to be connected with turbulence. Since that time, much work has been conducted on the analysis of turbulent flow, including estimating the Hausdorff dimension of attractors. Nevertheless, the fundamental question of whether singularities do indeed develop in finite time remains unanswered.

Exercises for Section 11.4

1. Use (43) to show (42b) provided that $\alpha > 1/2$ for $n = 2$, and $\alpha > 3/4$ for $n = 3$.

2. If $\vec{f}(x,t)$ in (37) satisfies additional regularity, show that the solution of Theorem 1 is a classical solution; that is, if $\vec{f}$ is $C^{1,\nu}$ with respect to both $x \in \overline{\Omega}$ and $0 \leq t \leq T$ for some $0 < \nu < 1$, show that $\vec{u} \in C^2(\overline{\Omega} \times [0,\tau])$.

3. If $V \subset X = X' \subset V'$ are Hilbert spaces and $u \in L^2((0,T), V)$ satisfies $u_t \in L^2((0,T), V')$, show that $u \in C([0,T], X)$ (up to a set of measure zero).

4. If $\vec{u}^j \rightharpoonup \vec{u}$ weakly in $L^2((0,T), V)$ and $\vec{u}_t^j \rightharpoonup v$ weak* in $L^1((0,T), V')$, show that $\vec{v} = \vec{u}_t$.

5. (a) If $u_i \to u$ in $L^2(\Omega)$, show that $u_i^2 \to u^2$ in $L^1(\Omega)$.
   (b) If u_i is bounded in $L^2(\Omega)$ and $u_i \to u$ in $L^1(\Omega)$, then $u_i \rightharpoonup u$ weakly in $L^2(\Omega)$.

6. For $n \geq 3$, use the Sobolev inequality to prove $\vec{u} \in L^2((0,T), V) \cap L^\infty((0,T), X) \Rightarrow \vec{u} \in L^4((0,T), L^q(\Omega, \mathbf{R}^n))$, where $q = 2n/(n-1)$. (For $n = 3, 4$ this provides an alternative a priori estimate for $\vec{u}$.)

7. *Proof of Uniqueness for $n = 2$.* Let $\vec{u}_1, \vec{u}_2 \in L^2((0,T), V) \cap L^\infty((0,T), X)$ be two solutions of (46) with $\vec{u}_1(0) = \vec{u}_2(0)$.
   (a) Show that $\langle B(\vec{u}_1) - B(\vec{u}_2), \vec{u}_1 - \vec{u}_2 \rangle = \{\vec{u}_1 - \vec{u}_2, \vec{u}_1, \vec{u}_1 - \vec{u}_2\}$.
   (b) Show that

$$\frac{1}{2}\frac{d}{dt}\|\vec{u}_1 - \vec{u}_2\|^2 + \nu\|\nabla(\vec{u}_1 - \vec{u}_2)\|^2 \leq C\|\vec{u}_1 - \vec{u}_2\| \cdot \|\nabla \vec{u}_1\| \cdot \|\nabla(\vec{u}_1 - \vec{u}_2)\|.$$

   (c) Show that

$$\frac{d}{dt}\|\vec{u}_1 - \vec{u}_2\|^2 \leq C\|\nabla \vec{u}_1\|^2 \cdot \|\vec{u}_1 - \vec{u}_2\|^2.$$

   (d) Conclude that $\|\vec{u}_1 - \vec{u}_2\| \equiv 0$ for all $t \in (0,T)$.

Further References for Chapter 11

A good reference for parabolic maximum principles and their applications is [Protter-Weinberger].

The presentation in this chapter of local and global existence borrows much from analytic semigroup theory, without engaging in the general context: sectorial operators, fractional powers A^α of infinitesimal generators, and regularity properties of the solution operator. A good reference for the general theory is [Henry].

For additional information on the evolutionary Navier-Stokes equations, see [Ladyzhenskaya, 1], [Temam, 1], or [Temam, 2]. Navier-Stokes is also discussed in [Lions], which also contains many other applications to nonlinear evolutionary equations. The issue of attractors for Navier-Stokes is a subject of much current research; see, for example, [Constantin-Foias-Temam].

For a nice collection of relatively contemporary articles on the subject of nonlinear diffusion, see [Ni-Peletier-Serrin]. Another interesting reference on nonlinear parabolic equations and reaction-diffusion systems is [Smoller], which contains more advanced topics such as the Conley index (which generalizes the Morse index).

12 Linear and Nonlinear Waves

In this chapter, we study linear and nonlinear wave equations. A basic technique is to reduce second-order time derivatives to first-order by converting the equation to a system. If we also reduce the spatial derivatives to first-order, then we naturally encounter symmetric hyperbolic systems, which we investigate first. Another approach, however, is to leave the spatial derivatives as second-order, similar to the heat equation, and view the evolution as a dynamical system. This latter approach is very useful for certain nonlinear wave equations, such as the semilinear Klein-Gordon equation, which arises in quantum field theory. We also consider the linear and nonlinear Schrödinger equations, which arise in physical models and display both diffusive and wave properties.

12.1 Symmetric Hyperbolic Systems

Let us consider the semilinear wave equation

$$u_{tt} - \Delta u = f(x, t, u) \quad \text{for } x \in \mathbf{R}^n, \text{ and } t > 0, \tag{1}$$

with pure initial conditions

$$u(x, 0) = g(x), \quad u_t(x, 0) = h(x) \quad \text{for } x \in \mathbf{R}^n. \tag{2}$$

Can we reduce this problem to a first-order system? Following the ideas of Example 8 in Section 2.2, let us introduce the vector

$$\vec{u} = \begin{pmatrix} u_0 \\ u_1 \\ \vdots \\ u_{n+1} \end{pmatrix} = \begin{pmatrix} u \\ u_{x_1} \\ \vdots \\ u_t \end{pmatrix}, \tag{3}$$

whose components satisfy

$$\begin{cases} (u_0)_t = u_{n+1} \\ (u_i)_t = (u_{n+1})_{x_i} \quad 1 \le i \le n \\ (u_{n+1})_t - \sum_{i=1}^{n}(u_i)_{x_i} = f(x, t, u_0). \end{cases} \tag{4}$$

Letting $N = n+2$, these equations may be written as

(5) $$A_0 \vec{u}_t + \sum_{i=1}^{n} A_i \vec{u}_{x_i} = \vec{f}$$

where the A_i are $N \times N$ symmetric matrices [A_0 is the identity matrix, whereas A_i for $1 \le i \le n$ consists of -1's in the (i, N) and (N, i) places and 0's everywhere else], and $\vec{f}$ is an N-vector-valued function [consisting of u_{n+1} as the first component, $f(x, t, u_0)$ as the last component, and 0's everywhere else]. We want to solve (5) with the initial condition

(6) $$\vec{u}(x, 0) = \begin{pmatrix} g(x) \\ g_{x_1}(x) \\ \vdots \\ h(x) \end{pmatrix}.$$

First-order systems of the general form (5) are called *symmetric hyperbolic* when the $N \times N$ matrices A_i are all symmetric. The A_i may depend on x and t, and $\vec{f}$ may depend on x, t and u; in this case (5) is at worst semilinear. More generally, the A_i may depend on $\vec{u}$ as well, in which case (5) is quasilinear. Of course, symmetric hyperbolic systems can occur independent of a reduction from a second-order equation. For example, *Maxwell's equations*

(7) $$\vec{E}_t - \operatorname{curl} \vec{H} = 0, \qquad \vec{H}_t + \operatorname{curl} \vec{H} = 0$$

for an electric field $\vec{E} = (E_1, E_2, E_3)$ and magnetic field $\vec{H} = (H_1, H_2, H_3)$ form a linear and homogeneous ($\vec{f} = 0$) symmetric hyperbolic system (cf. Exercise 2).

We first study linear symmetric hyperbolic systems. Certain a priori estimates for solutions may be used to obtain existence and uniqueness of solutions. The linear theory is then used to study the nonlinear theory. We always deal with equations in free space $\mathbf{R}^n$ to avoid dealing with boundary conditions, but bounded domains Ω are mentioned in some of the exercises.

a. Energy Estimates for Linear Systems

Let us consider the linear symmetric hyperbolic system

(8a) $$\vec{u}_t + \sum_{i=1}^{n} A_i(x, t) \vec{u}_{x_i} + B(x, t) \vec{u} = \vec{f}(x, t) \qquad \text{in } U_T = \mathbf{R}^n \times (0, T),$$

where $\vec{u}$ and $\vec{f}$ are N-vector-valued functions, and A_i and B are $N \times N$ matrix-valued functions; we assume that

(8b) $$\begin{cases} \text{all } A_i \text{ and } B \text{ have matrix elements} \in C([0, T], C_B^1(\mathbf{R}^n)), \\ \text{and } \vec{f} \text{ has vector components} \in L^2((0, T), H^{1,2}(\mathbf{R}^n)). \end{cases}$$

Solutions of (8a,b) satisfy certain energy estimates, which may be expressed in terms of first-order L^2-Sobolev norms. To simplify notation, as in Chapter 8 we write $\|\cdot\|_k$ in place of $\|\cdot\|_{k,2}$; in particular, $\|\cdot\|_0$ denotes the L^2-norm. All such norms are performed with respect to the spatial variables only, and we also frequently suppress the x-dependence in $\vec{u}$ to think of $\vec{u}(t)$ as an element of $L^2(\mathbf{R}^n)$.

Theorem 1 (Energy Estimate). *Suppose $\vec{u} \in C([0,T], H^{1,2}(\mathbf{R}^n, \mathbf{R}^N)) \cap C^1((0,T), L^2(\mathbf{R}^n, \mathbf{R}^N))$ satisfies (8a,b). Then $\vec{u}$ satisfies the energy estimate*

(9a) $$\max_{0 \le t \le T} \|\vec{u}(t)\|_1^2 \le e^{2\beta T} \left(\|\vec{u}(0)\|_1^2 + 2 \int_0^T \|\vec{f}(t)\|_1^2 dt \right).$$

Moreover, if $f \in C([0,T], H^{1,2}(\mathbf{R}^n))$, then

(9b) $$\max_{0 \le t \le T} \|\vec{u}_t(t)\|_0 \le C e^{\beta T} \left(\|\vec{u}(0)\|_1 + \max_{0 \le t \le T} \|\vec{f}(t)\|_1 \right).$$

In (9a,b), the constants C and β depend on the supremum of $|A_i|$, $|(A_i)_{x_j}|$, $|B|$, and $|B_{x_j}|$ in U_T.

Proof. Let us assume $n = 1$ and write $A_1(x,t)$ as $A(x,t)$; the generalization to $n > 1$ is purely formal. The symmetry of A implies

(10) $$2\vec{u} \cdot A\vec{u}_x = (\vec{u} \cdot A\vec{u})_x - \vec{u} \cdot A_x \vec{u}.$$

Now let us use (8a) and (10) to compute

(11) $$\frac{d}{dt} \int \frac{|\vec{u}|^2}{2} dx = \int \vec{u} \cdot \vec{u}_t \, dx = \int \left(\vec{u} \cdot \vec{f} - \vec{u} \cdot B\vec{u} - \vec{u} \cdot A\vec{u}_x \right) dx$$
$$= \int \left(\vec{u} \cdot \vec{f} - \vec{u} \cdot B\vec{u} + \frac{1}{2} \vec{u} \cdot A_x \vec{u} - \frac{1}{2}(\vec{u} \cdot A\vec{u})_x \right) dx.$$

By the density of $C_0^1(\mathbf{R})$ in $H^{1,2}(\mathbf{R})$, we may assume that $\vec{u}$ has compact x-support for each $t \in [0,T]$, so the last term in (11) vanishes, leaving

(12) $$\frac{d}{dt} \int \frac{|\vec{u}|^2}{2} dx = \int \left(\frac{1}{2} A_x - B \right) \vec{u} \cdot \vec{u} \, dx + \int \vec{u} \cdot \vec{f} \, dx$$
$$\le K_1 \int |\vec{u}|^2 \, dx + \int |\vec{f}|^2 \, dx,$$

where we have used the Cauchy-Schwarz inequality; the constant K_1 depends on $\sup_{U_T}(|B|, |A_x|)$. Now we can apply Gronwall's inequality (cf. Exercise 4) to conclude that

(13) $$\max_{0 \le t \le T} \|\vec{u}(t)\|_0^2 \le e^{2K_1 T} \left(\|\vec{u}(0)\|_0^2 + 2 \int_0^T \|\vec{f}(t)\|_0^2 dt \right).$$

If we differentiate (8a) with respect to x, we get $\vec{w}_t + A\vec{w}_x + B'\vec{w} = \vec{f'}$, where $\vec{w} = \vec{u}_x$, $B' = B + A_x$, and $\vec{f'} = \vec{f}_x - B_x\vec{u}$. Applying the preceding argument to $\vec{w}$ and using $\int \vec{w} \cdot \vec{f'}\,dx \le 2\int(|B_x\vec{u}|^2 + |\vec{f}_x|^2)\,dx \le C\int(|\vec{u}|^2 + |\vec{f}_x|^2)\,dx$, we get the following in place of (12):

$$(14)\quad \frac{d}{dt}\int \frac{|\vec{w}|^2}{2}\,dx = -\int\left(\frac{1}{2}A_x + B\right)|\vec{w}|^2\,dx - \int B_x\vec{w}\cdot\vec{u}\,dx + \int \vec{w}\cdot\vec{f}_x\,dx \le K_2\int(|\vec{w}|^2 + |\vec{u}|^2)\,dx + \int|\vec{f}_x|^2\,dx,$$

where K_2 depends on $\sup_{U_T}(|B|, |B_x|, |A_x|)$.

If we replace $\vec{w}$ in (14) by $\vec{u}_x$ and combine with (12), we obtain

$$(15)\quad \frac{d}{dt}\frac{\|\vec{u}(t)\|_1^2}{2} \le K_3\|\vec{u}(t)\|_1^2 + \|\vec{f}(t)\|_1^2.$$

We again apply Gronwall's inequality to conclude that

$$(16)\quad \max_{0\le t\le T}\|\vec{u}(t)\|_1^2 \le e^{2K_3 T}\left(\|\vec{u}(0)\|_1^2 + 2\int_0^T \|\vec{f}(t)\|_1^2\,dt\right),$$

which is (9a) with $\beta = K_3$.

Now let us assume that $f \in C([0,T], H^{1,2}(\mathbf{R}^n))$. In order to estimate $\|\vec{u}_t(t)\|_0$, we use (8a) to conclude that $\|\vec{u}_t(t)\|_0 \le \|\vec{f}(t)\|_0 + K_4\|\vec{u}(t)\|_1$, where K_4 depends on $\sup_{U_T}(|A|, |B|)$; apply (9a) to the last term to obtain

$$(17)\quad \|\vec{u}_t(t)\|_0 \le \|\vec{f}(t)\|_0 + K_4 e^{\beta T}\left(\|\vec{u}(0)\|_1^2 + 2\int_0^T\|\vec{f}(t)\|_1^2\,dt\right)^{1/2}.$$

Using $\int_0^T\|\vec{f}(t)\|_1^2\,dt \le T\max_{0\le t\le T}\|\vec{f}(t)\|_1^2$, $(a^2+b^2)^{1/2} \le |a|+|b|$, $\sqrt{T}e^{\beta T} \le e^{\beta' T}$, and $\|\vec{f}(t)\|_0 \le \|\vec{f}(t)\|_1$, the estimate (9b) follows easily. ♠

Under stronger differentiability assumptions on the coefficients in (8a), we can obtain L^2-estimates on the higher-order derivatives of solutions. Let us suppose for some integer $k \ge 1$ that

$$(19)\quad \begin{cases} A_i \text{ and } B \text{ have matrix elements in } C([0,T], C_B^k(\mathbf{R}^n)), \text{ and} \\ \vec{f} \text{ has vector components in } C([0,T], H^{k,2}(\mathbf{R}^n)). \end{cases}$$

Theorem 1' (Energy Estimate). *For some integer $k \ge 1$, suppose $\vec{u} \in C([0,T], H^{k,2}(\mathbf{R}^n, \mathbf{R}^N)) \cap C^1((0,T), H^{k-1,2}(\mathbf{R}^n, \mathbf{R}^N))$ satisfies (8a),(19). Then $\vec{u}$ satisfies the energy estimate*

$$(20)\quad \max_{0\le t\le T}(\|\vec{u}(t)\|_k + \|\vec{u}_t(t)\|_{k-1}) \le C_k e^{\beta_k T}\left(\|\vec{u}(0)\|_k + \max_{0\le t\le T}\|\vec{f}(t)\|_k\right),$$

where the constants C_k and β_k depend on the supremum of the A_i, B, and their spatial derivatives up to order k on U_T.

The proof easily generalizes from that of Theorem 1 and so is omitted.

Remark. The estimates remain true for more general systems in which $\vec{u}_t$ in (8a) is replaced by $A_0(x,t)\vec{u}_t$ as in (5) (cf. Exercise 5).

b. Existence for Linear Systems

The a priori estimates of the previous subsection were derived assuming $\vec{u}$ is a sufficiently regular solution of (8). In fact, these estimates may be used to *prove* the existence of a solution of (8) that also satisfies the initial condition

(21) $$\vec{u}(0) = \vec{g} \quad \text{for } x \in \mathbf{R}^n,$$

where $\vec{g} \in H^{1,2}(\mathbf{R}^n, \mathbf{R}^N)$. The idea is to approximate the problem (8), (21) by a simpler system for which a solution is known to exist and then use the a priori estimates to show that the approximate solutions converge to a solution of (8), (21). There are various approximation schemes available for this procedure, with various advantages and disadvantages. Let us mention a few that we do *not* pursue, as they have been at least mentioned earlier in this book.

(i) *Galerkin Method.* Replace $\vec{u}(x,t)$ by $\vec{u}^\ell(x,t) = \sum_{j=1}^\ell a_j(t)\vec{w}_j(x)$ where $\vec{w}_1, \vec{w}_2, \ldots$ is a dense set in the separable Hilbert space $H^{1,2}(\Omega)$; here $\Omega = \mathbf{R}^n$ for the pure initial value problem (8), (21), or more generally $\Omega \subset \mathbf{R}^n$ if we also impose boundary conditions. This replaces (8a) by a linear system of ODEs for the functions $a_1(t), a_2(t), \ldots$, and it only remains to show that, as $\ell \to \infty$, $\vec{u}^\ell$ converges to a solution $\vec{u}$ of (8), (21). (This method was used in Section 11.4 for the Navier-Stokes equations; for the application to hyperbolic systems, see [Ladyzhenskaya, 2].)

(ii) *Finite Differences.* Replace derivatives in (8a) by finite difference quotients [i.e., $\partial_{x_j}\vec{u}(x,t)$ is replaced by $h^{-1}(\vec{u}(x+he_j,t) - u(x,t))$, where $e_j = x_j/|x_j|$, and $\partial_t \vec{u}(x,t)$ is replaced by $\eta^{-1}(\vec{u}(x,t+\eta)-\vec{u}(x,t))$]. This replaces (8a) by a recursion relation that shows how to extend $\vec{u}(x,0)$ to $\vec{u}(x,\eta)$, and so on. It remains to show that this approximation converges as $h, \eta \to 0$. (Pure spatial finite differences were encountered in Section 8.2 in conjunction with linear elliptic equations; for the application to hyperbolic systems, see [John, 1] or [Ladyzhenskaya, 2].)

(iii) *Analytic Approximation.* Approximate the data [i.e., the coefficients in (8a) and $\vec{g}$ in the initial condition (21)] by real analytic functions. The Cauchy-Kovalevski theorem (cf. Section 2.1) then guarantees local existence of a real analytic solution. It may be shown that in fact the real analytic approximate solutions are defined on $(0,T)$ and converge to a solution of (8), (21). (For details, see [John, 1].)

Section 12.1: Symmetric Hyperbolic Systems

We pursue another approximation scheme called the *artificial* or *vanishing viscosity method*. The idea is to (i) introduce a viscosity term with $\epsilon > 0$ in (8a) to create a diffusive system:

(22) $\quad \vec{u}_t - \epsilon \Delta \vec{u} + \Sigma_{i=1}^n A_i(x,t)\, \vec{u}_{x_i} + B(x,t)\, \vec{u} = \vec{f}(x,t) \quad \text{in } U_T = \mathbf{R}^n \times (0,T);$

(ii) show that there is a solution $\vec{u}^\epsilon$ of the diffusive problem (22), (21); and (iii) let $\epsilon \to 0$. In order to invoke the existence theory of Chapter 11, however, we need to improve the regularity assumptions: for some $\alpha > 0$,

(23) $\quad \begin{cases} A_i \text{ and } B \text{ have matrix elements in } C^\alpha([0,T], C_B^k(\mathbf{R}^n)), \text{ and} \\ \vec{f} \text{ has vector components in } C^\alpha([0,T], H^{k,2}(\mathbf{R}^n)). \end{cases}$

Proposition 1. *Suppose $k \geq 1$, (23) is satisfied, and $\vec{g} \in H^{k,2}(\mathbf{R}^n, \mathbf{R}^N)$. Then for every $\epsilon > 0$, the problem (22), (21) admits a unique solution*

(24) $\quad \vec{u}^\epsilon \in C([0,T], H^{k,2}(\mathbf{R}^n, \mathbf{R}^N)) \cap C^1((0,T], H^{k-1,2}(\mathbf{R}^n, \mathbf{R}^N)).$

Moreover, $\vec{u}^\epsilon$ satisfies the estimate (20), independent of ϵ.

Proof. To find the solution $\vec{u}^\epsilon$, let us write (22) in the following form:

(25a) $\quad \vec{u}_t = \epsilon \Delta \vec{u} + \vec{F}(t, \vec{u}) \quad \text{in } U_T = \mathbf{R}^n \times (0,T), \text{ where}$

(25b) $\quad \vec{F}(t, \vec{u}) = \vec{f}(x, t, \vec{u}, \nabla \vec{u}) = \vec{f}(x,t) - \Sigma_{i=1}^n A_i(x,t)\, \vec{u}_{x_i} - B(x,t)\, \vec{u}.$

Observe that replacing t by $t' = \epsilon t$ has the effect of setting $\epsilon = 1$ [i.e., (25a) becomes $\vec{u}_{t'} = \Delta \vec{u} + \vec{F}(\epsilon^{-1} t', \vec{u})$]. This means that we may use the solution operator for the heat equation,

(26) $\quad S(t)\vec{g} = \int_{\mathbf{R}^n} K(x,y,t) \vec{g}\, dy = (4\pi t)^{-n/2} \int_{\mathbf{R}^n} e^{\frac{-|x-y|^2}{4t}} \vec{g}(y)\, dy,$

to write Duhamel's principle for (25a) as

(27) $\quad \vec{u}(t') = S(t')\vec{g} + \int_0^{t'} S(t'-s) \vec{F}(\epsilon^{-1} s, \vec{u}(s))\, ds.$

To invoke the contraction mapping principle, we observe that

(28a) $\quad \|S(t)\vec{g}\|_{k,2} \leq C_k t^{-1/2} \|\vec{g}\|_{k-1,2} \quad \text{for } \vec{g} \in C_0^k(\mathbf{R}^n),\ k \geq 1.$

Moreover, the linearity of $\vec{F}$ and the assumptions (23) imply that

(28b) $\quad \vec{F} \colon [0,T] \times H^{k,2}(\mathbf{R}^n, \mathbf{R}^N) \to H^{k-1,2}(\mathbf{R}^n, \mathbf{R}^N)$

is Hölder continuous in t and Lipschitz continuous in $\vec{u}$. Theorem 3 in Section 11.2 with $X = H^{k,2}(\mathbf{R}^n, \mathbf{R}^N)$ and $X' = H^{k-1,2}(\mathbf{R}^n, \mathbf{R}^N)$ then implies that (22), (21) have a unique solution $\vec{u}^\epsilon$ as in (24) with T replaced by τ_ϵ, where τ_ϵ depends on $\|\vec{g}\|_k$ (and the functions $A_i, B, \vec{f}$).

To show that $\vec{u}^\epsilon$ is in fact defined on $[0, T]$, we recall from Section 11.3 that we require an energy bound. But the energy bound (20) applies to $\vec{u}^\epsilon$, independently of ϵ, as may readily be seen by repeating the proof of Theorem 1, using $\epsilon \int |\nabla \vec{u}|^2 \, dx \geq 0$. This completes the proof. ♠

If we replace the assumptions (23) by the weaker assumptions (19), then (21), (22) will still have a *weak solution* in the following sense:

$$(29) \quad \langle \vec{u}_t, \vec{v} \rangle + \epsilon \langle \nabla \vec{u}, \nabla \vec{v} \rangle + \langle \Sigma_{i=1}^n A_i \vec{u}_{x_i}, \vec{v} \rangle + \langle B\vec{u}, \vec{v} \rangle = \langle \vec{f}, \vec{v} \rangle$$
for all $\vec{v} \in H^{1,2}(\mathbf{R}^n, \mathbf{R}^N)$ and almost every $t \in (0, T)$,

where $\langle , \rangle$ denotes the spatial L^2-inner product (cf. Section 9.2d). The following result is proved in Exercise 6, using duality techniques similar to the proof of Theorem 2.

Proposition 2. *Suppose $k \geq 1$, (19) is satisfied, and $\vec{g} \in H^{k,2}(\mathbf{R}^n, \mathbf{R}^N)$. Then for every $\epsilon > 0$, the problem (22),(21) admits a unique weak solution $\vec{u}^\epsilon \in L^\infty((0,T), H^{k,2}(\mathbf{R}^n, \mathbf{R}^N))$ with $\vec{u}_t^\epsilon \in L^\infty((0,T), H^{k-1,2}(\mathbf{R}^n, \mathbf{R}^N))$. Moreover, $\vec{u}^\epsilon$ satisfies the estimate (20), independent of ϵ.*

Remark. We may integrate $\vec{u}_t^\epsilon \in L^\infty((0,T), H^{k-1,2}(\mathbf{R}^n, \mathbf{R}^N))$ to show that $u^\epsilon \in C([0,T], H^{k-1,2}(\mathbf{R}^n, \mathbf{R}^N))$, so (21) makes sense.

Finally, we want to let $\epsilon \to 0$ and conclude that our symmetric hyperbolic system (8a) admits a weak solution; that is,

$$(30) \quad \langle \vec{u}_t, \vec{v} \rangle + \langle \Sigma_{i=1}^n A_i \vec{u}_{x_i}, \vec{v} \rangle + \langle B\vec{u}, \vec{v} \rangle = \langle \vec{f}, \vec{v} \rangle$$
for all $\vec{v} \in H^{1,2}(\mathbf{R}^n, \mathbf{R}^N)$ and almost every $t \in (0, T)$.

Theorem 2. *If (19) is satisfied and $\vec{g} \in H^{k,2}(\mathbf{R}^n, \mathbf{R}^N)$, then (8a), (21) admit a unique weak solution $\vec{u} \in L^\infty((0,T), H^{k,2}(\mathbf{R}^n, \mathbf{R}^N))$ with $\vec{u}_t \in L^\infty((0,T), H^{k-1,2}(\mathbf{R}^n, \mathbf{R}^N))$; $\vec{u}$ also satisfies (20).*

Proof. Let $Y^* = L^\infty((0,T), L^2(\mathbf{R}^n, \mathbf{R}^N))$, which is the dual space of $Y = L^1((0,T), L^2(\mathbf{R}^n, \mathbf{R}^N))$. The sequence $\{\vec{u}^\epsilon\}_{0 < \epsilon < 1}$ is uniformly bounded in Y^*, so by Theorem 3 of Section 6.3, there is a sequence $\{\vec{u}^{\epsilon_j}\}_{j=1}^\infty$ that converges weak* to $\vec{u} \in Y^*$; that is,

$$(31) \quad \int_0^T \langle \vec{u}^{\epsilon_j}, \vec{w} \rangle \, dt \to \int_0^T \langle \vec{u}, \vec{w} \rangle \, dt \quad \text{for all } \vec{w} \in Y.$$

But (20) also implies that $\{\partial_{x_i} \vec{u}^{\epsilon_j}\}_{j=1}^\infty$ is uniformly bounded in Y^*, so we may assume that it also converges weak* in Y^*; taking $\vec{w} = \partial_{x_i} \vec{w}_1$ in (31), where $\vec{w}_1 \in L^1((0,T), H^{1,2}(\mathbf{R}^n, \mathbf{R}^N))$, shows that $\partial_{x_i}\vec{u}$ is the weak* limit of $\{\partial_{x_i}\vec{u}^{\epsilon_j}\}_{j=1}^{\infty}$. Thus $\vec{u} \in L^{\infty}((0,T), H^{1,2}(\mathbf{R}^n, \mathbf{R}^N))$ and, in particular, $\int_0^T \langle \nabla \vec{u}^{\epsilon_j}, \nabla \vec{w}\rangle\, dt \to \int_0^T \langle \nabla \vec{u}, \nabla \vec{w}\rangle\, dt$ for all $\vec{w} \in L^1((0,T), H^{1,2}(\mathbf{R}^n, \mathbf{R}^N))$. Similarly, we conclude that $\vec{u} \in L^{\infty}((0,T), H^{k,2}(\mathbf{R}^n, \mathbf{R}^N))$.

Now (20) also shows that $\{\vec{u}_t^{\epsilon_j}\}_{j=1}^{\infty}$ is uniformly bounded in Y^*, so we may assume that $\vec{u}_t^{\epsilon_j} \overset{*}{\rightharpoonup} \vec{u}_t$ in Y^* (cf. Exercise 7). Taking spatial derivatives as before shows that in fact $\vec{u}_t \in L^{\infty}((0,T), H^{k-1,2}(\mathbf{R}^n, \mathbf{R}^N))$.

Finally, for $\vec{w} \in L^1((0,T), H^{1,2}(\mathbf{R}^n, \mathbf{R}^N))$ of the form $\vec{w} = \phi \vec{v}$, where $\phi \in C([0,T])$ and $\vec{v} \in H^{1,2}(\mathbf{R}^n, \mathbf{R}^N)$, we know by (29) that

$$\int_0^T \left(\epsilon_j \langle \nabla \vec{u}^{\epsilon_j}, \nabla \vec{w}\rangle + \langle \vec{u}_t^{\epsilon_j} + \Sigma_{i=1}^n A_i \vec{u}_{x_i}^{\epsilon_j} + B\vec{u}^{\epsilon_j}, \vec{w}\rangle\right) dt = \int_0^T \langle \vec{f}, \vec{w}\rangle\, dt.$$

But weak*-convergence enables us to take $\epsilon_j \to 0$ and obtain

$$\int_0^T \left(\langle \vec{u}_t, \vec{w}\rangle + \langle \Sigma_{i=1}^n A_i \vec{u}_{x_i}, \vec{w}\rangle + \langle B\vec{u}, \vec{w}\rangle\right) dt = \int_0^T \langle \vec{f}, \vec{w}\rangle\, dt.$$

Since $\phi \in C([0,T])$ was arbitrary, we obtain (30).

Uniqueness follows by taking the difference of two solutions and applying (20). ♠

Appealing to the Sobolev imbedding theorem, we obtain the following.

Corollary. *Suppose (19) is satisfied and $\vec{g} \in H^{k,2}(\mathbf{R}^n, \mathbf{R}^N)$ for some $k > 2 + \frac{n}{2}$. Then (8a), (21) admit a unique solution $\vec{u} \in C^1([0,T] \times \mathbf{R}^n, \mathbf{R}^N)$.*

Proof. Since $H^{k-1,2} \subset C^1$ for $k > 2 + \frac{n}{2}$, the solution of Theorem 2 satisfies $\vec{u}_t \in L^{\infty}((0,T), C^1(\mathbf{R}^n, \mathbf{R}^N))$; integrating with respect to t shows that $\vec{u} \in C([0,T], C^1(\mathbf{R}^n, \mathbf{R}^N))$, so $\vec{u}_{x_i}$ is continuous. The equation (8a) then shows that $\vec{u}_t$ is continuous. ♠

c. Local Existence for Quasilinear Systems

Let us now consider a quasilinear symmetric hyperbolic system (5) (i.e., the A_i and $\vec{f}$ are now allowed to depend on $\vec{u}$); we would like to show that (5), (6) admit a unique solution $\vec{u}$ in U_τ provided that $\tau > 0$ is sufficiently small. As usual, we prove local existence by means of the contraction mapping principle, which depends on a priori estimates as in Section 12.1a. In particular, let us recall from the proof of Theorem 1 that solutions $\vec{w}$ of

(32a) $$\vec{w}_t + \sum_{i=1}^n A_i(x,t)\vec{w}_{x_i} = \vec{f}(x,t)$$

satisfy the energy estimate

(32b) $$\frac{d}{dt}\frac{\|\vec{w}(t)\|_0^2}{2} \leq K_1 \|\vec{w}(t)\|_0^2 + \|\vec{f}(t)\|_0^2 \quad \text{for } 0 \leq t \leq T,$$
where $K_1 = \sup\{|(A_i)_{x_i}(x,t)|^2 : 1 \leq i \leq n,\ x \in \mathbf{R}^n,\ 0 \leq t \leq T\}$.

Moreover, an application of Gronwall's inequality (cf. Exercise 4) yields

$$\max_{0 \leq t \leq T} \|\vec{w}(t)\|_0^2 \leq e^{2K_1 T} \left(\|\vec{w}(0)\|_0^2 + 2 \int_0^T \|\vec{f}(t)\|_0^2 \right). \tag{32c}$$

To simplify the calculations somewhat, we assume in our quasilinear system (5) that $\vec{f} \equiv 0$ and that the $A_i = A_i(\vec{u})$ are C^∞ functions of $\vec{u} \in \mathbf{R}^N$ with no direct dependence on x or t:

$$\vec{u}_t + \sum_{i=1}^n A_i(\vec{u})\vec{u}_{x_i} = 0 \quad \text{for } 0 < t < \tau, \tag{33a}$$

$$\vec{u}(0) = \vec{g}. \tag{33b}$$

Systems of the form (33a) frequently arise as conservation laws (cf. Section 10.2 and the application to gas dynamics in Section 12.1d).

The idea is to consider the transformation $\mathcal{T}$ defined by $\vec{v} = \mathcal{T}\vec{u}$, where $\vec{u}$ is given and $\vec{v}$ satisfies the *linear* system

$$\vec{v}_t + \sum_{i=1}^n A_i(\vec{u})\vec{v}_{x_i} = 0 \quad \text{for } 0 < t < \tau, \tag{34a}$$

$$\vec{v}(0) = \vec{g}. \tag{34b}$$

We want to show that $\mathcal{T}$ has a fixed point, at least for small τ.

We consider

$$\vec{u} \in Y^{k,\tau} \equiv L^\infty((0,\tau), H^{k,2}(\mathbf{R}^n, \mathbf{R}^N)) \quad \text{with } k > 1 + (n/2), \tag{35}$$

so the Sobolev imbedding theorem implies that $\vec{u}(t), \vec{u}_{x_i}(t) \in C(\mathbf{R}^n, \mathbf{R}^N)$ for each $t \in [0, \tau]$. In fact, we assume that $\vec{u}$ satisfies

$$\|\vec{u}\|_{Y^{k,\tau}} \equiv \max_{0 \leq t \leq \tau} \|\vec{u}(t)\|_k \leq R. \tag{36}$$

In particular, (36) implies that

$$\begin{aligned} |\vec{u}(t)|_\infty &\equiv \sup_{x \in \mathbf{R}^n} |\vec{u}(x,t)| \leq C_0 R, \text{ and} \\ |\vec{u}_{x_i}(t)|_\infty &\equiv \sup_{x \in \mathbf{R}^n} |\vec{u}_{x_i}(x,t)| \leq C_1 R, \end{aligned} \tag{37}$$

where C_i is the imbedding constant for $H^{k-i,2}(\mathbf{R}^n) \subset C_B(\mathbf{R}^n)$.

Proposition 3. *If $\vec{u} \in Y^{k,\tau}$ with $k > 1 + (n/2)$ satisfies (36), then a (sufficiently smooth) solution $\vec{v}$ of (34a) must satisfy the energy estimate*

$$\frac{d}{dt}\frac{\|\vec{v}(t)\|_k^2}{2} \leq K_R \|\vec{v}(t)\|_k^2, \tag{38}$$

where $K_R < \infty$ for fixed R in (36).

Section 12.1: Symmetric Hyperbolic Systems

Proof. For ease of notation, let us write (34a) as if $n = 1$; that is, $\vec{v}_t + A(\vec{u})\vec{v}_x = 0$. Differentiation with respect to x successively ℓ times yields

(39)
$$(D_x^\ell \vec{v})_t + A(\vec{u})(D_x^\ell v)_x = \vec{f}^{(\ell)}(t)$$
$$\text{where } \vec{f}^{(\ell)}(t) \equiv A(\vec{u})D_x^\ell \vec{v}_x - D_x^\ell(A(\vec{u})v_x).$$

Application of (32b) with $\vec{w} = D_x^\ell \vec{v}$ and summing $0 \leq \ell \leq k$ yields

(40a)
$$\frac{d}{dt}\frac{\|\vec{v}(t)\|_k^2}{2} \leq K_1 \|\vec{v}(t)\|_k^2 + \sum_{\ell=1}^k \|\vec{f}^{(\ell)}(t)\|_0^2,$$

where we may use (37) to conclude that

(40b)
$$K_1 = \sup\{|(D_{\vec{u}}A(\vec{u}))\,\vec{u}_x| : (x,t) \in U_T\} \leq \beta_1(R),$$

where $\beta_1(R) < \infty$ for fixed R in (36). We want to estimate the last term in (40a). If $\ell = 1$, then $\|\vec{f}^{(1)}\|_0^2 = \|D_{\vec{u}}A(\vec{u})\vec{u}_x\vec{v}_x\|_0^2 \leq K_1 \|\vec{v}\|_1^2$. More generally, interpolation inequalities may be used (cf. Exercise 8) to show that

(41)
$$\sum_{\ell=1}^k \|\vec{f}^{(\ell)}\|_0^2 \leq K_2 \|\vec{v}\|_k^2,$$

where $K_2 = \beta_2(R) < \infty$ for fixed R in (36). Substitution of (41) in (40a) yields (38). ♠

Corollary. Suppose $k > 1 + (n/2)$, $\vec{g} \in H^k(\mathbf{R}^n, \mathbf{R}^N)$, and $\vec{u} \in Y^{k,\tau}$ satisfies (36). Then there exists a unique solution $\vec{v} \in Y^{k,\tau}$ of (34) that satisfies the energy estimates

(42a)
$$\max_{0 \leq t \leq \tau} \|\vec{v}(t)\|_k \leq e^{K_R \tau} \|\vec{v}(0)\|_k$$

(42b)
$$\max_{0 \leq t \leq \tau} \|\vec{v}_t(t)\|_{k-1} \leq C_R \|\vec{v}(0)\|_k,$$

where $K_R, C_R < \infty$ for fixed R in (36).

Proof. Select $\vec{u}_j \in C^\infty([0,\tau] \times \mathbf{R}^n, \mathbf{R}^N) \cap Y^{k,\tau}$ with $\vec{u}_j \to \vec{u}$ in $Y^{k,\tau}$. Theorem 2 may be applied to construct a unique solution $\vec{v}_j$ of (34), and $\vec{v}_j$ is sufficiently regular that (38) is satisfied. Gronwall's inequality and (34a) then yield (42a,b). Since $\vec{v}_j$ is uniformly bounded in $Y^{k,\tau}$, we may, as in the proof of Theorem 2, find the weak solution $\vec{v}$ corresponding to $\vec{u}$. Uniqueness follows as before. ♠

In order to study local existence for (33a,b), let us take

(43)
$$B_R^{k,\tau} = \{\vec{u} \in Y^{k,\tau} : \|\vec{u}\|_{Y^{k,\tau}} \leq R\}, \quad \text{where } R \geq 2\|\vec{g}\|_k.$$

It is clear from (42a) that $\mathcal{T}: B_R^{k,\tau} \to B_R^{k,\tau}$ provided that τ is sufficiently small. We want to show that $\mathcal{T}$ is a contraction on $B_R^{k,\tau}$ in the $Y^{0,\tau}$-norm:

(44) $\quad \|\mathcal{T}\vec{u}_1 - \mathcal{T}\vec{u}_2\|_{Y^{0,\tau}} \leq \frac{1}{2}\|\vec{u}_1 - \vec{u}_2\|_{Y^{0,\tau}} \quad$ for all $\vec{u}_1, \vec{u}_2 \in B_R^{k,\tau}$,

provided that τ is sufficiently small.

To prove (44), let $\vec{v}_j = \mathcal{T}\vec{u}_j$ and $\vec{w} = \vec{v}_1 - \vec{v}_2$. Then $\vec{w}$ satisfies the linear system

(45)
$$\vec{w}_t + \sum_{i=1}^n A_i(\vec{u}_1)\vec{w}_{x_i} = \vec{f}(x,t) \quad \text{and} \quad \vec{w}(0) = 0,$$
$$\text{where } \vec{f}(x,t) = \sum_{i=1}^n (A_i(\vec{u}_2) - A_i(\vec{u}_1))(\vec{v}_2)_{x_i}.$$

Now $\vec{u}_j \in B_R^{k,\tau}$ and (37) together with the mean value theorem imply $|A_i(\vec{u}_1) - A_i(\vec{u}_2)| \leq C_R|\vec{u}_1 - \vec{u}_2|$. Similarly, $\vec{v}_2 \in B_R^{k,\tau}$ together with (37) implies $(\vec{v}_2)_{x_i} \leq C_R$. Therefore, $\|\vec{f}(t)\|_0 \leq C_R \|\vec{u}_1 - \vec{u}_2\|_0$. If we apply (32c), we obtain

(46) $\quad \max_{0 \leq t \leq \tau} \|\mathcal{T}\vec{u}_1 - \mathcal{T}\vec{u}_2\|_0^2 \leq C_R e^{2K_1\tau}\tau \max_{0 \leq t \leq \tau} \|\vec{u}_1 - \vec{u}_2\|_0^2.$

Taking τ very small in (46) yields (44).

This means that $\mathcal{T}$ is a contraction on $B_R^{k,\tau}$ in the norm of $Y^{0,\tau}$. But this enables us to find $\vec{u}$ by the following iteration scheme. Let $\vec{u}_1 = \vec{g}$, and define $\vec{u}_{j+1} = \mathcal{T}\vec{u}_j$ for $j = 1, 2, \ldots$. As in Section 7.3a, $\vec{u}_j$ converges to a unique $\vec{u} \in Y^{0,\tau} = L^\infty((0,T), L^2(\mathbf{R}^n, \mathbf{R}^N))$.

However, for $k > 2 + (n/2)$, we would like to show that $\vec{u} \in C^1([0,\tau] \times \mathbf{R}^n, \mathbf{R}^N)$, and $\vec{u}$ is a classical solution of (33a,b). It suffices to show that $\vec{u} \in C^1([0,\tau] \times \overline{\Omega})$, where Ω is an arbitrary, smooth, bounded domain in $\mathbf{R}^n$. But (42b) implies that $\{\vec{u}_j(t)\}$ is equicontinuous in $C([0,T], H^{k-1,2}(\Omega))$. The compactness of the imbedding $H^{k-1,2}(\Omega) \subset C^1(\overline{\Omega})$ then implies the existence of a subsequence $\{\vec{u}_{j'}\}$ such that $\vec{u}_{j'} \to \vec{v} \in C([0,T], C^1(\overline{\Omega}))$. But $\vec{u}_j \to \vec{u}$ in $L^\infty((0,\tau), L^2(\Omega))$, so $\vec{u} = \vec{v} \in C([0,T], C^1(\overline{\Omega}))$. In addition, the equation $(\vec{u}_{j'} - \vec{u}_{j'-1})_t = \sum_{i=1}^n (A(\vec{u}_{j'-2})(\vec{u}_{j'-1})_{x_i} - A(\vec{u}_{j'-1})(\vec{u}_j)_{x_i})$ shows that $\vec{u}_{j'}$ also converges in $C^1([0,T], C(\overline{\Omega}))$. In the limit, $\vec{u}$ satisfies (33a,b).

It is not difficult to see that C^1 solutions of (33a,b) are unique (cf. Exercise 9), and so we obtain the following.

Theorem 3. *If $\vec{g} \in H^{k,2}(\mathbf{R}^n, \mathbf{R}^N)$ for some $k > 2 + (n/2)$, then there exists $\tau > 0$ such that (33a,b) admits a unique solution $\vec{u} \in C^1([0,\tau] \times \mathbf{R}^n, \mathbf{R}^N)$.*

This theorem easily generalizes to produce $\vec{u} \in C^m([0,\tau] \times \mathbf{R}^n, \mathbf{R}^N)$ when $\vec{g} \in H^{k,2}(\mathbf{R}^n, \mathbf{R}^N)$ with $k > m + 1 + (n/2)$.

d. Application to Gas Dynamics

Let us consider the Euler equations for the motion in $\mathbf{R}^n$ of a compressible, inviscid fluid such as a gas:

(47)
$$\begin{cases} \vec{v}_t + (\vec{v} \cdot \nabla)\vec{v} + \dfrac{1}{\rho}\nabla p = 0 \\ \rho_t + \operatorname{div}(\rho\vec{v}) = 0, \end{cases}$$

where $\vec{v}$ denotes the velocity field, ρ is the density, and p is the pressure. If we assume that the gas is isentropic [i.e., $p = p(\rho)$] and introduce

(48)
$$\vec{u} = \begin{pmatrix} \rho \\ \vec{v} \end{pmatrix} = \begin{pmatrix} \rho \\ v_1 \\ \vdots \\ v_n \end{pmatrix},$$

then we can write (47) as

(49)
$$\vec{u}_t + \sum_{i=0}^{n} A_i(\vec{u})\vec{u}_{x_i} = 0,$$

where $A_i(\vec{u})$ is the $(n+1) \times (n+1)$ matrix consisting of u_i down the main diagonal, ρ in the ith position of the first row, and $p'(\rho)/\rho$ in the ith position of the first column. For example, when $n = 2$ we have

$$A_1(\vec{u}) = \begin{pmatrix} u_1 & \rho & 0 \\ p'/\rho & u_1 & 0 \\ 0 & 0 & u_1 \end{pmatrix} \quad \text{and} \quad A_2(\vec{u}) = \begin{pmatrix} u_2 & 0 & \rho \\ 0 & u_1 & 0 \\ p'/\rho & 0 & u_1 \end{pmatrix}.$$

Now the system (49) is not symmetric but can be made so by using the change of dependent variables that was used in Section 10.3; that is, replace ρ by

(50)
$$h(\rho) \equiv \int_0^{\rho} \frac{c(r)}{r}\, dr, \quad \text{where } c(\rho) = \sqrt{dp/d\rho}.$$

In terms of the new variables, $h, v_1, \ldots, v_n$, (49) is of the form (33a) (cf. Exercise 10). As a consequence, we conclude from Theorem 3 that a compressible gas, which initially has smooth density and smooth velocity field, will remain so for at least a short period of time. Of course, after a time, it is possible for shocks to develop.

Exercises for Section 12.1

1. Consider the linear equation $u_{tt} = \sum_{i,j=1}^{n} a_{ij}(x)u_{x_i x_j} + \sum_{i=1}^{n} b_i(x)u_{x_i} + c(x)u_t + d(x)u = f(x)$, where the matrix $a_{ij}(x)$ is symmetric and uniformly elliptic. Show that this equation may be reduced to a first-order symmetric hyperbolic system.

2. Show that Maxwell's equations (7) may be written as a symmetric hyperbolic system of six equations in the six dependent variables.

3. Suppose $(\vec{E}, \vec{H})$ is a solution of Maxwell's equations (7) in a smooth domain $\Omega \subset \mathbf{R}^3$, and $\vec{E}$ satisfies the condition $\nu \times \vec{E} = 0$ on $\partial\Omega$.
   (a) Show that the energy $\mathcal{E}(t) = \int_\Omega (|\vec{E}|^2 + |\vec{H}|^2)\, dx\, dy\, dz$ is constant.
   (b) What sort of uniqueness theorem can you formulate from this?

4. Prove *Gronwall's inequality*: If $y \geq 0$ satisfies $dy/dt \leq ky(t) + h(t)$ for $0 \leq t \leq \tau$ where $k \geq 0$ is a constant and $h(t) \geq 0$, $h \in L^1((0,\tau))$, then $y(t) \leq e^{k\tau}[y(0) + \int_0^\tau h(s)ds]$ for all $t \in [0,\tau]$.

5. In place of (8a), consider the linear system $A_0 \vec{u}_t + \sum_{i=1}^n A_i \vec{u}_{x_i} + B\vec{u} = \vec{f}$, where A_0 is also a symmetric matrix function of (x,t) satisfying (8b). Prove that, as in Theorem 1, a solution $\vec{u}$ satisfies (9a,b).

6. Take $N = 1$ and suppose A_i, B, and f all satisfy (19). For $\mu > 0$, choose $A_i^\mu, B^\mu \in C^1([0,T], Z)$ where $Z = C_B^k(\mathbf{R}^n)$ such that $\max_{0 \leq t \leq T} \|A_i^\mu - A_i\|_Z < \eta$ and $\max_{0 \leq t \leq T} \|B^\mu - B\|_Z < \eta$; similarly, approximate f by $f^\mu \in C^1([0,\tau], H^{k,2}(\mathbf{R}^n))$. Let u^μ be the solution of

$$u_t = \Delta u - \sum_{i=1}^n A_i^\mu(x,t) u_{x_i} - B^\mu(x,t) u + f^\mu(x,t).$$

   (a) Show that there exists a sequence u^{μ_j} that converges weak* in $L^\infty((0,T), H^{k,2}(\mathbf{R}^n))$ as $\mu_j \to 0$ to a function u which satisfies (29).
   (b) Show that this weak solution u is unique.

7. If $u^j \overset{*}{\rightharpoonup} u$ in $L^\infty((0,T), L^2(\mathbf{R}^n))$ and also $u_t^j \overset{*}{\rightharpoonup} v$ in $L^\infty((0,T), L^2(\mathbf{R}^n))$, then show that $u_t = v$.

8. *Interpolation inequalities* may be used to bound lower derivatives in terms of higher ones. For example (cf. [Friedman, p. 24]),

$$\|D^j u\|_{L^p} \leq C \|D^m u\|_{L^r}^a \|u\|_{L^q}^{1-a}$$

holds for $u \in C_0^m(\mathbf{R}^n)$, where $j/m \leq a < 1$, $C = C(n, m, j, q, r, a)$, and

$$\frac{1}{p} = \frac{j}{n} + a\left(\frac{1}{r} - \frac{m}{n}\right) + (1-a)\frac{1}{q}.$$

Here D^j represents any derivatives of order j.
   (a) Show that $\|D^i u\|_{L^p} \leq C \|D^m u\|_{L^2}^{i/m} \|u\|_{L^\infty}^{1-i/m}$, for some $1 < p < \infty$.
   (b) Show that $\|D^\ell(u_1 u_2)\|_{L^2} \leq C(\|D^\ell u_1\|_{L^2} \|u_2\|_{L^\infty} + \|u_1\|_{L^\infty} \|D^\ell u\|_{L^2})$.
   (c) Prove the estimate (41).

9. Show that C^1 solutions of (33) are unique.

10. For $n = 2$, express (47) in terms of the variables h, v_1, and v_2 to see that the system is indeed symmetric hyperbolic.

12.2 Linear Wave Dynamics

The dynamical systems approach to wave equations involves a reduction to a first-order evolution $u_t = Au$ on a Banach space, as in Section 9.2. In order to view wave equations involving u_{tt} as a first-order evolution, let us take our cue from the reduction of a second-order ODE to a first-order system (cf. Example 1 of Section 9.2) and introduce

(51) $$\vec{u}(t) = \begin{pmatrix} u(t) \\ u_t(t) \end{pmatrix}$$

to reduce the order of the time derivatives from two to one.

In this section, we carry out this program for linear wave equations, both on $\mathbf{R}^n$ and on bounded domains. The goal is always to express the equation as a first-order evolution

(52) $$\begin{cases} \vec{u}_t = A\vec{u} \\ \vec{u}(0) = \vec{g} \end{cases}$$

on some Banach space X, and show that the resultant solution operator $S(t): X \to X$ forms a quasicontractive C^0-semigroup. The semigroup properties (i) $S(t+s) = S(t)S(s)$ and (ii) $S(0) = I$ follow from the existence and uniqueness of solutions to (52). It remains then to check the C^0-continuity, which means

(53a) $\qquad \|S(t)u - u\|_X \to 0 \qquad$ as $t \to 0$, for all $u \in X$,

and the quasicontractivity, which means to find $\beta \in \mathbf{R}$ so that for some $C > 0$,

(53b) $\qquad \|S(t)u\|_X \leq Ce^{\beta t}\|u\|_X \qquad$ for all $t > 0$ and $u \in X$.

In the next section, we apply this dynamical formulation to the study of semilinear wave equations.

a. The Wave Equation in $\mathbf{R}^n$

If we consider the initial value problem

(54) $$\begin{cases} u_{tt} = \Delta u & \text{for } x \in \mathbf{R}^n \text{ and } t > 0, \\ u(x,0) = g_0(x), \quad u_t(x,0) = g_1(x) & \text{for } x \in \mathbf{R}^n, \end{cases}$$

then (51) yields the first-order evolution (52), where

(55) $$A = \begin{pmatrix} 0 & 1 \\ \Delta & 0 \end{pmatrix} \qquad \text{and} \qquad \vec{g} = \begin{pmatrix} g_0 \\ g_1 \end{pmatrix}.$$

For our Banach space X, let us use

(56a) $\quad X = \begin{pmatrix} H^{1,2}(\mathbf{R}^n) \\ L^2(\mathbf{R}^n) \end{pmatrix} = \left\{ \vec{v} = \begin{pmatrix} v_0 \\ v_1 \end{pmatrix} : v_0 \in H^{1,2}(\mathbf{R}^n), v_1 \in L^2(\mathbf{R}^n) \right\}$

with the norm

(56b) $\quad\quad\quad\quad\quad\quad \|\vec{v}\|_X^2 = \|v_0\|_{1,2}^2 + \|v_1\|_2^2.$

Of course, A is not a bounded operator on X, but let us take as domain

(57) $\quad\quad\quad\quad\quad\quad D = \begin{pmatrix} H^{2,2}(\mathbf{R}^n) \\ H^{1,2}(\mathbf{R}^n) \end{pmatrix}.$

Then D is dense in X (using mollifiers as in Section 6.5), and $A: D \to X$ is a closed operator, as a consequence of the a priori estimate

(58) $\quad\quad \|u\|_{2,2} \leq C(\|\Delta u\|_2 + \|u\|_2) \quad \text{for } u \in H^{2,2}(\mathbf{R}^n)$

(cf. Exercise 11 in Section 8.1 and Exercise 1a at the end of this section).

Now we know how to solve (54), assuming, for example, that $g_0, g_1 \in C_0^\infty(\mathbf{R}^n)$; in fact, explicit formulas were given in Chapter 3. This means that there is a solution operator $S(t)$ for (52), so that $\vec{u}(t) = S(t)\vec{g}$ is the solution of (52), at least when $\vec{g} \in C_0^\infty(\mathbf{R}^n, \mathbf{R}^2)$. We want, however, to show that $S(t)$ is a bounded linear map on X satisfying (53a,b). To do so, we use the energy $\mathcal{E} = \mathcal{E}_u$:

(59) $\quad\quad \mathcal{E}(t) = \frac{1}{2}\int_{\mathbf{R}^n} (u_t^2 + |\nabla u|^2)\, dx = \frac{1}{2}\left(\|\vec{u}(t)\|_X^2 - \|u(t)\|_2^2 \right).$

Differentiating $\mathcal{E}$ with respect to t shows that the energy of the solution is a constant (cf. Section 3.3). In terms of the norms, we may write

(60) $\quad\quad 2\mathcal{E}(t) = \left(\|S(t)\vec{g}\|_X^2 - \|u(t)\|_2^2 \right) = \|\vec{g}\|_X^2 - \|g_0\|_2^2 = 2\mathcal{E}(0).$

Next we want to estimate the L^2-norm of $u(t)$. If we observe $u(t) = \int_0^t u_t(s)\, ds + g$, then we obtain $\|u(t)\|_2 \leq t\left(\max_{0 \leq s \leq t} \|u_t(s)\|_2 \right) + \|g_0\|_2$. But $\|u_t(s)\|_2^2 \leq 2\mathcal{E}(s) = 2\mathcal{E}(0) \leq \|\vec{g}\|_X^2$, so

(61) $\quad\quad\quad\quad\quad\quad \|u(t)\|_2 \leq t\|\vec{g}\|_X + \|g_0\|_2.$

Combining (60) and (61), we obtain $\|S(t)\vec{g}\|_X^2 = \|\vec{g}\|_X^2 + \|u(t)\|_2^2 - \|g_0\|_2^2 \leq \|\vec{g}\|_X^2 + t^2\|\vec{g}\|_X^2 + 2t\|\vec{g}\|_X \|g_0\|_2 \leq \|\vec{g}\|_X^2(1 + t^2 + 2t) = (1+t)^2 \|\vec{g}\|_X^2$ which we summarize as

(62) $\quad\quad\quad\quad\quad\quad \|S(t)\vec{g}\|_X \leq (1+t)\|\vec{g}\|_X.$

Now (62) was obtained under the assumption $g_0, g_1 \in C_0^\infty(\mathbf{R}^n)$. But density implies that (62) holds for all $\vec{g} \in X$, so $S(t): X \to X$ is a bounded operator with operator norm $\|S(t)\| \leq (1+t)$. In particular, (62) says that $S(t)$ is β-contractive for any $\beta > 0$.

It remains to verify C^0-continuity. But for $g_0, g_1 \in C_0^\infty(\mathbf{R}^n)$, we know from Section 3.2 that $u(t) \in C_0^\infty(\mathbf{R}^n)$ with $u(t) \to g_0$ and $u_t(t) \to g_1$ pointwise. For such $\vec{g}$, $\|S(t)\vec{g} - \vec{g}\|_X \to 0$ is certainly true; since such $\vec{g}$ are dense in X and $S(t)$ is bounded on X uniformly for $t \in [0, \epsilon]$, it is easy to deduce (53a) (cf. Exercise 2).

This means that we have proved the following.

Theorem 1. *The first-order evolution associated with the wave equation on $\mathbf{R}^n$ generates a C^0-semigroup that is β-contractive for all $\beta > 0$.*

From Section 9.2 we know that the C^0-semigroup $S(t)$ provides the solution to (52) whenever $\vec{g} \in D(A)$,; that is, we have the following.

Corollary. *For every $g_0 \in H^{2,2}(\mathbf{R}^n)$ and $g_1 \in H^{1,2}(\mathbf{R}^n)$, there exists a unique solution $u \in C([0, \infty), H^{1,2}(\mathbf{R}^n)) \cap C^1((0, \infty), L^2(\mathbf{R}^n))$ of (54).*

Remark. The hypotheses $g_0 \in H^{2,2}(\mathbf{R}^n)$ and $g_1 \in H^{1,2}(\mathbf{R}^n)$ are necessary to conclude that $u \in C([0, \infty), H^{1,2}(\mathbf{R}^n)) \cap C^1((0, \infty), L^2(\mathbf{R}^n))$; this loss of regularity was encountered in Section 3.2 and reflects the fact that the semigroup is *not* analytic. Of course, the solution will have more regularity if we assume more regularity of the initial conditions (cf. Exercise 3).

b. The Klein-Gordon Equation in $\mathbf{R}^n$

Let us consider the initial value problem for the *Klein-Gordon equation*:

$$\text{(63)} \quad \begin{cases} u_{tt} - \Delta u + m^2 u = 0 & \text{for } x \in \mathbf{R}^n \text{ and } t > 0, \\ u(x, 0) = g_0(x), \quad u_t(x, 0) = g_1(x) & \text{for } x \in \mathbf{R}^n. \end{cases}$$

The reduction (51) yields the first-order evolution (52) where

$$\text{(64)} \quad A = \begin{pmatrix} 0 & 1 \\ \Delta - m^2 & 0 \end{pmatrix} \quad \text{and} \quad \vec{g} = \begin{pmatrix} g_0 \\ g_1 \end{pmatrix}.$$

As in the case $m = 0$, we use the Banach space X of (56).

We assume that we know how to solve (63) at least for $g_0, g_1 \in C_0^\infty(\mathbf{R}^n)$ (cf. Exercise 8 in Section 3.1 and Exercise 5 in Section 3.2). It is easy to see that $m > 0$ implies that the solution operator $S(t)$ satisfies

$$\text{(65)} \quad C^{-1}\|\vec{g}\|_X \leq \|S(t)\vec{g}\|_X \leq C\|\vec{g}\|_X,$$

for some constant $C \geq 1$; in particular, $S(t)$ is a *0-contraction* (i.e., a quasicontraction with $\beta = 0$). In fact, $S(t)$ is an isometry if $m = 1$ or an equivalent norm is introduced on X (cf. Exercise 4b). In any case, the C^0-continuity follows as for $m = 0$.

Theorem 2. *The first-order evolution associated with the Klein-Gordon equation on $\mathbf{R}^n$ generates a C^0-semigroup that is 0-contractive.*

As for (54), if $g_0 \in H^{2,2}(\mathbf{R}^n)$ and $g_1 \in H^{1,2}(\mathbf{R}^n)$, then there exists a unique solution $u \in C([0,\infty), H^{1,2}(\mathbf{R}^n)) \cap C^1((0,\infty), L^2(\mathbf{R}^n))$ of (63).

c. Equations on Bounded Domains

Let Ω be a bounded domain in $\mathbf{R}^n$ with smooth boundary, and consider the wave equation with Dirichlet condition

$$\text{(66)} \quad \begin{cases} u_{tt} - \Delta u = 0 & \text{for } x \in \Omega \text{ and } t > 0, \\ u(x,0) = g_0(x), \quad u_t(x,0) = g_1(x) & \text{for } x \in \Omega, \\ u(x,t) = 0 & \text{for } x \in \partial\Omega \text{ and } t > 0. \end{cases}$$

(The introduction of $m^2 u$ as in the previous subsection no longer has a dramatic effect.) We again use (51) to reduce the problem to (52), but now

$$\text{(67)} \quad X = \begin{pmatrix} H_0^{1,2}(\Omega) \\ L^2(\Omega) \end{pmatrix} \quad \text{and} \quad D = \begin{pmatrix} H_0^{1,2}(\Omega) \cap H^{2,2}(\Omega) \\ H_0^{1,2}(\Omega) \end{pmatrix}.$$

As before, $A: D \to X$ is densely defined and closed (cf. Exercise 1b).

We can use the eigenvalues and eigenfunctions (λ_k, ϕ_k) of the Dirichlet Laplacian (cf. Section 7.2) to define explicitly the solution operator. Let us use the eigenfunction expansions for $g_0 \in H_0^{1,2}(\Omega)$ and $g_1 \in L^2(\Omega)$,

$$\text{(68)} \quad \begin{cases} g_0(x) = \sum_{k=1}^{\infty} a_k \phi_k(x), & \text{where} \quad \|\vec{g}_0\|_{1,2}^2 = \sum_{k=1}^{\infty}(\lambda_k + 1)a_k^2, \\ g_1(x) = \sum_{k=1}^{\infty} b_k \phi_k(x), & \text{where} \quad \|\vec{g}_1\|_{1,2}^2 = \sum_{k=1}^{\infty} b_k^2, \end{cases}$$

to write the solution of (66) as

$$\text{(69a)} \quad u(x,t) = \sum_{n=1}^{\infty} \left(a_k \cos \sqrt{\lambda_k} t + \frac{b_k}{\sqrt{\lambda_k}} \sin \sqrt{\lambda_k} t \right) \phi_k(x).$$

Differentiating with respect to t, we obtain

$$\text{(69b)} \quad u_t(x,t) = \sum_{n=1}^{\infty} \left(-a_k \sqrt{\lambda_k} \sin \sqrt{\lambda_k} t + b_k \cos \sqrt{\lambda_k} t \right) \phi_k(x).$$

This defines the solution operator $S(t): X \to X$, which satisfies

$$\text{(70)} \quad \|S(t)\vec{g}\|_X^2 \leq C \sum_{k=1}^{\infty} \left((\lambda_k + 1)a_k^2 + b_k^2 \right) = C\|\vec{g}\|_X^2.$$

In other words, $S(t)$ is a 0-contraction. Moreover, it is easy to check that (53a) holds for all $\vec{u} \in X$ (cf. Exercise 5).

Theorem 3. *The first-order evolution associated with the wave equation on a bounded domain Ω generates a C^0-semigroup that is 0-contractive.*

In particular, if $g_0 \in H_0^{1,2}(\Omega) \cap H^{2,2}(\Omega)$ and $g_1 \in H_0^{1,2}(\Omega)$, then (66) admits a unique solution $u \in C([0,\infty), H_0^{1,2}(\Omega) \cap H^{2,2}(\Omega)) \cap C^1((0,\infty), L^2(\Omega))$.

d. The Schrödinger Equation

Replacing t in the heat equation by it yields the Schrödinger equation

$$(71) \quad \begin{cases} u_t = i\Delta u & \text{for } x \in \mathbf{R}^n \text{ and } t > 0, \\ u(x,0) = g(x) & \text{for } x \in \mathbf{R}^n, \end{cases}$$

which arises in quantum mechanics. Notice that (71) is already in the form (52) with

$$(72) \quad A = i\Delta, \quad X = L^2(\mathbf{R}^n), \quad \text{and} \quad D = H^{2,2}(\mathbf{R}^n),$$

although we must now consider *complex-valued* functions u. Nevertheless, the a priori estimate (58) still holds, which shows that $A \colon D \to X$ is a closed operator.

The solution operator $S(t)$ is formally obtained by modifying the heat kernel:

$$(73) \quad u(x,t) = S(t)g = (4\pi i t)^{-n/2} \int_{\mathbf{R}^n} e^{\frac{-|x-y|^2}{4it}} g(y) \, dy.$$

If $g \in L^1(\mathbf{R}^n)$, then (73) is certainly well defined; in fact,

$$(74) \quad |u(x,t)| \leq (4\pi t)^{-n/2} \|g\|_1.$$

However, it is not immediately clear that $S(t) \colon X \to X$ is a bounded linear map. In fact, we want to show that $S(t)$ is an isometry on X:

$$(75) \quad \|S(t)g\|_2 = \|g\|_2 \quad \text{for all } t > 0.$$

Formally, this is certainly true; if we assume that $u(t) = S(t)g$ solves (71), then

$$(76) \quad \begin{aligned} \frac{d}{dt} \int_{\mathbf{R}^n} |u|^2 \, dx &= \frac{d}{dt} \int_{\mathbf{R}^n} u\bar{u} \, dx = \int_{\mathbf{R}^n} (u_t \bar{u} + u \bar{u}_t) \, dx \\ &= \int_{\mathbf{R}^n} (i\Delta u \bar{u} - iu\Delta \bar{u}) \, dx = \int_{\mathbf{R}^n} (-i\nabla u \cdot \nabla \bar{u} + i\nabla u \cdot \nabla \bar{u}) \, dx = 0. \end{aligned}$$

A formal proof of (75) is given in Exercise 7 for $n = 1$. In fact, it is also clear from Exercise 7 (and Exercise 2) that $S(t)$ satisfies (53a), and so we have the following.

Theorem 4. *The first-order evolution associated with Schrödinger's equation on $\mathbf{R}^n$ generates a C^0-semigroup of isometries on $X = L^2(\mathbf{R}^n)$.*

There is a theorem of functional analysis, called the *Riesz convexity theorem*, that enables us to interpolate between the two estimates (74) and (75). We obtain the L^p-$L^{p'}$ estimate where $p' = p/(p-1)$

(77) $\qquad \|S(t)g\|_{p'} = \|u(t)\|_{p'} \leq (4\pi t)^{(\frac{n}{2} - \frac{n}{p})} \|g\|_p \quad \text{for } 1 \leq p \leq 2.$

[See Exercise 8 for a further discussion of the derivation of (77).] This estimate is useful in studying semilinear Schrödinger equations.

Exercises for Section 12.2

1. (a) Use (58) to show that A given in (55) with domain (57) is a closed operator on the Banach space X of (56).
   (b) Use the elliptic estimates of Section 8.2 to verify that $A: D \to X$ defined by (55) and (67) is a closed operator.

2. If $\{S(t)\}_{0 \leq t \leq \epsilon}$ is a family of bounded linear operators on a Banach space X with $\|S(t)\| \leq M$ for all $0 \leq t \leq \epsilon$, and X_0 is a dense subset of X for which $g \in X_0$ implies $S(t)g \to g$ as $t \to 0$, show that $S(t)g \to g$ as $t \to 0$ holds for all $g \in X$.

3. *Improved Regularity.* If $g_0 \in H^{k+1,2}(\mathbf{R}^n)$ and $g_1 \in H^{k,2}(\mathbf{R}^n)$ for some integer $k > 1$, prove that (54) admits a unique solution $\vec{u} \in C([0,\infty), H^{k+1,2}(\mathbf{R}^n)) \cap C^1((0,\infty), L^2(\mathbf{R}^n))$.

4. (a) Verify (65) for the solution operator for (63) with $m > 0$.
   (b) Verify that (65) holds with $C = 1$ [i.e., $S(t)$ is an isometry] if $m = 1$ or an equivalent norm on X is introduced.

5. Use the eigenfunction expansions to show directly that the solution operator $S(t)$ defined by (69ab) satisfies (53a).

6. *Dissipation.* Let $\Omega \subset \mathbf{R}^n$ be a smooth bounded domain, and consider the equation $u_{tt} + \alpha u_t - \Delta u = 0$ for $x \in \Omega$ and $t > 0$, where $\alpha > 0$ so that αu_t corresponds to *dissipation*. Assume Dirichlet boundary condition $u(x,t) = 0$ for $x \in \partial\Omega$, and show that the associated first-order evolution $S(t)$ on $X = L^2(\Omega)$ is β-contractive for some $\beta < 0$.

7. This exercise establishes (75) for $n = 1$, where $S(t)$ is given in (73).
   (a) For $\phi(x) = e^{-b(x-a)^2}$ with $b > 0$, show by direct calculation that

   $$(S(t)\phi)(y) = (1 + 4ibt)^{-1/2} e^{-b(x-a)^2/(1+4ibt)}.$$

   (b) For ϕ as in (a), show that $\|\phi\|_2^2 = \sqrt{\pi/(2b)} = \|S(t)\phi\|_2^2$.
   (c) For ϕ as in (a), show that the Fourier transform of ϕ is given by

   $$\widehat{\phi}(\xi) = (2b)^{1/2} e^{-\frac{\xi^2}{4b} - ia\xi}.$$

(d) Let D be the collection of all linear combinations of functions ϕ as in (a). Show that D is dense in $L^2(\mathbf{R})$ by using (c) and the fact from Fourier analysis that the Fourier transform is an isometry on $L^2(\mathbf{R}^n)$.

8. The *Riesz convexity theorem* is the following. Suppose $T\colon L^{p_i}(\mathbf{R}^n) \to L^{q_i}(\mathbf{R}^n)$ is bounded with norm k_i for $i = 0, 1$. For $0 \leq \lambda \leq 1$, let $1/p_\lambda = \lambda/p_0 + (1-\lambda)/p_1$ and $1/q_\lambda = \lambda/q_0 + (1-\lambda)/q_1$. Then $T\colon L^{p_\lambda}(\mathbf{R}^n) \to L^{q_\lambda}(\mathbf{R}^n)$ is bounded with norm $k_\lambda \leq k_0^\lambda k_1^{(1-\lambda)}$. Use this to derive (77) from (74) and (75).

12.3 Semilinear Wave Dynamics

Semilinear wave equations of the form

(78) $$\begin{cases} u_{tt} - \Delta u + m^2 u = f(x, t, u) & \text{for } x \in \Omega \text{ and } t > 0, \\ u(x, 0) = g_0(x), \quad u_t(x, 0) = g_1(x) & \text{for } x \in \Omega, \end{cases}$$

where $m \geq 0$ and Ω is a domain in $\mathbf{R}^n$ (possibly $\mathbf{R}^n$ itself), may be studied by means of a nonlinear first-order evolution on the same Banach space X used for the linear evolution. For simplicity, we suppose f is independent of t. This leads us to consider the autonomous evolution:

(79) $$\begin{cases} \vec{u}_t = A\vec{u} + \vec{F}(\vec{u}) \\ \vec{u}(0) = \vec{g}, \end{cases}$$

where A and $\vec{g}$ are given by (64), and $F\colon \mathbf{R} \times X \to X$ is defined by

$$F(\vec{u})(x) = \begin{pmatrix} 0 \\ f(x, u(x)) \end{pmatrix}.$$

In order to solve (79) when A generates a semigroup $S(t)$, we use Duhamel's principle as in Section 9.2e to consider the integral equation

(80) $$\vec{u}(t) = S(t)\vec{g} + \int_0^t S(t-s) F(\vec{u}(s))\, ds.$$

We investigate local existence of a solution of (80) [i.e., a mild solution of (79)] and when this solution is a classical solution of (79) (i.e., C^1 in t). We then investigate global existence.

a. Local Existence

We assume that $A\colon D \to X$ is a closed operator generating a quasicontractive C^0-semigroup $S(t) = e^{tA}$ on the Banach space X. We also make the following assumptions on F:

(81a) $\quad \|F(\vec{u})\| \leq M_R \|\vec{u}\| \quad$ for all $\|\vec{u}\| \leq R$,

(81b) $\quad \|F(\vec{u}) - F(\vec{v})\|_X \leq M_R \|\vec{u} - \vec{v}\| \quad$ for all $\|\vec{u}\|, \|\vec{v}\| \leq R$.

We consider functions $\vec{u} \in C([0,T], X)$; for such a function, (81b) implies that $F(\vec{u}(s))$ and $S(t-s)F(\vec{u}(s))$ are both continuous functions of s (cf. Exercise 1a), so the integral in (80) exists as a limit of Riemann sums (cf. Exercise 10 in Section 9.2). We may therefore use Theorem 2 of Section 9.2 to conclude local existence of a mild solution $u \in C([0,\tau], X)$ when $\vec{g} \in D(A)$. Let us consider an example.

Example 1. Let $\Omega = \mathbf{R}^n$ and $f(u) = \gamma u^k$, where $k > 1$ is an integer. Then (78) becomes

$$(82) \quad \begin{cases} u_{tt} - \Delta u + m^2 u + \gamma u^k = 0 & \text{for } x \in \mathbf{R}^n \text{ and } t > 0, \\ u(x,0) = g_0(x), \quad u_t(x,0) = g_1(x) & \text{for } x \in \mathbf{R}^n. \end{cases}$$

For what values of k can we conclude local existence of a mild solution? Let X be given by (56). To verify (81a), we use the Sobolev imbedding inequality (cf. Section 6.4):

$$(83) \quad \|u\|_{2k} \le C\|u\|_{1,2} \quad \text{provided that} \quad \begin{cases} n = 1,2 \text{ and any } k > 1, \\ n = 3,4 \text{ and } 1 < k \le \dfrac{n}{n-2}. \end{cases}$$

For these values of k, we find $\|F(\vec{u})\| = \|\gamma u^k\|_2 = \gamma\|u\|_{2k}^k \le \gamma C^k \|u\|_{1,2}^k = \gamma C^k \|u\|_{1,2}^{k-1} \|u\|_{1,2} \le M_R \|\vec{u}\|$. We may also verify (81b):

$$\|u^k - v^k\|_2 = \|(u-v)(u^{k-1} + \cdots + v^{k-1})\|_2$$
$$\le \|u-v\|_2 (\|u^{k-1}\|_2 + \|u^{k-2}\|_2 \|v\|_2 + \cdots + \|v^{k-1}\|_2)$$
$$\le C\|u-v\|_{1,2}(\|u\|_{1,2} + \|v\|_{1,2})^{k-1}.$$

We conclude that (82) admits a unique mild solution $u \in C([0,\tau], X)$ provided that k satisfies the restrictions in (83), $g_0 \in H^{2,2}(\mathbf{R}^n)$, $g_1 \in H^{1,2}(\mathbf{R}^n)$, and τ is sufficiently small. ♣

It remains to verify that the mild solution $u \in C([0,\tau], X)$ in fact satisfies (79) in the classical sense; that is, $u \in C^1((0,\tau], X)$. For this to be true, we naturally need a more intimate relationship between F and A:

(84a) $\qquad\qquad F: D \to D,$

(84b) $\qquad\qquad \|AF(\vec{u})\| \le M_R \|A\vec{u}\|,$

(84c) $\qquad\qquad \|A(F(\vec{u}) - F(\vec{v}))\| \le M_R \|A(\vec{u} - \vec{v})\|,$

which are to hold for all $\vec{u}, \vec{v} \in D$ such that $\|\vec{u}\|, \|\vec{v}\|, \|A\vec{u}\|, \|A\vec{v}\| \le R$. As a consequence, $S(t-s)AF(\vec{u}(s))$ is a continuous function of s (cf. Exercise 1b). In fact, we now prove the following.

Section 12.3: Semilinear Wave Dynamics

Lemma. *Under the afore-mentioned hypotheses on A and F,*

$$(85) \qquad A\int_0^t S(t-s)F(\vec{u}(s))\,ds = \int_0^t S(t-s)AF(\vec{u}(s))\,ds.$$

Proof. Consider the Riemann sum $\sigma_N = \sum_{i=1}^N S(t-s_i)F(\vec{u}(s_i))\Delta s_i$, which is in $D(A)$ by the properties of F and A. Moreover,

$$A\sigma_N = \sum_{i=1}^n S(t-s_i)AF(\vec{u}(s_i))\Delta s_i \to \int_0^t S(t-s)AF(\vec{u}(s))\,ds,$$

by Exercise 1b. But then $\sigma_N \to \sigma = \int_0^t S(t-s)F(\vec{u}(s))\,ds$ and the fact that A is a closed operator together imply (85). ♠

Theorem 1. *Suppose $A: D \to X$ is a closed operator generating a β-contractive C^0-semigroup on the Banach space X, $\vec{g} \in D$, and F satisfies (81) and (84). Then for τ sufficiently small, (79) admits a unique classical solution $\vec{u} \in C^1([0,\tau], X)$. In fact, for any $a, b > 0$, $\tau = \tau_{a,b}$ may be chosen uniformly for initial conditions satisfying $\|\vec{g}\| \leq a$ and $\|A\vec{g}\| \leq b$.*

Proof. For continuous functions $\vec{u}: [0,T] \to X$ such that $\vec{u}(t) \in D$ and $A\vec{u}: [0,T] \to X$ is also continuous, let

$$(86) \qquad \|\vec{u}\|_{Z_T} = \max_{0 \leq t \leq T} \|\vec{u}(t)\| + \max_{0 \leq t \leq T} \|A\vec{u}(t)\|,$$

and let Z_T be the space of all such $\vec{u}: [0,T] \to D$ for which $\|\vec{u}\|_{Z_T} < \infty$. The closedness of A shows that Z_T is a Banach space with the norm (86). The standard contraction mapping argument (cf. Exercise 2) shows that (80) admits a unique solution $\vec{u} \in C([0, \tau_{a,b}], X)$ for all $\vec{g}$ satisfying $\|\vec{g}\| \leq a$ and $\|A\vec{g}\| \leq b$.

It only remains to show that $\vec{u}$ is differentiable. For this purpose, suppose $\|\vec{u}\|_{Z_\tau} \leq R$ and for $0 \leq t < t + h < \tau$ use (80) to write

$$(87) \quad \begin{aligned} \frac{\vec{u}(t+h) - \vec{u}(t)}{h} &= \left(\frac{S(h) - I}{h}\right)S(t)\vec{g} + \frac{1}{h}\int_t^{t+h} S(t+h-s)F(\vec{u}(s))\,ds \\ &\quad + \int_0^t S(t-s)\left(\frac{S(h) - I}{h}\right)F(\vec{u}(s))\,ds = I_1 + I_2 + I_3. \end{aligned}$$

Letting $h \to 0$, we find $I_1 \to AS(t)\vec{g}$. The integrand in I_2 is continuous in s by Exercise 1a, and so $I_2 \to S(0)F(\vec{u}(t)) = F(\vec{u}(t))$. The integrand in I_3 converges to $S(t-s)AF(\vec{u}(s))$ for each s and is bounded uniformly by $Ce^{\beta(t-s)}\|AF(\vec{u}(s))\| \leq Ce^{\beta(t-s)}M_R R$, where we have used (84b). By the dominated convergence theorem, $I_3 \to \int_0^t S(t-s)AF(\vec{u}(s))\,ds$. This means that $\vec{u}: [0,\tau) \to X$ admits right derivatives; a similar argument using

$0 < t - h < t \leq \tau$ shows that $\vec{u}\colon (0,\tau] \to X$ also admits left derivatives that agree with the right derivatives and satisfy (79):

$$\frac{d\vec{u}}{dt} = AS(t)\vec{g} + F(\vec{u}(t)) + A\int_0^t S(t-s)F(\vec{u}(s))\,ds = A\vec{u} + F(\vec{u}). \spadesuit$$

Example 1 (Revisited). Let us verify the hypotheses of Theorem 1 for (82). We have

(88) $$AF(\vec{u}) = AF\begin{pmatrix} u \\ u_t \end{pmatrix} = \begin{pmatrix} u^k \\ 0 \end{pmatrix},$$

so to verify (84a) we must show that $u \in H^{2,2}(\mathbf{R}^n)$ implies $u^k \in H^{1,2}(\mathbf{R}^n)$. This will be true for exactly the same values of k as before. In fact, applying (83) to ∇u in place of u enables us to compute

(89) $$\|\nabla(u^k)\|_2 \leq C\|u^{k-1}\|_{2k/(k-1)} \|\nabla u\|_{2k}$$
$$\leq C\|u\|_{2k}^{k-1} \|\nabla^2 u\|_{1,2} \leq C\|u\|_{1,2}^{k-1} \|u\|_{2,2}.$$

This verifies (84a). To check (84b), let us compute

(90) $$\|AF(\vec{u})\| = \|u^k\|_{1,2} \leq C(\|\nabla(u^k)\|_2 + \|u^k\|_2) \leq C\|u\|_{1,2}^{k-1} \|u\|_{2,2}.$$

Now, for $m > 0$ we may verify by Fourier analysis (cf. Exercise 3a)

(91) $$\|u\|_{2,2} \leq C_m \|(\Delta - m^2)u\|_2;$$

combined with (90) and

(92) $$\|A\vec{u}\|^2 = \|u_t\|_{1,2}^2 + \|(\Delta - m^2)u\|_2^2,$$

we obtain (84b). All these observations may also be used to verify (84c) (cf. Exercise 3b). We conclude that (82) admits a unique classical solution $\vec{u} \in C^1([0,\tau], X)$ provided that τ is sufficiently small. $\spadesuit$

b. Global Behavior for Conservative Systems

In this section we consider global existence for the semilinear wave equation (78). However, global existence for (78) does not hold in general: For example, when $m = 0$, $f(u) = \pm u^2$, and $n = 3$, then a mild solution exists for a short time (cf. Example 1 with $k = 2$), but it is known (cf. [John, 3]) that all solutions blow up in finite time.

As indicated in Section 9.2e, we might try to establish *global existence* by iterating our local existence argument; we obtain a solution $\vec{u}(t)$ of (78) on $0 \leq t \leq \tau + \tau_1 + \tau_2 + \cdots$. This will give us global existence provided that $\tau + \tau_1 + \tau_2 + \cdots$ diverges to infinity; for example, provided that $\tau_j \geq \epsilon > 0$ for all j. This latter condition is obtained by establishing an a priori bound on the solution [i.e., $\|\vec{u}(t)\| \leq M$]. This will give a lower bound on τ provided that the choice of τ depends on $\|\vec{g}\|$ but *not* on $\|A\vec{g}\|$. This, in turn, will be the case if we require the constant in (84b) to be independent of $\|A\vec{g}\|$:

(93) $$\|AF(\vec{u})\| \leq M_R \|A\vec{u}\| \quad \text{for all } \|\vec{u}\| \leq R.$$

In other words, we have the following result.

Theorem 2. *Suppose the hypotheses of Theorem 1 are satisfied and (93) is satisfied. Also suppose that, on every finite interval $[0,T]$ on which a strong solution $\vec{u}$ of (79) exists, the a priori bound $\|\vec{u}(t)\| \leq M_{\vec{g}}$ holds. Then (79) admits a unique classical solution defined for all $t > 0$.*

The a priori bounds on solutions are frequently obtained by means of the energy. Let us return to our example.

Example 1 (Revisited). For (82), our calculation in (90) to verify (84b) in fact shows that (93) is satisfied for all k satisfying the restrictions in (83). We therefore need only obtain an a priori bound on solutions.

As already observed, global existence does not always hold, so let us consider a specific example, the *cubic Klein-Gordon equation*, which arises in relativistic quantum mechanics:

$$(94) \quad \begin{cases} u_{tt} - \Delta u + m^2 u + \gamma u^3 = 0 & \text{in } \mathbf{R}^3 \times (0, \infty), \\ u(x,0) = g_0(x), \quad u_t(x,0) = g_1(x) & \text{for } x \in \mathbf{R}^3, \end{cases}$$

where $m > 0$ and $\gamma \geq 0$. As with the linear wave and Klein-Gordon equations, it is natural to define the *energy* of a solution of (93) to be

$$(95) \quad \mathcal{E}(t) = \int_{\mathbf{R}^3} \left(\frac{1}{2} u_t^2 + \frac{1}{2} |\nabla u|^2 + \frac{m^2}{2} u^2 + \frac{\gamma}{4} u^4 \right) dx.$$

We want to show this to be an invariant of the time evolution of (94), but we first must verify that $\mathcal{E}(t)$ is well defined and finite for our solution u; clearly this only requires verifying that $u \to \int u^4 \, dx$ is continuous for $u \in H_0^{1,2}(\mathbf{R}^n)$. To do so, we invoke the Sobolev inequality [cf. (35) of Section 6.4 with $p = r = 2$, $q = 4$, and $n = 3$]. We find $\|u\|_4 \leq C \|\nabla u\|_2^{3/4} \|u\|_2^{1/4}$, which shows that $\mathcal{E}(t)$ is indeed well defined and continuous on X.

For a strong solution of (82), we calculate

$$\frac{d\mathcal{E}}{dt} = \int_{\mathbf{R}^3} (u_t u_{tt} + \nabla u \cdot \nabla u_t + m^2 u u_t + \gamma u^3 u_t) \, dx$$

and then integrate by parts to show that $d\mathcal{E}/dt = 0$. This enables us to obtain our a priori bound:

$$(96) \quad \|\vec{u}(t)\|^2 = \int_{\mathbf{R}^3} (u_t^2 + |\nabla u|^2 + |u|^2) \, dx \leq C \mathcal{E}(t) = C \mathcal{E}(0).$$

As a consequence of Theorem 2, we conclude that the initial value problem (94) has a unique strong solution $u(x,t)$; that is, $u \in C^2([0,\infty), L^2(\mathbf{R}^3))$ and $u(t) \in H^{2,2}(\mathbf{R}^3)$ for all $t > 0$. ♣

Remark. The global existence depends strongly on the condition $\gamma \geq 0$; if $\gamma < 0$, then it may fail. Although our proof also strongly depended on the condition $m > 0$, the global existence also holds when $m = 0$; see [Reed].

Exercises for Section 12.3

1. Suppose $A: D \to X$ is a closed operator generating a quasicontractive semigroup $S(t) = e^{tA}$ on the Banach space X and $u \in C([0,T], X)$.
   (a) Suppose that $F: D \to X$ satisfies (81a,b). Show that both $F(u(s))$ and $S(t-s)F(u(s))$ are continuous functions of $s \in [0,t]$ when $t \leq T$.
   (b) Suppose that F satisfies (84a,b,c). Show that $AS(t-s)F(u(s)) = S(t-s)AF(u(s))$ is a continuous function of $s \in [0,t]$ when $t \leq T$.

2. In the proof of Theorem 1, show that (80) admits a unique solution $u \in C([0, \tau_{a,b}])$ for all $\vec{g}$ satisfying $\|\vec{g}\| \leq a$ and $\|A\vec{g}\| \leq b$.

3. (a) Use Fourier analysis on $\mathbf{R}^n$ (cf. Exercise 11 of Section 8.1) to verify (91).
   (b) Show that Example 1 satisfies (84c).

4. The semilinear equation $u_{tt} = \Delta u + \sin u$ is called the *sine-Gordon equation* (partly because of the rhyme with Klein-Gordon). Show that the initial value problem for the sine-Gordon equation has a global solution (defined for all $t > 0$).

5. Consider the nonlinear Schrödinger equation in the form $iu_t = \Delta u + f(u)$ in $\mathbf{R}^n$, where $f: \mathbf{C} \to \mathbf{C}$ is of the form $f(u) = g(|u|^2)u$ with $g: \mathbf{R}^+ \to \mathbf{R}$. Let $G(s) = \int_0^s g(\sigma)\,d\sigma$, $F(u) = -G(|u|^2)/2$, and define the energy of a solution u by
$$\mathcal{E}(t) = \int_{\mathbf{R}^n} \left(\frac{1}{2}|\nabla u|^2 + F(u) \right) dx.$$
   (a) For a solution u, show that $\|u\|_2$ is constant in t.
   (b) For a solution u, show that $\mathcal{E}$ is constant in t.

Further References for Chapter 12

For a more thorough treatment of nonlinear wave equations, the reference [Reed] treats the abstract questions of local and global existence, blow-up, and scattering theory for nonlinear Klein-Gordon and other equations (sine-Gordon, Schrödinger, etc.). A similar but more complete and up-to-date reference is [Strauss]. A specific analysis of blow-up for certain nonlinear wave equations may be found in [John, 2].

Adding a dissipation term to a nonlinear wave equation allows for the existence of attractors; see [Hale] or [Ladyzhenskaya, 3].

An important topic in the theory of nonlinear waves, which we have not had the opportunity to mention, is the subject of *solitons;* these arise in conjunction with such equations as the Korteweg de Vries equation; see [Whitham].

13 Nonlinear Elliptic Equations

In this chapter, we study the semilinear Laplace equation and other nonlinear elliptic equations in various settings. First, we use the implicit function theorem to study a nonlinear eigenvalue problem as a perturbation of the linear case; this leads to *bifurcation theory*. Then we use three different methods to investigate equations and solutions that are *not* merely perturbations of the linear case. The *method of sub- and supersolutions* (also called *monotone iteration*) is based on the elliptic maximum principle of Section 8.3 and is quite general provided that certain barrier functions can be found. A second method uses the *variational theory* of Chapter 7 to try to find a weak solution as the critical point of a nonlinear functional. Finally, the *Schauder fixed point theory* of Section 9.1 is used both for semilinear and quasilinear equations. As we shall see. of these three methods, the appropriate choice may depend on the particular equation under consideration.

13.1 Perturbations and Bifurcations

In this section, we apply the implicit function theorem to nonlinear problems that are small perturbations of linear problems for which solvability is already understood.

a. Nonlinear Eigenvalue Problems

Let us consider the nonlinear elliptic equation with parameter λ and Dirichlet boundary condition

(1)
$$\begin{cases} \Delta u + \lambda u + f(u) = 0 & \text{in } \Omega \\ u = 0 & \text{on } \partial\Omega, \end{cases}$$

where Ω is a smooth, bounded domain and f is a smooth function of the real variable u satisfying $f(0) = f'(0) = 0$. Notice that $u_\lambda \equiv 0$ is a solution of (1). Are there other (perhaps small) solutions for certain values of λ?

To invoke the implicit function theorem, let us take $X = \mathbf{R}$, $Y = H^{k+2,2}(\Omega) \cap H_0^1(\Omega)$, and $Z = H^{k,2}(\Omega)$ in Theorem 4 of Section 7.3. Consider $F: \mathbf{R} \times Y \to Z$ defined by $F(\lambda, u) = \Delta u + \lambda u + f(u)$. Notice that $F: \mathbf{R} \times Y \to Z$ is C^1, and even C^2, provided that k is sufficiently large (cf. Exercise 1). Moreover, $F(\lambda, 0) \equiv 0$, and we may easily compute $D_u F(\lambda, 0)v = \Delta v + \lambda v$ for all $v \in Y$. If we take a particular value λ_* that is *not* an eigenvalue for Δ, then the results of Section 8.4 show that $D_u F(\lambda_*, 0): Y \to Z$ is an isomorphism. The implicit function theorem provides a unique mapping $\lambda \to u_\lambda$, for λ near λ_*, such that $F(\lambda, u_\lambda) = 0$. But uniqueness implies that $u_\lambda \equiv 0$ is the only small solution of (1) for λ near λ_*.

On the other hand, if λ_* *is* an eigenvalue for Δ, then $D_u F(\lambda_*, 0): Y \to Z$ is *not* an isomorphism and there may be small nontrivial solutions u_λ for λ near λ_*. For example, let us consider the case where $\lambda_* = \lambda_1$ is the smallest eigenvalue for Δ, which we know to be a simple eigenvalue by Theorem 4 of Section 8.3. If there is a nontrivial branch of solutions (λ, u_λ) satisfying $u_\lambda \to 0$ as $\lambda \to \lambda_1$, then we may think of these nontrivial solutions as *bifurcating* at $\lambda = \lambda_1$ off the curve of trivial solutions (see Figure 1).

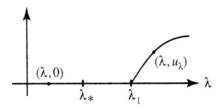

Figure 1. Bifurcation at $\lambda = \lambda_1$.

To handle the analysis near $(\lambda_1, 0)$, we introduce the following reduction method, which we present in a general setting.

b. The Method of Lyapunov-Schmidt

Suppose Y and Z are Hilbert spaces, and $F \in C^2(\mathbf{R} \times Y, Z)$ with

$$(2) \qquad F(\lambda, 0) = 0 \qquad \text{for all } \lambda \in \mathbf{R}.$$

For $\lambda = \lambda_*$, suppose that the linear operator $L = D_u F(\lambda_*, 0): Y \to Z$ is *not* invertible. We adopt the following hypothesis, which is valid for the application we have in mind:

(3) $\quad L: Y \to Z$ has finite-dimensional nullspace N_L and closed range R_L such that $R_L^\perp$ has finite dimension (i.e., R_L has finite deficiency).

If we let $k = \dim(R_L^\perp)$, then we may find $z_1, \ldots, z_k \in Z$ such that $z_1, \ldots, z_k$ form a basis for $R_L^\perp$. Let Z_k denote the linear span of $\{z_1, \ldots, z_k\}$, so that

$$(4) \qquad Z = Z_k \oplus R_L$$

is an orthogonal decomposition. Let $P: Z \to Z_k$ be the linear projection; that is, $P(\alpha_1 z_1 + \cdots \alpha_k z_k + \rho) = \alpha_1 z_1 + \cdots \alpha_k z_k$ for any $\rho \in R_L$ and $\alpha_j \in \mathbf{R}$.

Let $Q = I - P$ denote the projection $Q\colon Z \to R_L$. Notice that the equation that we are trying to solve,

(5) $$F(\lambda, u) = 0,$$

is equivalent to the two equations

(6a) $$PF(\lambda, u) = 0,$$
(6b) $$QF(\lambda, u) = 0.$$

Let us also define an orthogonal decomposition of Y by

(7) $$Y = V \oplus W, \qquad \text{where } V = N_L \text{ and } W = N_L^\perp.$$

Furthermore, let us write

(8) $$F(\lambda, u) = Lu + H(\lambda, u), \qquad \text{where } H(\lambda, u) = o(|\lambda| + \|u\|_Y).$$

For $u = v + w \in Y$, let us use $Lv = 0$ to write

(9) $$F(\lambda, v + w) = Lw + H(\lambda, v + w).$$

If we apply Q and use $Lw \in R_L$, then (6b) is equivalent to

(10) $$Lw + QH(\lambda, v + w) = 0.$$

Let $G(\lambda, v, w) = Lw + QH(\lambda, v + w)$ and observe that $G \in C^2(\mathbf{R} \times V \times W)$. Moreover, $G_w(\lambda_*, 0, 0)\colon w \to Lw + QH_u(\lambda_*, 0)w = Lw$, since $H(\lambda, u) = F(\lambda, u) - Lu$ implies $H_u(\lambda_*, 0) = F_u(\lambda_*, 0) - L = 0$. But $L\colon W \to R_L$ is injective and surjective, hence an isomorphism, so we can apply the implicit function theorem to solve (10) uniquely in a neighborhood of $(\lambda_*, 0, 0) \in \mathbf{R} \times V \times W$; that is,

(11) $$w = \gamma(\lambda, v)$$

satisfies (10). Substituting this into (6a), we obtain

(12) $$P(F(\lambda, v + \gamma(\lambda, v))) = 0.$$

This equation is called the *bifurcation equation*, because solving (12) for $(\lambda, v) \in \mathbf{R} \times V$ is equivalent to solving (5); that is, the existence of a bifurcating solution as in Figure 1. Moreover, (12) is a finite-dimensional problem, consisting of k equations in $1+\dim V$ unknowns. We only consider the special case when $k = 1 = \dim V$, so that (12) is just one equation in the two unknowns (λ, v).

c. Bifurcation from a Simple Eigenvalue

We now return to our consideration of (1) with $Y = H^{k+2,2}(\Omega) \cap H_0^{1,2}(\Omega)$, $Z = H^{k,2}(\Omega)$, and $F(\lambda, u) = \Delta u + \lambda u + f(u): \mathbf{R} \times Y \to Z$. We assume that λ_* is a simple eigenvalue for Δ on Ω; for example, we could take $\lambda_* = \lambda_1$. By Theorem 3 in Section 8.4, we know that $L = \Delta + \lambda_*$ has zero Fredholm index so the nullity and deficiency are both 1. This means that the bifurcation equation (12) is just one equation in two unknowns. In fact, let us pick $v_0 \in N_L$, $v_0 \neq 0$, so that $V = \{\mu v_0 : \mu \in \mathbf{R}\}$, and $R_L = \{f \in Z : \langle f, v_0 \rangle = 0\}$.

Now, for $\mu \neq 0$, define $G(\lambda, \mu, w) = \frac{1}{\mu} F(\lambda, \mu(v_0 + w))$, and in the limit as $\mu \to 0$ obtain $G(\lambda, 0, w) = D_u F(\lambda, 0)(v_0 + w)$. Then $G: \mathbf{R} \times \mathbf{R} \times W \to Z$ is continuous, and in fact C^1 (since F was C^2). Now we want to solve near $(\lambda_*, 0, 0)$ for (λ, w) in terms of μ, so we shall apply the implicit function theorem. Checking the hypotheses, we certainly have $G(\lambda_*, 0, 0) = L v_0 = 0$. Moreover, $D_\lambda G(\lambda_*, 0, 0) = \frac{\partial}{\partial \lambda}(D_u F(\lambda, 0) v_0)|_{\lambda = \lambda_*} = \frac{\partial}{\partial \lambda}(\Delta v_0 + \lambda v_0)|_{\lambda = \lambda_*} = v_0$, and $D_w G(\lambda_*, 0, 0) w = D_u F(\lambda_*, 0) w = L w$ for all $w \in W$. But $v_0 \perp Lw$, so $D_{(\lambda, w)} G(\lambda_*, 0, 0): \mathbf{R} \times W \to Z$ is an isomorphism. The implicit function theorem states that there is an $\epsilon > 0$ such that for $-\epsilon < \mu < \epsilon$ there is a map $\mu \to (\lambda_\mu, w_\mu)$ satisfying $G(\lambda_\mu, \mu, w_\mu) \equiv 0$. But this certainly implies for $\mu \neq 0$ that $F(\lambda_\mu, \mu(v_0 + w_\mu)) = 0$, establishing the desired curve of solutions $u_\lambda = \mu(v_0 + w_\mu)$ where $\lambda = \lambda_\mu$.

We have proved the following.

Theorem. *Suppose $f: \mathbf{R} \to \mathbf{R}$ and satisfies $f(0) = f'(0) = 0$. If λ_* is a simple eigenvalue for the Dirichlet problem for Δ in Ω, then a bifurcation of solutions for (1) occurs at λ_*.*

Bifurcation may also occur at a multiple eigenvalue; see [Ambrosetti-Prodi]. Bifurcation is important in the study of the stability of equilibrium solutions of reaction-diffusion equations; see [Sattinger] or [Smoller].

Exercises for Section 13.1

1. (a) If $f: \mathbf{R} \to \mathbf{R}$ is C^{k+2} and Ω is a smooth, bounded domain in $\mathbf{R}^n$, show that $\tilde{f}(u)(x) = f(u(x))$ defines a map $\tilde{f}: H^{k,2}(\Omega) \to H^{k,2}(\Omega)$ that is C^2 provided that $k > n/2$.
   (b) Use (a) to show that $F(\lambda, u) = \Delta u + \lambda u + f(u)$ defines a C^2-map $\mathbf{R} \times Y \to Z$, where $Y = H^{k+2,2}(\Omega) \cap H_0^{1,2}(\Omega)$ and $Z = H^{k,2}(\Omega)$.

2. Consider the nonlinear eigenvalue problem

$$(*) \qquad \Delta u + \lambda u + f(u) = 0 \qquad \text{on the torus } \mathbf{T}^n,$$

   where $f \in C^\infty$ with $f(0) = f'(0) = 0$.
   (a) Show that, for $\lambda < 0$, $u \equiv 0$ is the only small solution of $(*)$ (i.e., bifurcation does not occur).
   (b) Show that $\lambda = 0$ is a bifurcation point for $(*)$.

Additional exercises involving nonlinear eigenvalue problems will be found at the end of Section 13.2.

13.2 Method of Sub- and Supersolutions

In this section we consider a method for solving a nonlinear elliptic equation that only requires the construction of two *global barrier functions*, one called a *subsolution* and one a *supersolution,* that satisfy elliptic differential inequalities. It may not be an easy task to find these barriers, but inequalities are easier to solve than equations, and once the functions are found, then the maximum principle can be used to create a monotone sequence of functions converging to the solution. In this way, the method is relatively elementary and quite general: the only restrictions occur in the ability to solve the inequalities. Let us now describe the method in detail.

a. Barriers for Semilinear Equations

If Ω is a smooth, bounded domain in $\mathbf{R}^n$ with $n \geq 2$, let us consider the boundary value problem

$$\text{(13)} \quad \begin{cases} \Delta u = f(x, u) & \text{in } \Omega \\ u = g & \text{on } \partial\Omega, \end{cases}$$

where the regularity of $f(x, u)$ on $\overline{\Omega} \times \mathbf{R}$ and g on $\partial\Omega$ is to be specified.

A function $u_+ \in C^2(\Omega) \cap C(\overline{\Omega})$ is called a *supersolution* (or an *upper function*) for (13) if

$$\text{(14)} \quad \begin{cases} \Delta u_+ \leq f(x, u_+) & \text{in } \Omega \\ u_+ \geq g & \text{on } \partial\Omega. \end{cases}$$

Similarly, a *subsolution* (or *lower function*) for (13) is defined by reversing the inequalities in (14). We say that $u_\pm$ are *global barriers* if they also satisfy $u_+ \geq u_-$. We hope to find a solution u of (13) satisfying

$$\text{(15)} \quad u_+ \geq u \geq u_- \quad \text{in } \Omega.$$

For example, if $f(x, u) \equiv 0$ and $g \in C(\partial\Omega)$, then we know from Section 4.3 that (13) admits a unique solution $u \in C^2(\Omega) \cap C(\overline{\Omega})$; moreover, if $u_\pm$ satisfy (13), then we may use the maximum principle to conclude that (15) holds. However, the idea of this section is to use the global barriers and the maximum principle to *prove* the existence of a solution u.

In order to simplify the application of the a priori estimates of Section 8.3, we assume that our global barriers satsify $u_\pm \in C^{2,\alpha}(\overline{\Omega})$ for some $0 < \alpha < 1$. This also yields greater regularity up to the boundary for our solution u of (13). If we only assume $u_\pm \in C^2(\Omega) \cap C(\overline{\Omega})$, then the arguments must be modified by restricting the a priori estimates to compact subdomains of Ω.

In fact, if we assume that

$$\text{(16)} \quad f(x, u) \text{ is } C^\alpha \text{ in } x \in \overline{\Omega}, \text{ and } C^1 \text{ in } -\infty < u < \infty,$$

then it is not difficult to verify (see Exercise 1) that $u \in C^\alpha(\overline{\Omega})$ implies $\tilde{f}(x) \equiv f(x, u(x)) \in C^\alpha(\overline{\Omega})$. If, in addition, $u \in C^2(\Omega)$ and solves (13), then the regularity theory of Section 8.2 implies that $u \in C^{2,\alpha}(\overline{\Omega})$ as a consequence of the a priori estimate [cf. (46b) in Chapter 8]

$$(17) \qquad |v|_{2,\alpha} \le C(|\Delta v|_\alpha + |v|_{2,\alpha;\partial\Omega}),$$

where C is independent of $v \in C^{2,\alpha}(\overline{\Omega})$. For this reason, we need to assume that

$$(18) \qquad g \in C^{2,\alpha}(\partial\Omega),$$

by which we mean that g admits an extension as a $C^{2,\alpha}$-function to a neighborhood of $\partial\Omega$. Notice that the estimate (17) does not hold for $\alpha = 0$ (this is in fact the very reason to use the Hölder spaces); however, it is not difficult to prove the weaker estimate

$$(19) \qquad |v|_1 \le C(|\Delta v|_0 + |v|_{1;\partial\Omega}) \qquad \text{for all } v \in C^2(\overline{\Omega})$$

(cf. Exercise 2).

b. Monotone Iteration

Now let us assume that we have global barriers for (13); that is, $u_\pm \in C^{2,\alpha}(\overline{\Omega})$ satisfying (14) and $u_+ \ge u_-$ in Ω. Let us take $k \ge 0$ satisfying

$$(20) \qquad k \ge \max\left\{ \frac{\partial f(x,u)}{\partial u} : x \in \overline{\Omega},\ \max_{x \in \overline{\Omega}} u_+(x) \ge u \ge \min_{x \in \overline{\Omega}} u_-(x) \right\}.$$

This means that

$$(21) \qquad \begin{aligned} k(v - w) &\ge f(x, v) - f(x, w) \quad \text{whenever} \\ \max_{x \in \overline{\Omega}} u_+(x) &\ge v \ge w \ge \min_{x \in \overline{\Omega}} u_-(x). \end{aligned}$$

Let $L = -\Delta + k$. Since $k \ge 0$ the weak elliptic maximum principle of Section 8.3a shows that for $w \in C^2(\Omega) \cap C(\overline{\Omega})$

$$(22) \quad Lw \ge 0 \text{ in } \Omega \quad \text{and} \quad w \ge 0 \text{ on } \partial\Omega \quad \Rightarrow \quad w \ge 0 \text{ in } \Omega.$$

We want to inductively define a sequence $u_m \in C^{2,\alpha}(\overline{\Omega})$ that is monotonic,

$$(23) \qquad u_0 \ge u_1 \ge \cdots \ge u_- \qquad \text{in } \Omega,$$

and satisfies (for $m \ge 1$)

$$(24) \qquad \begin{cases} Lu_m = -f(x, u_{m-1}) + ku_{m-1} & \text{in } \Omega \\ u_m = g & \text{on } \partial\Omega. \end{cases}$$

Section 13.2: Method of Sub- and Supersolutions 387

Indeed, let $u_0 = u_+$, and then find the unique $u_1 \in C^{2,\alpha}(\overline{\Omega})$ satisfying (24) with $m = 1$ (cf. Theorem 5 in Section 8.2 and the Remark following Theorem 2 in Section 8.4). But $Lu_0 \equiv -\Delta u_+ + ku_+ \geq -f(x, u_+) + ku_+ = Lu_1$ in Ω, and $u_0 \equiv u_+ \geq g = u_1$ on $\partial\Omega$, imply by (22) that $u_0 \geq u_1$. We may also use (21) to conclude that $-f(x, u_+) + ku_+ \geq -f(x, u_-) + ku_-$, which implies that $Lu_1 \geq Lu_-$ in Ω; since $u_1 = g \geq u_-$ on $\partial\Omega$, we may again use (22) to conclude that $u_1 \geq u_-$ in Ω. We have verified $u_0 \geq u_1 \geq u_-$ in Ω, which is the first step in (23).

At each step of the iteration defined by (24), the argument to prove (23) is the same; namely use (24) together with (21) and the definition of the subsolution u_- to conclude that

$$\begin{cases} Lu_{m-1} \geq Lu_m \geq Lu_- & \text{in } \Omega \\ u_{m-1} \geq u_m \geq u_- & \text{on } \partial\Omega, \end{cases}$$

and then use (22) to conclude that $u_{m-1} \geq u_m \geq u_-$ in Ω.

Thus we have defined a monotonically decreasing sequence $\{u_m\}$ that is bounded below by u_-. Using (19), we find that $\{u_m\}$ is bounded in $C^1(\overline{\Omega})$; thus by the Arzela-Ascoli theorem (cf. Example 5 in Section 6.3), a subsequence converges to some function $u \in C^0(\overline{\Omega})$. But since $\{u_m\}$ is monotone, we must have $u_m \to u$ in $C^0(\overline{\Omega})$; in particular,

(25) $\qquad |u_m - u_n|_0 \to 0 \qquad \text{as } m, n \to \infty.$

To improve (25) to convergence in $C^{2,\alpha}(\overline{\Omega})$, let us write

(26) $\Delta(u_m - u_n) = f(x, u_{m-1}) + k(u_m - u_{m-1}) - f(x, u_{n-1}) - k(u_n - u_{n-1}).$

The right-hand side of (26) tends to zero in the $C^0(\overline{\Omega})$-norm as $m, n \to \infty$, so by (19)

(27) $\qquad |u_m - u_n|_1 \to 0 \qquad \text{as } m, n \to \infty.$

But if we use (16), then (27) implies

(28) $\qquad |f(x, u_m) - f(x, u_n)|_\alpha \to 0 \qquad \text{as } m, n \to \infty.$

Applying (17) to $v = u_m - u_n$ and using (26) yields

$$|u_m - u_n|_{2,\alpha} \leq C(|f(x, u_m) - f(x, u_n)|_\alpha + k|u_m - u_n|_\alpha + k|u_{m-1} - u_{n-1}|_\alpha).$$

Using $|v|_\alpha \leq C|v|_1$, (27), and (28), we conclude that $|u_m - u_n|_{2,\alpha} \to 0$, and hence $u \in C^{2,\alpha}(\overline{\Omega})$ is a solution of (13).

We may formulate this result as follows.

Theorem 1. *Suppose $f(x,u)$ satisfies (16), $g(x)$ satisfies (18), and $u_\pm \in C^{2,\alpha}(\overline{\Omega})$ are global barriers for (13). Then (13) admits a solution $u \in C^{2,\alpha}(\overline{\Omega})$ satisfying $u_+ \geq u \geq u_-$.*

c. Application with Uniformly Bounded $f(x,u)$

Theorem 1 reduces the solution of (13) to finding global barriers. One case in which we may easily construct such barriers is when $f(x,u)$ is uniformly bounded on $\overline{\Omega} \times \mathbf{R}$; that is,

(29) $\qquad |f(x,u)| \leq M \quad \text{for all } x \in \overline{\Omega} \text{ and } -\infty < u < \infty.$

Without loss of generality, we may assume that $g \equiv 0$ in (13); this is achieved by the substitution $v = u - g$, where g has been extended to $\overline{\Omega}$. Then $\Delta v = f(x, v+g) + \Delta g \equiv \tilde{f}(x,v)$, where $\tilde{f}$ is again uniformly bounded.

With $g \equiv 0$ in (13), we may take $u_\pm \in C^\infty(\overline{\Omega})$ to be the solutions of

$$\begin{cases} \Delta u = \mp M & \text{in } \Omega \\ u = 0 & \text{on } \partial\Omega. \end{cases}$$

By the maximum principle $u_+ \geq 0 \geq u_-$, so Theorem 1 implies that we have a solution $u \in C^{2,\alpha}(\overline{\Omega})$ of (13) with $g \equiv 0$. This yields the following result (for general g).

Theorem 2. *If $f(x,u)$ satisfies (16) and (29), and $g(x)$ satisfies (18), then there is a solution $u \in C^{2,\alpha}(\overline{\Omega})$ of (13).*

We cannot assert that the solution u in Theorem 2 is unique. However, if w is any other solution of (13) with $g \equiv 0$, then $\Delta u_+ = -M \leq f(x,w) = \Delta w$ implies by the maximum principle that $u_+ \geq w$; inductively we find $u_m \geq w$ for all m and hence $u \geq w$ in Ω. In other words, our monotone iteration scheme has produced the *maximal solution* of (13). [If we had constructed a monotone increasing sequence by taking $u_0 = u_-$, then the limiting solution u would be the *minimal solution* of (13).]

d. Application with $f(x,u)$, Nondecreasing in u

In order to guarantee uniqueness of the solution of (13), we require f to be nondecreasing in u:

(30) $\qquad \dfrac{\partial f(x,u)}{\partial u} \geq 0 \quad \text{for } x \in \overline{\Omega} \text{ and } -\infty < u < \infty.$

As a consequence of (30), we obtain uniqueness of solutions of (13); that is, *if u_1 and u_2 are two solutions of (13), then $u_1 \equiv u_2$*. To prove this uniqueness, we use the mean value theorem to write

(31) $\qquad f(x,u_1) - f(x,u_2) = \dfrac{\partial f}{\partial u}(x, \theta(x))\,(u_1 - u_2),$

where $\theta(x)$ is some function with values between $u_1(x)$ and $u_2(x)$. If we let $w = u_1 - u_2$, then $w(x) = 0$ for $x \in \partial\Omega$, and $\Delta w + c(x)w = 0$ for $x \in \Omega$, where $c(x) = -\partial f(x, \theta(x))/\partial u \leq 0$ in Ω. By the maximum principle $w(x) \leq 0$; that is, $u_1 \leq u_2$; interchanging u_1 and u_2 shows that $u_1 \equiv u_2$.

Another consequence of (30) is the existence of a priori bounds for solutions of (13). In fact, if we let $u_1 = u$ and $u_2 = 0$ in (31), we obtain

$$f(x, u) = \frac{\partial f}{\partial u}(x, \theta(x))u + f(x, 0),$$

so that (13) with $g \equiv 0$ may be written

$$(32) \quad \begin{cases} \Delta u + c(x)u = f(x, 0) & \text{in } \Omega \\ u = 0 & \text{on } \partial\Omega \end{cases}$$

with $c(x) = -\partial f(x, \theta(x))/\partial u \leq 0$ as before. But we may then apply Theorem 2 of Section 8.2 to conclude

$$(33) \quad \max_{x \in \overline{\Omega}} |u(x)| \leq C \max_{x \in \overline{\Omega}} |f(x, 0)| = C_0$$

holds for every solution u of (13), where C_0 depends on Ω and f but not on u.

The a priori bound (33) enables us to reduce the case (30) to the uniformly bounded case (29). Namely, let us modify the function $f(x, u)$:

$$(34) \quad f_0(x, u) = \begin{cases} f(x, u) & \text{for } |u| \leq C_0 \\ f(x, C_0) & \text{for } |u| > C_0. \end{cases}$$

A solution of (13) with $g \equiv 0$ and f satisfying (30) must satisfy the same equation with f replaced by f_0. But we may then apply Theorem 2. Since $f(x, u)$ satisfies (30) if and only if $\tilde{f}(x, u) = f(x, u + g) + \Delta g$ does, we may extend this solvability to $g \not\equiv 0$:

Theorem 3. *Suppose $f(x, u)$ satisfies (16) and (30), and g satisfies (18). Then (13) has a unique solution $u \in C^{2,\alpha}(\overline{\Omega})$.*

As an example, let us consider a semilinear equation that essentially coincides with the conformal scalar curvature equation in dimension two.

Example. Suppose $\Omega \subset \mathbf{R}^n$ is a smooth, bounded domain, and $K(x)$ is a smooth function on $\overline{\Omega}$ satisfying $-C_0 \leq K(x) \leq 0$. Consider the problem

$$(35) \quad \begin{cases} \Delta u + K(x)e^u = 0 & \text{for } x \in \Omega \\ u = g & \text{for } x \in \partial\Omega, \end{cases}$$

where g satisfies (18). Notice that $f(x, u) = -K(x)e^u$ satisfies (30), so Theorem 3 implies the existence of a unique solution $u \in C^{2,\alpha}(\overline{\Omega})$. ♣

Exercises for Section 13.2

1. Assuming that $f(x, u)$ is C^1 in $x \in \overline{\Omega}$ and $-\infty < u < \infty$, and assuming that $u \in C^\alpha(\overline{\Omega})$ for some $0 < \alpha < 1$, prove that $g(x) \equiv f(x, u(x)) \in C^\alpha(\overline{\Omega})$.

2. Use a domain potential to prove (19).

3. Consider (35) where K is allowed to become positive on a set $\Omega_+ \subset\subset \Omega$: $-C_0 \leq K(x) \leq$ for $x \in \Omega \backslash \Omega_+$ and $-C_0 \leq K(x) \leq \epsilon$ for $x \in \Omega_+$. Assuming $g \equiv 0$ and $\epsilon > 0$ is sufficiently small, show that a solution $u \in C^{2,\alpha}(\overline{\Omega})$ exists.

4. With K and Ω as in the example in this section, show that

$$\begin{cases} \Delta u + K(x)e^u = 0 & \text{for } x \in \Omega \\ u \to +\infty & \text{as } x \to \partial\Omega \end{cases}$$

admits a solution $u \in C^2(\Omega)$.

5. If $K(x)$ and $f(x)$ are smooth functions satisfying $K(x), f(x) \leq 0$ in the bounded domain $\Omega \subset \mathbf{R}^n$, show that

$$\begin{cases} \Delta u + K(x)u^2 = f(x) & \text{in } \Omega \\ u = 0 & \text{on } \partial\Omega \end{cases}$$

admits a unique smooth solution u.

6. For a smooth, bounded domain Ω, consider the nonlinear eigenvalue problem

$$\begin{cases} \Delta u + \lambda f(x, u) = 0 & \text{in } \Omega \\ u = 0 & \text{on } \partial\Omega, \end{cases}$$

where $f(x, u)$ satisfies (16), $f(x, u) \geq 0$ for all x, u, and $f(x, u) > 0$ for $u > 0$. We are interested in nonnegative solutions $u \in C^2(\Omega) \cap C(\overline{\Omega})$.
(a) If u is a nonnegative solution, then either u is positive in Ω or $u \equiv 0$.
(b) If $f(x, 0) > 0$, show that there exists $\lambda_1 > 0$ for which a positive solution u_1 exists.
(c) If $f(x, 0) > 0$ and $\lambda_1 > 0$ admits a positive solution u_1, show that there is a positive solution u_λ for every $0 < \lambda < \lambda_1$.

7. Let $\Omega \subset \mathbf{R}^n$ be a smooth, bounded domain for which the first Dirichlet eigenvalue λ_1 satsifies $0 < \lambda_1 < 1$. Show that
$$\begin{cases} \Delta u + u(1-u) = 0 & \text{in } \Omega \\ u = 0 & \text{on } \partial\Omega, \end{cases}$$
admits a solution u that is positive in Ω.

8. *Semilinear Elliptic Comparison Principle.* Suppose that
$L = \sum_{ij} a_{ij}(x) \partial^2/\partial x_i \partial x_j + \sum_k b_k(x) \partial/\partial x_k$
is uniformly elliptic in a bounded domain Ω, $f(x, u, \xi)$ is nondecreasing in $u \in \mathbf{R}$, and $u, v \in C^2(\Omega) \cap C(\overline{\Omega})$ satisfy $u \leq v$ on $\partial\Omega$.
(a) If $Lu - f(x, u, \nabla u) \geq Lv - f(x, v, \nabla v)$ in Ω, then $u \leq v$ in Ω.
(b) If $Lu - f(x, u, \nabla u) > Lv - f(x, v, \nabla v)$ in Ω, then $u < v$ in Ω.

9. For a smooth, bounded domain Ω, consider the nonlinear eigenvalue problem
$$\begin{cases} \Delta u + \lambda u - u^3 = 0 & \text{in } \Omega \\ u = 0 & \text{on } \partial\Omega. \end{cases}$$

Let λ_1 be the principal Dirichlet eigenvalue for the Laplacian on Ω. In Section 10.1, we saw that there is no nontrivial solution for $\lambda < \lambda_1$, but that λ_1 is a bifurcation point. Show that for every $\lambda > \lambda_1$, there exist at least two nontrivial solutions, one positive and one negative.

13.3 The Variational Method

In this section we consider certain cases of (13) in which the method of sub- and supersolutions fails, but for which a solution may be found as the critical point of a nonlinear functional on a Banach space. Unless the functional is convex, it may have several critical points, so we cannot expect such equations to have unique solutions in general.

a. A Semilinear Equation and Weak Solutions

Consider the boundary value problem

(36) $$\begin{cases} \Delta u + a(x) u^\sigma = 0 & \text{in } \Omega \\ u = 0 & \text{on } \partial\Omega, \end{cases}$$

where Ω is a smooth, bounded domain in $\mathbf{R}^n$ with $n \geq 3$, $\sigma > 1$ is a constant, and $a \in C^\infty(\overline{\Omega})$. This equation certainly has the trivial solution $u \equiv 0$. Does it have any other solutions?

Notice that an immediate question arises as to the meaning of the nonlinear term $a(x) u^\sigma$ when σ is not an integer: If u becomes negative, then how is u^σ defined? For this reason, we should consider instead the equation

$$(36') \qquad \begin{cases} \Delta u + a(x)|u|^{\sigma-1}u = 0 & \text{in } \Omega \\ u = 0 & \text{on } \partial\Omega. \end{cases}$$

If $a \leq 0$ in Ω, then the maximum principle shows that the trivial solution of (36') is unique (i.e., there is no nontrivial solution). We could then replace the homogeneous boundary condition by $u = g$ on $\partial\Omega$, or consider the more general equation $\Delta u + a(x)|u|^{\sigma-1}u = f(x)$; the method of sub- and supersolutions might be useful in solving such problems.

In this section, however, we consider the opposite sign condition,

$$(37) \qquad a(x) > 0 \quad \text{for all} \quad x \in \overline{\Omega}.$$

In this case, we no longer have an a priori bound on solutions, and the method of sub- and supersolutions fails. However, our experience with eigenvalue problems in Section 8.3 suggests that a nontrivial solution may exist: If $\sigma = 1$ and $a(x) \equiv \lambda_1$, then $u(x) = u_1(x)$ is a *positive* solution. This we leads us to consider the existence of nontrivial positive solutions of (36'); notice that such a solution will also solve (36).

We try to find a *weak solution* of (36'); that is, $u \in H_0^{1,2}(\Omega)$ must satisfy

$$(38) \qquad \int \nabla u \cdot \nabla v \, dx = \int a(x)|u|^{\sigma-1} u v \, dx \quad \text{for all } v \in H_0^{1,2}(\Omega).$$

Define the functional

$$(39) \qquad J(u) = \int_\Omega \left(\frac{1}{2}|\nabla u|^2 - \frac{a(x)|u|^{\sigma+1}}{\sigma+1} \right) dx \quad \text{for } u \in H_0^{1,2}(\Omega).$$

If we can show that J is a C^1 functional on $X = H_0^{1,2}(\Omega)$, for which (38) is the Euler-Lagrange equation, then critical points of J correspond to weak solutions of (36') (cf. Section 7.1). Since $F(u) = (1/2)\int |\nabla u|^2 \, dx$ was already studied in Example 1 of Section 7.1, we may focus on $\int a(x)|u|^{\sigma+1} \, dx$. We encounter the following with $q = \sigma + 1$:

(40a) $\quad \|u\|_q \leq C_q \|u\|_{1,2} \quad$ for all $u \in H_0^{1,2}(\Omega)$, if $1 \leq q \leq \dfrac{2n}{n-2}$,

(40b) $\quad H_0^{1,2}(\Omega) \subset L^q(\Omega) \quad$ is compact for $1 \leq q < \dfrac{2n}{n-2}$

(cf. Theorem 1 of Section 6.4 and Theorem 4 of Section 6.5).

Section 13.3: The Variational Method

Proposition 1. If $1 < \sigma \leq \frac{n+2}{n-2}$, then $G(u) = \int a(x)|u|^{\sigma+1}\,dx$ defines a C^1 functional on $H_0^{1,2}(\Omega)$ with derivative given by

$$G'(u)v = (\sigma+1)\int a(x)|u|^{\sigma-1} uv\,dx \qquad \text{for all } v \in H_0^{1,2}(\Omega).$$

Proof. Let $X = H_0^{1,2}(\Omega)$; by (40a), $G(u)$ is well defined and finite for $u \in X$. Now let us compute $G'(u)$:

$$G'(u)v = \lim_{\epsilon \to 0} \frac{\int_\Omega a(x)[|u+\epsilon v|^{\sigma+1} - |u|^{\sigma+1}]\,dx}{\epsilon}.$$

To evaluate this limit, let us apply Taylor's theorem with remainder to $f \in C^2[a,b]$ in order to write

(41) $$f(b) = f(a) + f'(a)(b-a) + \frac{f''(c)}{2}(b-a)^2$$

for some $c \in (a,b)$. If we let $f(t) = |t|^{\sigma+1}$, then we may calculate $f'(t) = (\sigma+1)|t|^{\sigma-1}t$ and $f''(t) = \sigma(\sigma+1)|t|^{\sigma-1}$. Substitution in (41) with $a = u$ and $b = u + \epsilon v$ yields

(42) $$|u+\epsilon v|^{\sigma+1} = |u|^{\sigma+1} + (\sigma+1)\epsilon|u|^{\sigma-1}uv + \frac{\sigma(\sigma+1)}{2}\epsilon^2 |u+\theta v|^{\sigma-1} v^2,$$

where $0 < \theta(x) < \epsilon$. Thus

$$\frac{\int_\Omega a(x)[|u+\epsilon v|^{\sigma+1} - |u|^{\sigma+1}]\,dx}{\epsilon} = (\sigma+1)\int_\Omega a(x)|u|^{\sigma-1}uv\,dx + O(\epsilon).$$

Letting $\epsilon \to 0$, we obtain our desired formula for $G'(u)$.

Next we wish to show that G is differentiable (cf. Section 7.1a). Using (42) with $\epsilon = 1$, we obtain

(43) $$G(u+v) - G(u) - G'(u)v = \frac{\sigma(\sigma+1)}{2}\int_\Omega a(x)|u+\theta v|^{\sigma-1} v^2\,dx.$$

Now the elementary inequality

(44) $$(a+b)^r \leq 2^r(a^r + b^r),$$

which holds for all $a,b,r > 0$, together with $0 < \theta(x) < 1$ imply $|u + \theta v|^{\sigma-1} \leq (|u| + \theta|v|)^{\sigma-1} \leq C(|u|^{\sigma-1} + |v|^{\sigma-1})$. We conclude that

(45) $$\int_\Omega a(x)|u+\theta v|^{\sigma-1} v^2\,dx \leq C\left(\int_\Omega |u|^{\sigma-1} v^2\,dx + \int_\Omega |v|^{\sigma+1}\,dx\right).$$

Now $2 < \sigma + 1 = q \leq \frac{2n}{n-2}$ in (40a) implies

$$(46) \qquad \int_{\Omega} |v|^{\sigma+1} \, dx < C \, \|v\|_X^{\sigma+1}.$$

Similarly, we can first use the Hölder inequality and then (40a) to obtain

$$(47) \qquad \int_{\Omega} |u|^{\sigma-1} v^2 \, dx \leq \left(\int_{\Omega} |u|^{(\sigma-1)p} \, dx \right)^{1/p} \left(\int_{\Omega} |v|^{2p'} \, dx \right)^{1/p'}$$
$$\leq C \, \|u\|_X^{\sigma-1} \|v\|_X^2;$$

this is valid provided that we can choose p and p' satisfying the three conditions

$$(48) \qquad \frac{1}{p} + \frac{1}{p'} = 1, \quad (\sigma-1)p \leq \frac{2n}{n-2}, \quad 2p' \leq \frac{2n}{n-2}.$$

It is straightforward to verify that (48) can be achieved provided that $\sigma \leq \frac{n+2}{n-2}$. Combining (43) through (47), we obtain

$$G(u+v) - G(u) - G'(u)v \leq C \left(\|u\|_X^{\sigma-1} + \|v\|_X^{\sigma-1} \right) \|v\|_X^2 = o(\|v\|_X)$$

as $\|v\|_X \to 0$, proving that G is indeed differentiable.

A somewhat similar calculation shows that G' is in fact continuous (Exercise 1), so G is C^1. ♠

We conclude that critical points of the functional J in (39) are indeed weak solutions of (36′); it remains to show that such critical points exist. Here we encounter a difficulty in that J does not have a global minimum nor a global maximum. In fact, if we choose smooth "bump" functions u_k with shrinking support but increasing maximum point, we can arrange $\|u_k\|_{\sigma+1} = 1$ and $\|\nabla u\|_2 \to \infty$, so $J(u_k) \to +\infty$; on the other hand if we fix $v \in X$ and let $u_k = kv$, then $\sigma + 1 > 2$ implies $J(u_k) \to -\infty$. Thus we must either look for local extrema or saddle points, or use a constrained minimization procedure somewhat analogous to that used for the eigenvalue problems of Section 7.2. We use both methods to prove the following result.

Theorem 1. *If $1 < \sigma < \frac{n+2}{n-2}$, then (36′) admits a nontrivial weak solution $u \in H_0^{1,2}(\Omega)$.*

In fact, we show that the weak solution is positive and C^2:

Theorem 2. *If $1 < \sigma < \frac{n+2}{n-2}$, then (36′) admits a positive solution $u \in C^2(\overline{\Omega})$, which is therefore a solution of (36).*

Incidentally, the restriction $\sigma < \frac{n+2}{n-2}$ in these theorems is not just an artifice of the variational method used to solve (36). In fact, for $\sigma \geq \frac{n+2}{n-2}$, a result of Pohozaev shows that, when Ω is a ball in $\mathbf{R}^n$, the only solution of (36) is $u \equiv 0$; this is outlined in Exercise 2. The case $\sigma = (n+2)/(n-2)$ is often called the critical case since it leads to $q = 2n/(n-2)$ in the Sobolev imbedding, so we have (40a) but *not* the compactness (40b). This is exactly the case that arises in differential geometry in connection with the conformal scalar curvature equation for $n \geq 3$.

b. Application of Lagrange Multipliers

We apply the method of Lagrange multipliers to prove Theorem 1. Let us define the following functionals on $u \in X = H_0^{1,2}(\Omega)$,

(49a) $$F(u) = \frac{1}{2} \int_\Omega |\nabla u|^2 \, dx \quad \text{and}$$

(49b) $$G(u) = \int_\Omega a(x)|u|^{\sigma+1} \, dx - 1,$$

and try to minimize $F(u)$ on the constraint set $\mathcal{C}$ defined by

(50) $$\mathcal{C} = \{u \in X : G(u) = 0\}.$$

Proposition 2. *If $1 < \sigma < \frac{n+2}{n-2}$, then F attains its minimum on $\mathcal{C}$.*

Proof. Let $I = \inf\{F(u) : u \in \mathcal{C}\} \geq 0$ and let $u_j \in \mathcal{C}$ be a minimizing sequence: $F(u_j) \leq I+1$ and $F(u_j) \to I$. With $q = 2$ in (40b), we conclude that we may find a subsequence (which we continue to denote $\{u_j\}$) and a nonnegative $u_0 \in X = H_0^{1,2}(\Omega)$ such that $u_j \to u_0$ in $L^2(\Omega)$ and $u_j \rightharpoonup u_0$ weakly in X. As in Section 7.1b, the weak lower semicontinuity of F implies $F(u_0) = I$. With $q = \sigma+1$ we may also use (40b) to conclude that a subsequence (again denoted $\{u_j\}$) satisfies $u_j \to u_0$ in $L^q(\Omega)$. In particular, $G(u_j) \to G(u_0)$; but $G(u_j) \equiv 0$ implies $G(u_0) = 0$; that is, $u_0 \in \mathcal{C}$. In other words, F attains its minimum on $\mathcal{C}$ at $u_0 \in \mathcal{C}$. ♠

Now let us calculate the Euler-Lagrange equation:

$$F'(u_0)v = \int_\Omega \nabla u_0 \cdot \nabla v \, dx = \mu(\sigma+1) \int_\Omega a(x)|u_0|^{\sigma-1} u_0 v \, dx = \mu G'(u_0)v$$

for all $v \in H_0^{1,2}(\Omega)$. If we let $v = u_0$, we obtain

$$\int_\Omega |\nabla u_0|^2 \, dx = \mu(\sigma+1) \int_\Omega a(x)|u_0|^{\sigma+1} \, dx,$$

which shows that the Lagrange multiplier μ is positive (if $\mu = 0$, then $|\nabla u_0| \equiv 0$, and hence $u_0 \equiv 0$, contradicting $u_0 \in \mathcal{C}$). In other words, u_0 is a nonnegative weak solution of

$$\Delta u_0 + \lambda a(x)|u_0|^{\sigma-1} u_0 = 0, \quad \text{where } \lambda = \mu(\sigma+1).$$

Finally, we may let $u = \tau u_0$, where $\tau = \lambda^{1/(\sigma-1)} = (\mu(\sigma+1))^{1/(\sigma-1)}$ to obtain a weak solution of (36′). This proves Theorem 1.

c. Application of the Mountain Pass Theorem

We now give an alternative proof of Theorem 1 by showing that J in (39) has a saddle-type critical point on $X = H_0^{1,2}(\Omega)$. Throughout this discussion, we use the Poincaré inequality (Theorem 1 in Section 6.2) to consider $\|\nabla u\|_2$ as the norm on X. Moreover, as a consequence of (40a),

$$(51) \quad \int a(x)|u(x)|^{\sigma+1}\,dx \leq C_{a,\sigma}\|\nabla u\|_2^{\sigma+1}.$$

In order to apply Theorem 2 of Section 7.1 to the functional J in (39), we must verify its hypotheses. Clearly, (i) $J(0) = 0$. Next let us show that (ii) $J|_{\partial B_\rho(0)} \geq \alpha$ for some $\rho, \alpha > 0$. For all u satisfying $\|\nabla u\|_2 = \rho$, (51) implies that

$$J(u) = \frac{\|\nabla u\|_2^2}{2} - \frac{1}{\sigma+1}\int a(x)|u|^{\sigma+1}\,dx \geq \frac{\rho^2}{2} - \frac{C_{\sigma,a}}{\sigma+1}\rho^{\sigma+1} \geq \alpha > 0,$$

provided that we take ρ sufficiently small; this verifies (ii). We verify (iii) $J(\overline{u}) < 0$ for some $\overline{u} \in X \setminus B_\rho(0)$ by taking $\overline{u} = kv$ with k sufficiently large.

Finally, we must verify (iv) the Palais-Smale condition. Suppose that $\{u_m\} \in X$ is a Palais-Smale sequence; that is,

(52a) $\qquad\qquad J(u_m)$ is bounded, and

(52b) $\qquad\qquad J'(u_m) \to 0$ in X^*.

We must show that $\{u_m\}$ contains a convergent subsequence. Now (52a) states $J(u_m) \leq C < \infty$, which implies that

$$(53) \quad \frac{1}{2}\|\nabla u_m\|_2^2 \leq C + \frac{1}{\sigma+1}\int a(x)|u_m|^{\sigma+1}\,dx,$$

whereas (52b) implies that, for any $\epsilon > 0$,

$$(54) \quad \left|\int \nabla u_m \nabla \phi\,dx - \int a(x)|u_m|^{\sigma-1}u_m\phi\,dx\right| \leq \epsilon\|\nabla \phi\|_2 \quad \text{for all } \phi \in X,$$

provided that m is sufficiently large. But if we take $\epsilon = 1$ and $\phi = u_m$ in (54), then for m sufficiently large we have

$$\int a(x)|u_m|^{\sigma+1}\,dx \leq \|\nabla u_m\|_2^2 + \|\nabla u_m\|_2,$$

which, combined with (53), yields

$$(55) \quad \frac{1}{2}\|\nabla u_m\|_2^2 \leq C + \frac{1}{\sigma+1}\left(\|\nabla u_m\|_2^2 + \|\nabla u_m\|_2\right).$$

Section 13.3: The Variational Method   397

But $\sigma+1 > 2$ together with the inequality $\|\nabla u_m\|_2 \le \epsilon \|\nabla u_m\|_2^2 + C_\epsilon (\text{vol}\,\Omega)^2$ enables us to conclude from (55) that $\{u_m\}$ is bounded in X.

At this point, we know that (50) implies that $\{u_m\}$ is bounded in X; also $J'(u) = L + K(u)$, where $L = -\Delta\colon X \to X^*$ is an isomorphism (by Theorem 1 of Section 6.3), and $K(u)v = -\int a(x)|u|^{\sigma-1} uv\,dx$. But with $p = \sigma+1$ and $p' = 1+(1/\sigma)$, $u \to a|u|^{\sigma-1}u$ is bounded $L^p(\Omega) \to L^{p'}(\Omega)$, and $L^{p'}(\Omega) \subset X^*$ is compact by (40b); from this it is not difficult to conclude that J is Palais-Smale (cf. Exercise 11 in Section 7.1). We conclude from Theorem 2 of Section 7.1 that J has a critical point in X, which proves Theorem 1.

d. Regularity and Positivity.

The proof of Theorem 2 proceeds in three steps: (i) use a bootstrap argument to show that any weak solution of (36') must in fact be in $C^2(\overline{\Omega})$, (ii) modify the variational argument in order to produce a *nonnegative* weak solution [which by (i) is in fact C^2], and (iii) use a maximum principle argument to show that the nonnegative solution is in fact positive.

Step (i) uses the Sobolev imbedding theorems of Chapter 6 and the elliptic estimates of Chapter 8 to improve gradually the regularity of a weak solution $u \in H_0^{1,2}(\Omega)$. In particular, $H_0^{1,2}(\Omega) \subset L^{q_1}(\Omega)$ for $q_1 = 2n/(n-2)$ implies that $a(x)|u|^{\sigma-1}u \in L^{p_1}(\Omega)$, where $p_1 = q_1/\sigma$, and then Theorem 5 of Section 8.2 implies that $u \in H^{2,p_1}(\Omega)$. If $2p_1 > n$, then Theorem 6 of Section 6.5 implies that $H^{2,p_1}(\Omega) \subset C^\alpha(\overline{\Omega})$ for some $\alpha > 0$, and hence $a(x)|u|^{\sigma-1}u \in C^\beta(\overline{\Omega})$ for some $0 < \beta \le \alpha$; then Theorem 5 of Section 8.2 shows that $u \in C^{2,\beta}(\overline{\Omega}) \subset C^2(\overline{\Omega})$ (i.e., u is a classical solution). If, on the other hand, $2p_1 < n$, then Theorem 6 of Section 6.5 only yields $H^{2,p_1}(\Omega) \subset L^{q_2}(\Omega)$ where $q_2 = np_1/(n-2p_1)$; however, it is easy to verify that $q_2 > q_1$, so this process has produced some improvement in the regularity of u, and so we can repeat the process until we eventually get $2p_k > n$ and we appeal to Theorem 5 of Section 8.2 to conclude $u \in C^{2,\beta}(\overline{\Omega}) \subset C^2(\overline{\Omega})$. (If $2p_1 = n$, then a slight modification is necessary; cf. Exercise 3.)

For step (ii), the Lagrange multiplier argument may be modified easily by observing that the functionals F and G in (49) are both invariant if u is replaced by $|u|$; thus the minimizing sequence may be chosen with $u_j \ge 0$, and since the limit u_0 is continuous, u_0 must also be nonnegative. On the other hand, for the mountain pass argument, simply replace the functional J by

$$\tilde{J}(u) = \int_\Omega \frac{1}{2}|\nabla u|^2 - a(x,u)\,dx,$$

where

$$a(x,u) = \begin{cases} (\sigma+1)^{-1} a(x)|u|^{\sigma+1} & \text{if } u \ge 0 \\ 0 & \text{if } u < 0. \end{cases}$$

The mountain pass argument still applies to produce a critical point u. If u is negative somewhere, let $U = \{x \in \Omega : u(x) < 0\}$. Then, by the definition of $a(x,u)$, $\Delta u = 0$ in U with $u = 0$ on ∂U. But the maximum principle then implies $u \equiv 0$ in U; that is, $U = \emptyset$.

Step (iii) is then an immediate consequence of the strong maximum principle (cf. Theorem 3 of Section 8.3).

Exercises for Section 13.3

1. Show that G as in Theorem 1 is C^1 on $X = H_0^{1,2}(\Omega)$; that is, show that $G' : X \to X^*$ is a continuous mapping by showing that

$$\frac{|(G'(u+w) - G'(u))v|}{\|v\|_X} = o(\|w\|_X), \quad \text{as } \|w\|_X \to 0.$$

2. This exercise shows that the restriction $1 < \sigma < (n+2)/(n-2)$ is necessary for the existence of positive solutions of (36) by considering

$$(*) \quad \begin{cases} \Delta u + f(u) = 0 & \text{in } \Omega \\ u = 0 & \text{on } \partial \Omega, \end{cases}$$

where $f(u)$ is continuous in $u \in \mathbf{R}$.
(a) If $F(u) = \int_0^u f(t)\, dt$, use integration by parts to show

$$n \int_\Omega F(u)\, dx + \int_\Omega f(u) \sum_{i=1}^n x_i \frac{\partial u}{\partial x_i}\, dx = 0,$$

for any $u \in C^1(\overline{\Omega})$ with $u = 0$ on $\partial\Omega$.
(b) If $u \in C^2(\Omega) \cap C(\overline{\Omega})$ satisfies (*), then prove *Pohozaev's identity*

$$\frac{n-2}{2} \int_\Omega |\nabla u|^2\, dx - n \int_\Omega F(u)\, dx + \frac{1}{2} \int_{\partial\Omega} \left(\frac{\partial u}{\partial \nu}\right)^2 (x \cdot \nu)\, ds = 0,$$

where ν is the exterior unit normal.
(c) When Ω is a ball in $\mathbf{R}^n$, show that (*) admits no positive solution $u \in C^2(\Omega) \cap C(\overline{\Omega})$ when $f(u) = u^\sigma$ with $\sigma \geq (n+2)/(n-2)$.

3. Suppose $\Omega \subset \mathbf{R}^3$ is an arbitrary domain, and $u \in H_0^{1,2}(\Omega)$ is a weak solution of $\Delta u + a(x)u^4 = 0$ in Ω, where $a \in C^\infty(\Omega)$. Show that $u \in C^\infty(\Omega)$.

4. Suppose Ω is a smooth, bounded domain in $\mathbf{R}^2$, and $a \in C^\infty(\overline{\Omega})$. For any $\sigma > 1$, show that (36) admits a positive solution $u \in C^\infty(\overline{\Omega})$.

5. If $a(x)$ is a smooth function that is positive in the bounded domain Ω, and $1 < q < (n+2)/(n-2)$, consider the equation

$$\begin{cases} \Delta u + a(x)u^q = \mu u & \text{in } \Omega \\ u = 0 & \text{on } \partial\Omega. \end{cases}$$

a) Show that there is a positive solution u provided that $\mu > -\lambda_1$, where λ_1 is the first Dirichlet eigenvalue for Δ on Ω.

b) If $\mu = -\lambda_1$, show that there can be *no* weak solution u that is positive in Ω.

13.4 Fixed Point Methods

In this section we explore the use of the Schauder fixed point theorem of Section 9.1 to solve semilinear and quasilinear elliptic equations. This method does not require the construction of global barriers but generally requires special properties of the equation to guarantee the existence of a priori estimates. We first illustrate the method for semilinear equations.

a. Semilinear Equations

Let us consider (13) again, where $f(x,u)$ satisfies (16) and g satisfies (18), for fixed $0 < \alpha < 1$. Let $X = C^\alpha(\overline{\Omega})$. For $u \in X$ we saw (in Exercise 1 of Section 13.2) that $\tilde{f}(x) \equiv f(x,u(x)) \in C^\alpha(\overline{\Omega})$, so the existence and regularity theory for the Laplacian (cf. Theorem 2 in Section 6.5 and the corollary to Theorem 5 in Section 8.2) shows that

$$(56) \quad \begin{cases} \Delta v = \tilde{f}(x) & \text{in } \Omega \\ v = g & \text{on } \partial\Omega \end{cases}$$

admits a unique solution $v \in C^{2,\alpha}(\overline{\Omega})$. Thus, if we let $v = Tu$, we have defined a map $T: X \to X$ that is continuous: If $u_j \to u$ in X, then $\tilde{f}_j(x) = f(x, u_j(x)) \to \tilde{f}(x) = f(x, u(x))$ in X and by (17) we have $|Tu_j - Tu|_\alpha \le C|Tu_j - Tu|_{2,\alpha} \le C|\tilde{f}_j - \tilde{f}|_\alpha$, so $Tu_j \to Tu$ in X.

Now suppose we can find a fixed point $u \in X$ of T (i.e., $u = Tu$). Then $u \in C^{2,\alpha}(\overline{\Omega})$ and by (56) we find that u solves (13). To find a fixed point of T, we may try to use one of the fixed point theorems of Section 9.1. Let us consider first the example that was treated in Section 13.2 by the method of sub- and superfunctions.

Example 1 (Uniformly Bounded f(x, u)). Suppose that the function $f(x,u)$ satisfies (29). Then we may use $C^1(\overline{\Omega}) \subset C^\alpha(\overline{\Omega})$ and (19) to conclude that every solution v of (56) satisfies

$$(57) \quad |v|_\alpha \le C_1|v|_1 \le C_2(|\Delta v|_0 + |g|_{1;\partial\Omega}) = C_2(|f(x,u)|_0 + |g|_{1;\partial\Omega}) \le C^*.$$

Therefore, let us define $A = \{u \in X : |u|_\alpha \le C^*\}$, where C^* is given in (57). A is clearly closed and convex in X, and $T: A \to A$, with $T(A) \subset B$, where $B = \{u \in C^1(\overline{\Omega}) : |u|_1 \le C^* C_1^{-1}\}$. But B is closed and bounded in $C^1(\overline{\Omega})$, and $C^1(\overline{\Omega}) \subset C^\alpha(\overline{\Omega})$ is compact by the Arzela-Ascoli theorem [cf. (70) in Section 6.5], so B is a compact subset of X. Thus $\overline{T(A)}$ is also compact in X. Therefore, we may apply Theorem 3 of Section 9.1 to conclude that T

has a fixed point $u \in A$; that is, u solves (13). We have therefore produced an alternative proof of Theorem 2 of Section 13.2. ♣

Example 2 (f(x, u) Nondecreasing in u). Now let us replace the uniform boundedness of $f(x, u)$ by the property (30) that f be nondecreasing in u. Recall that (30) implies the uniqueness of solutions and a priori bounds for solutions of (13). As in the method of sub- and supersolutions, we may now reduce this case to the previous example. We have therefore produced an alternative proof of Theorem 3 of Section 13.2. ♣

b. Quasilinear Equations

Let us consider the Dirichlet problem

(58a)
$$\begin{cases} Qu = f(x, u, \nabla u) & \text{in } \Omega \\ u = g & \text{on } \partial\Omega, \end{cases}$$

where Qu denotes the quasilinear operator

(58b)
$$Qu = \sum_{i,j=1}^{n} a_{ij}(x, u, \nabla u)\partial_i \partial_j u, \quad \text{with } \partial_i = \partial/\partial x_i.$$

We assume that Q is uniformly elliptic in Ω; that is,

(58c)
$$\sum_{i,j=1}^{n} a_{ij}(x, u, \zeta)\xi_i \xi_j \geq \epsilon |\xi|^2 \quad \text{for all } x \in \overline{\Omega},\ u \in \mathbf{R},\ \xi, \zeta \in \mathbf{R}^n,$$

and the coefficients $a_{ij}(x, u, \zeta)$ and $f(x, u, \zeta)$ are all smooth with respect to $u \in \mathbf{R}$ and $\zeta \in \mathbf{R}^n$. Moreover, for some $0 < \alpha < 1$, we assume that

(58d) $\qquad a_{ij}, f \in C^\alpha$ with respect to $x \in \overline{\Omega}$, and $g \in C^{2,\alpha}(\partial\Omega)$.

Let $X = C^{1,\alpha}(\overline{\Omega})$. Given $u \in X$, let us define a function $\tilde{f} = \tilde{f}_u \in C^\alpha(\overline{\Omega})$ and a linear operator $L = L_u$ by

(59a)
$$\tilde{f}(x) = f(x, u, \nabla u) \quad \text{and} \quad L = \sum_{i,j=1}^{n} a_{ij}(x, u, \nabla u)\partial_i \partial_j.$$

Using the results of Section 8.4 (especially the Remark after Theorem 2), there is a unique solution $v \in C^{2,\alpha}(\overline{\Omega})$ of the linear equation

(59b)
$$\begin{cases} Lv = \tilde{f} & \text{in } \Omega \\ v = g & \text{on } \partial\Omega. \end{cases}$$

As in the semilinear case, this defines a continuous map $T: X \to X$. If we could show that T maps the ball $A = \{u \in X : |u|_{1,\alpha} \leq C_1\}$ into itself, then we might hope to use Schauder fixed point theory to conclude the existence of a fixed point $u \in X$ of T. Such a fixed point will actually lie in $C^{2,\alpha}(\overline{\Omega})$ and will be a solution of (58a).

Instead, we prepare to use the Leray-Schauder fixed point theorem by considering the one-parameter family of equations

(60) $$\begin{cases} Qu = tf(x, u, \nabla u) & \text{in } \Omega \\ u = tg & \text{on } \partial\Omega, \end{cases}$$

for $0 \leq t \leq 1$. The following result reduces the solvability of (58a,b,c,d) to the existence of a priori bounds on solutions of (60).

Theorem 1. *If Ω is a smooth, bounded domain, then (58a,b,c,d) admits a solution $u \in C^{2,\alpha}(\overline{\Omega})$ provided that every solution $u \in C^{2,\alpha}(\overline{\Omega})$ of $(60)_t$ satisfies*

(61) $|u|_{1,\alpha} \leq C$,   *where C is independent of $t \in [0,1]$ and u.*

Proof. As observed previously, (17) shows that $T: X \to X$ is continuous, in fact with $T(X) \subset C^{2,\alpha}(\overline{\Omega})$. The compactness of the imbedding $C^{2,\alpha}(\overline{\Omega}) \subset C^{1,\alpha}(\overline{\Omega}) = X$ shows that $T: X \to X$ is a compact map on X, so the theorem follows immediately from Theorem 4 of Section 9.1. ♠

For each t, (60) is of the form (58), so we focus on obtaining a priori bounds for solutions of (58). In particular, this requires establishing an a priori bound on ∇u on $\partial\Omega$; that is, a *boundary gradient estimate*. This is usually achieved by means of local barriers, which we shall describe next.

c. Local Barriers and Boundary Gradient Estimates

In order to obtain the a priori bound (61) for solutions of (58), as desired for Theorem 1, we must bound both $|u(x)|$ and $|\nabla u(x)|$ in Ω. The idea is to use the maximum principle to extend bounds obtained on the boundary of Ω into its interior. An important tool for doing so is a *comparison principle*, which we have encountered for linear and semilinear elliptic equations (cf. Exercise 2 in Section 8.3 and Exercise 8 in Section 13.2), as well as semilinear parabolic equations (cf. Section 11.1). For this purpose, we now add the following hypotheses on (58a):

(62a)   the $a_{ij} = a_{ij}(x, \zeta)$ in (58c) are independent of u;
(62b) $f(x, u, \zeta)$ is nondecreasing in $u \in \mathbf{R}$ for each $\zeta \in \mathbf{R}^n$.

Moreover, it is convenient to define

(63) $$Q_1 u = Qu - f(x, u, \nabla u).$$

Proposition 1 (Comparison Principle). *Assume that Ω is a bounded domain, and Q and f satisfy (58b,c,d) and (62a,b). Let $u, v \in C^2(\Omega) \cap C(\overline{\Omega})$ satisfy $u \leq v$ on $\partial \Omega$.*
 (i) *If $Q_1 u \geq Q_1 v$ in Ω, then $u \leq v$ in Ω.*
 (ii) *If $Q_1 u > Q_1 v$ in Ω, then $u < v$ in Ω.*

Proof. To prove (i), we write

$$
\begin{aligned}
Qu - Qv &= \sum_{i,j} a_{ij}(x, \nabla u)\partial_i\partial_j u - \sum_{i,j} a_{ij}(x, \nabla v)\partial_i\partial_j v \\
&= \sum_{i,j} a_{ij}(x, \nabla u)\partial_i\partial_j (u-v) + \sum_{i,j} (a_{ij}(x, \nabla u) - a_{ij}(x, \nabla v))\partial_i\partial_j v \\
&\geq f(x, u, \nabla u) - f(x, v, \nabla v) \\
&= f(x, u, \nabla u) - f(x, u, \nabla v) + f(x, u, \nabla v) - f(x, v, \nabla v).
\end{aligned}
\tag{64}
$$

Use the mean value theorem to write

$$
\begin{aligned}
f(x, u, \nabla u) - f(x, u, \nabla v) &= \nabla_\zeta f(x, u, \theta(x)) \cdot (\nabla u - \nabla v), \text{ and} \\
a_{ij}(x, \nabla u) - a_{ij}(x, \nabla v) &= \nabla_\zeta a_{ij}(x, \theta_{ij}(x)) \cdot (\nabla u - \nabla v),
\end{aligned}
\tag{65}
$$

where $\theta(x)$ and $\theta_{ij}(x)$ are functions with values between $\nabla u(x)$ and $\nabla v(x)$. If we let $w = u - v$ and $\Omega^+ = \{x \in \Omega^+ : w(x) > 0\}$, then (62b) implies $f(x, u, \nabla v) - f(x, v, \nabla v) \geq 0$ in Ω^+, and so (64) and (65) combine to yield

$$
\begin{cases}
\sum_{i,j} \tilde{a}_{ij}(x)\partial_i\partial_j w + \sum_k b_k(x)\partial_k w \geq 0 & \text{in } \Omega^+ \\
w \leq 0 & \text{on } \partial\Omega^+,
\end{cases}
\tag{66}
$$

where $\tilde{a}_{ij}(x) = a_{ij}(x, \nabla u)$ and $b_k(x)$ involves the $\partial f/\partial \zeta$, $\partial a_{ij}/\partial \zeta$, and $\partial_i\partial_j v$. By the weak maximum principle (Theorem 1 in Section 8.3), $\Omega^+ = \emptyset$, proving that $u \leq v$ in Ω. The same argument using the strong maximum principle (Theorem 3 in Section 8.3) proves (ii). ♠

With this proposition in mind, we make the following definition.

Definition. For $x_0 \in \partial\Omega$, we say that $w^\pm$ are *local barriers* at x_0 if, for some neighborhood U of x_0, $w^\pm \in C^2(\Omega) \cap C^1(U \cap \overline{\Omega})$ satisfy
 (i) $\pm Q_1 w^\pm < 0$ in $U \cap \Omega$;
 (ii) $w^\pm(x_0) = u(x_0)$; and
 (iii) $w^-(x) \leq u(x) \leq w^+(x)$ for $x \in \partial(U \cap \Omega)$.

To see how local barriers yield a boundary gradient estimate, let us assume that Q and f are as in Proposition 1, so that (i), (ii), and (iii) imply that

$$
w^-(x) - w^-(x_0) \leq u(x) - u(x_0) \leq w^+(x) - w^+(x_0) \quad \text{for all } x \in U \cap \Omega.
\tag{67}
$$

Assuming that $\partial\Omega$ is smooth, we can let ν denote the exterior unit normal and conclude from (67) that

$$(68) \qquad \frac{\partial w^-}{\partial \nu}(x_0) \leq \frac{\partial u}{\partial \nu}(x_0) \leq \frac{\partial w^+}{\partial \nu}(x_0).$$

This means that $|\partial u/\partial \nu|$ is bounded at x_0 by $\max\{|\partial w^+/\partial \nu|, |\partial w^-/\partial \nu|\}$. If local barriers can be found at each $x_0 \in \partial\Omega$ with $|\partial w^\pm/\partial \nu|$ bounded independently of $x_0 \in \partial\Omega$, then we have a global bound on $|\partial u/\partial \nu|$ on $\partial\Omega$. Since tangential boundary derivatives of $u = g$ are bounded by $|g|_{1;\partial\Omega}$, we get the following result.

Theorem 2. *Suppose $u \in C^2(\Omega) \cap C^1(\overline{\Omega})$ is a solution of (58a,b,c,d), where (62) is satisfied. Suppose also that at each point $x_0 \in \partial\Omega$, there exist local barriers $w^\pm$ with $|\partial w^\pm/\partial \nu| \leq M$. Then there exists a constant C, depending on M and $|g|_{1;\partial\Omega}$, so that we have the following boundary gradient bound:*

$$(69) \qquad |\nabla u(x)| \leq C \qquad \text{for all } x \in \partial\Omega.$$

d. Hölder Estimates of De Giorgi and Nash

In order to improve (69) to an estimate in the norms $|\cdot|_{1,\alpha}$, we need to appeal to some important estimates for uniformly elliptic linear equations which, for simplicity, we assume to be in divergence form and homogeneous:

$$(70) \qquad Lu = \sum_{i,j=1}^n \partial_i(a_{ij}(x)\partial_j u) = 0 \qquad \text{in } \Omega.$$

If $a_{ij} \in C^{1,\alpha}(\Omega)$ and $u \in C^2(\Omega)$ solves (70), then the linear elliptic estimates of Section 8.2 show $u \in C^{2,\alpha}(\Omega)$. But suppose we assume less regularity of a_{ij}, and only that u is a *weak* solution of (70); that is,

$$(71a) \qquad \int_\Omega \sum_{i,j=1}^n a_{ij}(x)\, \partial_j u\, \partial_i \phi\, dx = 0 \qquad \text{for all } \phi \in C_0^1(\Omega).$$

Can we still conclude that u is continuous? Hölder continuous? We need to assume that the $a_{ij}(x)$ are measurable, uniformly bounded, and uniformly elliptic in Ω; that is,

$$(71b) \qquad \lambda|\xi|^2 \leq \sum_{i,j=1}^n a_{ij}(x)\xi_i \xi_j \leq \Lambda|\xi|^2 \qquad \text{for all } x \in \Omega \text{ and } \xi \in \mathbf{R}^n.$$

The following result is a generalization of the Harnack inequality of Section 4.2 (Theorem 6); however, its proof is quite technical, so we not reproduce it here; see [Gilbarg-Trudinger] or [Ladyzhenskaya-Ural'tseva].

Proposition 2 (Weak Harnack Inequality). Suppose $u \in H^{1,2}(\Omega)$ is a nonnegative solution (71a,b). Suppose also that $y \in \Omega$ and $R > 0$ satisfy $B_{2R}(y) \subset \Omega$. Then for every $0 < \alpha < 1$, there is a constant C depending on R, α, λ, and Λ, but independent of u, such that

(72) $$\sup_{x \in B_{\alpha R}(y)} u(x) \leq C \inf_{x \in B_{\alpha R}(y)} u(x).$$

As a consequence of (72), we now show that solutions $u \in H^{1,2}(\Omega)$ of (71a,b) are Hölder continuous.

Corollary. Suppose $u \in H^{1,2}(\Omega)$ is a weak solution of (71a,b); then $u \in C^\alpha(\Omega)$ for some $0 < \alpha < 1$. Moreover, for $\overline{\Omega'} \subset \Omega$, there is a constant C, depending on $\operatorname{dist}(\overline{\Omega'}, \partial\Omega)$ but independent of u, such that

(73) $$|u|_{\alpha;\overline{\Omega'}} \leq C \|u\|_{\infty;\Omega}.$$

Proof. Let $y \in \Omega$, and choose $R > 0$ so that $B_{2R}(y) \subset \Omega$. Let us define the *oscillation*, $\omega(R)$, of the solution u in $B_R(y)$ by

$$\omega(R) = M(R) - m(R), \text{ where}$$
$$M(R) = \sup\{u(x) : x \in B_R(y)\} \text{ and}$$
$$m(R) = \inf\{u(x) : x \in B_R(y)\}.$$

If we define $v(x) = u(x) - m(R)$ and $w(x) = M(R) - u(x)$, then these are nonnegative solutions of (71a) in $B_R(y)$. Let us take $\alpha = 1/2$ in (72), and let $c = C(\alpha)$, to conclude that $M(R/2) - m(R) \leq c(m(R/2) - m(R))$ and $M(R) - m(R/2) \leq c(M(R) - M(R/2))$. Adding these two inequalities, we find $\omega(R) + \omega(R/2) \leq c(\omega(R) - \omega(R/2))$, or

$$\omega(R/2) \leq \frac{c-1}{c+1}\omega(R).$$

If we let $\beta = (c-1)/(c+1)$, then by induction we have

$$\omega(2^{-k}R) \leq \beta^k \omega(R), \quad \text{where } \beta \in (0,1).$$

Now $\omega(r)$ is increasing in r, so $R\,2^{-(k+1)} \leq r \leq R\,2^{-k}$ implies that $\omega(r) \leq \beta^k \omega(R)$. But we then have

$$\beta^{k+1} \leq \left(\frac{r}{R}\right)^{-\log\beta/\log 2} \leq \beta^k.$$

This implies that

(74) $$\omega(r) \leq \beta^{-1}(r/R)^{-\log\beta/\log 2}\omega(R).$$

Since $y \in \Omega$ was arbitrary, this estimate shows that $u \in C^\alpha(\Omega)$ with $\alpha = -\log\beta/\log 2$. The estimate (73) easily follows from (74). ♠

As might be expected, we cannot take $\Omega' = \Omega$ in the estimate (73) without additional regularity assumptions on $\partial\Omega$. For a proof of the following, see [Gilbarg-Trudinger] or [Ladyzhenskaya-Ural'tseva].

Theorem 3 (Hölder Regularity of Weak Solutions). *Suppose $\partial\Omega \in C^1$ and $u \in H^{1,2}(\Omega)$ is a solution of (71). Then u is Hölder continuous in $\overline{\Omega}$. Moreover, there is a constant C depending on λ, Λ, and $\|u\|_\infty$ such that*

$$|u|_{\alpha;\overline{\Omega}} \leq C(\lambda, \Lambda, \|u\|_\infty). \tag{75}$$

We next see how to use all of the aforementioned ingredients to obtain the a priori bound (61) and conclude the existence of a solution to a quasilinear equation.

e. Application to the Minimal Surface Equation

Let us consider the minimal surface equation in a smooth, bounded domain Ω. Recall that the equation may be written in divergence form as

$$\sum_{i=1}^{n} \partial_i [(1 + |\nabla u|^2)^{-1/2} \partial_i u] = 0 \quad \text{in } \Omega. \tag{76}$$

Let us consider the Dirichlet problem and express the problem in the form of (58a); that is,

$$\begin{cases} Qu \equiv (1 + |\nabla u|^2)\Delta u - \sum_{i,j=1}^{n} u_i u_j u_{ij} = 0 & \text{in } \Omega \\ u = g & \text{on } \partial\Omega. \end{cases} \tag{77}$$

Here and throughout this section, subscripts are used to denote partial derivatives; that is, $u_i = \partial_i u$ and $u_{ij} = \partial_i \partial_j u$. Notice that (77) satisfies (62a,b).

We construct local barriers for (77) that are defined in a neighborhood of $\partial\Omega$ and do not depend on the choice of $x_0 \in \partial\Omega$. Let $d(x) = \text{dist}(x, \partial\Omega)$, and let us assume that $g \in C^{2,\alpha}(\partial\Omega)$ has been extended to $g \in C^{2,\alpha}(\overline{\Omega})$, with $\max_{x \in \partial\Omega} |g(x)| = \max_{x \in \Omega} |g(x)|$. This can be done because, for $d_0 > 0$ but sufficiently small, there is a diffeomorphism

$$U_{d_0} \equiv \{x \in \Omega : 0 < d(x) < d_0\} \approx \partial\Omega \times (0, d_0), \tag{78}$$

given by $x \to (x_0, t)$ where $x_0 \in \partial\Omega$ is the closest point to x and $t = d(x)$; so the values of g on $\partial\Omega$ may be extended independent of $t \in (0, d_0)$ and then multiplied by a smooth function $\phi(t)$ that vanishes for $t \geq d_0/2$. In fact, the smoothness of $\partial\Omega$ implies that $d(x)$ is a smooth function in U_{d_0} that satisfies

$$|\nabla d| = 1. \tag{79}$$

We look for our local barrier in the form $w(x) = g(x) + \psi(d(x))$, where $\psi(t)$ is a C^2 function in an interval $[0, \epsilon]$; both $\psi(t)$ and ϵ are to be determined.

If we use (79), which by differentiation also implies $\sum_i^n d_i d_{ij} = 0$ for any j, to calculate $Q(w)$, we get (after cancellations and grouping terms according to the ψ factor)

(80)
$$\begin{aligned} Q(w) = &(1 + |\nabla g|^2)\Delta g - \sum_{i,j} g_i g_j g_{ij} \\ &+ \psi'\{2\nabla g \cdot \nabla d \Delta g + (1 + |\nabla g|^2)\Delta d - \sum_{i,j}(2d_i g_j g_{ij} + d_{ij} g_i g_j)\} \\ &+ (\psi')^2\{\Delta g - \sum_{i,j} d_i d_j g_{ij} + 2\nabla g \cdot \nabla d \Delta d\} \\ &+ \psi''\{1 + |\nabla g|^2 - (\nabla d \cdot \nabla g)^2\} + (\psi')^3 \Delta d. \end{aligned}$$

We now impose the following conditions on $\psi(t)$:

(81) $\qquad \psi(0) = 0, \quad \psi'(t) \geq 1, \quad \text{and} \quad \psi''(t) < 0.$

Using $g, d \in C^2(\overline{U_{d_0}})$, we can bound the derivatives of g and d by constants to obtain from (80)

(82) $\qquad Q(w) \leq C_1 + \psi' C_2 + (\psi')^2 C_3 + \psi'' + (\psi')^3 \Delta d;$

here we have also used $\psi'' < 0$ and $|\nabla g|^2 - (\nabla g \cdot \nabla d)^2 \geq 0$, which follows from (79) and the Cauchy-Schwartz inequality. Moreover, using $\psi' \geq 1$, we may dominate $C_1 + C_2 \psi'$ by $C(\psi')^2$ to conclude that

(83) $\qquad Q(w) < C(\psi')^2 + \psi'' + (\psi')^3 \Delta d.$

In order to remove the term $(\psi')^3 \Delta d$, we place a convexity condition on $\partial \Omega$:

(84) $\qquad \partial \Omega$ has nonnegative mean curvature.

This condition is certainly satisfied when Ω is a *convex* domain, but more generally, (84) means that the *average of the principal curvatures* is nonnegative (cf. Section 7.3e). To exploit this condition, we need to use the following important properties relating the mean curvature to $d(x)$:
  (i) At $x \in \Omega$ with $d(x) = t \in (0, d_0)$, we have $-\Delta d(x) = n\, h(x)$, where $h(x)$ is the mean curvature of the hypersurface $\Sigma_t = \{y \in \Omega : \text{dist}(y, \partial \Omega) = t\}$.
  (ii) As x moves away from $\partial \Omega$ along the normal to $\partial \Omega$, $\Delta d(x)$ decreases.

For proofs of these properties of $d(x)$, see [Gilbarg-Trudinger] or [Giusti]. We simply observe that, with (84), the properties imply that $\Delta d \leq 0$ in U_{d_0}.

As a consequence, we now have

(85) $$Q(w) < C(\psi')^2 + \psi''.$$

But if we take

(86) $$\psi(t) = \frac{1}{C}\log(1+\beta t),$$

then we see that $\psi(0) = 0$, $\psi''(t) < 0$, and $C(\psi')^2 + \psi'' = 0$; moreover, if we take $\epsilon = 1/\beta$ with β sufficiently large, then we can also arrange $\psi'(t) \geq 1$ for $0 \leq t \leq \epsilon$.

If we define $w^+ = w$, then we have an upper barrier. Similarly, if we replace g by $-g$ and follow the preceding construction, then setting $w^- = -w$ produces a lower barrier. We conclude the following.

Lemma. *If Ω satisfies (84), then every solution u of (77) must satisfy the estimate (69), where C depends on $|g|_2$ but not on the solution u.*

We are now in a position to prove the following.

Theorem 4. *Suppose Ω is a smooth, bounded domain with $\partial\Omega$ having nonnegative mean curvature. Then for any $g \in C^{2,\alpha}(\partial\Omega)$, there is a unique solution $u \in C^{2,\alpha}(\overline{\Omega})$ of (77).*

Proof. Uniqueness follows from Proposition 1. To prove existence using Theorem 1, we must show that a solution $u \in C^{2,\alpha}(\overline{\Omega})$ of (77) with g replaced by tg satisfies

(87) $\quad |u|_{1,\alpha} \leq C,\quad$ where C is independent of $t \in [0,1]$ and u.

We assume that $t = 1$; the derivation of (87) clearly extends to $0 \leq t \leq 1$. We proceed in several steps, estimating $|u|$, $|\nabla u|$, and finally $|\nabla u|_\alpha$ in terms of the boundary data.

Step 1. By the weak maximum principle (cf. Theorem 1 in Section 8.3),

(88) $$\max_{x\in\overline{\Omega}} |u(x)| \leq \max_{x\in\partial\Omega} |g(x)|.$$

Step 2. In order to estimate $\max_{\overline{\Omega}} |\nabla u|$ in terms of boundary data, we need to use the divergence form of the equation, so let us write (76) as $\operatorname{div} \vec{a}(\nabla u) = 0$, where $\vec{a}(\nabla u)$ is the vector-valued function whose components are given by

$$a_i(\nabla u) = (1+|\nabla u|^2)^{-1/2}\partial_i u.$$

We may view u as a weak solution, so

$$\int_\Omega \vec{a}(\nabla u)\cdot \nabla\phi\,dx = 0 \quad \text{for all } \phi \in C_0^1(\Omega).$$

If we replace ϕ by $\partial_k \phi$ and integrate by parts, we obtain

$$\int_\Omega \sum_{i,j=1}^n \frac{\partial a_i}{\partial \zeta_j}(\nabla u)(\partial_j \partial_k u)\partial_i \phi \, dx = 0.$$

Therefore, $w = \partial_k u \in C^1(\overline{\Omega})$ is a weak solution of

(89)
$$\sum_{i,j=1}^n \partial_i a_{ij}(\nabla u)\partial_j w = 0, \quad \text{where}$$

$$a_{ij}(\zeta) = \frac{\partial a_i}{\partial \zeta_j}(\zeta) = \frac{(1+|\zeta|^2)\delta_{ij} - \zeta_i \zeta_j}{(1+|\zeta|^2)^{3/2}}.$$

But if we let $\tilde{a}_{ij}(x) = a_{ij}(\nabla u(x))$, then w is a weak solution of a linear elliptic equation in divergence form, so we may apply the generalized weak maximum principle (cf. Exercise 8 in Section 8.3) to conclude that $|w|$ attains its maximum on $\partial\Omega$. Applying this for each $k = 1, \ldots, n$, we conclude that

(90)
$$\max_{\overline{\Omega}} |\nabla u| \leq \max_{\partial\Omega} |\nabla u| \leq C,$$

where C depends on $|g|_2$ but not on u.

Step 3. We want to estimate $[\nabla u]_{\alpha;\overline{\Omega}}$. Notice that (89) is of the form (70), where the constants λ and Λ in (71b) depend on ∇u. But, by (90), we may take λ and Λ independent of u, and so applying (75) to $w = \partial_k u$, we obtain

(91) $\qquad |w|_{\alpha;\overline{\Omega}} \leq C, \qquad$ where C is independent of u.

This implies (87) and completes the proof. ♠

Remark. The condition (84) is not only a sufficient condition but is also a *necessary* condition for the solvability of (77) for every $g \in C^{2,\alpha}(\partial\Omega)$ (cf. [Gilbarg-Trudinger] or [Giusti]). However, if (84) fails, then it may still be possible to solve (77) for some functions g (cf. Exercise 2).

Exercises for Section 13.4

1. If $a(x)$ and $f(x)$ are smooth functions satisfying $a(x) \leq 0$ in the bounded domain $\Omega \subset \mathbf{R}^n$ and $q > 1$ is an odd integer, show that

$$\begin{cases} \Delta u + a(x)u^q + \lambda u = f(x) & \text{in } \Omega \\ \qquad\qquad\qquad\quad u = 0 & \text{on } \partial\Omega \end{cases}$$

admits a smooth solution u whenever $\lambda < \lambda_1$, where λ_1 is the first Dirichlet eigenvalue for Δ in Ω.

2. If Ω does not satisfy the convexity assumption (84), then (77) may still be solvable under certain conditions on the *boundary manifold*

$$\Gamma = \{(x, z) \in \partial\Omega \times \mathbf{R} : z = g(x)\}.$$

For example, Γ satisfies the *bounded slope condition* if
  (i) For each $P \in \Gamma$, there exist planes $\pi_P^+(x)$ and $\pi_P^-(x)$ passing through P that satisfy $\pi_P^+(x) \geq g(x) \geq \pi_P^-(x)$ for all $x \in \partial\Omega$;
  (ii) The slopes of these planes are uniformly bounded independent of P (i.e., $|\nabla \pi_P^\pm| \leq M$ for all $P \in \Gamma$).
In this case, prove that (77) admits a unique solution $u \in C^{2,\alpha}(\overline{\Omega})$.

3. Consider the *prescribed mean curvature equation*

$$(*) \qquad \sum_{i=1}^{n} \partial_i[(1 + |\nabla u|^2)^{-1/2} \partial_i u] = n\, h(x),$$

which was encountered briefly in Section 7.3. Let us assume that the bounded domain Ω is *uniformly convex;* that is, there is a fixed $R > 0$ such that, for every $x_0 \in \partial\Omega$, there is a ball $B = B_R(y) \supset \Omega$ such that $x_0 \in \partial B$. If h satisfies the condition

$$\sup_{x \in \Omega} |h(x)| \leq \frac{n-1}{nR},$$

show that every solution $u \in C^2(\overline{\Omega})$ of $(*)$ satisfies the boundary gradient estimate (69), where C depends only on $u|_{\partial\Omega}$.

Further References for Chapter 13

References on nonlinear elliptic equations tend to specialize to specific types of equations and/or applications. For connections with bifurcation and stability theory, see [Sattinger]. For conformal scalar curvature type or Monge-Ampere type, see [Aubin]. For more about topological and variational methods, see [Nirenberg, 1], [Nirenberg, 2], or [Struwe]. For quasilinear elliptic equations, see [Ladyzhenskaya-Ural'tseva] or [Gilbarg-Trudinger], the latter especially for equations of mean curvature type. More techniques and applications can be found in [Zeidler, vols. I, II].

It should also be observed that the minimal surface equation was studied here as an application of fixed point methods. However, it may also (and perhaps more naturally) be studied by variational methods. In fact, the minimal surfaces that are studied by means of the minimal surface equation are special surfaces realized as graphs rather than in parametric form; for this reason the graph solutions are often called nonparametric minimal surfaces. For more on parametric and nonparametric minimal surfaces, see [Giusti].

Appendix on Physics

In this Appendix, we collect some background on physical principles and how they are used to derive some of the important equations and systems discussed in this book.

A.1 Physical Principles and PDEs

The PDEs of mathematical physics are derived from physical principles such as Newton's law (balancing forces and momentum) and conservation laws (such as conservation of mass). In the process, simplifying assumptions frequently allow nonlinear equations to be replaced by linear ones. In this section we restrict our attention to scalar-dependent variables.

a. Vibrating Strings and Membranes

First let us consider a flexible string of length L that is under tension T (a force directed along the string) and that undergoes small displacements from its equilibrium position. We suppose that the equilibrium position is $0 \leq x \leq L$ on the x-axis and that the displacements occur in a vertical plane (with no horizontal component); consequently, the displacement at a given $x \in [0, L]$ and time t is given by $u(x,t)$. (See Figure 1.) Let ρ denote the linear density (mass per unit length) of the string, which we allow to vary along the string; that is, $\rho(x)$ depends on $x \in [0, L]$. We allow the tension to vary with position and time; that is, $T(x,t)$.

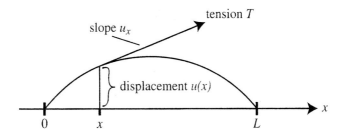

Figure 1. Vibrating string.

At any point, the motion of the string is due to the vertical component of the force, which in turn depends on the slope of the string: simple trignometry shows that the vertical component of the force is just $Tu_x/\sqrt{1+u_x^2}$. Now if we consider the portion of the string between x_1 and x_2, then Newton's law ($F = ma$) gives

$$\left. \frac{Tu_x}{\sqrt{1+u_x^2}} \right|_{x_1}^{x_2} = \int_{x_1}^{x_2} \rho u_{tt}\, dx;$$

from which we obtain the nonlinear PDE

$$\frac{\partial}{\partial x}\left(\frac{Tu_x}{\sqrt{1+u_x^2}}\right) = \rho\, u_{tt}, \tag{1}$$

This is rather complicated, but if we assume that u_x is very small (due to our assumption that the displacement u is small), then the binomial theorem $\sqrt{1+u_x^2} = 1 + \frac{1}{2}u_x^2 + \cdots$ suggests that we approximate $\sqrt{1+u_x^2}$ by 1. If we make this approximation in (1), we get the linear equation

$$(Tu_x)_x = \rho\, u_{tt}. \tag{2}$$

Finally, let us assume that the density ρ and tension T are both constants; this is a natural assumption when the string is *homogeneous* (so ρ is independent of x) and *perfectly elastic* (so that the tension depends only on the local stretching of the string and so for small displacements may be assumed constant). Then we can rewrite (2) as

$$u_{tt} = c^2 u_{xx}, \qquad \text{where } c^2 = T/\rho. \tag{3}$$

This is the form of the one-dimensional wave equation that we study in Chapter 3.

Now let us consider the two-dimensional case of a flexible membrane stretched over a planar region Ω. Again we assume that the displacement is entirely vertical, and so may be denoted by $u(x, y, t)$ for a given point $(x, y) \in \Omega$ and time t. (See Figure 2.) If we assume that the membrane is homogeneous and perfectly elastic, then the density ρ (mass per unit area) and tension T may both be assumed constant.

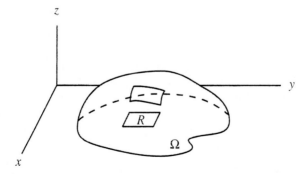

Figure 2. Vibrating membrane.

The vertical force on the membrane again depends on its slope at boundary points, which now is determined by the gradient, ∇u. In fact, if we let R denote a small subregion of Ω (e.g., a rectangle), then at every point of the boundary ∂R, the tensile force is given by $(\nabla u/\sqrt{1+|\nabla u|^2})\cdot \vec{\nu}$, where $\vec{\nu}$ deontes the exterior unit normal to ∂R. Thus, Newton's law implies that

$$\text{(4)} \qquad \int_{\partial R} T \frac{\nabla u}{\sqrt{1+|\nabla u|^2}} \cdot \vec{\nu}\, dS = \int_R \rho u_{tt}\, dA.$$

To obtain a PDE from this, let us first use the approximation (for small displacements) $\sqrt{1+|\nabla u|^2} \approx 1$ and then apply the divergence theorem to write

$$\int_R (\operatorname{div}(T\nabla u) - \rho u_{tt})\, dA = 0.$$

Because R was an arbitrary small region in Ω, we conclude that

$$\text{(5)} \qquad u_{tt} = c^2 \operatorname{div} \nabla u = c^2 \Delta u, \qquad \text{where } c^2 = T/\rho.$$

This is the form of the two-dimensional wave equation that we study in Chapter 3.

We can also obtain from (4) the quasilinear equation for a minimal surface. Namely, let us suppose that T is constant and u is in steady state so that $u_{tt} = 0$. Then, applying the divergence theorem, we obtain

$$\text{(6)} \qquad \operatorname{div}\left(\frac{\nabla u}{\sqrt{1+|\nabla u|^2}}\right) = 0.$$

For $n=3$, equation (5) is shown in Section 3.5 to model the propagation of sound and light waves.

b. Diffusion

Let us consider the one-dimensional diffusion of a substance, such as a dye, through a fluid medium, such as water. Let $u(x,t)$ denote the dye concentration (mass per unit length) at the point x and time t. Between two points $x_1 < x_2$, the total mass of the dye is $m(t) = \int_{x_1}^{x_2} u(x,t)\, dx$. We assume that there are no internal sources or sinks in $[x_1, x_2]$, so by conservation of mass, $m(t)$ can only change due to the dye entering or exiting the interval through the endpoints:

$$\text{(7)} \qquad \frac{dm}{dt} = \int_{x_1}^{x_2} u_t(x,t)\, dx = q_1(t) - q_2(t),$$

where $q_j(t)$ is the rate (mass per unit time) at which the dye flows past x_j. (*Note:* $q_j > 0$ means that substance is moving from left to right at x_j.) Now let us invoke *Fick's law*, which states that the flow rate at any point is proportional to the gradient of the concentration:

$$\text{(8)} \qquad q = -k\partial_x u;$$

k is called the *diffusivity constant*. Combining (7) and (8), we obtain

$$\int_{x_1}^{x_2} u_t(x,t)\,dx = -ku_x(x_1,t) + ku_x(x_2,t) = k\int_{x_1}^{x_2} u_{xx}\,dx.$$

But this may be rewritten as $\int_{x_1}^{x_2}(u_t - ku_{xx})\,dx = 0$, and since x_1 and x_2 are arbitrary, we must have

(9) $$u_t = ku_{xx},$$

which is the one-dimensional diffusion equation that we study in Chapter 5. If the fluid has nonconstant diffusivity $k(x)$, or if additional sources (or sinks) of the dye exist, then we would get the more general equation

(10) $$u_t = (ku_x)_x + f(x,t).$$

It is interesting that heat diffuses through a conducting material in a similar manner. In that case, $u(x,t)$ denotes the temperature at the point x and time t, conservation of mass is replaced by conservation of energy, $q(x,t)$ is called the heat flow (or flux), and relationship (8) is known as *Fourier's law*. But in the end, we find that (9) [or more generally (10)] governs the one-dimensional diffusion of heat.

Now let us consider diffusion in dimensions two or three; to be specific, we consider the diffusion of heat. Let Ω be the domain in which the diffusion takes place, and let R denote a small subregion of Ω. The total amount of heat energy in R depends on the physical properties of the conducting material and is given by

$$H(t) = \int_R c(x)\rho(x)u(x,t)\,dx,$$

where c is the *specific heat*, which is the heat energy required to raise a unit of mass one unit in temperature, ρ is the mass density, and u is the temperature. The change in the heat energy in R is given by

$$\frac{dH}{dt} = \int_R c(x)\rho(x)u_t(x,t)\,dx,$$

which by conservation of energy (assuming no internal heat sources or sinks) must also be given by

$$\frac{dH}{dt} = -\int_{\partial R} \vec{\Phi}\cdot\vec{\nu}\,dS,$$

where $\vec{\Phi}$ is the *heat flux* across the boundary ∂R, and $\vec{\nu}$ is the unit exterior normal to ∂R. Now *Fourier's law* states that

$$\vec{\Phi} = -\kappa\nabla u,$$

for some proportionality factor κ (which could depend on x). But now we can invoke the divergrence theorem:

$$\int_R c(x)\rho(x)u_t(x,t)\,dx = \int_{\partial R} \kappa\nabla u \cdot \vec{\nu}\,dS = \int_R \operatorname{div}(\kappa\nabla u)\,dx.$$

Because R was an arbitrary subregion of Ω, we may conclude that the following PDE holds in Ω:

(11) $$c\,\rho\,u_t = \operatorname{div}(\kappa\nabla u).$$

In case c, ρ, and k are all constants, (11) simplifies to

(12) $$u_t = k\Delta u,$$

where $k = \kappa/(c\rho)$; this is the heat equation that we study in Chapter 5.

A.2 Fields, Tensors, and Systems of PDEs

Physical applications frequently involve several unknown quantities. These may occur as apparently separate unknowns, or as a vector, or as an even more complicated object such as a matrix. In this section, we discuss two such situations: electromagnetism and fluid mechanics. We generally keep our arrow notation for vectors (i.e. $\vec{u}$, $\vec{V}$, etc.), but occasionally it is more convenient to denote vectors in lowercase boldface **u** and matrices in uppercase boldface **T**.

a. Curl and Its Properties

If $\vec{V}(x) = \langle V_1(x), V_2(x), V_3(x) \rangle$ is a smooth vector field on a domain $\Omega \subset \mathbf{R}^3$, recall that the *curl* of $\vec{V}$ is another vector field on Ω defined by

(13) $$\operatorname{curl}\vec{V} = \nabla \times \vec{V} = \det\begin{pmatrix} \mathbf{e}_1 & \mathbf{e}_2 & \mathbf{e}_3 \\ \partial_1 & \partial_2 & \partial_3 \\ V_1 & V_2 & V_3 \end{pmatrix},$$

where $\mathbf{e}_1, \mathbf{e}_2, \mathbf{e}_3$ denote the usual unit vectors in the coordinate directions x_1, x_2, x_3, respectively, and the determinant is computed formally by expansion along the first row. A vector field $\vec{V}$ satisfying $\operatorname{curl}\vec{V} = 0$ is called *irrotational*; the Stokes theorem of vector calculus shows that an irrotational vector field in a simply connected domain is a gradient field (i.e. $\vec{V} = \nabla\phi$ for some function ϕ).

The following identities are easily verified by direct computation:

(14a) $$\operatorname{curl}(\nabla\phi) = 0,$$
(14b) $$\operatorname{div}(\operatorname{curl}\vec{V}) = 0,$$
(14c) $$\operatorname{curl}(\operatorname{curl}\vec{V}) = \nabla(\operatorname{div}\vec{V}) - \Delta\vec{V},$$

where ϕ is a function and $\vec{V}$ is a vector field, both sufficiently smooth.

b. Maxwell's Equations

Let us turn to the equations of electromagnetism. Charged particles create an electric field $\vec{E}$, and moving particles create a magnetic field $\vec{H}$, both of which are functions of $x \in \mathbf{R}^3$ and time t. The interaction of these fields is governed by

(15a) $\qquad \partial_t \vec{E} = c \operatorname{curl} \vec{H} - 4\pi \vec{J} \qquad$ (Ampere's law),

(15b) $\qquad \partial_t \vec{H} = -c \operatorname{curl} \vec{E} \qquad$ (Faraday's law),

where c is the velocity of light and $\vec{J}$ is the electric current density. (We have assumed that the medium is perfectly conducting so as to identify $\vec{E}$ and $\vec{H}$ with dielectric displacement $\vec{D}$ and magnetic induction $\vec{B}$, respectively; in the general case, we have $\partial_t \vec{D} = c \operatorname{curl} \vec{H} - 4\pi \vec{J}$ and $\partial_t \vec{B} = -c \operatorname{curl} \vec{E}$.) The following scalar equations are usually paired with these vector equations:

(16a) $\qquad \operatorname{div} \vec{E} = 4\pi \rho \qquad$ (Coulomb's law),

(16b) $\qquad \operatorname{div} \vec{H} = 0 \qquad$ (absence of magnetic monopoles),

where ρ is the charge density. Notice that (15a) and (16a) together imply

$$\rho_t + \operatorname{div} \vec{J} = 0,$$

which is just consevration of charge. On the other hand, if the fields are static ($\partial_t \vec{E} = 0 = \partial_t \vec{H}$), we may recover the basic equations of magnetostatics, namely $c \operatorname{curl} \vec{H} = 4\pi \vec{J}$ and $\operatorname{div} \vec{H} = 0$, as well as the basic equations of electrostatics, namely $\operatorname{div} \vec{E} = 4\pi \rho$ and $\operatorname{curl} \vec{E} = 0$ (the latter equation implies that $\vec{E} = -\nabla u$ for some function u). The equations (15a,b) and (16a,b) collectively are called *Maxwell's equations*.

Let us now impose a constitutive relation between $\vec{J}$ and $\vec{E}$:

(16c) $\qquad\qquad\qquad \vec{J} = \sigma \vec{E},$

where σ is some constant depending on the medium. For example, in a vacuum we have $\sigma = 0$, so that $\vec{J} \equiv 0$; of course, in a vacuum we would also have $\rho \equiv 0$, so that equations (15a,b)–(16a,b) become homogeneous.

Since (15ab) represents a time evolution, we need initial conditions:

(17) $\qquad\qquad \vec{E}(x,0) = \vec{E}_0(x), \quad \vec{H}(x,0) = \vec{H}_0(x).$

Obviously, we can also obtain from (17) and (16a) the initial value for ρ: $\rho(x,0) = (4\pi)^{-1} \operatorname{div} \vec{E}_0(x)$. In fact, using (16c), we can obtain $\rho(x,t)$ for all t: take the divergence of (15a) and use (14b) to obtain $\partial_t \rho + 4\pi\sigma\rho = 0$, which is easily solved to find

$$\rho(x,t) = \rho(x,0) e^{-4\pi\sigma t}.$$

Thus we may view (15)–(17) as an initial value problem for a system of PDEs that we must solve for $\vec{E}$ and $\vec{H}$.

c. Ideal Fluids and Euler's Equations

Consider a fluid (such as a gas or water) moving in a region Ω of n-dimensional space. Physically, of course, we only consider $n = 1, 2,$ or 3, but here we do not need to make this restriction. At any point $x \in \Omega$ and time t, let $\vec{u}(x,t)$ denote the velocity vector of the fluid; if we fix t, then we can consider $\vec{u}(\cdot, t)$ as a vector field on Ω. We want to discover the equations that govern $\vec{u}$.

First we invoke *conservation of mass*. This requires the introduction of the fluid density; for this exposition, we assume $n = 3$ so that the mass density $\rho(x,t)$ has units of mass per unit volume. If we consider a small subregion $R \subset \Omega$, the total fluid mass in R at time t is given by $m_R(t) = \int_R \rho(x,t)\,dV$, and conservation of mass tells us that the total mass can change only due to fluid entering or exiting the region:

$$\frac{d}{dt}\int_R \rho(x,t)\,dV = -\int_{\partial R} \rho\vec{u}\cdot\vec{\nu}\,dA,$$

where $\vec{\nu}$ is the exterior unit normal on the boundary ∂R. If we apply the divergence theorem, we can write this as

$$\int_R \left(\frac{\partial \rho}{\partial t} + \operatorname{div}(\rho\vec{u})\right) dV = 0.$$

Since R is arbitrary, we obtain the PDE for conservation of mass:

$$(18) \qquad \frac{\partial \rho}{\partial t} + \operatorname{div}(\rho\vec{u}) = 0 \quad \text{in } \Omega.$$

If the fluid is *incompressible*, then obviously $\rho = \text{const}$ and $\operatorname{div}\vec{u} = 0$.

Now let us see what Newton's law (balancing forces and the change in momentum) tells us. Here we assume that the fluid is *ideal*: If S is a surface in the fluid with a chosen unit normal $\vec{\nu}$, then the force across S per unit area is given by $p(x,t)\vec{\nu}$, where $p(x,t)$ is the pressure. (This absence of tangential forces excludes most viscous fluids; we see how to amend this assumption in Section A.2d.) Consider a region R in the fluid with exterior unit normal $\vec{\nu}$ on its boundary ∂R. The total force on R due to pressure on ∂R is $-\int_{\partial R} p\vec{\nu}\,dS$ (negative because $\vec{\nu}$ points outward). According to Newton's law, we have

$$(19) \qquad \frac{d}{dt}\int_R \rho\vec{u}\,dx + \int_{\partial R} \rho\vec{u}(\vec{u}\cdot\vec{\nu})\,dS + \int_{\partial R} p\vec{\nu}\,dS = 0,$$

where the first term is the change of momentum within R and the second term is the flux of momentum across ∂R. We want to use the divergence theorem to convert the boundary integrals to domain integrals. Using u^j as the components of $\vec{u}$, we obtain

$$\int_{\partial R} \rho u^j (\vec{u} \cdot \vec{\nu}) \, dS = \int_R \operatorname{div}(\rho u^j \vec{u}) \, dx = \int_R u^j \operatorname{div}(\rho \vec{u}) + \rho \vec{u} \cdot \nabla u^j \, dx.$$

If we let $\mathbf{e_j}$ denote the unit vector in the jth direction, we obtain

$$\int_{\partial R} p \nu^j \, dS = \int_{\partial R} p \mathbf{e_j} \cdot \vec{\nu} \, dS = \int_R \operatorname{div}(p \mathbf{e_j}) \, dx = \int_R \frac{\partial p}{\partial x^j} \, dx.$$

Plugging these into (19) and using the fact that R was arbitrary, we obtain the following PDE:

$$(\rho \vec{u})_t + \vec{u} \operatorname{div}(\rho \vec{u}) + \rho \vec{u} \cdot \nabla \vec{u} + \nabla p = 0 \quad \text{in } \Omega.$$

But if we use (18), this last PDE is easily simplified to the following form of conservation of momentum, usually called *Euler's equations*:

$$(20) \qquad \vec{u}_t + (\vec{u} \cdot \nabla) \vec{u} + \frac{1}{\rho} \nabla p = 0 \quad \text{in } \Omega.$$

The term $\vec{u}_t + (\vec{u} \cdot \nabla) \vec{u}$ is called the *material derivative* and is often denoted by $D u / D t$.

In order to solve (18) and (20), we also need to impose initial and boundary conditions (when $\partial \Omega \neq \emptyset$). Now a natural boundary condition is $\vec{u} \cdot \vec{\nu} = 0$ on $\partial \Omega$, assuming that the fluid is not allowed to flow across the boundary. But there is another difficulty: (18) and (20) together form a system of four equations involving five unknown quantities, $\vec{u}$, ρ, and p. One more equation is required to specify the fluid motion completely. In general, this is provided by conservation of energy, although in practice we can use instead a simplifying assumption about the fluid's properties. For example, a fluid is *isentropic* if its pressure is a known function of its density: $p = p(\rho)$. More specifically, gases are often *polytropic*, which means that $p = A \rho^\gamma$ for positive constants A and $\gamma > 1$.

d. Cartesian Vector Fields and Tensors

In the two previous subsections, we encountered various vector fields in $\mathbf{R}^n$: electric fields, magnetic fields, and fluid velocity fields. Such a vector field $\vec{v}$ can be represented in terms of components $(v^1, \ldots, v^n)$ in the standard orthonormal basis of vectors $\mathbf{e}_1, \ldots, \mathbf{e}_n$ associated with the coordinates $x^1, \ldots, x^n$ of $\mathbf{R}^n$: $\vec{v} = v^1 \mathbf{e}_1 + \cdots + v^n \mathbf{e}_n$. (The reason for indexing the v^i and x^i with superscripts will become clear later.) But now suppose we have another orthonormal basis $\bar{\mathbf{e}}_1, \ldots, \bar{\mathbf{e}}_n$ for $\mathbf{R}^n$, and we write $\vec{v} = \bar{v}^1 \bar{\mathbf{e}}_1 + \cdots + \bar{v}^n \bar{\mathbf{e}}_n$. How are the coefficients v^i and $\bar{v}^i$ related?

Expressing each $\bar{\mathbf{e}}_j$ in terms of the standard basis defines an $n \times n$ matrix $\mathbf{A} = (a_j^i)$, namely $\bar{\mathbf{e}}_j = \sum_{i=1}^n a_j^i \mathbf{e}_i$. It is now a simple matter to compute $\vec{v} = \sum_i v^i \mathbf{e}_i = \sum_j \bar{v}^j \bar{\mathbf{e}}_j = \sum_{i,j} \bar{v}^j a_j^i \mathbf{e}_i$, so that

$$(21) \qquad v^i = \sum_{j=1}^n a_j^i \bar{v}^j = a_j^i \bar{v}^j,$$

where in the last expression we have invoked the *Einstein summation rule* (i.e. automatically sum over indices that appear both as superscripts and subscripts). (This rule was the reason for indexing the v^i with superscripts.) If we denote by $\mathbf{v}$ and $\bar{\mathbf{v}}$ the column vectors of coefficients in the $\mathbf{e}$-basis and $\bar{\mathbf{e}}$-basis, respectively, we can write (21) as $\mathbf{v} = \mathbf{A}\bar{\mathbf{v}}$ or $\bar{\mathbf{v}} = \mathbf{A}^T \mathbf{v}$, where T denotes transpose. This means that $\mathbf{v} = \mathbf{A}\mathbf{A}^T \mathbf{v}$ and $\bar{\mathbf{v}} = \mathbf{A}^T \mathbf{A} \bar{\mathbf{v}}$ for all column vectors $\mathbf{v}$ and $\bar{\mathbf{v}}$, so $\mathbf{A}\mathbf{A}^T = \mathbf{A}^T\mathbf{A} = \mathbf{I}$ (i.e. $\mathbf{A}$ is an orthogonal matrix).

Let us now consider a velocity field $\vec{u}$ and note that it is natural to view $\vec{u}$ as a first-order differential operator: If $f(x)$ is a smooth function, then $\vec{u}(f)(x)$ represents the rate of change of f at x with respect to the flow determined by $\vec{u}$. Moreover, if we use our coordinates $x^1, \ldots, x^n$ to determine the components $u^1, \ldots, u^n$ for $\vec{u}$, then $\vec{u}(f) = \sum_{j=1}^n u^j \partial f / \partial x^j$. [To verify this, let $x(t)$ denote an integral curve for $\vec{u}$ and use the chain rule to differentiate $f(x(t))$ with respect to t.] This suggests that we write

$$(22) \qquad \vec{u} = \sum_{i=1}^n u^i \frac{\partial}{\partial x^i}$$

and consider the u^i to be the coefficients for $\vec{u}$ with respect to the basis $\partial / \partial x^1, \ldots \partial / \partial x^n$. Now suppose we have a second system $\bar{x}^1, \ldots, \bar{x}^n$ of orthonormal coordinates that we use to determine corresponding coefficients $\bar{u}^1, \ldots, \bar{u}^n$ for $\vec{u}$. Using the chain rule (and summation rule), we obtain

$$\vec{u}(f) = u^i \frac{\partial f}{\partial x^i} = \bar{u}^j \frac{\partial f}{\partial \bar{x}^j} = \bar{u}^j \frac{\partial x^i}{\partial \bar{x}^j} \frac{\partial f}{\partial x^i}.$$

We conclude that the components u^i and $\bar{u}^i$ must satisfy the following transformation rule:

$$(23) \qquad u^i = \frac{\partial x^i}{\partial \bar{x}^j} \bar{u}^j,$$

as a specific instance of (21); in particular, $a_j^i = \partial x^i / \partial \bar{x}^j$.

On the other hand, if $f(x)$ is a smooth function, then we may consider its differential

$$(24) \qquad df = \sum_{i=1}^n \frac{\partial f}{\partial x^i} dx^i$$

as a vector field with coefficients $v_i = \partial f/\partial x^i$ computed with respect to the basis $dx^1, \ldots, dx^n$. If we change bases by considering the coordinate system $\bar{x}^1, \ldots, \bar{x}^n$ (and employ the summation convention), we obtain

$$v_i dx^i = \bar{v}_j d\bar{x}^j = \frac{\partial f}{\partial \bar{x}^j} \frac{\partial \bar{x}^j}{\partial x^i} dx^i,$$

so that the coefficients v_i satisfy the transformation rule

(25) $$v_i = \frac{\partial \bar{x}^j}{\partial x^i} \bar{v}_j \quad \left(\text{or, equivalently,} \quad \bar{v}_j = \frac{\partial x^i}{\partial \bar{x}^j} v_i \right).$$

Now (25) looks different from (23), but if we use the matrix $\mathbf{A} = (a^i_j)$, we find that (25) can be written $\bar{\mathbf{v}} = \mathbf{A}^T \mathbf{v}$, or as $\mathbf{v} = (\mathbf{A}^T)^{-1}\bar{\mathbf{v}}$. But by orthogonality, $(\mathbf{A}^T)^{-1} = \mathbf{A}$, so (25) agrees with (23).

Vector fields u^i that satisfy the transformation rule (23) are called *first-order tensor fields* or *1-tensors*. Now suppose that u^i and v^j are two 1-tensors expressed with respect to a given basis $\mathbf{e}_i$, and let us form the matrix-valued function $\mathbf{T} = \mathbf{T}(x)$ by

(26) $$T^{ij} = u^i v^j \quad \text{for } i,j = 1, \ldots, n.$$

With respect to a second basis $\bar{\mathbf{e}}_j = \sum_i a^i_j \mathbf{e}_i$, the terms $\bar{T}^{ij} = \bar{u}^i \bar{v}^j$ obviously satisfy the following transformation law:

(27) $$T^{ij} = a^i_k a^j_\ell \bar{T}^{k\ell}.$$

Any $n \times n$ matrix-valued function $\mathbf{T}$ that satisfies (27) is called a *second-order tensor field* or *2-tensor*; the special second-order tensor field defined by (26) is called the *tensor product* of $\vec{u}$ and $\vec{v}$ and is denoted by $\vec{u} \otimes \vec{v}$. Higher-order tensors are defined analogously.

The tensors in this subsection are called *Cartesian* because we only consider orthogonal changes of bases; as we have seen, this makes the transformation rules (23) and (25) the same. (When general curvilinear changes of coordinates are considered, these transformation laws do not agree, leading to a distinction betweeen covariant and contravariant vector tensor fields.)

e. Viscous Stress and Newtonian Fluids

For an ideal fluid in $\mathbf{R}^3$, we saw that the force on a surface element S is directed along its normal $\vec{\nu}$. But in viscous fluids, the stress force on S is likely to have tangential components. At each point of S we assume that the stress force (per unit area) is of the form $\mathbf{T}\vec{\nu}$, where $\mathbf{T}$ acts linearly on $\vec{\nu}$ (at that point):

$$(\mathbf{T}\vec{\nu})_i = \sum_{j=1}^{3} T_{ij} \nu_j.$$

where T_{ij} denotes the ith component of the force per unit area exerted on a small surface element perpendicular to $\mathbf{e}_j$, and $\vec{\nu} = \nu_1 \mathbf{e}_1 + \nu_2 \mathbf{e}_2 + \nu_3 \mathbf{e}_3$. The T_{ij} satisfy the transformation law for a 2-tensor, and $\mathbf{T}$ is called the *stress tensor*. Let us write

$$(28) \qquad \mathbf{T} = -p\mathbf{I} + \mathbf{\Sigma},$$

where p is the pressure and $\mathbf{I}$ is the identity matrix; the 2-tensor $\mathbf{\Sigma}$ is called the *viscous stress tensor*. In an ideal fluid, we have $\mathbf{\Sigma} = 0$, but in most situations the stress tensor is at least symmetric, $\Sigma_{ij} = \Sigma_{ji}$.

In particular, we call the fluid *Newtonian* if the viscous stress tensor is in the following form:

$$(29) \qquad \Sigma_{ij} = \lambda (\operatorname{div} \vec{u}) \delta_{ij} + 2\mu D_{ij},$$

where λ and μ are constants called the *coefficients of viscosity*, δ_{ij} is the Kronecker delta, and $\mathbf{D}$ is the *deformation tensor* (or *rate of strain tensor*), defined by

$$D_{ij} = D_{ji} = \frac{1}{2}\left[\frac{\partial u_i}{\partial x_j} + \frac{\partial u_j}{\partial x_i}\right].$$

Some words of explanation about (29) are in order. To begin with, if the fluid is incompressible, then $\operatorname{div} \vec{u} = 0$ and we see that *the viscous stress is proportional to the deformation*; thus with either zero viscosity ($\mu = 0$) or zero deformation ($\mathbf{D} = 0$) we have zero viscous stress ($\mathbf{\Sigma} = 0$). For one-dimensional considerations, this proportionality is called *Newton's law of viscosity*, which explains why fluids satisfying (29) are called Newtonian. The tensor $\mathbf{D}$ is sometimes called the rate of deformation tensor due to the fact that $\vec{u}$ is already a rate. On the other hand, in elasticity theory a deformation $\vec{\xi}$ leads to a tensor $\mathbf{E}$ defined by

$$E_{jk} \equiv \frac{1}{2}\left(\frac{\partial \xi_j}{\partial x_k} + \frac{\partial \xi_k}{\partial x_j}\right)$$

and called the *(linearized) strain tensor*; in our case, the velocity field $\vec{u} = d\vec{\xi}/dt$, which explains the terminology rate of strain tensor for $\mathbf{D}$.

The assumption (29) is not unreasonable but eliminates certain viscoelastic (or non-Newtonian) fluids, such as polymers, and many organic fluids, such as blood.

f. The Navier-Stokes Equations

If we apply Newton's law with the more general stress tensor (28), we obtain

$$(30) \qquad \frac{d}{dt}\int_R \rho \vec{u}\, dx + \int_{\partial R} \rho \vec{u}(\vec{u}\cdot\vec{\nu})\, dS + \int_{\partial R} p\vec{\nu}\, dS = \int_{\partial R} \mathbf{\Sigma}\vec{\nu}\, dS.$$

Applying (29) and the divergence theorem to the last term in (30) yields

$$\int_{\partial R} \Sigma_{ij}\nu_j \, dS = \int_R \frac{\partial}{\partial x_j}(\lambda \operatorname{div} \mathbf{u}\, \delta_{ij} + 2\mu D_{ij})\, dx$$
$$= \int_R \left[\lambda \frac{\partial}{\partial x_i}(\operatorname{div} \mathbf{u}) + \mu \frac{\partial}{\partial x_j}\left(\frac{\partial u_i}{\partial x_j} + \frac{\partial u_j}{\partial x_i}\right)\right] dx.$$

Summing on j we obtain

$$\int_{\partial R} \mathbf{\Sigma}\vec{\nu}\, dS = \int_R [(\lambda + \mu)\nabla(\operatorname{div}\vec{u}) + \mu\Delta\vec{u}]\, dx.$$

Combined with the calculation already performed on (19), we see that a Newtonian viscous fluid satisfies the following second-order PDE:

(31) $\qquad \rho\vec{u}_t + \rho(\vec{u}\cdot\nabla)\vec{u} + \nabla p = (\lambda + \mu)\nabla(\operatorname{div}\vec{u}) + \mu\Delta\vec{u} \quad \text{in } \Omega,$

called the *Navier-Stokes equations*.

In the special case that the fluid is incompressible, we have $\operatorname{div}\vec{u} = 0$ and $\rho = \text{const}$. In that case, the equations (31) simplify considerably to the *Navier-Stokes equations for incompressible flow*:

(32) $\qquad \vec{u}_t + (\vec{u}\cdot\nabla)\vec{u} + \frac{1}{\rho}\nabla p = \nu\Delta\vec{u} \quad \text{in } \Omega,$

where $\nu = \mu/\rho$ is called the *kinematic viscosity*.

If $\partial\Omega \neq \emptyset$, then either (31) or (32) must be supplemented with a boundary condition. Due to the viscous nature of the fluid, it is natural to impose the nonslip boundary condition, $\vec{u} = 0$ on $\partial\Omega$. (This is different from the boundary condition $\vec{u}\cdot\vec{\nu} = 0$ used for Euler's equations.) If there are external forces $\vec{f}$, then (32) becomes a nonhomogeneous equation.

Further References for the Appendix

For additional information on the derivation of the equations of mathematical physics and fluid dynamics, see [Sobolev] and [Aris]. For additional information on electromagnetism, see [Jackson].

Hints and Solutions for Selected Exercises

We collect here hints and solutions for many of the exercises in the book. In some cases, a complete solution is given; for others, only hints are provided. The hints sometimes give the outline for a complete solution, while in other cases the hints are cryptic and require additional thought. You should use these hints as a last resort after being stuck on a problem for a while.

Hints for Exercises in Chapter 1

Section 1.1

1. Let $\chi = (x(t), y(t), z(t))$ and define $\varphi(t) = z(t) - u(x(t), y(t))$. Show that $\varphi(t) \equiv 0$.
2. Use Exercise 1 to conclude that the characteristic curve through $P \in S_1 \cap S_2$ is contained in both S_1 and S_2.
3. Consider the Cauchy problem for curves that intersect Γ.
4. (a) $u(x, y) = \frac{1}{2} y^2 + x^2 e^{-2y}$ exists for all x, y.
5. (a) $u(x, y, z) = h(x e^{-z}, y e^{-z}) e^z$ exists for all x, y, z.
6. (a) $u(x, y) = (3y + 1)^{1/3}$ exists for all x, y.
7. (a) $u = f((x + u)/(y + u))$. (b) $u = y f(x^2 y + y^3 + u^2 y)$.

Section 1.2

1. $z = h(x - a(z)y)$ defines $z = u(x, y)$ implicitly.
2. Show that characteristics intersect at the point $(1, u_0^{-1})$; then consider the cases $y < u_0^{-1}$ and $y > u_0^{-1}$ separately.
3. The shock curve is given by $x = \xi(y) = 1 - (1 + u_0 y)^{1/2}$. The solution is $u(x, y) = 0$ for $x < \xi(y)$, and $u(x, y) = u_0(x - 1)/(1 + u_0 y)$ for $x > \xi(y)$.
4. The solution involves the fan given by $u(x, y) = (x - x_0)/y$ if $u_l y < x - x_0 < u_r y$.
5. A simple wave solution is $u(x, y) = (x/y)^{1/3}$.
7. (b) Assuming a car has length ℓ, then $\rho(x, 0) = 1/\ell$ for $|x - x_0| \leq \ell/2$, $\rho(x, 0) = 0$ for $|x - x_0| > \ell/2$ describes a single car centered at x_0. [Alternatively, $\rho(x, 0) = \delta_{x_0}(x)$, the delta distribution of Section 2.3.]
8. The shock curve is $x = \xi(t) = -\frac{1}{2} ct$.
9. There is no shock. The solution is continuous, and involves a fan, $\rho(x, t) = \rho_{\max}(ct - x)/2ct$ when $-ct < x < ct$.

Section 1.3

2. $u(x, y) = \pm y + \frac{1}{2}(1 - x^2)$.

3. It is possible to solve the equation if $0 < a \le \frac{1}{4}$. If $0 < a < \frac{1}{4}$, there are two solutions: $u(x, y) = (\sqrt{a}x \pm \frac{\sqrt{1-4a}}{2}y)^2$. But for $a = \frac{1}{4}$, there is a unique solution: $u(x, y) = \frac{1}{4}x^2$.

6. (b) Use $w = a$, for example, in which case $u(x, y) = -\frac{1}{4}(x + y)^2$.

7. (b) For example, use $w(a) = a$ to obtain $u(x, t) = (x + 1)^2/4t$, or $w(a) = a^2$ to obtain $u(x, t) = x^2/4(t - 1)$.

8. The solutions are $u(x, y) = \pm c^{-1}(x^2 + y^2)^{1/2}$.

10. (a) $u(x, y, z) = k + (\alpha x + \beta y + z)/\sqrt{\alpha^2 + \beta^2 + 1}$.

Hints for Exercises in Chapter 2

Section 2.1

2. The heat equation is in normal form on the t-axis *not* on the x-axis.

3. $u(x, y) = \sum_{k=0}^{\infty} \frac{1}{(2k)!} 2^k e^x y^{2k}$.

4. $u(x, y) = \frac{\pi}{4}x + \frac{1}{\sqrt{2}}y$.

5. The solution itself is a polynomal (which converges everywhere):
$$u(x, t) = \sum_{j,k=0}^{2j+k=n} \frac{(2j+k)!}{j!k!} a_{2j+k} t^j x^k.$$

6. Formally, a solution is given by
$$u(x, t) \sim \sum_{j,k=0}^{\infty} \frac{(j+2k)!}{j!k!}(ix)^j(-t)^k.$$

7. The explicit solution is $u(x, t) = \frac{1}{k^2}\sin(kx)\sinh(ky)$.

Section 2.2

1. (a) $u_{\mu\eta} = 0$, where $\mu = y - 2x$ and $\eta = y - 3x$.

2. (a) $u(x, y) = f(y - x - \cos(x)) + g(y + x - \cos(x))$.

6. The reduced system solution is $v_1(x, t) = -x + 3t$, $v_2(x, t) = x + t$, which corresponds to the solution of the initial problem $u_1(x, t) = x + 5t$, $u_2(x, t) = -4t$.

7. (a) decouples, but (b) does not.

8. Let $\mu = x - \frac{t}{\sqrt{LC}}$, $\eta = x + \frac{t}{\sqrt{LC}}$, and $k = R/L$. Then

$$I(x,t) = \frac{1}{2}[I_0(\mu) + I_0(\eta) + \sqrt{\frac{C}{L}}V_0(\mu) - \sqrt{\frac{C}{L}}V_0(\eta)]e^{-kt}$$

$$V(x,t) = \frac{1}{2}[\sqrt{\frac{L}{C}}I_0(\mu) - \sqrt{\frac{L}{C}}I_0(\eta) + V_0(\mu) + V_0(\eta)]e^{-kt}.$$

9. Use $u_0 = u$, $u_1 = u_x$, $u_2 = u_y$ to reduce to a first-order system that is hyperbolic with eigenvalues $\lambda = 0 \pm 1$. Your answer should agree with the d'Alembert formula of Section 3.1.

Section 2.3

1. (a) $L^* v = \sum_{|\alpha| \leq m}(-1)^{|\alpha|} D^\alpha(\bar{a}_\alpha v)$.
   (b) $L^* \vec{v} = \sum_{|\alpha| \leq m}(-1)^{|\alpha|} D^\alpha(\bar{a}_\alpha^T \vec{v})$, where $\bar{a}_\alpha^T$ denotes the conjugate transpose matrix for a_α.

3. (a) A weak solution must satisfy $\int u(v_t + cv_x)\,dx = 0$ for all $v \in C_0^1(\Omega)$. Changing variables to $\mu = x - ct$ and $\eta = x + ct$, we have that $u(x,t) = f(\mu)$ is a weak solution since $\int_\Omega f(\mu) v_\eta(\mu,\eta)\,d\mu\,d\eta = 0$ for all $v \in C_0^1(\mathbf{R}^2)$. (b) Take $f(\mu) = f(x - ct)$ with discontinuous f. (c) No.

5. (b) Consider the sequence $\xi_k = (k, 0, \cdots, 0)$.

6. (b) Consider the distribution $G = \delta'$ on $\mathbf{R}^1$ and $f(x) = |x|$.

7. (b) $\langle a\partial_\nu \delta_\Gamma, v\rangle = -\int_\Gamma \nabla(a(z)v(z))\cdot\nu(z)\,dz$.

8. Use the mean value theorem.

10. (a) Use e^{-ax} as an integrating factor.

11. (a) Apply the Lemma of this section to show that $u \in C(\mathbf{R})$.
    (b) Write $u(x) = \frac{1}{2}\int_{-\infty}^x (x-y)f(y)\,dy + \frac{1}{2}\int_x^\infty (y-x)f(y)\,dy$, and show that $h^{-1}[u(x+h) - u(x)] \to \frac{1}{2}\int_{-\infty}^x f(y)\,dy - \frac{1}{2}\int_x^\infty f(y)\,dy$. Finally, use the hypotheses on f to verify $u' \in C(\mathbf{R})$.
    (c) Use (b) to compute u''.

12. (a) Use $u = u_p + u_h$ to find $u(x) = u_0 - \frac{1}{2}\int_0^\infty |y|f(y)\,dy + (u_0' + \frac{1}{2}\int_0^\infty f(y)\,dy)x + \frac{1}{2}\int_0^\infty |x-y|f(y)\,dy$.

13. (b) Use $v_{tt} - c^2 v_{xx} = -4c^2 v_{\mu\eta}$. One fundamental solution is $F_1(x,t) = (2c)^{-1}H(ct+x)H(ct-x)$, having support in $-ct < x < ct$ and $t > 0$.

15. Partition $\Omega = (a,b) \times (c,d)$ by rectangles (i.e., introduce $a = x_0 < x_1 < \cdots < x_m = b$ and $c = y_0 < y_1 < \cdots < y_n = d$), and approximate the integral over Ω as a Riemann sum, using difference quotients in place of v_y and v_x. Recombine terms to perform a discrete integration by parts, and obtain difference quotients involving u. Then use the condition of Section 1.2 to evaluate the limit as the partition increment shrinks to zero.

Hints for Exercises in Chapter 3

Section 3.1
1. (a) $u(x,t) = x^3 + 3c^2xt^2 + c^{-1}\sin(x)\sin(ct)$.
2. The Fourier series solution is
$$u(x,t) = \frac{2}{\pi}\sum_{n=1}^{\infty}\frac{1-\cos(n\pi)}{n^2}\sin(nx)\sin(nt).$$
Using the parallelogram rule, the solution is given by $u(x,t) = t$ in region R_1, $u(x,t) = x$ in region R_2, $u(x,t) = \pi - x$ in region R_3, $u(x,t) = \pi - t$ in region R_4, and so on.
3. (a) The Fourier series solution is
$$u(x,t) = \frac{\pi}{2} + \frac{2}{\pi}\sum_{k=1}^{\infty}\frac{1}{k^2}[\cos(k\pi) - 1]\cos(kx)\sin(kt).$$
(b) Use the parallelogram rule for u_x in R_1 and R_2. The solution is not unique; it may be C^0, but it is not C^1.
4. You need to show that the solution given by d'Alembert's formula satisfies the boundary condition $u(0,t) = 0$.
5. Consider the function $U(x,t) = u(x,t) - \alpha(t)$. Modify the method in Exercise 4 for the equation U satisfies.
6. Use the ideas in the Remark ending this section to solve this problem. You should use a particular solution depending on x!
7. The solution is given by (9) with $L = \pi$.
 (a) $a_n(t) = c_n \sin \lambda_n t + d_n \cos \lambda_n t$, where $\lambda_n = (c^2 n^2 + m^2)^{1/2}$.
 (b) If $c^2 < m^2$, then $a_1(t)$ is either trivial or unbounded.

Section 3.2
1. (a) $|\alpha|^2 = c^{-2}$.
 (b) $\nabla g(x) = \alpha F'(\alpha \cdot x) = -\alpha h(x)$.
 (c) $u(x,t) = 1 + x_1 - x_2 \pm c\sqrt{2}t$.
2. $u(x,t) = x^2 + y^2 + 2t^2$.
3. The solution is given by
$$u(x,t) = (4\pi)^{-1}\int_0^t\int_{|\xi|=1}(t-s)f(x+c(t-s)\xi, s)\,dS_\xi\,ds.$$
We will need f to be C^2 in x and C^0 in t.
4. $u(x,y,t) = \sin(\pi x/a)\sin(2\pi y/b)\cos(\pi(4a^2+b^2)^{1/2}t/ab)$.

5. Let $u(x_1, x_2, x_3, t) = \cos(\frac{m}{c}x_3) v(x_1, x_2, t)$, and show that u satisfies a three-dimensional wave equation with initial conditions. Compute u from (37), and then use this to compute v.

6. (a) Estimate the area of intersection of the sphere of radius ct with the support of the functions g and h.
   (b) No. This is a consequence of Huygens's principle and may be proved by using (39).

7. (a) First show by induction on $k \geq 1$ that
$$\partial_r^2 (r^{-1}\partial_r)^{k-1}(r^{2k-1}\phi(r)) = (r^{-1}\partial_r)^k(r^{2k}\partial_r\phi(r))$$
for any C^{k+1}-function $\phi(r)$. Then replace $\phi(r)$ by $M_u(x, r, t)$ and use (28) to conclude (32).

   (b) First show by induction on $k \geq 1$ that
$$(r^{-1}\partial_r)^{k-1}(r^{2k-1}\phi(r)) = \sum_{j=0}^{k-1} a_j^k r^{j+1} \partial_r^{j+1} \phi(r)$$
for any C^{k-1}-function $\phi(r)$, where the a_j^k are independent of ϕ and $a_0^k = 1 \cdot 3 \cdots (2k-1)$. This implies that $V^x(0,t) = 0 = G^x(0) = H^x(0)$, so we may use (35).

Section 3.3

1. Differentiate $\mathcal{E}_\Omega(t)$ with respect to t, and then use the boundary condition to show that $\int_\Omega \sum u_{x_i} u_{x_i t}\, dx = -\int_\Omega u_t \Delta u\, dx$.

3. Expand $||\nu_{n+1} \nabla u - u_t \langle \nu_1, \cdots, \nu_n \rangle||^2 \geq 0$, where $||\cdot||$ denotes the length of the n-vector inside, then use (48).

4. (a) The energy integral is $\mathcal{E}(t) = \int (|u_t|^2 + c^2|\nabla u|^2 + q(x)u^2)\, dx$ and may be used to define global and local energy. (b) Duplicate the proof of Theorem 2 for the energy defined in (a).

5. (a) Duplicate the proof of Theorem 2 for the equation in this exercise. The divergence of the vector field is now nonpositive instead of zero.

Section 3.4

1. $u(x, t) = Ae^{i(k \cdot x - \omega t)}$, where $k = \langle k_1, \cdots, k_n \rangle$ is an n-vector and $x \cdot k$ is the dot product; $w(k) = \pm (c^2|k|^2 + m^2)^{1/2}$.
2. (a) $w(k) = \pm \gamma k^2$.
3. $w(k) = -ik^2$.
4. Try $u(x, t) = A\cos(kx - \omega t)$.
5. The initial conditions must satisfy $\pm\sqrt{k^2 + \lambda}\, g'(x) = kh(x)$.

Hints for Exercises in Chapter 4

Section 4.1

2. You should get a series of the form $a_o y + \sum_{k=1}^{\infty} a_k \cos(kx) \sinh(ky)$.

3. Let $w = u_1 - u_2$ and use $\int_\Omega |\nabla w|^2 \, dx = \int_{\partial \Omega} w \frac{\partial w}{\partial \nu} \, dS$.

4. (a) Use separation of variables in polar coordinates where $\beta(\vartheta) = c_0 + \sum_{k=1}^{\infty} [c_k \cos(k\vartheta) + d_k \sin(k\vartheta)]$. (b) Consider $u(r, \theta) = r(a \cos \theta + b \sin \theta)$.

5. Use (9) with either Dirichlet or Neumann boundary condition to conclude that $\int (qu^2 + |\nabla u|^2) \, dx = 0$ when $u = u_1 - u_2$.

7. (a) Apply the maximum principle to u on $\Omega_+ = \{x : u(x) > 0\}$ and to $-u$ on $\Omega_- = \{x : u(x) < 0\}$. (b) Use part (a) on the domains $\{x : 1 < |x| < R\}$ and let $R \to \infty$.

8. Let $v(x) = \exp(-\alpha|x - x_1|^2) - \exp(-\alpha\epsilon^2)$ for $x \in B_\epsilon(x_1)$, where $\alpha > 0$. For α sufficiently large, show that $\Delta v < 0$ in $B_{\epsilon/2}(x_0)$. For $\eta > 0$ sufficiently small, show that $u + \eta v \leq u(x_0)$ on $B_\epsilon(x_1) \cap B_{\epsilon/2}(x_0)$. Conclude that $\partial(u + \eta v)/\partial \nu \geq 0$ at x_0. Compute $\partial v/\partial \nu$ at x_0.

Section 4.2

1. (a) Show $|x - \xi|^2 = 1 + r^2 - 2r\cos(\theta - \phi)$ when $\xi = (r \cos \theta, r \sin \theta)$ and $x = (\cos \phi, \sin \phi)$.

2. Generalize the proof of (iii) in the Proposition of this section.

3. (b) Fix $x \neq \xi$ in Ω and let $\Omega_\epsilon = \Omega \setminus (B_\epsilon(x) \cap B_\epsilon(\xi))$, where $\epsilon > 0$ is sufficiently small. Apply (7) in Ω_ϵ with $u(y) = G(x, y)$ and $v(y) = G(y, \xi)$; then let $\epsilon \to 0$.

4. (a) Fix $\xi \in \Omega$ and apply the maximum principle to $G(x, \xi)$ in $\Omega_\epsilon = \{x \in \Omega : |x - \xi| > \epsilon\}$.

5. Apply (33) to $\Omega_R = \{x \in \mathbf{R}^n_+ : |x| < R\}$, and show that the integral over $|x| = R$ vanishes as $R \to \infty$.

7. Consider $\overline{B_r(\xi)} \subset \Omega$ and v harmonic in $B_r(\xi)$ with $v = u$ on $\partial B_r(\xi)$; apply Exercise 9 in Section 4.1.

8. Define $\tilde{u}(x_1, \cdots, x_n)$ by $\tilde{u}(x_1, \ldots, x_n) = -u(x_1, \ldots, -x_n)$ if $x_n \leq 0$ and $\tilde{u} = u$ if $x_n \geq 0$. Use the preceding exercise to show that $\tilde{u}$ is harmonic.

9. Use the preceding exercise and Liouville's theorem.

10. (a) Use the fact that $a - |\xi| \leq |x - \xi| \leq a + |\xi|$ for all x with $|x| = a$ and all $\xi \in B_a(0)$. (b) Prove the inequality for a ball first. Then use a compactness argument.

12. Use $n^k = (1 + \cdots 1)^k = \sum_{|\alpha|=k}(|\alpha|!/\alpha!)$ to show that $|\alpha|! \leq n^{|\alpha|}\alpha!$ for each $\alpha = (\alpha_1, \ldots, \alpha_n)$.

Section 4.3

1. Suppose $\Delta u < 0$ on $\Omega' = \overline{B_a(\xi)} \subset \Omega$, and apply (31) in Ω' to $G(x,\xi) = K(x,\xi) - \Psi(a)$, which is negative in Ω' and vanishes on $\partial\Omega'$.
2. If $B_\epsilon(\xi)$ is an exterior sphere at $z \in \partial\Omega$, use $Q_z(x) = -\log(|x - \xi|/\epsilon)$ (when $n = 2$).
3. Without loss of generality, we can assume that the boundary point is 0, and nearby $\partial\Omega$ is given by a graph $x_n = f(x_1, \cdots, x_{n-1})$ that has $\mathbf{R}^{n-1}$ as its tangent plane. The C^2 hypothesis ensures that all the second-order derivatives of f are continuous, hence bounded, at 0.
4. Assume that $z = 0$ and $\theta_0 = 0$. Seek the barrier in the form $Q(r,\theta) = r^\lambda Y(\theta)$, where λ is a certain eigenvalue for ∂_θ^2.

Section 4.4

1. (c) The smallest frequencies are $\sqrt{5}/2b$ and $\sqrt{2}/b$.
2. Use separation of variables: $u(x,y,z) = X(x)Y(y)Z(z)$.
3. For each fixed t, we can decompose $f(x,t)$ by eigenfunctions to obtain $f(x,t) = \sum f_n(t)\varphi_n(x)$. From Duhamel's principle, we then have
$$u(x,t) = \sum_{n=1}^{\infty} \frac{1}{\sqrt{\lambda_n}} \varphi_n(x) \int_0^t f_n(s) \sin(\sqrt{\lambda_n}(t-s))\,ds.$$

4. (a)
$$u(x,t) = \sum_{n=1}^{\infty} \frac{a_n}{\sqrt{\lambda_n}} \frac{\omega \sin(\sqrt{\lambda_n}t) - \sqrt{\lambda_n}\sin(\omega t)}{\omega^2 - \lambda_n} \phi_n(x),$$
where $a_n = \int_\Omega A(x)\phi_n(x)\,dx$.
7. Multiply the PDE in (62) by u and integrate over Ω.

Hints for Exercises in Chapter 5

Section 5.1

2. Consider $u(x,t) = v(x,t) + w(x)$ with $\Delta w = 0$ in Ω, $w = h$ on $\partial\Omega$.
3. Assume that $f(x,t)$ admits an eigenfunction expansion of the form $f(x,t) = \sum_{n=1}^{\infty} a_n(t)\varphi_n(x)$, which converges absolutely and uniformly on $\overline{U_T}$ for every $T > 0$. Then $u(x,t) = \sum_{n=1}^{\infty} \int_0^t a_n(s)e^{\lambda_n(s-t)}ds\phi_n(x)$.
4. Assume that $h(x,t)$ extends to $x \in \overline{\Omega}$ such that the eigenfunction expansion $h(x,t) = \sum a_n(t)\varphi_n(x)$ converges absolutely and uniformly on $\overline{U_T}$, then use Exercise 3 on $u(x,t) = v(x,t) + h(x,t)$. (You should discuss the issue of convergence for this solution.)
5. The eigenvalues are $\lambda_{mn} = \pi^2(m^2 + n^2)/a^2$ and the eigenfunctions are $\phi_{mn}(x,y) = \cos(m\pi x/a)\cos(n\pi y/a)$, for $m, n = 1, 2, \ldots$.

Hints for Exercises in Chapter 6

6. Apply the weak maximum principle to $v(x,t) = e^{-ct} u(x,t)$.
7. (a) Integrate by parts. (b) Consider the energy for the difference between two solutions.

Section 5.2

1. (a) Direct calculations. (b) and (c) Use the substitution $z = (x-y)/2\sqrt{t}$. (d) Write $u(x,t) - g(x) = \int K(x,y,t)(g(y) - g(x))\,dy$ and imitate the proof of Theorem 3 in Section 4.2.
5. (a) Use the following trick to prove that $u(x,t) \leq m$. Pick $x_0 \in \mathbf{R}^n$ and show that $v_{x_0}(x,t) = u(x,t) - \epsilon(2nt + |x-x_0|^2) \leq m$ for all x and $\epsilon > 0$. Let $\epsilon \to 0$ to conclude that $u(x_0, t) \leq m$ for all $x_0 \in \mathbf{R}^n$.
7. For both parts, use (19) after extending g to all of $\mathbf{R}$.
8. During integration by parts, the fact that v is allowed to be nonzero at $t = 0$ is related to the fact that u vanishes at $t = 0$.
10. Similar to Theorem 2.
11. Consider the function $v(x,t) = e^t u(x,t)$.

Section 5.3

1. Modify the proof of the theorem in this section.
2. (a) Assume that $u(x,t) = t^\alpha w(x/\sqrt{t})$, and show that $\int u(x,t) = 1$ for all $t > 0$ implies that $\alpha = -n/2$. (b) Solve the ODE in (43a) for all A.
3. Use (38b) to conclude that $\beta = 1$ and $u_t + au_x = t^{\alpha-1}[\alpha w(z) + (a-z)w'(z)] = 0$; then solve $w'(z) = \alpha(z-a)^{-1}w(z)$.
6. Use (38b) and compare powers of t to conclude that $\alpha - 1 = \alpha\gamma - 2\beta$ and $\alpha w - \beta z \cdot \nabla_z w = \Delta(w^\gamma)$. Now assume that w is radial: $w(z) = w(r)$, where $r = |z|$, and denote dw/dr by w'. Then w satisfies $\alpha w - \beta r w' = (w^\gamma)'' + (n-1)r^{-1}(w^\gamma)'$. Choose $\alpha = -n\beta$ and multiply by r^{n-1} to conclude that $(r^{n-1}(w^\gamma)')' + \beta(r^n w)' = 0$. Integrate to find $r^{n-1}(w^\gamma)' + \beta r^n w = A$, where the constant $A = 0$ if we assume $w, w' \to 0$ as $r \to \infty$. Conclude that $w^{\gamma-2}w' = -\beta r$, which may be integrated to find $w^{\gamma-1} = C - (\gamma-1)\beta r^2/(2\gamma)$.

Hints for Exercises in Chapter 6

Section 6.1

1. (a) Apply the triangle inequality to $x = (x-y) + y$. (b) Use (a). (c) Expand $\|x - \lambda y\|^2 \geq 0$ and let $\lambda = \|x\|/\|y\|$.
3. Apply the Hölder inequality to $\int |u|^q\,dx = \int |u|^{q\lambda}|u|^{q(1-\lambda)}\,dx$.
4. (b) Let A_j be a Cauchy sequence in ℓ^2; that is, $A_j = \{a_n^{(j)}\}$ satisfies $\sum_{n=1}^\infty (a_n^{(j)} - a_n^{(k)})^2 \to 0$ as $j,k \to \infty$. By the completeness of the real numbers, $a_n^{(j)}$ converges to some a_n as $j \to \infty$. Show that $\sum a_n^2 < \infty$;

that is, $A = \{a_n\} \in \ell^2$.

5. (a) If $\phi \in C_0^1(\Omega)$, use integration by parts to compute $\langle u, \partial\phi/\partial x\rangle$.

6. Use the cutoff functions $\phi_k(x) = \phi(k^{-1}|x|)$ for $k \geq 1$, where we have fixed $\phi \in C^\infty([0,\infty))$ satisfying $\phi(t) = 1$ if $0 \leq t \leq 1$, and $\phi(t) = 0$ if $2 \leq t < \infty$.

7. No; consider $u_j \in C(\overline{\Omega})$ with $\|u_j\|_p = 1$ and $u_j(x_0) = j$.

10. If $u = v + w = v_1 + w_1$, where $v, v_1 \in Y$ and $w, w_1 \in Y^\perp$, then $v - v_1 = w - w_1 \in Y \cap Y^\perp$, so $v - v_1 = 0 = w_1 - w$ by Exercise 8b.

13. Use Exercise 8 to write $X = \overline{S} \oplus S^\perp$. Extend f to $\overline{S}$ by continuity, and let $F = 0$ on $S^\perp$.

14. (a) Expand $\|u - \sum_{n=1}^N \alpha_n x_n\|^2$. (b) Expand $\|u - \sum_{n=1}^N c_n x_n\|^2$.

15. For any $x \in X$, define $F_x(y) = B(x,y)$ for all $y \in X$ and show that $F_x: X \to \mathbf{R}$ is a bounded linear functional. By Theorem 3 (Riesz representation), there is a unique $z \in X$ for which $L_x(y) = \langle y, z\rangle$ for all $y \in X$. Define $Ax = z$, and show that A is indeed a bounded linear operator on X.

17. Show that $R \subset (N^*)^\perp$ and recall that $(N^*)^\perp$ is closed; then use the projection theorem to show that $y \notin \overline{R}$ implies $y \notin (N^*)^\perp$.

Section 6.2

1. For $v_n \in C_0^1(\Omega)$ with $\|v_n - v\|_{1,2} \to 0$, use continuity of the L^2-inner product as well as (14).

4. (a) Show that $\lambda_1 \geq C^{-1}$, where C is the constant in Theorem 1. (b) Prove that $B(u,v) = \int(\nabla u \cdot \nabla v - cuv)\,dx$ is positive [cf. (25)].

6. For any $\vec{\phi} \in C_0^2(\Omega, \mathbf{R}^3)$, use (14b) in the Appendix to conclude that $\langle \vec{g}, \operatorname{curl} \vec{\phi}\rangle = 0$, and then integrate by parts to conclude $\operatorname{curl} \vec{g} \equiv 0$ in Ω. Use the simple-connectivity of Ω and Stokes' theorem to conclude that $\vec{\phi} = \nabla p$ for some scalar potential function $p \in C^2(\Omega)$.

Section 6.3

1. (c) If $\{F_n\}$ is Cauchy in the operator norm, verify that each $\{F_n u\}$ is a Cauchy sequence of real numbers, and use this to obtain a limit operator F.

2. If $\iota X \neq X^{**}$, then there exists a nonzero $y \in X^{***}$ such that $y(x) = 0$ for all $x \in \iota X$.

5. Let $\{x_n\}_1^\infty$ be a sequence in S, and let A be the set $\{x_n : n = 1, 2, \ldots\}$. Consider the cases where A is finite and infinite separately.

7. If $\{y_k\}$ is dense in X, then $\{x_n\}$ coverges weakly in X if and only if $\langle y_k, x_n\rangle$ converges as $n \to \infty$, for every k.

8. (a) First show that $C_0(\Omega)$ is separable (using polynomials with rational coefficients), then use the density of $C_0(\Omega)$ in $L^p(\Omega)$.

9. Let $\{x_k\}_{k=1}^{\infty}$ be a dense sequence in X. Since $F_j(x_1)$ is a bounded set in $\mathbf{R}$, extract a convergent subsequence, $F_{j_1}(x_1)$. From the subsequence, extract a subsequence $\{F_{j_2}\}$ such that $\{F_{j_2}(x_2)\}$ converges. Proceed inductively. A diagonalization argument produces a subsequence $\{F_{j_j}\}$ such that $\{F_{j_j}(x_k)\}$ converges for every k. Show that this guarantees that $\{F_{j_j}(x)\}$ converges for every $x \in X$.

Section 6.4

3. Use Exercise 3 of Section 6.1.
4. For b, s, t fixed, consider $f(a) = \frac{a^s}{s} - ab + \frac{b^t}{t}$ as a function of $a \in (0, \infty)$.
5. (a) See the hint for the preceding problem.
7. Find $1 \leq s < n$ for which $H_0^{1,p}(\Omega) \subset H_0^{1,s}(\Omega) \subset L^q(\Omega)$.

Section 6.5

3. (a) Use the divergence theorem to write $w_{ij}(x)$ as

$$\int_{\tilde{\Omega}} (\Gamma_{ij}(x-y)) \frac{(f(y) - f(x))}{|x-y|^\alpha} dy - f(x) \int_{\partial \tilde{\Omega}} \partial_j K(x-y) \nu_i(y) \, d\mathbf{S}_y,$$

where $\Gamma_{ij} = |x|^\alpha \partial_i \partial_j K(x) = O(|x|^{2-n+\alpha})$.

4. (c) For $u \in H_0^{1,p}(\Omega)$, let $v \in C_0^1(\Omega)$ satisfy $\|u - v\|_{1,p} < \epsilon/2$. Use (65) to show that $\|v - v_h\|_{1,p} \to 0$ as $h \to 0$, so for h is sufficiently small, $v_h \in C_0^\infty(\Omega)$ and $\|v - v_h\|_{1,p} < \epsilon/2$.

5. Approximate u by $\phi \in C_0(\Omega)$ and use the fact that $S_z \phi \to \phi$ uniformly in Ω as $|z| \to 0$.

7. (a) Show that $\max_{x \in \bar{\Omega}}\{|\nabla u(x)| + |u(x)|\} \leq M$ for every $u \in A$ implies that for some $\delta > 0$, $|u(x) - u(y)| \leq (M+1)|x-y|$ whenever $|x-y| < \delta$; that is, A is equicontinuous. (c) Use the following:

$$\frac{|\phi(x) - \phi(y)|}{|x-y|^\alpha} = \left(\frac{|\phi(x) - \phi(y)|}{|x-y|^\beta}\right)^{\alpha/\beta} |\phi(x) - \phi(y)|^{1-\frac{\alpha}{\beta}}.$$

9. Treat cases $p > n$ and $1 \leq p \leq n$ separately; for the latter case observe that $np \geq n > n - p$ and use Exercise 3 in Section 6.1.

10. Use $\|T^*u\|^2 \leq \|u\| \cdot \|TT^*u\|$ to show that $\{u_m\}$ bounded $\Rightarrow \{T^*u_m\}$ bounded $\Rightarrow \{TT^*u_m\}$ contains a convergent subsequence.

13. $H_0^{2,p}(\Omega) \subset H_0^{1,r}(\Omega)$ for $r = np/(n-p)$ and $H_0^{1,r}(\Omega) \subset\subset L^q(\Omega)$ for $q < nr/(n-r) = np/(n-2p)$.

14. (i) If $2p > n > p$, then must take $m = 0$ and use $q = np/(n-p)$ in $H_0^{2,p} \subset H_0^{1,q} \subset\subset C^0$. (ii) If $n = p$, then $H_0^{2,p} \subset H_0^{1,q}$ for all $1 \leq q < \infty$ so take $q > n$ for $H_0^{1,q} \subset\subset C^0$. (iii) If $n < p$, then take $m = 1$ and bound $\|\nabla u\|_\infty$ in terms of $\|\nabla u\|_{1,p}$.

Section 6.6

1. $\Rightarrow$: If $x_j \in D$ converges to $x \in X$, then the boundedness of T shows that $\{Tx_n\}$ converges, and the closedness of T shows that $x \in D$.

2. $\Rightarrow$: Let $x_j \in D$ converge to x in $\|\cdot\|_g$; then $\{x_j\}$ and $\{Tx_j\}$ are both convergent in X, so the closedness of T implies $x \in D$.

3. Recall that $S: \tilde{S} \to X$ is closed $\Leftrightarrow G(S)$ is closed in $X \times X$, so $T: D \to X$ is closeable $\Leftrightarrow G(T)$ is a submanifold of a closed linear manifold in $X \times X$ that is a graph $\Leftrightarrow \overline{G(T)}$ is a graph.

4. Suffices to show: Let T be closed and densely defined; if $T^{-1}: X \to X$ exists and is bounded, then $(T^*)^{-1}$ does too, and $(T^*)^{-1} = (T^{-1})^*$. In fact, we need only show that $T^*(T^{-1})^* v = v$ for all $v \in X$ and $(T^{-1})^* T^* u = u$ for all $u \in D$.

5. (a) If $|\lambda| > \|T\|$, then $(T - \lambda)^{-1} = -\lambda^{-1} \sum_{n=0}^{\infty} (T/\lambda)^n$ converges, so $\lambda \in \rho(T)$. (c) If $\sigma(T) = \emptyset$, then $f(\lambda)$ is a bounded analytic function.

6. Let $z = \lambda^{-1}$, so $(T - z^{-1})^{-1} = -z \sum_{n=0}^{\infty} T^n z^n$ has radius of convergence R given (by power series theory) by $R^{-1} = \limsup_{n \to \infty} \|T^n\|^{1/n}$; conclude that $\sigma^{\text{rad}}(T) = \limsup_{n \to \infty} \|T^n\|^{1/n}$. On the other hand, for $\lambda \in \mathbb{C}$, $T^n - \lambda^n = (T - \lambda)(\lambda^{n-1} + \lambda^{n-2} T + \cdots + T^{n-1})$, so $(T - \lambda)^{-1} = (T^n - \lambda^n)^{-1}(\lambda^{n-1} + \cdots T^{n-1})$; in particular, $\lambda \in \sigma(T) \Rightarrow \lambda^n \in \sigma(T^n) \Rightarrow |\lambda| < \|T^n\|^{1/n}$ [by Exercise 5(a)], so $\sigma^{\text{rad}}(T) \le \liminf_{n \to \infty} \|T^n\|^{1/n}$.

8. $\Rightarrow$ is clear. Conversely, it is easy to conclude $\langle Ax, x \rangle = 0$ for all $x \in D$ where $A = T - T^*$, and we want to conclude $A \equiv 0$. Expansion of $\langle A(\alpha x + \beta y), (\alpha x + \beta y) \rangle = 0$ shows that $\alpha \bar{\beta} \langle Ax, y \rangle + \bar{\alpha} \beta \langle Ay, x \rangle = 0$ for all $x, y \in D$. Take $\alpha = \beta = 1$ to conclude that $\langle Ax, y \rangle + \langle Ay, x \rangle = 0$, and $\alpha = i$, $\beta = 1$ to conclude that $\langle Ax, y \rangle - \langle Ay, x \rangle = 0$; hence $\langle Ax, y \rangle = 0$. Approximate $y = Ax$ by $y_j \in D$ to conclude that $\|Ax\| = 0$.

9. Use compactness to find $\phi_n \in X$ with $\|\phi_n\| = 1$ and $T\phi_n \to y$, where $y \in \mathcal{H}$ satisfies $\|y\| = \|T\|$. Use self-adjointness and the Cauchy-Schwarz inequality to show that $\langle T^2 y, y \rangle = \|T^2 y\| \cdot \|y\|$, and hence $T^2 y = \mu y$, where $\mu = \|T\|^2$. If $y + \|T\|^{-1} Ty = 0$, then y is an eigenvector with eigenvalue $-\|T\|$; otherwise, $y + \|T\|^{-1} Ky$ is an eigenvector with eigenvalue $\|T\|$.

12. See the proof of Lemma 3 in Section 8.4.

13. Since $T - \lambda: X \to R_{T-\lambda}$ is 1-1, onto, and has closed range (Exercise 12), we only need verify that $R_{T-\lambda} = X$ in order to conclude (by the closed graph theorem) that $(T - \lambda)^{-1}$ is continuous. If $X_1 \equiv (T - \lambda)X$ is a proper subset of X, then in general we may define $X_{n+1} \equiv (T - \lambda)X_n$, which is a propoer subset of X_n. Pick $y_n \in X_n$ with $\|y_n\| = 1$ and $\text{dist}(y_n, X_{n+1}) = 1$. If $n > m$, then $\lambda^{-1}(Ty_m - y_n) = y_m + (-y_n + \lambda^{-1}((T - \lambda)y_m - (T - \lambda)y_n)) = y_m - y$, where $y \in X_{m+1}$, and hence $\|y_m - y\| \ge 1$. We conclude that $\|\lambda^{-1}(Ty_m - Ty_n)\| \ge 1$, contradicting the compactness of T.

14. In light of Exercise 13, it suffices to show that the following cannot occur: There is a sequence of linearly independent vectors $\{x_n\}_{n=1}^{\infty}$ such that $Tx_n = \lambda_n x_n$ and $\lambda_n \to \lambda \neq 0$. Otherwise, let X_n be the linear span of $\{x_1, \ldots, x_n\}$. Note that dist $(x_n, X_{n-1}) = 1$. If $n > m$, then $\lambda_n^{-1} Tx_n - \lambda_m^{-1} Tx_m = x_n - x_m$, so that $\|\lambda_n^{-1} Tx_n - \lambda_m^{-1} Tx_m\| \geq 1$, in contradiction of the compactness of T and $\lambda_n \to \lambda \neq 0$.

15. Observe that for $K = (T - \lambda_0)^{-1}$, $Kx = \mu x \Leftrightarrow (\frac{1}{\mu} + \lambda_0)x = Tx$.

Hints for Exercises in Chapter 7

Section 7.1

3. (a) First observe that $\langle u/\|u\|, u_j \rangle \leq \|u_j\|$ provided that $\|u\| \neq 0$. (b) See Example 4 of Section 6.3.

4. (a) Square $|a| \leq |a+b| + |b|$ and use $2a \cdot b \leq |a|^2 + |b|^2$.

5. (a) $\int_\Omega (\nabla u \cdot \nabla v + fv)\, dx = 0$ for $u, v \in C^2(\Omega) \cap C^1(\overline{\Omega})$. (b) Use compactness $H^{1,2}(\Omega) \subset L^2(\Omega)$.

7. (a) Use the second variation. (b) Use the Taylor series and Legendre convexity of L to obtain $L(p,x) \geq L(q,x) + \nabla_p L(q,x) \cdot (p-q) + \frac{\epsilon}{2}|p-q|^2$.

11. Let $\{x_m\}$ be Palais Smale. By (ii), $\{x_m\}$ is bounded. Moreover, $J'(u_m) = Lu_m + K(u_m) \to 0$ implies that $u_m = o(1) - L^{-1}K(u_m)$. Since $L^{-1}K(u_m)$ contains a convergent subsequence, so does u_m.

Section 7.2

3. First assume that $f \in Y = H^{2,2}(\Omega) \cap H_0^{1,2}(\Omega)$.

4. Generalize the proof of Theorem 2.

6. For $\hat{\Omega}$ you will encounter the half-integral Bessel function $J_{1/2}(z) = (\frac{2}{\pi z})^{1/2} \sin z$.

Section 7.3

1. Use $d(\bar{x}_1, \bar{x}_2) = d(T_1\bar{x}_1, T_2\bar{x}_2) \leq d(T_1\bar{x}_1, T_1\bar{x}_2) + d(T_2\bar{x}_2, T_1\bar{x}_2)$, and $d((T_1\bar{x}_1, T_1\bar{x}_2) \leq \alpha_1 d(\bar{x}_1, \bar{x}_2)$ to obtain $(1 - \alpha_1)d(\bar{x}_1, \bar{x}_2) \leq \epsilon$.

3. See Exercise 1.

5. (a) $G(z+w) - G(w) = T_{z+w}(G(z+w)) - T_w(G(w))$. (b) Show that $G'(y) = (F'(G(y)))^{-1}$ as maps $Z \to X$.

7. Show that $\langle F'(u)v, \phi \rangle = -\int_\Omega \frac{\nabla v \cdot \nabla \phi}{(1+|\nabla u|^2)^{1/2}}\, dx + \int_\Omega \frac{(\nabla u \cdot \nabla v)(\nabla u \cdot \nabla \phi)}{(1+|\nabla u|^2)^{3/2}}\, dx$.

8. Using Hölder spaces, assume that $\phi \in C^{2,\alpha}(\overline{\Omega})$ where $0 < \alpha < 1$, and let $X = \{u \in C^{2,\alpha}(\overline{\Omega}) : u = 0 \text{ on } \partial\Omega\}$ and $Y = C^\alpha(\overline{\Omega})$. Using Sobolev spaces, assume that $\phi \in H^{k,2}(\Omega)$ where $k - 1 > n/2$, and let $X = \{u \in H^{k,2}(\Omega) : u - \phi \in H_0^{1,2}(\Omega)\}$ and $Y = H^{k-2,2}(\Omega)$. In either case, apply Theorem 2 to $F: X \to Y$.

Hints for Exercises in Chapter 8

Section 8.1

5. Use $\|u\|_1 \leq \|u\|_2 \leq C(\|\Delta u\|_0 + \|u\|_0)$.

6. Use $(1+m^2)^{k+1} \leq \eta(1+m^2)^{k+2} + C_\eta(1+m^2)^k$.

7. (i) In computing $\langle \phi^2 u, u\rangle_{\ell+1} - \langle \phi u, \phi u\rangle_{\ell+1}$, all terms involving $(D^\gamma u)^2$ with $|\gamma| = \ell+1$ cancel, enabling the estimate by $\|u\|_\ell \|u\|_{\ell+1}$ instead of $\|u\|_{\ell+1}^2$. The case $\ell = 0$ is particularly simple: $|\langle \phi^2 u, u\rangle_{\ell+1} - \langle \phi u, \phi u\rangle_{\ell+1}| = |\int [\nabla(\phi^2 u) \cdot \nabla u - \nabla(\phi u) \cdot \nabla(\phi u)]\,dx| = \int u^2 |\nabla \phi|^2\,dx \leq C\|u\|_0^2 \leq C\|u\|_0 \|u\|_1$, where $C = \max |\nabla \phi|^2$. Similarly, with $\ell = 1$, a calculation of $\partial_i \partial_j (\phi^2 u)\partial_i \partial_j u - (\partial_i \partial_j(\phi u))^2$ only involves terms such as u^2, $u\partial_i$, $u\partial_i \partial_j u$, $\partial_i u \partial_j u$, or $\partial_i u \partial_i \partial_j u$ (multiplied by derivatives of ϕ, but not $(\partial_i \partial_j u)^2$); hence $|\langle \phi^2 u, u\rangle_2 - \langle \phi u, \phi u\rangle_2| \leq C\|u\|_1 \|u\|_2$. (ii) The case $k=0$ is trivial: $\|\phi Lu\|_0^2 + \|\psi Lu\|_0^2 = \int_{\mathbf{T}^n}(\phi^2 + \psi^2)|Lu|^2 dx$. For $k=1$, we compute $\int |\nabla(\phi Lu)|^2 + |\nabla(\psi Lu)|^2\,dx = \int (\phi^2 + \psi^2)|\nabla(Lu)|^2\,dx + C \int(|Lu|^2 + |Lu||\nabla(Lu)|)\,dx$.

8. (a) $u = \sum u_\alpha e^{i\alpha \cdot x}$ with $|\alpha|^2 u_\alpha = 0 \Rightarrow u_\alpha = 0$ if $\alpha \neq 0$.

9. Use $u_0 = 0$, $\|u\|_1^2 = \sum(1+|\alpha|^2)u_\alpha^2$, and $\|\Delta u\|^2 = \sum \alpha^4 u_\alpha^2$.

Section 8.2

2. (a) For any $\Omega' \subset\subset \Omega$, $f \in H^k(\Omega')$ for all k. (b) For any $\Omega' \subset\subset \Omega$, choose $\Omega' = \Omega_{k+1} \subset\subset \cdots \subset\subset \Omega_1 \subset\subset \Omega$ as in the proof of Theorem 1. Use $u \in H^1(\Omega)$ and $\Delta u = -\lambda u$ to conclude by Theorem 1 that $u \in H^3(\Omega_1)$; repeat to conclude that $u \in H^5(\Omega_2)$, $u \in H^7(\Omega_3)$, and so on.

4. Write $u = v + g$, where $v \in C^2(\Omega) \cap C(\overline{\Omega})$ satisfies $\Delta v = f - \Delta g$ in Ω and $v = 0$ on $\partial \Omega$.

6. $\delta_j^h(a_\alpha(x)D^\alpha u(x)) = h^{-1}(a_\alpha(x+he_j)D^\alpha u(x+he_j) - a_\alpha(x)D^\alpha u(x)) = h^{-1}(a_\alpha(x+he_j)D^\alpha u(x+he_j) - a_\alpha(x)D^\alpha u(x+he_j) + a_\alpha(x)D^\alpha u(x+he_j) - a_\alpha(x)D^\alpha u(x)) = (\delta_j^h a_\alpha)(x)D^\alpha u^h(x) + a_\alpha(x)D^\alpha(\delta_j^h u(x))$.

7. If $2 \leq \ell \leq k+1$ and $u \in H^\ell(\Omega)$ satisfies $Lu = f$, then differentiate the equation to show that, for $|\alpha| = \ell - 1$, $D^\alpha u \in H^1(\Omega)$ is a weak solution of an equation like (35); apply (37) to show that $D^\alpha u \in H^2(\Omega')$.

8. In $\Omega = \{x \in \mathbf{R}^2 : |x| < 1\}$ consider the harmonic functions $u_m(r,\theta) = r^m \cos m\theta$.

Section 8.3

2. (a) If u_1, u_2 are two solutions, let $w = u_1 - u_2$, and suppose there is an open set $\Omega' \subset \Omega$ with $w(x) > 0$ for $x \in \Omega'$, and $w(x) = 0$ for $x \in \partial \Omega'$. Then (49) shows that $w(x) \leq 0$ for all $x \in \Omega'$, a contradiction.

3. Repeat the proof of Theorem 2 with $w(x) = M(e^{\alpha d} - e^{\alpha x_1}) + N$, where α is sufficiently large.

4. To prove (iii), consider the operator $L-c^+$ where $c^+(x) = \max(c(x), 0)$.
   (i) If $\sup u = M \geq 0$, then OK. Otherwise, let $v = u - M$, so $\sup v = 0$. (ii) If $u \leq 0$, let $c^+(x) = \max(c(x), 0)$ and apply (ii) to $L - c^+$: $(L - c^+)u = Lu - c^+ u \geq 0 + 0$ and $c - c^+ \leq 0$.

5. (a) If u is not constant, then it must achieve its maximum on $\overline{\Omega}$ at $x_0 \in \partial \Omega$, where (54) holds; but this violates the boundary condition $\partial u/\partial \nu = 0$.

6. (a) is the same. (b) Suppose u is not constant and $u(x_0) = M$ for some $x_0 \in \partial \Omega$. If $x_0 \in \partial \Omega \setminus (0, 0)$, then (54) holds; if $x_0 = (0, 0)$, then $D_v u(x_0) > 0$ for all directions $v \in \overline{Q_1}$.

8. (a) Suppose $x_0 \in \Omega$ exists such that $u(x_0) > m \equiv \sup_{\partial \Omega} u(x)$. Then there exists $\mu > 0$ and $\Omega' \subset\subset \Omega$ such that $v(x) \equiv u(x) - m - \mu$ satisfies $v(x) > 0$ for $x \in \Omega'$ and $v(x) = 0$ for $x \in \partial \Omega'$. Extend v by zero to $\Omega \setminus \Omega'$, so $v \in H_0^{1,2}(\Omega)$. If we approximate v by $\phi \in C_0^1(\Omega')$, we can use $\int \sum a_{ij} \partial_j \partial_i \phi \, dx \leq 0$ to conclude that $\int_{\Omega'} \sum a_{ij} \partial_j u \partial_i v \, dx \leq 0$. But $\partial_j u = \partial_j v$ in Ω', so $\int_{\Omega'} \sum a_{ij} \partial_j v \partial_i v \, dx \leq 0$. This implies $\nabla v \equiv 0$ in Ω'; since $v = 0$ om $\partial \Omega'$, we conclude that $v \equiv 0$ in Ω', a contradiction.

Section 8.4

2. (a) If $v = Au \in R_A$, define $Bv = u$; if $v \in R_A^\perp$, define $Bv = 0$.
   (b) Notice that $N_A \subset N_{B_1 A}$ and $R_{AB_2} \subset R_A$. By Proposition 1, $\dim N_{B_1 A} < \infty$ and $\dim R_{AB_2}^\perp < \infty$.

3. Use Exercise 2a to find $B \colon \mathcal{H} \to \mathcal{H}$ such that $AB - I = K_1$ and $BA - I = K_2$ are compact. Then $B(A + K) - I = K_2 + BK$ is compact (recall Exercise 8 in Section 6.5); similarly, $(A + K)B - I = K_1 + KB$ is compact; use Exercise 2b.

4. As for real numbers, the inverse of $I - A$ is given by a geometric series: $(I - A)^{-1} = I + A + A^2 + A^3 + \cdots$, which converges since $\sum_{k=0}^\infty \|A^k\| \leq \sum_{k=0}^\infty \|A\|^k < \infty$.

5. (a) Use Exercise 2a to find $C \colon \mathcal{H} \to \mathcal{H}$ such that $AC - I = K_1$ and $CA - I = K_2$ are compact. Then $CB = C(A - (A - B)) = I + K_2 - C(A - B)$. Assume $\epsilon < \|C\|^{-1}$, so $I - C(A - B)$ is invertible by Exercise 4. Hence $(I - C(A - B))^{-1} CB = I + K_3$, where K_3 is compact; so $(I - C(A - B))^{-1} C$ is a left Fredholm inverse for B. Similarly, construct a right Fredholm inverse, and invoke Exercise 2b.
   (b) Let $V = N_A^\perp$, $W = R_{A_2}^\perp$, and $X = V \oplus W$. For a bounded linear operator $B \colon \mathcal{H} \to \mathcal{H}$, define $\tilde{B} \colon X \to \mathcal{H}$ by $\tilde{B}(v+w) = Bv + w$. If $B = A$, then $\tilde{B} \colon X \to \mathcal{H}$ is an isomorphism; show this remains true if $A - B$ is sufficiently small (cf. Exercise 4). Hence B maps V isomorphically onto a closed subspace BV of $\mathcal{H}$, and $\dim((BV)^\perp) = \dim W = \operatorname{def}(A)$. We can write $\mathcal{H} = N_B \oplus Z \oplus V$, where $\dim Z = \dim BZ < \infty$ and B is 1-1 on $Z \oplus V$. Show $\operatorname{def}(B) = \dim W - \dim Z$ and $\operatorname{nul}(B) = \dim N_A - \dim Z$.

6. Consider $A + tK$ for $0 \leq t \leq 1$ and use Exercise 5b.

Hints for Exercises in Chapter 9

Section 9.1

1. (a) A shift operator $f(x) = x + v$ for $X = \mathbf{R}^n$ and some fixed $v \in \mathbf{R}^n$.
   (b) A rotation $f(e^{i\theta}) = e^{i(\theta+\phi)}$ for $\theta \in S^1 \subset \mathbf{R}^n$ and some fixed $\phi \in S^1$.
2. Show that $\overline{\mathrm{co}}(T(A))$ is a closed, compact subset of A; apply Theorem 1 to $T: \overline{\mathrm{co}}(T(A)) \to \overline{\mathrm{co}}(T(A))$.
4. Apply Theorem 2 with $A = \{u \in C^0(\overline{\Omega}) : \|u\|_{1,\infty} \le 1\}$.
5. (a) Assume by density that $\vec{u}, \vec{v}, \vec{w} \in C_0^1(\Omega, \mathbf{R}^n)$, and use div $\vec{u} = 0$ to show $\int_\Omega (\vec{u} \cdot \nabla)(\vec{v} \cdot \vec{w})\, dx = 0$. (c) Write $\int_\Omega \nabla(T\vec{w}_m - T\vec{w}_n) : \nabla \vec{v}\, dx = \{\vec{w}_n, \vec{w}_n - \vec{w}_m, \vec{v}\} + \{\vec{w}_n - \vec{w}_m, \vec{w}_m, \vec{v}\}$, and apply (10b).

Section 9.2

1. (b) $\dot{\mathcal{E}}(t) = \dot{y}\ddot{y} + ky\dot{y} = -c\dot{y}^2$. (c) With $\mu = \sqrt{4k - c^2}/2$:

$$\mathbf{x}(t) = \begin{pmatrix} y \\ \dot{y} \end{pmatrix} = e^{-ct/2} \begin{pmatrix} \frac{c}{2\mu}\sin\mu t + \cos\mu t & \frac{1}{\mu}\sin\mu t \\ -\frac{k}{\mu}\sin\mu t & -\frac{c}{2\mu}\sin\mu t + \cos\mu t \end{pmatrix} \mathbf{x_0}.$$

3. (a) Yes. (b) No: try a periodic step function.
4. (b) $S(h)u_n - u_n = \int_0^h (S(s)u_n)_s\, ds = \int_0^h S(s)Au_n\, ds$. Letting $n \to \infty$, we obtain $S(h)u - u = \int_0^h S(s)w\, ds$. Divide by h and let $h \to 0$ to obtain $u \in D(A)$ and $Au = w$.
5. (a) For a partition $0 = s_0 < s_1 < s_2 < \cdots < s_N = t$, let $\Delta s_i = s_i - s_{i-1}$, and form the partial sum $\Sigma_N = \sum_{i=1}^N f(s_i)\Delta s_i \in X$. Show that Σ_N converges in X as $N \to \infty$; this defines $F(t)$.
6. (a) Let $\omega = \inf_{t>0} \phi(t)/t$ and let $\ell > \omega$. There exists $t_0 > 0$ with $\phi(t_0)/t_0 < \ell$. For any $t \ge 0$, write $t = nt_0 + r$ where $n \in \mathbf{N}$ and $0 \le r < t_0$. Subadditivity implies $\phi(t)/t \le (n\phi(t_0) + \phi(r))/t$, and letting $t \to \infty$, we obtain $\limsup_{t\to\infty} \phi(t)/t \le \phi(t_0)/t_0 < \ell$. (b) Use the semigroup property to show that $\log\|S(t)\|$ is subadditive.
9. Assume that $S(t)$ is β-contractive (cf. Exercise 6). For $0 \le s_1 < s_2 \le t$, $\|h(s_1) - h(s_2)\| \le \|S(t - s_1)(f(s_1) - f(s_2))\| + \|(S(t - s_1) - S(t - s_2))f(s_2)\| \le Ce^{\beta(t-s_1)}\|f(s_1) - f(s_2)\| + \|S(t-s_2)(S(s_2-s_1) - I)f(s_2)\|$.
10. Write the integral as a limit of Riemann sums, and use the closedness of $A: D(A) \to X$.
11. Let $w(t) = Au_1(t)$ with u_1 as in the proof of Theorem 2'. Show that $w(t)$ is *uniformly* Hölder continuous (with exponent α) for $t \ge 0$. [However, $Au_2(t)$ is only Hölder continuous (with exponent α) for $t \ge \epsilon > 0$.]
12. (a) See Section 5.2, Exercise 9. (b) Look at the eigenfunction expansion and use $ze^{-z} \le C$ for $0 \le z < \infty$.

13. Extend u by 0 to $t < 0$ and $t > T$, then let $u_h = \rho_h \star u$ where $\rho_h(t)$ is the usual mollifier (cf. Section 6.5.b). Then $u'_h = \rho_h \star u'$ on $(h, T-h)$. Moreover, $u_h \to u$ and $u'_h \to u'$ in $L^p((0,T), Z)$. Since $u_h(t) = u_h(s) + \int_s^t u'_h(\tau)\,d\tau$, we obtain (ii) upon letting $h \to 0$. Since $t \to \int_0^t u'(\tau)\,d\tau$ is continuous, (i) also follows.

Hints for Exercises in Chapter 10

Section 10.1

1. See the Remark after the Theorem in this section.
3. $[\![D']\!]$ and hence τ are independent of t.
4. See Exercise 1 for regularity and Exercise 3 for global existence.
5. Follow the steps of the proof of the Theorem, using (7) in place of linearity to verify that the integral operator $\mathcal{T}$ is contractive.
6. Notice that $(v-w)_t = (v-w)^2$, $(v-w)(0) = 1$. Integrate.

Section 10.2

1. Review Sections 1.1 and 1.2.
2. (a) The system for $v_1 = u_x$, $v_2 = u_t$ is strictly hyperbolic with $\lambda_\pm = \pm(1+u_x)$. (b) Using $\lambda = 1+v$, find $V_1 = \theta$ and $V_2 = -\theta - \theta^2/2$, where θ satisfies $\theta_t + (1+\theta)\theta_x$. Thus θ is implicitly given by $\theta = h'(x - (1+\theta)t)$; u satisfies $u_t + u_x + u_x^2/2$, which may be solved to find
$$u(x,t) = h(x-(1+\theta)t) + \frac{t}{2}\left(h'(x-(1+\theta)t)\right)^2.$$
   (c) The solution is defined for $t > -1/2$.
3. No. (Only if a conservation law decouples.)
4. A 1-shock is given by $\xi(t) = -w_l^{-2}t\sqrt{14/3}$ and a 2-shock is given by $\xi(t) = w_l^{-2}t\sqrt{14/3}$.
6. (a) $\lambda_\pm = v \pm \sqrt{h}$. (b) Simple waves: $u = (v, h) = (\pm\theta, \frac{1}{2}\theta^2 + \text{const})$.
7. (a) There can be no 1-shock. For a 2-shock, we need $v_r < v_l + \sqrt{h_l}$; then find $v_r - v_\ell = -\sqrt{2h_\ell}$. (b) The path of the 2-shock is $x(t) = v_\ell t$.

Section 10.3

1. Recall that a left eigenvector of A is an eigenvector of A^T, and that $\langle v, Aw \rangle = \langle A^T v, w \rangle$. So, letting $w = v_1$ we have $\langle v, \lambda_1 v_1 \rangle = \langle A^T v, v_1 \rangle = 0$. Hence $A^T v$ is perpendicular to v_1 and therefore a multiple of v.
2. (b) A condition is $v_x(x, 0) \geq k\, w(x, 0)^{-\beta}\, |w_x(x, 0)|$.
5. (a) The Riemann invariants are $r = v + 2\sqrt{h}$ and $s = v - 2\sqrt{h}$.

Hints for Exercises in Chapter 11

Section 11.1

1. Follow the outline of the proof in Section 5.1.
3. To prove (iii), consider $v(x,t) = u(x,t)e^{-\alpha t}$ with $\alpha \geq 0$ sufficiently large.
4. See the proof of Theorem 3 in Section 8.3c.
7. Yes. Let $w(x,t) = v(x,t) - u(x,t)$ and $W(x,t) = e^{-\lambda t}w(x,t)$ with λ sufficiently large.

Section 11.2

1. (a) Apply the mean value theorem to $f(u)$. (b) Use the Remark after Theorem 3 in Section 5.2.
3. Use the global Lipschitz condition to show that we can use the iteration of Theorem 3 in Section 9.2.
5. (b) $\|A^\alpha e^{-At}\| \leq \sup_{j \geq 1} \lambda_j^\alpha e^{-\lambda_j t} = O(t^{-\alpha})$ as $t \to 0$.
7. (b) $\|\mathcal{T}(u(t)) - \mathcal{T}(v(t))\|' \leq \int_0^t \|S(t-s)(F(u(s)) - F(v(s)))\|' ds \leq CM_R \|u-v\|_Y \int_0^t (t-s)^{-1/2} ds \leq 2CT^{1/2} M_R \|u-v\|_Y$.
10. Show that $\|u\|_\alpha, \|v\|_\alpha \leq R$ implies that $\|F(u) - F(v)\|_{1,2}^2 \leq C_R \|u-v\|_\alpha$. Suffices to show that $\|f'(u)u_i - f'(v)v_i\|_2 \leq C_R \|u-v\|_\alpha$. But $\|f'(u)u_i - f'(v)v_i\|_2 \leq \|f'(u)u_i - f'(u)v_i\|_2 + \|f'(u)v_i - f'(v)v_i\|_2 \leq C_R \|u-v\|_{1,2} + C_R \|u-v\|_2 \|v_i\|_2$, and $\|u\|_{1,2} \leq C\|u\|_\alpha$. Yes, this may be extended to all k.

Section 11.3

1. Global in time; use Theorem 2.
2. The solution blows up at least by $t = 2/\sqrt{\epsilon}$.
7. Use eigenvalues and eigenfunction expansions of the Laplacian with Neumann boundary conditions (as in Exercise 5 of Section 7.2) to prove
$$\mu_1 = \inf \left\{ \frac{\int (\Delta w)^2 \, dx}{\int |\nabla w|^2 \, dx} : w \in C^2(\Omega), \frac{\partial w}{\partial \nu} = 0 \text{ on } \partial\Omega \right\}.$$
8. Show that $\frac{d}{dt}(e^{-\int_0^t \phi(s)ds} y(t)) \leq 0$ and integrate.

Section 11.4

1. See Example 2 of Section 11.2.
2. Replace $X = \tilde{L}^2(\Omega, \mathbf{R}^n)$ by $X = X_\alpha$ as in Section 11.2.
3. As in Exercise 13 of Section 9.2, use a mollifier to write $u_h = \rho_h \star u$. Then we can write $\frac{d}{dt} \|u_h(t)\|_X^2 = 2\langle u_h'(t), u_h(t) \rangle_X$. Integrating, we obtain for $0 \leq s, t \leq T$, $\|u_h(t)\|_X^2 = \|u_h(s)\|_X^2 + 2\int_s^t \langle u_h'(\tau), u_h(\tau) \rangle_X \, d\tau \leq$

$\frac{1}{T}\int_0^T \|u_h(\tau)\|_X^2\,d\tau + 2\int_0^T \|u_h'(\tau)\|_V \|u_h(\tau)\|_{V'}\,d\tau$. Using the Cauchy-Schwartz inequality, we obtain a bound on $\max_{t\in[0,T]} \|u_h(t)\|_X^2$. These estimates applied to $u_{h_1} - u_{h_2}$ show that $u_h \to u \in C([0,T], X)$.

4. See the hint for Exercise 7 in Section 12.1.

5. (a) Write $u_i^2 - u^2 = u_i^2 - u_i u + u_i u - u^2$ and apply the Cauchy-Schwarz inequality. (b) Boundedness implies that u_j has a subsequence that is weakly convergent in L^2; use $u_j \to u$ in L^1 to conclude that the original sequence is weakly convergent in L^2 to u.

7. (a) $\langle B(\vec{u}_1) - B(\vec{u}_2), \vec{u}_1 - \vec{u}_2\rangle = \{\vec{u}_1 - \vec{u}_2, \vec{u}_1, \vec{u}_1 - \vec{u}_2\} + \{\vec{u}_2, \vec{u}_1 - \vec{u}_2, \vec{u}_1 - \vec{u}_2\}$. (b) Subtract (47) for $\vec{u}_2$ from (47) for $\vec{u}_1$ and apply to $\vec{u}_1 - \vec{u}_2$; then use (a). (c) Apply Young's inequality to estimate the RHS of (b) by $\nu\|\nabla(\vec{u}_1-\vec{u}_2)\|^2 + C\|\nabla\vec{u}_1\|^2 \cdot \|\vec{u}_1-\vec{u}_2\|^2$. (d) Use Gronwall's inequality.

Hints for Exercises in Chapter 12

Section 12.1

3. (a) Use div$(\vec{H} \times \vec{E}) = \vec{E}\cdot\text{curl}\,\vec{H} - \vec{H}\cdot\text{curl}\,\vec{E}$.

4. Integrate $[e^{-kt}y(t)]' \le h(t)$.

7. Let $\phi \in C_0^1(0,T)$ and $w \in L^2(\mathbf{R}^n)$. Using distributional derivatives, $\int_0^T \langle u_t^j, \phi w\rangle\,dt = -\int_0^T \langle u^j, \phi' w\rangle\,dt \to -\int_0^T \langle u, \phi' w\rangle\,dt = \int_0^T \langle u_t, \phi w\rangle\,dt$.

8. (a) $p = 2m/i$; (b) Write $\|D^\ell(u_1 u_2)\|_{L^2} \le C\sum_{i+j=\ell}\|D^i u_1 D^j u_2\|_{L^2} \le C\sum_{i+j=\ell}\|D^i u_1\|_{L^p}\|D^j u_2\|_{L^q}$ where $p^{-1} + q^{-1} = 2^{-1}$, and apply (a).

Section 12.2

2. For $g_0 \in X_0$, use $\|S(t)g - g\| \le \|S(t)\|\,\|g - g_0\| + \|S(t)g_0 - g_0\| + \|g - g_0\|$.

5. Use the eigenfunction expansion to reduce to a damped oscillator as in Section 9.2a.

7. (d) Suppose $\widehat{\psi}(\xi) \in L^2$ is orthogonal to all functions $\phi = \phi_{b,a}$ where $b > 0$ is fixed but $a \in \mathbf{R}$ varies. Show that $e^{-|\xi|^2/4b}\widehat{\psi}(\xi) \in L^1$ has Fourier transform 0; conclude that $\widehat{\psi}(\xi) = 0$.

Section 12.3

1. (a) Use $\|F(u(s+h)) - F(u(s))\| \le M_R\|u(s+h) - u(s)\|$, where $R = \max\{\|u(s)\| : 0 \le s \le t\}$, and the continuity of u to show that $F(u(s))$ is continuous. Similarly, $\|S(t-(s+h))F(u(s+h)) - S(t-s)F(u(s))\| = \|S(t-(s+h))[F(u(s+h)) - F(u(s)) + F(u(s)) - S(h)F(u(s))]\| \le Ce^{\beta(t-(s+h))}(\|F(u(s+h)) - F(u(s))\| + \|(I - S(h))F(u(s))\|)$.

2. Show that $\mathcal{T}(\vec{u}(t)) = \int_0^t S(t-s)F(\vec{u}(s))\,ds$ has a fixed point on a closed ball in Z_T; see Theorem 2 in Section 9.2.

Hints for Exercises in Chapter 13

Section 13.1

1. (a) See the proof of Theorem 4 in Section 7.3.
2. (a) Use the maximum principle to arrive at a contradiction.

Section 13.2

2. (i) If $v \in C^2(\overline{\Omega})$ is harmonic, then $|v|_1 \leq |v|_{1;\partial\Omega}$ is proved by applying the weak maximum principle to v and $\partial v/\partial x$. (ii) If w is the domain potential of $f \in C(\overline{\Omega})$, then $|w|_1 < C|f|_0$.
3. Apply Theorem 1 with $u_+(x) = u_1(x) =$ the principal eigenfuncton and $u_-(x) = \alpha + \beta|x|^2$, where $\alpha < 0$ and $\beta > 0$ are constants.
4. Let $g_k = k$ on $\partial\Omega$ and u_k be the corresponding solution of (35). Show u_k is uniformly bounded on each fixed compact domain $\Omega_1 \subset\subset \Omega$.
5. Use the solution v of $\Delta v = f$ as one barrier.
6. (b) Try a quadratic function as an upper solution.
9. Use a constant function and the first eigenfunction to construct positive barriers.

Section 13.3

2. (a) Write $F(u) = F(u)\partial_{x_i}x_i$. (b) Apply the divergence theorem to the vector field $P(x) = (x \cdot \nabla u)\nabla u$ and use (a).
3. Use the fact that $H_0^{1,q} \subset L^{q_1}(\Omega)$ with $q_1 = 6$, so $p_1 = 3/2$ and $a(x)u^4 \in L^{3/2}(\Omega)$. Therefore, $u \in H^{2,3/2}(\Omega) \subset L^{q_2}(\Omega)$ for $q_2 < \infty$ (note that $n = 2p_1$). Repeat with any $p_2 < \infty$ to get $u \in H^{2,p_2} \subset C^\alpha$. But then $a(x)u^4 \in C^\alpha$, and we can bootstrap to $u \in C^\infty$.
4. Use linear elliptic regularity theory (Chapter 8) and the previous Hint.
5. Examine the minimum value of $\int_\Omega (|\nabla u|^2 - \lambda_1 u^2 - a(x)|u|^{q+1})\,dx$ over all $u \in C^\infty(\Omega)$ with $u = 0$ on $\partial\Omega$.

Section 13.4

1. Use solutions of $\Delta u + \lambda u = \kappa$ with κ an appropriately chosen constant to construct local barriers.
2. Use the planes $\pi_P^\pm$ as local barriers in order to establish the a priori bound (69).
3. Modify the barrier construction for the minimal surface equation but instead of using d as the distance function to the boundary, use $d(x) = \text{dist}(x, \partial B)$, the distance to the enclosing ball. Notice that you must now extend the boundary values g of solution u to be defined on the entire ball B.

References

[Adams] R. Adams, *Sobolev Spaces*, Academic Press, New York, 1975.

[ADN] S. Agmon, A. Douglis, and L. Nirenberg, Estimates near the boundary for solutions of elliptic partial differential equations satisfying general boundary conditions, I *Comm. Pure Appl. Math.* **12**, (1959) 623–727; II *Comm. Pure Appl. Math.* **17**, (1964) 35–92.

[Ambrosetti-Rabinowitz] A. Ambrosetti and P. Rabinowitz, Dual variational methods in critical point theory and applications, *Journal of Functional Analysis* **14** (1973), 349–381.

[Ambrosetti-Prodi] A. Ambrosetti and G. Prodi, *A Primer of Nonlinear Analysis*, Cambridge University Press, Cambridge, 1993.

[Aris] Aris, *Vectors, Tensors, and the Basic Equations of Fluid Mechanics*, Prentice Hall, Englewood Cliffs, NJ, 1962.

[Arnold] V. I. Arnold, *Catastrophe Theory*, Springer-Verlag, Berlin, 1992.

[Aubin] T. Aubin, *Nonlinear Analysis on Manifolds, Monge-Ampère Equations*, Springer-Verlag, New York, 1982.

[BJS] L. Bers, F. John, and M. Schechter, *Partial Differential Equations*, John Wiley & Sons, New York, 1964.

[Bluman-Kumei] G. W. Bluman and S. Kumei, *Symmetries and Differential Equations*, Springer-Verlag, New York, 1989.

[Chorin-Marsden] J. Chorin and J. E. Marsden, *A Mathematical Introduction to Fluid Mechanics*, 3rd edition, Springer-Verlag, New York, 1993.

[Chow-Hale] S.-N. Chow and J. Hale, *Methods of Bifurcation Theory*, Springer-Verlag, New York, 1982.

[Churchill] R. V. Churchill, *Fourier Series and Boundary-Value Problems*, McGraw-Hill, New York, 1963.

[Constantin-Foias-Temam] P. Constantin, C. Foias, and R. Temam, *Attractors Representing Turbulent Flows*, American Mathematical Society, Providence, RI, 1985.

[Conway] J.Conway, *A Course in Functional Analysis,* 2nd edition, Springer Verlag, New York, 1990.

[Courant-Hilbert] R. Courant and D. Hilbert, *Methods of Mathematical Physics*, Interscience Publishers, New York, 1953 (vol. I), 1962 (vol. II).

[Dieudonné] J. Dieudonné, *Foundations of Modern Analysis*, Academic Press, New York, 1960.

[DiPerna] R. DiPerna, Decay and asymptotic behavior of solutions to nonlinear hyperbolic systems of conservation laws, *Indiana Univ. Math. J.* **24** (1975), 1047–1071.

[Doob] J. L. Doob, *Classical Potential Theory and Its Probabilistic Counterpart*, Springer-Verlag, New York, 1984.

[Evans] L. C. Evans, *Partial Differential Equations*, Berkeley Mathematics Lecture Notes (vols. 3A & 3B), 1993.

[Folland] G. Folland, *Introduction to Partial Differential Equations*, Princeton University Press, 1976.

[Freedman] D. Freedman, *Brownian Motion and Diffusion*, Holden-Day, San Francisco, 1971.

[Friedman] A. Friedman, *Partial Differential Equations*, Holt, Rinehart, and Wilson, New York, 1969.

[Galdi] G. Galdi, *Introduction to the Mathematical Theory of the Navier-Stokes Equations* (vols. I & II), Springer-Verlag, New York, 1994.

[Garabedian] P. Garabedian, *Partial Differential Equations*, John Wiley & Son, New York, 1964.

[Gelfand-Shilov] I. M. Gelfand and G. E. Shilov, *Generalized Functions*, vol. 1 (transl., E. Saletan), Academic Press, New York, 1964.

[Gilbarg-Trudinger] D. Gilbarg and N. Trudinger, *Elliptic Partial Differential Equations of Second Order*, 2nd edition, Springer-Verlag, New York, 1983.

[Giusti] E. Giusti, *Minimal Surfaces and Functions of Bounded Variation*, Birkhauser, Boston, 1984.

[Goldstein] J. A. Goldstein, *Semigroups of Linear operators and Applications*, Oxford University Press, New York, 1985.

[Guenther-Lee] R. B. Guenther and J. W. Lee, *Partial Differential Equations of Mathematical Physics and Integral Equations*, Prentice Hall, Englewood Cliffs, NJ, 1988.

[Guckenheimer-Holmes] J. Guckenheimer and P. Holmes, *Nonlinear Oscillations, Dynamical Systems, and Bifurcations of Vector Fields*, Springer-Verlag, New York, 1983.

[Hale] J. Hale, *Asymptotic Behavior of Dissipative Systems*, American Mathematical Society, Providence, RI, 1988.

[Henry] D. Henry, *Geometric Theory of Semilinear Parabolic Equations*, Lecture Notes in Mathematics #840, Springer-Verlag, Berlin, 1974.

[Hewitt-Ross] E. Hewitt and K. A. Ross, *Abstract Harmonic Analysis* (vol. I), Springer-Verlag, New York, 1963.

[Hille-Phillips] E. Hille and R. S. Phillips, *Functional Analysis and Semi-Groups*, American Mathematical Society, Providence, RI 1957.

References

[Hirsch-Smale] M. Hirsch and S. Smale, *Differential Equations, Dynamical Systems, and Linear Algebra*, Academic Press, New York, 1974.

[Hörmander,1] L. Hörmander, *Linear Partial Differential Operators*, Springer-Verlag, New York, 1964.

[Hörmander,2] L. Hörmander, *The Analysis of Linear Partial Differential Operators* (vols. I–III), Springer-Verlag, New York, 1983.

[Jackson] J.D. Jackson, *Classical Electrodynamics*, John Wiley & Sons, New York, 1975.

[John, 1] F. John, *Partial Differential Equations*, 4th edition, Springer-Verlag, New York, 1982.

[John, 2] F. John, *Nonlinear Wave Equations, Formation of Singularities* University Lecture Series, American Mathematical Society, Providence, 1990.

[John, 3] F. John, Blow-up of solutions of nonlinear wave equations in three space dimensions, *Manuscripta Math.* **28**, (1979), 235-268.

[Kato] T. Kato, *Perturbation Theory for Linear Operators*, 2nd edition, Springer-Verlag, New York, 1976.

[Kellogg] O. Kellogg, *Foundations of Potential Theory*, Springer-Verlag, New York, 1967.

[Kevorkian] J. Kevorkian, *Partial Differential Equations, Analytical Solution Techniques*, Wadsworth & Brooks/Cole, 1990.

[Ladyzhenskaya, 1] O. A. Ladyzhenskaya, *The Mathematical Theory of Viscuous Incompresible Flows*, 2nd edition, Gordon & Breach, 1969.

[Ladyzhenskaya, 2] O. A. Ladyzhenskaya, *The Boundary Value Problems of Mathematical Physics*, Springer-Verlag, New York, 1985.

[Ladyzhenskaya, 3] O. A. Ladyzhenskaya, *Attractors for Semigroups & Evolution Equations*, Cambridge University Press, Cambridge, 1991.

[Ladyzhenskaya-Ural'tseva] O. A. Ladyzhenskaya and N. N. Ural'tseva, *Linear and Quasilinear Elliptic Equations*, Academic Press, New York, 1968.

[Landkof] N. S. Landkof, *Foundations of Modern Potential Theory*, Springer-Verlag, Berlin, 1972.

[Lax] P. Lax, *Hyperbolic Systems of Conservation Laws and the Mathematical Theory of Shocks*, CBMS no. 11, SIAM, 1973.

[Lions] J. L. Lions, *Quelques Méthodes de Résolution des Problèmes aux Limites Non Linéaires*, Dunod, Paris, 1969.

[Lions-Magenes] J. L. Lions and E. Magenes, *Nonhomogeneous Boundary Value Problems & Applications*, Springer-Verlag, New York, 1972.

[Logan] J. D. Logan, *An Introduction to Nonlinear Partial Differential Equations*, Wiley-Interscience, New York, 1994.

[Luneburg] R. K. Luneburg, *Mathematical Theory of Optics*, Univ. of Calif. Press, Berkeley & Los Angeles, 1966.

[Majda] A. Majda, *Compressible Fluid Flow and Systems of Conservation Laws in Several Space Variables*, Applied Math. Sciences, #53, Springer-Verlag, 1984.

[Massey] W. S. Massey, *A Basic Course in Algebraic Topology*, Springer-Verlag, New York, 1991.

[Meyers-Serrin] N. Meyers and J. Serrin, H=W, *Proc. Nat. Acad. Sci. USA* **51** (1964), 1055–1056.

[Nehari] Z. Nehari, *Conformal Mapping*, McGraw-Hill, New York, 1952.

[Ni-Peletier-Serrin] W. M. Ni, L. A. Peletier, and J. Serrin, *Nonlinear Diffusion Equations and Their Equilibrium States, I and II* (MSRI vol. 12 and 13), Springer-Verlag, New York, 1988.

[Nirenberg, 1] L. Nirenberg, *Topics in Nonlinear Functional Analysis*, Lecture Notes, New York University, 1973.

[Nirenberg, 2] L. Nirenberg, Variational and topological methods in nonlinear problems, *Bulletin American Mathematical Society (N.S.)* **4** (1981), 267–302.

[Olver] P. J. Olver, *Applications of Lie Groups to Differential Equations*, Springer-Verlag, New York, 1986.

[Palais] R. Palais, *Seminar on the Atiyah-Singer Index Theorem*, Princeton University Press, Princeton, 1965.

[Pazy] A. Pazy, *Semigroups of Linear Operators and Applications to Partial Differential Equations*, Springer-Verlag, New York, 1983.

[Pinsky] M. Pinsky, *Partial Differential Equations and Boundary-Value Problems with Applications*, McGraw-Hill, New York, 1991.

[Protter-Weinberger] M. Protter and H. Weinberger, *Maximum Principles in Differential Equations*, Springer-Verlag, New York, 1984.

[Rabinowitz] P. Rabinowitz, *Minimax Methods in Critical Point Theory w/ Applications to Differential Equations*, American Mathematical Society, Providence, RI, 1986.

[Reed] M. Reed, *Abstract Nonlinear Wave Equations*, Lecture Notes in Mathematics, vol. 507, Springer-Verlag, 1976.

[Reed-Simon] M. Reed and B. Simon, *Methods of Mathematical Physics*, Academic Press, New York, 1972 (vol. I), 1975 (vol. II).

[Renardy-Rogers] M. Renardy and R. C. Rogers, *An Introduction to Partial Differential Equations*, Springer-Verlag, New York, 1993.

[Roe] J. Roe, *Elliptic Operators, Topology and Asymptotic Methods*, Longman Scientific & Technical, New York, 1988.

[Royden] H. Royden, *Real Analysis*, 3rd edition, Macmillan, New York, 1988.

[Rudin,1] W. Rudin, *Principles of Mathematical Analysis*, 2nd edition, McGraw-Hill, New York, 1964.

[Rudin,2] W. Rudin, *Functional Analysis*, McGraw-Hill, New York, 1973.

[Rund] H. Rund, *The Hamilton-Jacobi Theory in the Calculus of Variations*, Van Nostrand, London, 1966.

[Sattinger] D. Sattinger, *Topics in Stability and Bifurcation Theory*, Lecture Notes in Mathematics, vol. 309, Springer-Verlag, New York, 1973.

[Schwartz] L. Schwartz, *Théorie des Distributions*, Hermann, Paris, 1966.

[Smoller] J. Smoller, *Shock Waves and Reaction-Diffusion Equations*, Springer-Verlag, New York, 1983.

[Sobolev] S. L. Sobolev, *Partial Differential Equations of Mathematical Physics*, Pergamon Press, Toronto, 1964.

[Sokolnikoff-Redheffer] I. S. Sokolnikoff and R. M. Redheffer, *Mathematics of Physics and Modern Engineering*, McGraw-Hill, 1966.

[Stein-Weiss] E. Stein and G. Weiss, *Introduction to Fourier Analysis on Euclidean Spaces*, Princeton University Press, Princeton, NJ, 1971.

[Strauss] W. Strauss, *Nonlinear Wave Equations*, CBMS no. 73, American Mathematical Society, Providence, 1989.

[Struwe] M. Struwe, *Variational Methods,* Springer-Verlag, Berlin, 1990.

[Talenti] G. Talenti, Best constant in the Sobolev inequality, *Ann. Mat. Pura Appl.* **110** (1976), 353-372.

[Taylor] M. Taylor, *Pseudodifferential Operators,* Princeton Univ. Press, Princeton, NJ, 1981.

[Temam, 1] R. Temam, *Navier-Stokes Equations, Theory and Numerical Analysis*, 2nd edition, North-Holland Pub. Co., Amsterdam, 1979.

[Temam, 2] R. Temam, *Navier-Stokes Equations and Nonlinear Functional Analysis*, SIAM, Philadelphia, 1983.

[Temam, 3] R. Temam, *Infinite-Dimensional Dynamical Systems in Mathematics and Physics*, Springer-Verlag, New York, 1988.

[Treves,1] F. Treves, *Basic Linear Partial Differential Equations*, Academic Press, New York, 1975.

[Treves,2] F. Treves, *Topological Vector Spaces, Distributions and Kernels*, Academic Press, New York, 1967.

[Troutman] J. Troutman, *Variational Calculus with Elementary Convexity Theory,* Springer-Verlag, New York, 1983.

[Wermer] J. Wermer, *Potential Theory,* Lecture Notes in Mathematics, vol. 408, Springer-Verlag, Berlin, 1974.

[Whitham] G. Whitham, *Linear and Nonlinear Waves*, Wiley-Interscience, New York, 1974.

[Widder] D. V. Widder, *The Heat Equation*, Academic Press, New York, 1975.

[Yoshida] K. Yoshida, *Functional Analysis* 2nd edition, Springer-Verlag, New York, 1968.

[Zeidler] E. Zeidler, *Nonlinear Functional Analysis and Its Applications I, II*, Springer-Verlag, New York, 1986.

Index

absolute extrema, 220
adjoint
    differential operator, 61
    Hilbert space, 173, 210
admissible set, 219
analytic approximation, 360
Alaoglu's theorem, 185
analytic semigroup, 291
a priori estimates
    on a torus, 248ff
    on domains, 253ff
approximation to the identity, 196
artificial viscosity, 361
Arzela-Ascoli theorem, 8, 123, 184
asymptotically stable, 285
attractor, 285
autonomous system, 285
average value, 204
axiom of choice, 169, 172

β-contraction, 284, 289
Banach space, 163
barriers, 130
    local, 265, 329, 402
    global, 385
basis, 163
Beltrami equation, 53, 59
Bessel's equation, 137
Bessel's inequality, 172
best constant, 193
bifurcation equation, 383
bifurcation theory, 381-384
biharmonic equation, 5, 228
blow up, 6, 11
boundary, 7
boundary conditions, 2
boundary gradient estimate, 401
boundary manifold, 408
boundary value problem, 2
boundary point lemma, 264
bounded
    bilinear form, 172, 178
    linear functional, 167
    linear operator, 167, 208
    slope condition, 408
Boussinesq equation, 99
Brouwer fixed point theorem, 277
Burgers' equation 1, 23

C^0-semigroup, 288
C^1-flow, 299
calculus of variations, 219
canonical form, 52
Cartesian tensors, 419
Cauchy data, 17, 43-44
Cauchy problem, 13, 43-44, 75
Cauchy-Riemann operator, 68

Cauchy-Schwarz inequality. 163
center manifold, 286
    subspace, 284
characteristic
    curves, 12, 15, 32, 50, 55
    equations, 14, 33, 55
    strips, 33
    hypersurface, 60
    vector, 60
characteristics
    (see "characteristic curves")
classical solution, 68, 288
closed convex hull, 277, 282
closed graph theorem, 209
closed operator, 209
closed subspace, 168
coefficients of viscosity, 420
coercive
    bilinear form, 178, 295
    functional, 221
compact map, 200
compact resolvent, 213, 274
compact support, 7
compactness, 183
comparison principles/theorems,
    linear elliptic, 266
    quasilinear elliptic, 401
    semilinear elliptic, 391
    semilinear diffusion, 331
compatibility condition, 79, 136
complete collection, 163
complete integral, 35
complete metric space, 162-163
completely continuous
    (see "compact map")
compressible fluids, 6
cone property, uniform, 201
conformal scalar curvature
    equation, 4
conservation law(s), 26, 306
conservation of energy, 91, 96, 417
conservation of mass, 6, 416
conservative system, 285
contact discontinuity, 316
continuous dependence on data, 47
    1-dim wave equation, 76
    Laplace equation, 109
    heat equation, 145
continuous flow, 299
continuous semigroup, 284
contraction, 238
contraction mapping principle, 238
convergence in mean, 134
convex functional, 224
    hull, 282
    set, 277
convolution
    of functions, 66
    of distributions, 67
critical point

of a functional, 218
of a flow, 284, 285
cubic Klein-Gordon equation, 379
cubic Schrödinger equation, 4

d'Alembert's formula, 75
damped harmonic oscillator, 284
damped pendulum, 286
damping (see "dissipation")
Darboux equation, 85
data, 2
deficiency, 271
deformation tensor, 420
delta function, 64
derivative
    of a functional, 217, 218
    of a mapping, 237
descent, method of, 88
difference quotients, 255
differentiability (Frechét)
    of a functional, 217, 218
    of a mapping, 237
diffusion, 142, 412
diffusive wave, 99
dimension, 163
Dirac delta function
    (see "delta function")
Dirichlet problem, 106
discrete wave train, 96
dispersion, 95
dispersive wave, 96
dissipation, 3, 4, 96, 285, 374
distribution, 64-66
distribution solution, 66
distributional
    derivative, 65
    integral, 66
divergence form, 5, 178
divergence-free, 140, 176
divergence theorem, 8
domain, 7
domain (of a linear operator), 208
domain of dependence, 76, 87, 88, 92, 98
domain potential, 113, 195
double layer potential, 114
dual space, 182
Duhamel's principle, 70, 81, 151, 291

eigenfunctions, 132
eigenvalues, 132
eikonal equation, 2, 37
Einstein summation rule, 418
electrostatics, 113
elliptic, 50, 73, 249
    uniformly elliptic, 178, 253, 400
energy, 91, 285
energy estimates, 296, 358
entropy condition, 28, 313
envelope, 32
equation of state, 6

equicontinuous, 184
equilibrium solution, 285
equivalent norms, 174
error function, 158
essentially bounded, 164
Eulerian coordinates, 308
Euler-Poisson-Darboux equation, 86
Euler-Lagrange equations, 218
Euler's equations, 5, 417
evolution equations, 2
evolution operator 284, 289
expansion in eigenfunctions,
    heat equation, 142-144
    Laplace equation, 132-135, 233
    wave equation (1-dim), 78
    wave equation (n-dim), 135-6
extension operator, 203
exterior cone condition, 131
exterior sphere condition, 131

fan (see "rarefaction wave")
Fick's law, 412
field, 5, 103, 414
finite differences, 360
finite propagation speed, 76
first variation, 219
fixed point
    of Brouwer, 277
    of contraction map, 238
    of flows, 284, 285
    of Schauder theory, 277-279
flexible beam equation, 99
focusing, 88
Fourier
    coefficients, 172, 245
    series, 78, 245
    transform, 146
Fourier's law, 413
Fredholm operator, 271
Fredholm index, 271
frequency, 95, 136
function spaces, 162
functional, 217
fundamental solution, 68
    heat operator, 150
    initial value problem, 70
    Laplace operator, 111-2

Galerkin's method, 301, 350, 360
gas dynamics, 367
Gaussian kernel, 149
general solutions for first-order equations
    quasilinear equations, 21
    nonlinear equations, 36
genuinely nonlinear, 313, 314
geometric optics, 2, 36-41
gradient catastrophe, 6
graph area functional, 227
graph norm, 214, 337
Green's function, 117

Index

Green's identities, 107
Gronwall's inequality, 346, 368

Hadamard's example, 47
Hahn-Banach theorem, 169
Hamiltonian system, 285
harmonic function, 103
Harnack inequality, 122
heat equation, 1, 142, 414
heat flux, 413
heat kernel, 149
Heaviside function, 64
Helmholtz decomposition, 140
Hilbert-Schmidt theorem, 214
Hilbert space, 163
Hille-Yoshida theorem, 290
hodograph transformation, 323
Hölder continuity, 194
Hölder inequality, 164
   generalized, 171
Holmgren uniqueness theorem, 48
homogeneous fluids, 416
Hopf maximum principle, 264
Huygens' principle, 89-90
hyperbolic equations, 50
hyperbolic systems, 56, 302
   strictly hyperbolic, 56, 302
   symmetric hyperbolic, 56

ideal fluid, 416
implicit function theorem, 9, 240
incompressible fluids, 6, 416
infinite-dimensional space, 162
infinitesimal generator, 289
initial conditions, 2
initial/boundary value problem, 2, 77
inner product, 163
integrable function, 7
integral curves, 13
integral equations, 126, 133
integral surface, 13, 32
invariant subspaces, 284
inverse Fourier transform, 147
inverse function theorem, 9, 239
isentropic gas, 6, 417
isometrically isomorphic, 182
inviscid Burgers' equation, 1, 23

jump condition, 26, 314
jump discontinuity, 26, 63

k-shock, 314
KdV equation (see "Korteweg de Vries")
kinematic viscosity, 421
Kirkhoff's formula, 87
Klein-Gordon equation, 3, 99, 371
Korteweg de Vries equation, 5, 99

L^2-derivatives, 166
Lagrange multipliers, 228
Lagrangian, 227

Lagrangian coordinates, 309
Laplace equation, 1, 2, 103
Laplacian (see "Laplace equation")
Lax-Milgram theorem, 179
Lebesgue measure, 164
left-shift operator, 276
Legendre convexity condition, 227
Leray-Schauder theorem, 279
Lewy example, 48
Liapunov function, 285
light cone, 38
light ray, 37
linear
   equations, 1
   functional, 168
   operator, 167, 208
linearization, 286
linearized stability analysis (LSA), 286
linearly degenerate, 316
Liouville's theorem, 123
Lipschitz continuous, 207
local existence, 286
locally finite partition of unity, 10
locally Lipschitz continuos, 298
log, 9
lower function, 263-4, 385

material derivative, 417
max, 9
maximin, 235-6
maximum principles
   heat equation, 144
   parabolic equations, 328
   subharmonic functions, 109, 127
   uniformly elliptic equations:
      weak, 261, strong, 264,
      generalized weak, 268
Maxwell's equations, 5, 99, 357, 415
mean curvature, 242
mean value property, 109
measurable sets and functions, 164
method of characteristics
   nonlinear first-order equations, 32
   quasilinear first-order equations, 13
method of continuity, 269
method of Lagrange, 21
mild solutions for evolution equations
   linear, 292
   nonlinear, 298
minimal surface equation, 4, 59, 227, 242, 404
Minkowski inequality, 164
mollifier, 196
Monge cone, 31
Monge-Ampère equation, 5, 59
monotone iteration, 386
mountain pass, 225
   theorem 226
multi-index, 43
multiplicity of an eigenvalue, 134

Navier-Stokes equations, 6
  derivation of, 421
  linearized (see Stokes)
  local existence, 346
  stationary (forced), 279
  weak solutions, 348
Neumann problem, 106, 142
Newtonian fluid, 420
Newton's law of viscosity, 420
nodal sets, 136
noncharacteristic, 14, 34, 44
nonhomogeneous equations, 2, 68
nonslip boundary condition, 175, 421
norm, 162
normal form, 44
normed vector space, 162
nullity, 271

ODE, 2
operator norm, 167
orbit, 284
order of an equation, 1
orthonormal, 163
oscillation, 403

p-system, 307
Palais-Smale condition/sequence, 226
parabolic, 50, 328
parabolic boundary, 144
partition of unity, 9
parallelogram rule, 77
PDE, 1
periodic functions, 245
Perron's method, 127
perturbation theory, 240
phase velocity, 95
Pohozaev's identity, 398
Poincaré inequality, 174, 204
point spectrum, 212
Poisson equation, 3, 103
Poisson integral formula, 118
Poisson kernel, 118
polytropic, 290, 417
porous medium equation, 5, 158
positive bilinear form, 178
potential energy, 113, 343
potential function, 3, 103, 113
prescribed mean curvature, 242, 408
principal curvatures, 242
principal eigenvalue, 134, 266
principal part, 49, 55, 60
principal symbol, 50, 55, 60
principally linear, 49
projected characteristic curve, 16
propagation of singularities, 62
propagation speed, 37, 74
pseudo-differential operators, 126

quasicontraction, 289
quasilinear, 1

range of influence, 76, 87, 88
Rankine-Hugoniot jump condition, 314
rarefaction wave, 28, 310
real analytic, 45, 124
reflection, method of, 118, 120
reflection principle, 125
reflexive, 182
regular set, 211
regularization, 196
representation theorem
  of Riesz, 169
  of potential theory, 114
resolvent, 212
resolvent set, 211
Riemann function, 325
Riemann problem, 307, 316-9
Riesz convexity theorem, 374, 375
Riesz representation theorem, 169, 182
Riesz-Fischer theorem, 164
Robin boundary condition, 107, 142

saddle point, 225
scale invariance, 156
Schauder fixed point theory, 277-9
Schrödinger's equation, 3, 99, 373
Schwartz class, 147
second variation, 221
self-adjoint
  differential operator, 62, 179
  operator on a Hilbert space, 213
semi-Fredholm operator, 273
semigroups, 283ff
semilinear equations, 1
separable, 184
separation of variables, 77, 104
sequentially compact, 184
shallow-water waves, 308
sharp signals, 87, 89, 90
shock condition (of Lax), 314
shocks, 24, 25, 313
side conditions, 2
similarity, 156
simple eigenvalue, 134, 266, 384
simple wave, 28, 309
simply-connected, 7
sine-Gordon equation, 4, 380
single layer potential, 114
singular integral operators, 126
smooth approximations of functions, 196, 202
Sobolev inequality, 186
Sobolev spaces, 165
solitons, 380
solution operator, 284, 289
spacelike, 38
specific heat, 413
spectral theory, 134, 211-214
spectrum, 134, 212
speed of sound, 102, 324
spherical mean, 83

stable manifold, 286
stable subspace, 284
steady-state, 2
Stirling's formula, 126
Stokes equations (free), 159
    stationary (forced), 175ff
strain tensor, 420
stress tensor, 420
strict contraction, 238
strict convexity (conservation laws), 313
strictly convex functional, 224
strictly hyperbolic, 56
strip condition, 34
strong maximum principle
    elliptic, 264
    parabolic, 328
strong solution, 206
strongly continuous semigroup, 288
Sturm-Liouville theory, 105
subadditive function, 300
subfunction, 128
subharmonic function, 126
subsolution, 263, 385
subspace, 168
successive approximations, 238
sup, 9
supersolution, 385
support
    of a characteristic strip, 32
    of a distribution, 65
    of a function, 7
symmetric hyperbolic, 56, 357

telegrapher's equation/system, 3, 57, 99
tensor fields, 419
tensor product, 419
test functions, 61, 65
third boundary value problem, 107
timelike, 38
totally bounded, 183
trace of a matrix, 267
trace spaces, 276
traffic flow, 28
trajectory, 284
transmission conditions, 62-64
transport equation, 1, 12
trilinear form, 280
type of an equation, 50
type of a semigroup, 300

uniform cone property, 201
uniform ellipticity, 178, 253, 400
uniform wave, 95
uniformly bounded, 184
uniformly convex domain, 408
unique solution, 2
unstable manifold, 286
unstable subspace, 284
upper function, 385

vanishing viscosity, 361

variational method, 126, 217, 219
viscous fluids, 6, 420
viscous stress tensor, 420

wave equation, 1, 74ff., 411-2
wave front, 36
wave number, 95
weak convergence, 183
weak* convergence, 183
weak derivative, 61
weak maximum principles
    elliptic, 261
    heat equation, 144
    Laplace equation, 103
    parabolic, 328
weak solutions
    first-order equations, 25
    linear equations, 61
    linear evolution equation, 294
    Navier-Stokes equations, 348
    Poisson equation, 173
    Stokes equations, 175
    symmetric hyperbolic systems, 362
    wave equation (1-dim), 76
weak uniqueness, 268
weakly lower semi-continuous, 221
weakly sequentially compact, 184
well posed problem, 2, 47
Weierstrass approximation theorem, 185

Young's inequality, 187

Index of Symbols

A^α, 336-337
$B_r(x)$, 7
$C(\Omega)$, $C^k(\Omega)$, 7
$C(\overline{\Omega})$, $C^k(\overline{\Omega})$, 7
$C_B(\Omega)$, $C_B^k(\Omega)$, 7
$C_0(\Omega)$, $C_0^k(\Omega)$, 7
$C(\Omega, \mathbf{R}^N)$, 7
$C^\alpha(\Omega)$, $C^\alpha(\overline{\Omega})$, 194
$C^{m,\alpha}(\Omega)$, $C^{m,\alpha}(\overline{\Omega})$, 195
$C_B^{m,\alpha}(\overline{\Omega})$, 205
$C(X, Z)$, 237
$C^1(X, Z)$, 238
$C^\alpha([0, T], X)$, 292
$C^{2;1}(U)$, 144
$C^{0,1}(\Omega)$, 207

D, $\text{dom}(T)$, 208
∂_x, $\partial/\partial x$, 7

$G(T)$, 208

$H_0^{1,2}(\Omega)$, 165
$H^{1,2}(\Omega)$, 166
$H_0^{1,p}(\Omega)$, $H^{1,p}(\Omega)$, 166
$H^{-1,2}(\Omega)$, 183
$H_0^{k,p}(\Omega)$, $H^{k,p}(\Omega)$, 205
$H^k(\mathbf{T}^n)$, $H^{k,2}(\mathbf{T}^n)$, 247
$H^{1,p}((0,T), Z)$, 295

$L^1(\Omega)$, $L^1_{\text{loc}}(\Omega)$, 7
$L^\infty(\Omega)$, $L^p(\Omega)$, 164
$L^2(\mathbf{T}^n)$, 247
$L^p((0,T), Z)$, 295

$\mathbf{N}$, 7
N_T, 209
$O(\cdot)$, $o(\cdot)$, 9
R_T, 209
$\rho(T)$, 211
$\mathcal{S}$, 147
$\text{supp } f$, 7
$\sigma(T)$, $\sigma_p(T)$, 212
$\sigma^{rad}(T)$, 212
$W^{1,2}(\Omega)$, $W^{1,p}(\Omega)$, 166
X_α, 337
$\mathbf{Z}$, 7
$\hat{}$, $\vee$, 146-147
$\|\cdot\|_\infty$, $\|\cdot\|_p$, 163-164
$\langle\cdot,\cdot\rangle_1$, $\|\cdot\|_{1,2}$, 165
$\perp$, 171
$|\cdot|_{1,2}$, $(\cdot,\cdot)_1$, 173
$\|\cdot\|_{-1,2}$, 183
$(\vec{u}\cdot\nabla)\vec{u}$, 6
$\nabla\vec{u} : \nabla\vec{v}$, 176
$\subset\subset$, 253